Advances in Science, Technology & Innovation

IEREK Interdisciplinary Series for Sustainable Development

Advances in Science, Technology & Innovation (ASTI) is a series of peer-reviewed books based on important emerging research that redefines the current disciplinary boundaries in science, technology and innovation (STI) in order to develop integrated concepts for sustainable development. It not only discusses the progress made towards securing more resources, allocating smarter solutions, and rebalancing the relationship between nature and people, but also provides in-depth insights from comprehensive research that addresses the **17 sustainable development goals (SDGs)** as set out by the UN for 2030.

The series draws on the best research papers from various IEREK and other international conferences to promote the creation and development of viable solutions for a **sustainable future and a positive societal** transformation with the help of integrated and innovative science-based approaches. Including interdisciplinary contributions, it presents innovative approaches and highlights how they can best support both economic and sustainable development, through better use of data, more effective institutions, and global, local and individual action, for the welfare of all societies.

The series particularly features conceptual and empirical contributions from various interrelated fields of science, technology and innovation, with an emphasis on digital transformation, that focus on providing practical solutions to **ensure food, water and energy security to achieve the SDGs**. It also presents new case studies offering concrete examples of how to resolve sustainable urbanization and environmental issues in different regions of the world.

The series is intended for professionals in research and teaching, consultancies and industry, and government and international organizations. Published in collaboration with IEREK, the Springer ASTI series will acquaint readers with essential new studies in STI for sustainable development.

ASTI series has now been accepted for Scopus (September 2020). All content published in this series will start appearing on the Scopus site in early 2021.

Ilaria Pigliautile · Cristina Piselli ·
Hirushie Pramuditha Karunathilake ·
Claudia Fabiani
Editors

Urban Resilience, Livability, and Climate Adaptation

Health, Environmental Dynamics, and Societal Well-Being

A culmination of selected research papers from the International Conference on Health and Environmental Resilience and Livability in Cities (HERL)—2nd Edition—University of Perugia 2023.

Springer

Editors
Ilaria Pigliautile
Department of Engineering
EAPLAB at CIRIAF—Interuniversity Research Center
University of Perugia
Perugia, Italy

Hirushie Pramuditha Karunathilake
Division of Physical and Computational Sciences
University of Pittsburgh
Bradford PA, PA, USA

Cristina Piselli
Department of Architecture (DIDA)
University of Florence
Florence, Italy

Claudia Fabiani
Department of Engineering
EAPLAB at CIRIAF—Interuniversity Research Center
University of Perugia
Perugia, Italy

ISSN 2522-8714 ISSN 2522-8722 (electronic)
Advances in Science, Technology & Innovation
IEREK Interdisciplinary Series for Sustainable Development
ISBN 978-3-031-54910-6 ISBN 978-3-031-54911-3 (eBook)
https://doi.org/10.1007/978-3-031-54911-3

This Springer imprint is published by the registered company Springer Nature Switzerland AG
The registered company address is: Gewerbestrasse 11, 6330 Cham, Switzerland

Scientific Committee

The Editors warmly thank all the Reviewers who have contributed their authority to the double-blind review process, to ensure the quality of this publication.

Preface

Enhancing the resilience and livability of urban spaces while adapting to the growing challenges of climate change has become a topic of global concern in recent times. The urban environments face severe challenges due to the growing population and mass migration, deteriorating infrastructure, changing climatic and weather patterns, and economic stresses. To achieve the long-term vision for climate change mitigation and adaptation, cities' adaptation and resilience are requirements that cannot be further postponed. Indeed, environmental challenges associated with climate change, e.g., heat waves, urban overheating, and air pollution, negatively impact human well-being and health in indoor and outdoor environments. The specific microclimate of cities is characterized by the phenomenon known as the Urban Heat Island (UHI) which further exacerbates the need for building energy use to maintain comfortable indoor conditions thus contributing to the same climate change. More specifically, UHI stands for an urbanized area that experiences significantly warmer temperatures than the rural surroundings, due to the presence of artificial infrastructures—more than natural ones—and human activities.

UHIs are where the majority of our buildings are located, which account for about 36% of total global energy use and the largest contribution of global GHG emissions. Moreover, about 68% of the world population is expected to live in cities by 2050, and the highest share of this future urban population growth will occur in the global south. Therefore, cities are the major contributors to climate change, thus providing fertile ground for the implementation of mitigation and adaptation strategies. In particular, cities in developing countries, which are now the hotspots of urban growth, have a high potential for transformation. We have to rethink the design and planning of our smart cities to take action in efforts to face the harmful effects of and to be resilient to climate change, UHI, and the associated phenomena from multiple frontiers.

However, the analysis and achievement of health, well-being, and livability in our cities have to be not only dependent on the development of technical solutions and innovations. We should also pay the needed attention to the social and environmental dimensions of urban development and sustainability. Research needs to place human needs and experiences at the center of urban development, also through people and community engagement. Health and well-being require a focus on environmental sustainability and climate resilience to ensure that cities are livable and safe for all residents, even in the face of climate change. Furthermore, research needs to be contextualized to the local boundary conditions to meet the specific needs in each context. To this aim, we should develop and take advantage of methods and innovations that are replicable, but also applicable from a local perspective for the analysis and design of tailored and human-centric mitigation strategies. Therefore, with this book, we hope to foster a productive multi-disciplinary discussion that may identify which are the most urgent research lines toward sustainable, efficient, healthy, and just cities.

In this panorama, the book offers valuable insights into the above-mentioned debate based on a range of applications and case studies presented in the key papers from the 2nd International Conference on Health & Environmental Resilience and Livability in Cities (HERL) of IEREK. This book fosters a productive multi-disciplinary discussion that may identify which are the most urgent research lines toward sustainable, efficient, healthy, and just cities. It adds to this topic by collecting perspectives from various countries across the world. More in detail, the book is composed of three parts. Part One is on "Urban Green Infrastructure and Climate Mitigation", containing five chapters. Chapter "Effects of Street Tree Configuration on Urban Heat Island Mitigation" focuses on the effect of street tree configuration on UHI mitigation, and Chapter "The Butterfly Effect Toward a Nature Smart Society" addresses the development of a nature-smart society. Chapter "Spatial Relationship Between Diet Diversity, the Food Environment and Transport: A Case Study in South Africa" presents a South African case study on the spatial relationships between food, environment, and transport, while Chapter "Spatial Analysis of Noise Contour Maps Based on Traffic Speed Using Predictor-LimA Software" discusses spatial analysis of noise contour maps based on traffic speed. Chapter "Critical Factors Affecting the Design and Use of Elevated Urban Spaces: The Sky Garden, London" explores the critical factors affecting the design and use of elevated urban spaces.

Part Two consisting of six chapters, explores a different dimension, centering on the "Human Experience and Well-Being in Urban Environments". Chapters "Assessment of the Impact of the Office Environment on Productivity Based on Employees' Satisfaction" and "Impact of Streetscapes on Anxiety: A Physiological Evidence" discuss how office environments and cityscapes have an effect on the satisfaction and anxiety of individuals, respectively. Chapters "Envisioning Cities of the Future: A Malaysia Youth Perspective Using Concept Mapping" and "Topia—The City, the Livable Place to Be" focus on future cities and the design factors that affect their use and livability. Chapter "Evaluating Daylighting Performance Within Existing Schools Expansion in Egypt" analyzes the daylighting performance of school buildings, whereas Chapter "Revisiting Scientific Theories, Towards Human Well-Being-Oriented Built Environments" more generally discusses the scientific theories on human well-being in the built environments.

Part Three and final part is on the "Adaptation, Livelihood, and Social Dynamics", and comprises six more chapters. The part discusses how communities and social groups need to adapt to climate-resilient, healthy, and livable environments. Chapter "Assessing Community Adaptive Capacity for Climate Resilient Development: A Complex Adaptive Systems Case Study Analysis of Mathare Valley, Nairobi, Kenya" is on the adaptive capacity of communities for climate-resilient development. Chapter "Assessing Street Vendors Liveability—A Comparative Study in Hosur and Vellore" focuses on street vendors' livability, while Chapter "Urban Refugees' Impact on the Urban Fabric Form of the City Structure: The Case Study of Urban Syrian Refugees in the Sixth of October City" assesses the impact of refugees on the urban fabric form. Chapter "Restorative Streets for Healthy Cities: A Critical Review Dissenting the Conventional Narrative of Restorative Environments" discusses the narratives on restorative environments for healthy cities. The final Chapters "Evaluating Users' Satisfaction on Urban Railway Based on Service Quality Model: The Study on KLIA Express in Malaysia" and "Sport as a Tool for the Development of Healthy and Sustainable Cities: A Strategic Documentation Review" analyze tourist satisfaction based on transport systems and the means of using sport as a tool for healthy and sustainable cities, respectively.

Overall, all of these contributions focus on different aspects of the development of livable, healthy, and resilient urban environments, where long-term planning and effective decision-making based on scientific evidence are needed to ensure that global urban centers can successfully withstand the pressures of the present. Further, it is highlighted that urban planning and management need to be a participatory and inclusive process where the needs and challenges of different stakeholder groups need to be taken into account. The content in the individual chapters explores specific case studies spanning different geographic locations, social contexts, and climatic conditions.

All in all, this volume is meant to provide a means for sharing knowledge, experiences, and ideas on how to build sustainable, resilient, and livable cities in the face of climate change. It highlights the latest research and practices in the field by high-profile scientists, policymakers, and practitioners who share their insights and experiences on the conference themes. Therefore, we believe that it will provide a valuable resource for academics, practitioners, policymakers, and other stakeholders interested in the intersection of smart cities, health, and well-being in the climate change era. We believe that by working together and sharing knowledge, we can build cities that are not only smart but also healthy, resilient, and livable for all residents.

Florence, Italy — Cristina Piselli
Bradford, USA — Hirushie Pramuditha Karunathilake
Perugia, Italy — Ilaria Pigliautile

Introduction

In the relentless march of time, the entwined forces of climate change and escalating urbanization pose formidable threats to the very essence of cities—their livability. As humanity confronts the challenges of an evolving world, the focal point of this discourse rests on the shoulders of urban resilience, a linchpin in the quest to construct a livable future for generations to come. The urgent imperative to grapple with these challenges has resulted in a paradigm shift in the realm of urban development, necessitating a concerted focus on adaptation and resilience.

Cities worldwide face complex challenges, including environmental degradation, continuous population growth due to mass migrations, aging infrastructure, and unpredictable climate patterns. The urgency to tackle these issues head-on is underscored by the imperativeness of urban adaptation and resilience becoming the guiding principles in the evolution of urban landscapes. Within urban environments, which pulsate as hubs of human activities and diversity, a dual role unfolds. On the one hand, these environments stand as significant contributors to the anthropogenic factors influencing climate change. Simultaneously, they are highly vulnerable to the effects of climate change, facing issues like intense heat waves and harmful air pollution. This affects the well-being of residents both indoors and in the city. The Urban Heat Island (UHI) phenomenon emerges as a poignant illustration, characterized by markedly elevated temperatures in urbanized areas. This not only disrupts the comfort and well-being of citizens but also casts a long shadow over global energy consumption and greenhouse gas emissions.

With cities currently accounting for a staggering 36% of global energy usage, the urgency to address climate-related issues is acutely apparent. As projections foresee 68% of the world's population dwelling in urban areas by 2050, the imperative for decisive action becomes ever more pressing. It is within the expanding urban landscapes of developing countries, undergoing unprecedented expansion, that the potential for transformative actions becomes ripe. To meet the challenges posed by climate change, the UHI phenomenon, and their associated harms, it is paramount to transcend the traditional boundaries of smart city design and planning. This transcendence, however, cannot be confined to the realm of technical innovations alone. A holistic approach is indispensable, one that directs its focus not only on the technological but also on the social and environmental dimensions of urban development. In this intricate dance for the future of cities, engaging communities and relevant stakeholders becomes not just a consideration but a cornerstone. Prioritizing environmental sustainability emerges as a pivotal step in the quest for cities that defy the odds—remaining not just habitable but safe for all inhabitants, resilient in the face of the ever-mounting challenges posed by a changing climate.

In the face of these complex challenges worldwide, this book serves as a guide, offering valuable insights from the 2nd International Conference on Health & Environmental Resilience and Livability in Cities (HERL) organized by IEREK. The book is divided into three parts, exploring essential themes for creating cities that are sustainable, efficient, healthy, and equitable.

The initial part, titled "Urban Green Infrastructure and Climate Mitigation", conducts a comprehensive examination of nature's potential in shaping dynamic urban spaces amidst the continually changing climate. This segment delves into the complexities of microclimate analysis and acoustic quality assessments, providing a nuanced understanding of how nature can be a fundamental element in enhancing urban resilience.

A significant emphasis is placed on microclimate analysis, delving into the detailed climatic conditions of specific urban areas. This exploration aims to unravel the localized impacts of urban green infrastructure, addressing factors such as temperature variations, wind patterns, and moisture levels. The goal is to illustrate how strategically incorporating green elements can serve as a moderator for climate conditions. This analytical depth is particularly valuable for urban planners, offering insights that go beyond generic solutions to tailor interventions that align with the distinctive challenges presented by individual urban environments. Simultaneously, the part extends its exploration to acoustic quality assessment, recognizing the often-overlooked aspect of soundscapes in urban life. Given the prevalent issue of noise pollution in urban areas, this investigation meticulously examines how the integration of green spaces can influence acoustic landscapes. These green elements serve as natural buffers, absorbing and mitigating sound. This nuanced approach acknowledges that the vibrancy of urban spaces is not solely visual but also extends to the auditory experience. It underscores the importance of fostering a harmonious soundscape for the well-being of urban dwellers.

Part Two, "Human Experience and Well-Being in Urban Environments", marks a definitive shift toward a human-centric paradigm in urban planning and design. Here, the spotlight is directed at cultivating environments that advocate the well-being of urban dwellers. The part becomes a proponent of salutogenic design, a meticulous approach aimed at crafting spaces that not only provide to the functional needs of inhabitants but actively contribute to their health, happiness, and overall well-being.

This part goes beyond the traditional architectural considerations and zoning regulations. It delves into the psychological and physiological dimensions of human interaction with the urban environment. Salutogenic environments prioritize health promotion rather than simply preventing illness, recognizing the profound impact that surroundings have on mental and physical wellness. Whether within the sheltered confines of buildings or amidst the expansive urban sprawl, the emphasis is on creating spaces that nurture a sense of community, encourage physical activity, and provide moments of respite. This part explores design interventions that enhance air and water quality, incorporate green spaces, and promote accessibility to amenities that contribute to a healthier lifestyle. The importance of natural light, greenery, and communal areas takes center stage, acknowledging their pivotal role in fostering not just aesthetically pleasing spaces but also those that actively contribute to the holistic well-being of residents.

Part Three and final part, "Adaptation, Livelihood, and Social Dynamics", the focus turns toward a comprehensive investigation into the adaptive capacities of diverse urban communities and stakeholders. This part becomes a nuanced exploration of the intricate social fabric that weaves through city life, acknowledging the pivotal roles played by individuals ranging from street vendors and refugees to tourists. Each group, although seemingly peripheral, emerges as a dynamic force shaping the very essence of the cityscape.

The investigation delves into the adaptive strategies employed by these diverse communities in response to the multifaceted challenges posed by urban living. From the resourcefulness of street vendors navigating economic uncertainties to the resilience of refugees seeking stability in unfamiliar urban environments, each narrative unfolds as a thread in the intricate tapestry of urban life. Moreover, the transient yet impactful presence of tourists introduces a dynamic layer, influencing not only the economic landscape but also shaping the cultural and social dynamics of the city. This part recognizes that the resilience of these communities is not merely a passive response but a proactive force influencing the city's ability to navigate challenges. Their adaptive capacities become a linchpin, a critical element for the city's overall ability to weather the storm of urban complexities. By understanding and harnessing the adaptive capabilities of these diverse groups, urban planners and policymakers can cultivate more inclusive and responsive urban environments.

This collection serves as a catalyst, igniting a transformative multi-disciplinary dialogue that transcends traditional boundaries in urban development. It acts as a crucible where urgent research directions are not only identified but also passionately pursued, aiming to sculpt urban environments that are sustainable, efficient, healthy, and equitable. The diverse perspectives encapsulated in this compilation create a mosaic of global insights, contributing significantly to the ongoing worldwide conversation on the imperative task of building urban spaces that can withstand current challenges while evolving into resilient fortresses of health and livability for all inhabitants.

The significance of this collection lies in its ability to foster collaboration and exchange of knowledge across borders. It recognizes that solutions for urban challenges cannot be one-size-fits-all but must be crafted through a lens that incorporates the varied experiences and contexts of different regions. As readers delve into the subsequent chapters, they embark on a journey that goes beyond geographical confines. This is a journey where shared knowledge and experiences become the driving force, propelling the vision of cities that surpass the conventional definitions of smartness and habitability. The narrative emphasizes not only the need for cities to be intelligent in their design but resilient in the face of a changing climate and adaptable to the evolving urban landscape. The aspiration is not merely to create habitable spaces but to cultivate environments that actively contribute to the health and well-being of all residents. It signifies a commitment to crafting urban spaces that are inclusive, responsive, and considerate of the diverse needs of the global population. In essence, this collection emerges as a beacon, guiding the way toward a future where cities transcend mere functionality, evolving into thriving hubs that embody the ideals of resilience, health, and livability for every individual within their bounds.

Ilaria Pigliautile
Claudia Fabiani
Cristina Piselli

Contents

Adaptation, Livelihood, and Social Dynamics

Urban Green Infrastructure and Climate Mitigation

Effects of Street Tree Configuration on Urban Heat Island Mitigation

Sachini Jayasinghe, Varuni Jayasooriya, and Sandun Dassanayake

Abstract

Urban Heat Island is increasingly affecting cities in developing countries with rapid urbanization. Uncomfortable thermal environments are present in the urban streets of tropical towns due to the need for more urban vegetation. A well-planned street tree configuration can positively influence the thermal environment within metropolitan regions. This study intends to explore the tree configuration impacts on the roadside thermal environment in urban canyons of selected sites in Colombo, Sri Lanka. Field data was collected to identify street canyons' heat patterns considering existing roadside greenery configurations. Suitable tree-planting scenarios to regulate the microclimatic conditions were also investigated. Three different roadside tree configurations in Colombo and its suburbs were considered. Climate data, tree, and surrounding attribute data were measured. The 3D microclimate simulation software ENVI-met Lite version was used to identify heat distribution patterns and carry out scenario simulations. The results indicated that the heat distribution in the roadside area is reduced when the tree configuration of continuous tree rows with densely foliated vegetation is applied.

S. Jayasinghe · V. Jayasooriya (✉)
Department of Forestry and Environmental Science, University of Sri Jayewardenepura, Nugegoda, Sri Lanka
e-mail: varuni.jayasooriya@sjp.ac.lk

S. Dassanayake
Department of Decision Sciences, University of Moratuwa, Moratuwa, Sri Lanka
e-mail: sandund@uom.lk

Keywords

Tree configurations · Street canyons · Thermal environment · Urban Heat Island · ENVI-met

1 Introduction

Accelerating urbanization in developing tropical cities intensifies the undesirable microclimatic and environmental consequences, like the Urban Heat Island (UHI) effect, lowering the quality of life for urban residents. The causes that increase UHI effects include low-albedo and dark-hued urban surfaces, paved and impermeable pavements, and road surfaces (Senevirathne et al., 2021). The increased temperature situation is elevated in a tropical country such as Sri Lanka due to year-round sunlight; solar radiation is high under cloudless sky conditions, and, due to the proximity to the equator, solar elevation is consistently high as well (Halwatura & Ranasinghe & Halwatura, 2012). Most of the asphalt-paved roadways in Colombo, Sri Lanka, a tropical and humid city, are directly exposed to the sun, promoting more significant absorption of solar radiation and heat discharge at night.

Moreover, the roadside vegetation scarcity and sparse urban vegetation configurations (Aram et al., 2019) alter the urban morphology affecting wind movement, shading effects, and compactness, also increasing UHI effects. Recent studies (Aram et al., 2019; Srivanit & Jareemit, 2020) show the potential of roadside vegetation cover and the configuration of trees in mitigating the UHI effect and drastic thermal environment alterations. Remote sensing research has demonstrated that different tree configurations affect the thermal environment in a significant way. According to Aram et al. (2019), clustered tree configuration had improved cooling effects compared to dispersed tree configuration. In Bangalore, India, sections of streets

I. Pigliautile et al. (eds.), *Urban Resilience, Livability, and Climate Adaptation*,
Advances in Science, Technology & Innovation, https://doi.org/10.1007/978-3-031-54911-3_1

abundant with trees had an average ambient air temperature of 5.6 °C lower than in other segments (Vailshery et al., 2013). Based on a study conducted in a subtropical region, the air temperatures on tree-aligned roadways and boulevards were 1–2.5 °C lower during the day's warmest hours (15:00 h) than on non-vegetated roads (Shashua-Bar & Hoffman, 2004). The temperature-mitigating benefits of varied tree coverage rates on outdoor thermal conditions also have been investigated in several other studies (Klemm et al., 2015).

Many researchers have used the microclimate model ENVI-met to investigate the thermal environment and microclimate. Among various computational modeling software and Energy Balance Models (EBM) such as RayMan, TEB-Veg, SOLWEIG, and green-CTTC, and Computational Fluid Dynamic (CFD)-based models like OpenFOAM, FLUENT, and PHOENICS, ENVI-met is considered to be one of the most effective CFD-based holistic 3D microclimate models used vastly in vegetation thermal effect simulation studies in the world. Field data were utilized by Yang et al. (2013) to assess the model of ENVI-met based on the thermal performance of several surface subtypes. It has been applied to study the effects of urban greening on outdoor thermal comfort and local meteorological parameters. Lee et al. (2016) estimated how trees and grasslands influence the microclimate of a typical residential neighborhood in Friburg, Germany, while Salata et al. (2016) examined the significant impact of vegetation on the moderation of urban heat island in Toulouse, France. Multiple studies have shown that ENVI-met correctly simulates the thermal surroundings of various climates (Forouzandeh, 2021; Srivanit & Jareemit, 2020; Yang et al., 2013).

In the Sri Lankan context, the ENVI-met model has been applied to investigate the effects of green infrastructure on tropical Sri Lankan urban settings as a UHI adaptation approach (Herath et al., 2018). The same model has also been employed to evaluate the effects of different urban design alternatives on air and surface temperatures and outdoor comfort conditions in Colombo (Emmanuel et al., 2007). Local ENVI-met studies relating vegetation and UHI mitigation primarily focus on green infrastructure effects on UHI and thermal comfort levels and the "Building and thermal environment" aspect, not specifically on the roadside tree configurations aspect. The influence of urban green infrastructure (green walls, roofs, etc.) on microclimatic conditions was researched by Herath et al. (2018). Emmanuel et al. (2007) explored the "Building and thermal environment" aspect showing how the aspect ratios of street canyons and different building configurations affect the air temperature mitigation while exploring the added benefits of shade provided by buildings in comparison to established UHI reduction methods using a densely urbanized road in Colombo. It was done by building density modification, albedo enhancement, and shading level adjustment.

Only modest efforts had been adopted in local scientific research to identify improved street tree configurations intended to enhance the urban cooling effect and mitigation of UHI in Colombo. Inspired by the approaches employed in recent similar studies, this study aims to present a model-based identification of improved street tree configurations suitable for Colombo (selected sites) by deriving heat distribution patterns within street canyons incorporating the current roadside tree configurations via the CFD-based 3D microclimatic model 'ENVI-met,' along with validation through field measurements. Hence, this study will investigate the effects of roadside tree configurations on the roadside thermal environments in urban canyons of the selected sites.

2 Materials and Methods

2.1 Data Collection

The major locations around which the study was based are Colombo, Sri Lanka, and its suburb of Ratmalana, representing three different roadside tree configurations in varied environments of urban canyons. These cities represented a typical hot, humid, humid tropical climate as encompassed in the study scope. A tropical monsoon climate, or Köppen climatic type Am, can be found in Colombo. From March to April, the average high temperature is approximately 31 °C ("Colombo weather," October 2009). Rainfall in Colombo is around 2,500 mm (98 in) a year, on average ("World Weather Information Service—Colombo." World Meteorological Organization, February 2017).

2.1.1 Case Study

Three roadside tree configurations (Configurations A, B, and C) in Colombo and Ratmalana were considered, and three sites were selected accordingly. The selected three sites were mainly based on the street tree configurations and also on the nature of the surrounding urban environment, i.e., whether the area is in a heavily urbanized region with institutionally pre-planned vegetation and buildings, whether the site is in a more residential environment with suburban characteristics, or whether the area is in a moderately urban environment representing a main road which facilitates vehicle fleets diverting to/from several directions. The tree configurations observed in these three urban environments represented the most common ones in Colombo and Ratmalana.

Figure 1.1 shows Google Earth and real site images of the selected tree configurations A, B, and C sites. Tree Configuration A consisted of single-lane urban roads with an abundance of roadside trees with high crown density on either side of the road, which are spaced more or less at equal distances. Tree crown shapes of 'Spreading,' 'Layered,' and 'Broad' forms were predominant in this configuration. The site selected for Configuration A representation was Horton's Place, Colombo 07. Tree Configuration B consisted of non-uniformly spaced 'wayside' trees which were individual/isolated trees on a single-lane road near a residential area. The trees were of 'Irregular,' 'Open' crown shapes and were of moderate height. The site selected for Configuration B was Mallikarama Road, Ratmalana. Configuration C represented a very low or minimal presence of roadside trees in the urban road site, and buildings on both sides flanked the road. The site selected for Configuration C was Barnes Place, Colombo 07.

The data collection process commenced on the August 14, 2021, and was done for ten months until June 2022. The onsite data was collected from 11:00 to 14:00 on clear, sunny days. This time was selected to collect different microclimatic data within the peak daytime hours with maximum UHI effect in the tropical climate considering the shadowing effect of trees in the daytime (Srivanit & Hokao et al., 2013; Senevirathne et al., 2021). Clear and cloudless days were selected for data collection, as the ENVI-met simulation model, which was the microclimate model used in the study for thermal environment simulations, does not take the cloudiness of the sky into account, and thus simulation of sunny days may be more realistic (Taleghani et al., 2015). The climate data measurements were done 1.5 m above the ground level (pedestrian height). The study was confined to the urban climate's Urban Canopy Layer (UCL) as the study's focus is on climate-sensitive urban roadside space. It examines the causes and consequences of changing morphology at the street level. In all three sites, the data was collected within a 100 m road stretch, and this particular measurement stretch was selected after carefully examining the surrounding road area ranging 200–300 m of the sites and after confirming that similar tree configuration attributes and repeated urban morphology were present in that particular region. Five points were located 20 m apart on each side of the road for data collection, resulting in 10 measurement points per site. At each of these measurement points, the climate data was measured by taking three replicates of the four parameters with a 5-s interval between each replicate. The average of the three replicates was then calculated. The three sites measured the climate parameter at 11 a.m., 12 p.m., 1 p.m., and 2 p.m.

Table 1.1 shows the data types collected during onsite data collection. A list of the specification details for the instruments utilized for data collection is shown in Table 1.2.

Regarding the tree attribute data, the following are considered. In measuring the tree crown spread measurements, the average crown spreads of the roadside trees identified in the sites were measured by the following method (Eq. 1):

$$A(\textit{Average crown spread of trees}) = (\textit{Longest spread} + \textit{Longest cross spread})/2 \quad (1)$$

Other information on the considered roadside trees of each site was also recorded; whether the tree is cylindric/heart-shaped/spherical, the size of the trunk (large/medium/small), and whether the tree crown is visually

Fig. 1.1 Real street images (above) and Google Earth images (below) of selected sites

Table 1.1 Data types collected

Climate data	Tree attribute data	Data on adjacent environment
Ambient temperature (Ta)	Tree height	Road length and width, Road and pavement materials
Relative Humidity (RH)	Tree-spacing distances	Ground conditions
Wind Speed (WS)	Trunk height	Surrounding building attributes
Wind Direction (WD)	Crown shape	Roughness length

Table 1.2 Instruments used

Instrument	Measurement parameter	Accuracy
Digital anemometer AS816	Wind velocity	±5%
CO2/Temp/RH Data Logger	Ambient temperature	±0.6 °C
	Relative humidity	±3%
Wind Vane Mk2	Wind direction	
Forestry pro II laser range finder/hypsometer	Tree height/trunk height	±0.3 m
Fiberglass measuring tape	Distance measures/tree crown diameter	±0.1 m/s or ±5%

dense or sparse based on the foliage and tree crown spread measurements.

The data on the adjacent environment collected includes the surrounding building attributes such as the building wall material (concrete/glass, etc.), roof material (terracotta/tile, etc.), building height, and ground and soil conditions of the pavements and the roads (loamy soil/asphalt/concrete, etc.).

2.2 ENVI-Met Simulations

The 3D simulation software ENVI-met Lite 4.0 version was used to identify heat distribution patterns within the selected sites. Model simulations were conducted in ENVI-met in each location using the data collected. The simple forcing tool was used during the simulations, which enabled to force or set climatic variables of relative humidity and temperature over 24 h throughout the simulation. Through the simple forcing feature, the measured onsite meteorological data of Ambient temperature, Relative humidity, Wind speed, and Wind direction were input in forcing the model for the simulation for all the sites in consideration.

Major input variables entered for simulations on ENVI-met are shown in Table 1.3.

Table 1.3 Major input parameters for ENVI-met simulations

Metrological input variables	Ambient temperature (Ta) and Relative humidity (RH)	Measured average data and EPW data
	Horizontal solar and infrared radiation	EPW data
	Specific humidity (2500 m)	7 g/kg*
Total simulation time (h)	Per simulation	1
Wind speed at 10 m height (m/s)—WS		2.5*
Building	Concrete walls	Thermal conductivity—1.6 W/(m·K)*
	Roofs—Terracotta	Thermal conductivity—0.81 W/(m·K)*
	Roofs—Tile	Thermal conductivity—0.84 W/(m·K)*
Vegetation		Differing according to the site
Roughness length (m)	Urban environment	0.01

*Model default

The model simulations required two files for functioning: an Urban configuration file in which the study area is displayed with the vegetation, receptors, buildings, surfaces, and soil, vegetation, soil, and surfaces, and a Climate configuration file with the initialization values and microclimate data of Ta and RH. The general procedure used in the simulations of the three sites is as follows. In preparation for the urban configuration file, a.INX area file was created in the 'Spaces' feature of ENVI-met, using a previously prepared bitmap file of the study site (with located vegetation, buildings, and road) using QGIS software (an open-source, cross-platform GIS application). A grid resolution of 5 m*5 m*5 m was used for digitizing the study area with the model dimensions of 50*50*40, the maximum model size in ENVI-met Lite. The model domain was rotated by the respective degree belonging to each study site to align with the location's North axis. Other major model domain settings were adjusted accordingly with the site in consideration. In Model Location, the location's name, location coordinates, and reference time zone were adjusted.

In digitizing the area after adjusting the model domain settings, the vegetation, buildings, soils and surfaces, and receptors were defined with the help of the background georeferenced bitmap image. The roadside buildings were drawn in with the collected building data of building geometry (building height, roof, and wall materials). The soils and surfaces of the area were also retrieved from the model's database. The following soil profiles with their respective thermal conductivities are used: For the roads: Asphalt Road (0.17–0.2 W/m °K), for the pavements; concrete pavement gray (2.25 W/m °K), for the natural surfaces; Loamy Soil (0.47–2.56 W/m °K). In the representation of vegetation, the most resembling vegetation types were selected from the model database. The vegetation which was to be included can be chosen from the database according to the shape of the tree crown (Cylindric/Heart-shaped/Spherical), Height of the tree (Medium—15 m/ Small—5 m /Large—25 m), and Trunk height (Medium/ Small/Large). The most appropriate vegetation was selected in accordance with the field data collected to represent the actual area better. Receptors were also defined at the site to better identify the assigned measurement points.

After editing the materials and digitizing the area, the climate configuration file was created in the ENVI-guide feature. The forcing file for the simulation was created with an Energy Plus Weather File (EPW file) containing general yearly weather data of Colombo-Ratmalana. This and the onsite data were useful in forcing the Ta and RH values over 24 h. Default values derived by ENVI-met were adopted for Specific humidity (at 2500 m height), and Roughness length (0.01 m) showcasing an urban environment was adopted (roughness length at measurement site—roughness of exterior conditions). Table 1.3 represents the major input parameters included for the ENVI-met simulations. Thus, they were run after creating the respective simulations' urban configuration and climate files. The simulation results and outputs (Atmosphere files) obtained can be used to create 2D or 3D heat distribution maps of the study areas for data analysis.

2.3 Tree Configuration Scenario Analysis

To identify suitable roadside urban forest planning scenarios to regulate the microclimatic conditions within the selected urban canyons, different greening modification scenarios were developed for the selected sites of the three tree configurations, scenario simulations were run, and the results obtained were compared.

Scenario simulations are done for the three Tree Configurations A, B, and C sites, according to the following pre-selected roadside urban forest planting scenarios:

Scenario 1—Base case (original/actual site conditions)
Scenario 2—Control case (no vegetation)
Scenario 3—Widely spaced tree row (sparsely foliated—low LAD)
Scenario 4—Continuous tree row (densely foliated—high LAD).

When adding the vegetation in the scenarios, a constant deciduous tree from the model database was considered in all the simulations: Cylindric, medium trunk, medium height (15 m).

3 Results

The onsite collection of data on the parameters of Ta and RH values for the experimental road sites belonging to the three tree configurations was done for ten months, from August 2021 to June 2022. The below-mentioned results relate to a data set collected on the 26th of January 2022 during the 11 a.m. hour window for Configuration A and a data set obtained on January 31, 2022, during the 11 a.m. hour window for Configuration C.

The selected days for result interpretation were clear, sunny days devoid of precipitation for the day or the previous two days. ENVI-met model simulations were done for the hours and dates of Horton's Place, Colombo 07, of Tree Configuration A, and Barnes Place, Colombo 07, of Tree Configuration C.

The primary input data for Configuration A simulation: Average Ambient Temperature (Ta)—34.2 °C, Average Relative Humidity (RH)—65.7%, Average Wind Speed at 10 m height (WS)—2.5 m/s, Wind Direction (WD)—SW-NE, and Roughness Length—0.01 m.

The primary input data for Configuration C simulation: Ta—33.6 °C, RH—68.9%, WS—2.5 m/s, WD—NW-SE, and Roughness Length—0.01 m.

In identifying the heat distribution patterns within the street canyon of the selected site, the output files (atmosphere files) generated by the model simulation results were analyzed using LEONARDO. The 2D heat distribution maps for Ambient temperature and Relative humidity of the study areas obtained as a result are shown in Figs. 1.2, 1.3, 1.4, and 1.5. The white polygons in the heat maps represent the adjacent buildings on either side of the considered road site.

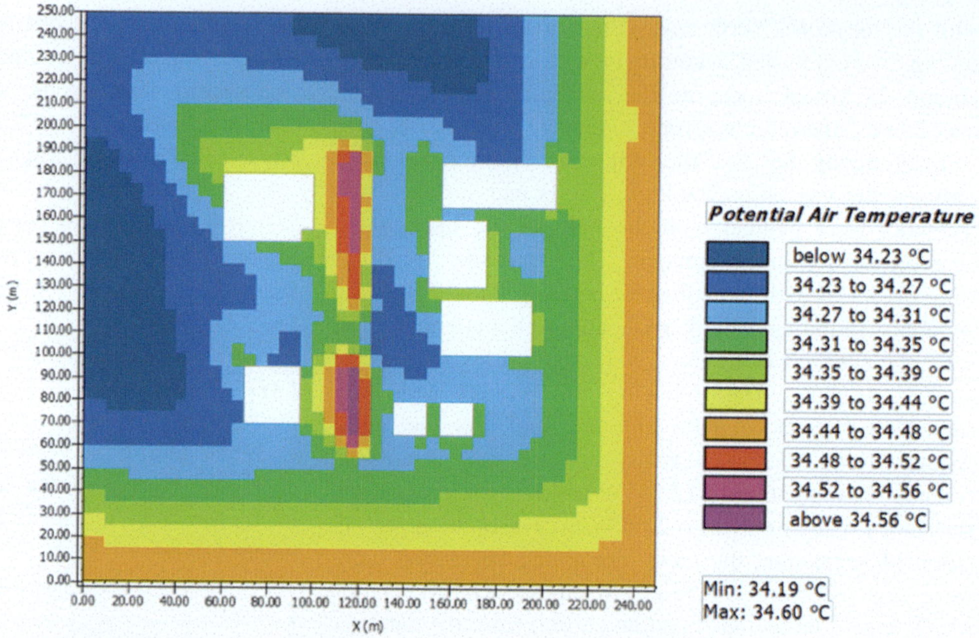

Fig. 1.2 Ambient temperature distribution—Horton's place (11 a.m.)

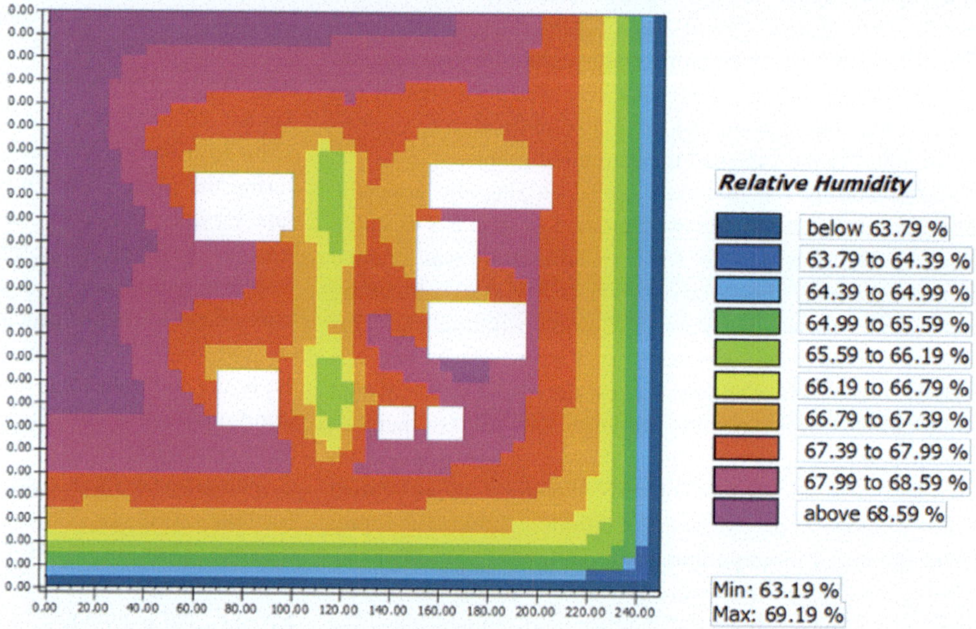

Fig. 1.3 Relative humidity distribution—Horton's place (11 a.m.)

3.1 Tree Configurations A and C

Configuration A:

See Figs. 1.2 and 1.3.

Configuration C:

See Figs. 1.4 and 1.5.

In the study site of Configuration A, an approximately continuous tree row configuration on both roadsides was present with predominating densely foliated trees. According to the results, the value ranges for the ambient temperature distributions of the roadside sites of Configuration A—Horton's place and Configuration C—Barnes place for the considered dates were not significantly different. The values were in the same range of 33–34 °C for both sites at 11 a.m. A higher relative humidity value distribution (Figs. 1.4, 1.5) resulted in the Configuration C site (67.84–69.13%) when compared with that of the Configuration A site (65.59–67.99%). In Configuration A—Horton's place, near the receptor where a large, over 25 m tall densely foliated tree and other associated trees

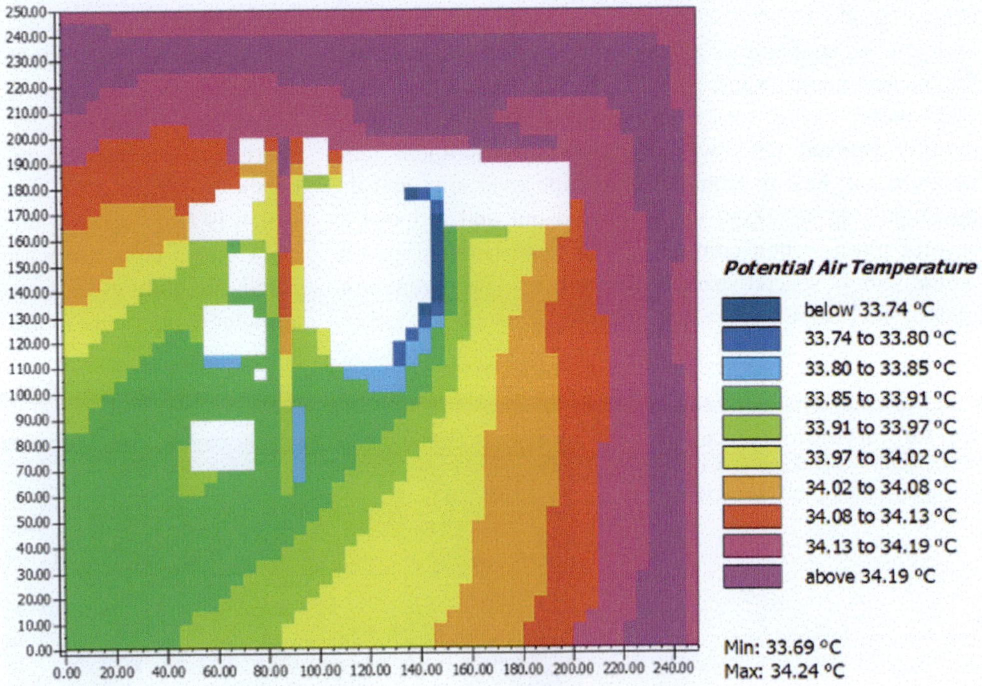

Fig. 1.4 Ambient temperature distribution—Barnes place (11 a.m.)

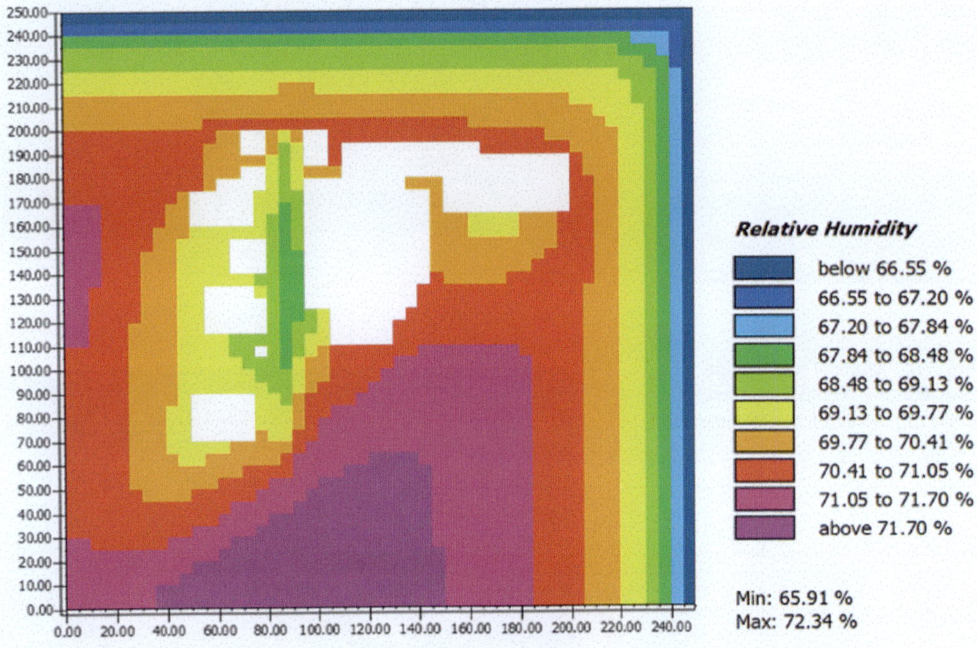

Fig. 1.5 Relative humidity distribution—Barnes place (11 a.m.)

are present, the lowest ambient temperature was observed (below 34.23 °C). The dark blue color distribution to the right side of the road area demonstrates this.

The following ambient temperature heat distribution maps were generated for the scenario comparison, considering four scenarios Mallikarama Road of Tree Configuration B for a particular data set collected on February 3, 2022, at 11 a.m.

3.2 Tree Configuration B—Scenario Comparison

For the scenario comparison, for all four scenarios considered, the following input parameters were made constant in the model according to the actual data obtained: Ta—33.2 °C, RH—72.8%, WS—2.5 m/s, WD—SW-NE, and Roughness Length—0.01 m. When the Base case

(Fig. 1.6) and Control case (Fig. 1.7) were compared, the ambient air temperature at several points on the road and the measurement points were higher in the control case. In Scenario 4 (Fig. 1.9), where a continuous tree row with densely foliated trees was positioned, the ambient temperature and heat distribution above the road area and the pavement regions seem to be thinned out and reduced at several places when compared with Scenario 3 (Fig. 1.8) where widely spaced sparsely foliated tree rows were simulated. In Scenario 3, when sparsely foliated trees were spaced with wide spaces in-between each other at either side of the road, the ambient temperature values averaged 34.04–34.16 °C and above 34.16 °C in the roadside area. In Scenario 4, when densely foliated trees were placed with short distances in-between, the ambient temperature in most of the roadside areas had decreased to 33.66–34.04 °C, which is more cooling than in Scenario 3. However, when the Control case (Scenario 2) and Scenario 3 were compared, the findings indicated no discernible difference in the ambient temperature in the roadside region.

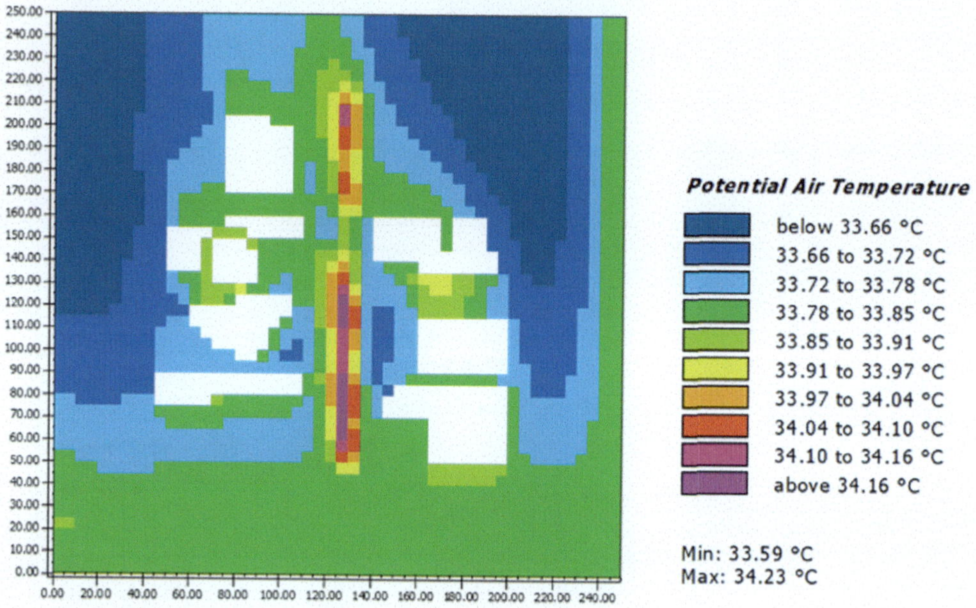

Fig. 1.6 Ambient temperature distribution—Mallikarama road (11 a.m.)—base case; Scenario 1

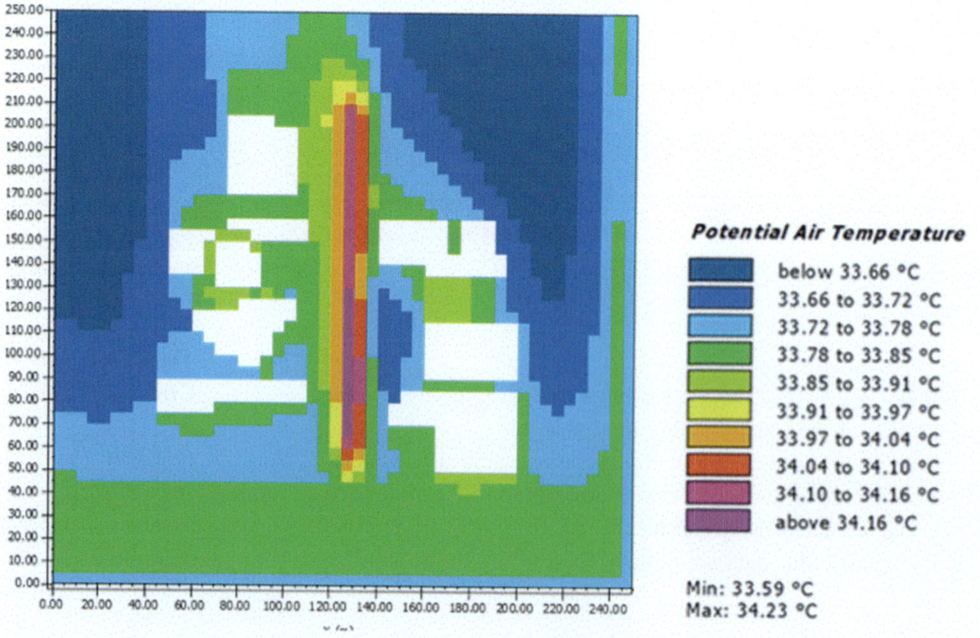

Fig. 1.7 Ambient temperature distribution—Mallikarama road (11 a.m.)—control case; Scenario 2

4 Discussion

The present study evaluated the effects of roadside tree configuration on urban thermal conditions. Evaluating the scenario simulations of the selected site derived using ENVI-met and under the current conditions—according to the results of the comparison between the base scenario (original site) conditions and the control scenario (no vegetation), the base scenario showed cooler regions on the roadside and road area when compared with the control scenario in which any roadside vegetation was included. This result can demonstrate the effect of roadside greening, which reduces the ambient temperature of the considered area. Additionally, the ENVI-met results presented a Ta difference of 0.25 °C between areas with and without greenery in Scenarios 1 (Fig. 1.6) and 2 (Fig. 1.7), respectively. This reduction in ambient temperature and cooling effect could be explained by the shading effects formed by the branches and leaves of roadside trees, which reduces the solar irradiance that enters the region beneath the canopy of the trees.

Furthermore, the evapotranspiration of trees helps lower temperatures in roadside vegetation areas (Srivanit & Hokao, 2013). Moreover, surface temperatures increase when the sun's angle with the surface decreases (Forouzandeh, 2021). During the data collection period section from July to December, the sun did not travel directly over Sri Lanka, which may have reduced the solar angle. Compared to some of the earlier research carried out in comparable tropical nations, this may account for the present study's lower ambient temperature ranges during peak hours.

For the Tree Configuration B site (Mallikarama Road), the base condition (Scenario 1) was compared and contrasted with two greening strategies and a control case without vegetation. This comparison considers only ambient temperature variations and is done to analyze the effect of these modifications to the tree configurations on mitigating the ambient temperature in the study area. For the comparison to be fair and to distinguish hot from cold sections along the roadside area, the temperature range is the same for all four scenarios. The density of buildings, roads, and pavement materials remains constant in all four scenarios, and only the tree configurations are altered. When the heat maps for Scenario 3 (Fig. 1.8) with widely spaced sparsely foliated tree rows on either side of the road are compared with Scenario 4 (Fig. 1.9) with continuous densely foliated tree rows on either side of the road, the ambient temperature distribution has decreased in Scenario 4. This shows that dense green areas with densely foliated trees in close proximity contribute to the cooling effect, which keeps the roadside area cooler during peak hours. The densely foliated aspect of the tree canopy, produced by dense and compacted clusters of leaves which block incoming solar radiation, results in lower surface temperatures and, as a result, lower ambient temperatures (Shinzato et al., 2019). In contrast, hot spots were identified over most of the road areas, as shown with the red-pink color gradient in Fig. 1.8 of Scenario 3, with sparsely foliated trees set widely apart.

Some general inconsistencies in model results and less significant ambient temperature differences among different simulated scenarios may be due to some limitations in the moderately simplified calculation methods related to the tree models and the static wind and cloud settings in the simple forcing function of the ENVI-met model (Yang et al., 2021). The use of model-assumed values instead of using measured data owing to the unavailability of appropriate scientific measuring instruments, including the specific surrounding building thermal attributes relating to specific heat capacity, emissivity, the absorption coefficient of walls, thermal conductivity, and the tree attributes (LAD, foliage albedo, and RAD) (Simon et al., 2018), may have caused irregularities in some ENVI-met output results. Model uncertainty may also have been introduced due to choosing typical (default) vegetation to represent the study area's roadside vegetation systems (Wu & Chen, 2017).

The present study can be thus useful in the identification of heat distribution patterns within street canyons considering selected roadside greenery configurations and determining suitable roadside urban tree planting scenarios to regulate the microclimatic conditions within the urban canyons of Colombo and suburbs, backed by field measurements and the application of the numerical modeling software of ENVI-met. Further research will be required to establish more solid conclusions by considering specific aspects of the roadside thermal environment, such as the surrounding building heat interactions and urban settings, including the aspect ratio differences, the impact derived from pavement and urban surface material variations on road thermal climate, and the impact of locally available tree species specifics (leaf size, branching arrangement, and other micro-scale characteristics) for analyzing their cooling potentials. Moreover, the data is preferable to be supported by an inclusive model calibration and validation through data collection over a more extended period by incorporating continuous daily measurements and mobile measurements capturing the daytime and nighttime impacts of street-side trees on UHI, further including multiple data sites covering a majority of tree and street configurations of Colombo and suburbs.

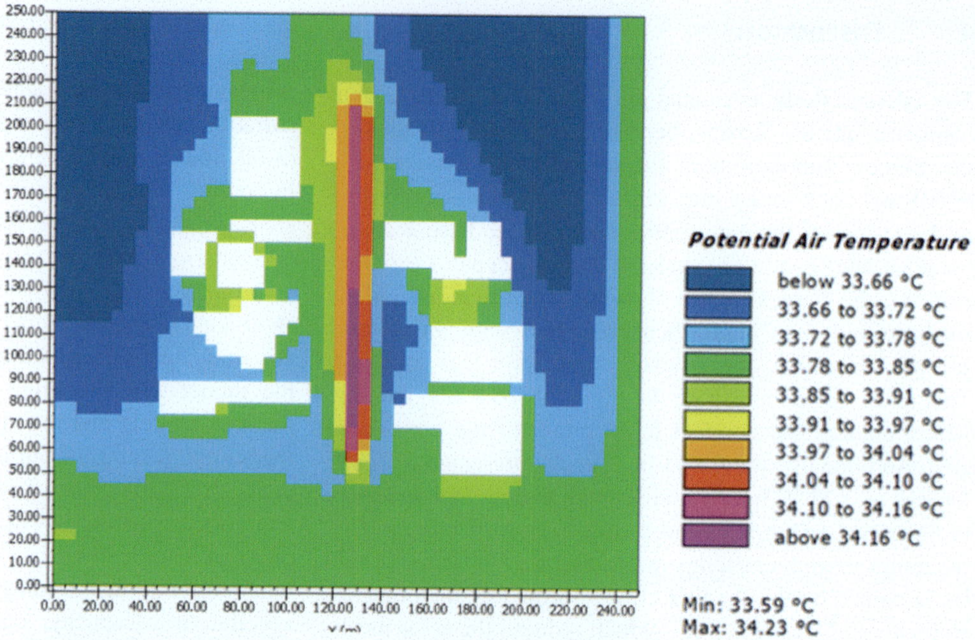

Fig. 1.8 Ambient temperature distribution—Mallikarama road (11 a.m.)—widely spaced tree row (sparsely foliated—low LAD); Scenario 3

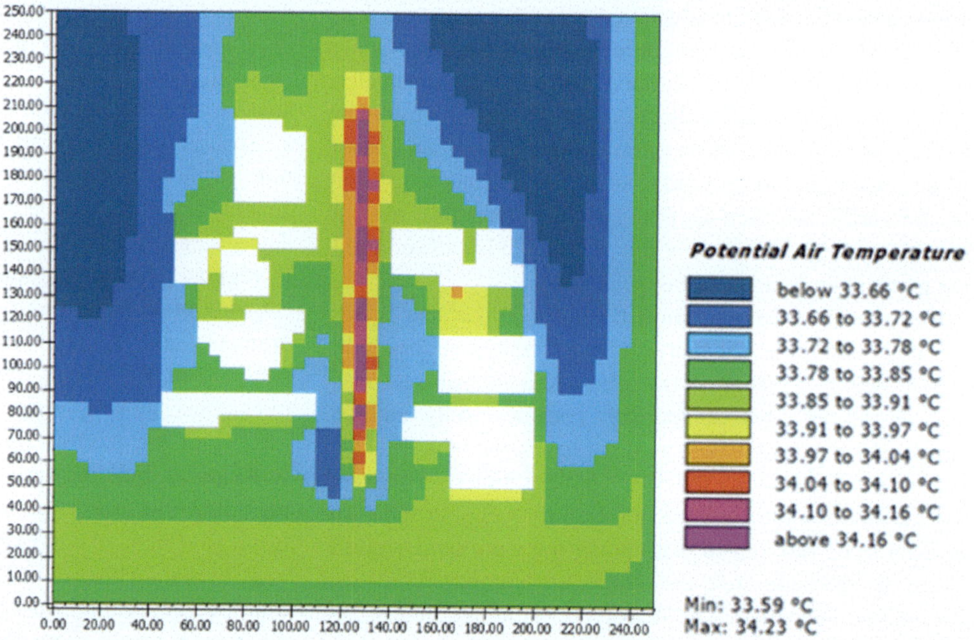

Fig. 1.9 Ambient temperature distribution—Mallikarama road (11 a.m.)—continuous tree row (densely foliated—high LAD); Scenario 4

5 Conclusion

The present study evaluated the effects of roadside tree configurations on the roadside thermal environment in urban canyons of selected sites in Colombo, Sri Lanka, using the 3D microclimate model ENVI-met. The analyzed results and scenarios relating to the effects of street trees on variations in relative humidity and ambient temperature of the immediate thermal surroundings reveal the following conclusions. When comparing the ambient temperature distributions between Tree Configuration A—Horton's place and Tree Configuration C—Barnes place for the selected dates, as shown in the results section, they are not significantly different from each other, and the values are in the same range of 33–34 °C for both the sites at the hour of 11 a.m. This may result from the data of the two tree configurations belonging to two different days contributing to slight climatic differences. A higher relative humidity value distribution is observed in the Tree Configuration C site (without vegetation) compared to that of the Tree Configuration A site.

The following observations are evident in the scenario comparison in the selected site of Tree Configuration B. The absence of trees caused an increase in ambient temperature in the thermal environment. The ambient temperature and heat distribution above the road area and the pavement regions are thinned out and reduced at several places when the tree configuration of a continuous tree row with densely foliated trees (Scenario 4) is used instead of the widely spaced tree row with sparsely foliated tree configuration (Scenario 3), with an overall simulated ambient temperature reduction of 0.5 °C in Scenario 4 when compared to Scenario 3. Based on the aforementioned findings, various roadside tree layouts in urban streets have varying effects on the microclimate at the pedestrian level, underscoring the necessity to formulate urban tree planting recommendations based on these aspects without resorting to ad-hoc roadside planting designs and systems.

References

Aram, F., García, E. H., Solgi, E., & Mansournia, S. (2019). Urban green space cooling effect in cities. *Heliyon, 5*(4), e01339.

Emmanuel, R., Rosenlund, H., & Johansson, E. (2007). Urban shading—A design option for the tropics? A study in Colombo, Sri Lanka. *International Journal of Climatology: A Journal of the Royal Meteorological Society, 27*(14), 1995–2004.

Forouzandeh, A. (2021). Prediction of surface temperature of building surrounding envelopes using holistic microclimate ENVI-met model. *Sustainable Cities and Society, 70*, 102878.

Herath, H. M. P. I. K., Halwatura, R. U., & Jayasinghe, G. Y. (2018). Evaluation of green infrastructure effects on tropical Sri Lankan urban context as an urban heat island adaptation strategy. *Urban Forestry & Urban Greening, 29*, 212–222.

Klemm, W., Heusinkveld, B. G., Lenzholzer, S., & van Hove, B. (2015). Street greenery and its physical and psychological impact on thermal comfort. *Landscape and Urban Planning, 138*, 87–98.

Lee, H., Mayer, H., & Chen, L. (2016). Contribution of trees and grasslands to the mitigation of human heat stress in a residential district of Freiburg, Southwest Germany. *Landscape and Urban Planning, 148*, 37–50.

Monthly weather summary—February 2017 (2017) World Meteorological Organization. Retrieved from https://public.wmo.int/en/media/news-from-members/monthly-weather-summary-%E2%80%93-february-2017

Ranasinghe, A. W. L. H., & Halwatura, R. U. (2012). User comfort on urban roads.

Salata, F., Golasi, I., de Lieto Vollaro, R., & de Lieto Vollaro, A. (2016). Urban microclimate and outdoor thermal comfort. A proper procedure to fit ENVI-met simulation outputs to experimental data. *Sustainable Cities and Society, 26*, 318–343.

Senevirathne, D. M., Jayasooriya, V. M., Dassanayake, S. M., & Muthukumaran, S. (2021). Effects of pavement texture and colour on Urban Heat Islands: An experimental study in tropical climate. *Urban Climate, 40*, 101024.

Shashua-Bar, L., & Hoffman, M. E. (2004). Quantitative evaluation of passive cooling of the UCL microclimate in hot regions in summer, case study: Urban streets and courtyards with trees. *Building and Environment, 39*(9), 1087–1099.

Shinzato, P., Simon, H., Silva Duarte, D. H., & Bruse, M. (2019). Calibration process and parametrization of tropical plants using ENVI-met V4–Sao Paulo case study. *Architectural Science Review, 62*(2), 112–125.

Simon, H., Lindén, J., Hoffmann, D., Braun, P., Bruse, M., & Esper, J. (2018). Modeling transpiration and leaf temperature of urban trees–a case study evaluating the microclimate model ENVI-met against measurement data. *Landscape and Urban Planning, 174*, 33–40.

Srivanit, M., & Hokao, K. (2013). Evaluating the cooling effects of greening for improving the outdoor thermal environment at an institutional campus in the summer. *Building and Environment, 66*, 158–172.

Srivanit, M., & Jareemit, D. (2020). Modeling the influences of layouts of residential townhouses and tree-planting patterns on outdoor thermal comfort in Bangkok suburb. *Journal of Building Engineering, 30*, 101262.

Taleghani, M., Kleerekoper, L., Tenpierik, M., & Van Den Dobbelsteen, A. (2015). Outdoor thermal comfort within five different urban forms in the Netherlands. *Building and Environment, 83*, 65–78.

Vailshery, L. S., Jaganmohan, M., & Nagendra, H. (2013). Effect of street trees on microclimate and air pollution in a tropical city. *Urban Forestry & Urban Greening, 12*(3), 408–415.

Wu, Z., & Chen, L. (2017). Optimizing the spatial arrangement of trees in residential neighborhoods for better cooling effects: Integrating modeling with in-situ measurements. *Landscape and Urban Planning, 167*, 463–472.

Yang, J., Hu, X., Feng, H., & Marvin, S. (2021). Verifying an ENVI-met simulation of the thermal environment of Yanzhong Square Park in Shanghai. *Urban Forestry & Urban Greening, 66*, 127384.

Yang, X., Zhao, L., Bruse, M., & Meng, Q. (2013). Evaluation of a microclimate model for predicting the thermal behavior of different ground surfaces. *Building and Environment, 60*, 93–104.

The Butterfly Effect Toward a Nature Smart Society

Anne Stenros and M. Arch

Abstract

We are living in times of great transformation. In a world of multiple crises, a drastic change is underway, and nothing we are used to will stay the same. We need to unleash our creative energies and actively seek for what's the Next and the New Normal in the future. This transformation is clearing the way for a new society to emerge. Only through embracing change and understanding the very essence of human creative capacity and human-nature connection can we build a future for the betterment of humans and the planet. A Nature Smart Society is the way forward, based on the different types of human needs: sense of trust, sense of consistency, sense of belonging, sense of being valued, sense of being heard, and sense of meaning and purpose. To support all levels of human needs is to create a more resilient culture and a more humane society. A Nature Smart Society is a safe and healthy place for all to flourish in all aspects of our lives: to live, love, learn, work, and play well. It is a natural human habitat of wellbeing, wealth, health, harmony, hope, and happiness as a future goal for humanity and the planet.

Keywords

Creative industries · Speculative architecture · Climate emergency · Gen Z · Regenerative architecture · Chief heat officer · Green infrastructure · Civic engagement · Placemaking · Inclusive city · Participatory planning · Net zero economy

A. Stenros (✉) · M. Arch
Romanian-American University, București, Romania
e-mail: anne.stenros@kolumbus.fi

1 Introduction: To See or not to See

Over ten years ago, in 2010, there was a global conference in Shanghai, titled *Partnership for Urban Innovation*, hosted by Cisco. The focus of the conference was on the urban development of cities and their future innovation potential. After the conference, I wrote an essay titled "The Butterfly Effect—Future Trends in Urban Transformation" (Stenros, 2010). Originally, the butterfly effect concept referred to weather conditions, but later it has also been used as a broad term for any situation where a small change is supposed to cause significantly larger consequences. In the introduction of my essay, I stated:

"We usually tend to think that the butterfly effect is something unexpected that happen—in a negative, dramatic sense. What if… a butterfly effect works for the positive purpose, in a good sense—and creates unexpected, yet desired snowball effect for the positive transformation. In that case, the butterfly effect is all about scaling. Scaling for example the social impact—maximizing good in our environment… Creating a circle of happiness which will affect to the well-being and support the pursuit of good life—for all and everyone" (Stenros, 2010).

In my presentation and later in the paper, I also drafted *The Butterfly Model* with the principles of betterness in the built environment: "Betterness in the built environment is about maximizing the good, i.e., the positive impact of architecture and landscape architecture in the urban environment. It begins from the sustainable values and responsible choices. It stands for creating the sense of place—*Genius Loci*—for people to share memories and mementos. In its way, it integrates art with architecture in order to create inspiring all-encompassing experiences. It respects the human scale, thus representing the value set of caring ethics. It shows how cultivating greenery and supporting biodiversity is a source of wellbeing through experiencing the natural cycle of renewal. The strength of the

I. Pigliautile et al. (eds.), *Urban Resilience, Livability, and Climate Adaptation*,
Advances in Science, Technology & Innovation, https://doi.org/10.1007/978-3-031-54911-3_2

Butterfly Model is a constant strive for the betterness of the quality of human life.

The actions taken are following the principle of the soft path as a model of growth" (Stenros, 2010); see Fig. 2.1.

What struck me today—a decade later—is the accuracy of the future vision of urban city transformation we discussed and envisioned at the conference in Shanghai and, further, as a summary in my paper in 2010. All these major trends of urbanity—environmental and social sustainability, biodiversity, partnerships for innovation, citizen involvement, and the rise of creative industries—also show the spectrum of today's lens to look at the city of tomorrow. What if… the audience would have been more responsive? What if… decision makers and influencers had been more visionary? What if… the business world had been more open to new ideas? What if…?

It took a decade to finalize my thoughts about the betterness of the built environment and the future of urbanity within the concept of *The Nature Smart City* (Stenros, 2022). However, the caterpillar poised to transform into the butterfly was already in existence and underway in my presentation at the urban summit in Shanghai in 2010. "It is a city which is both preserving its heritage and creating future at the same time. It is a living entity with a healthy ecosystem of economy, equality, and environment" (Stenros, 2010).

Three years later, in 2013, I conducted a scenario planning process titled *Future Flow 2030* for KONE Corporation. Together with the German Fraunhofer Research Institute and the Finnish VTT Technical Research Center, we created six future scenarios based on the socioeconomic analysis of megatrends. The scenarios clearly showed six plausible paths toward future society, globally; see Fig. 2.2.

Looking back at those times, it's evident the seeds of the future were already hidden in the present. It also tells me how we can see, yet not understand, what we have seen. These examples of futuring or making the future (visible) never got the attention they should have. The future narratives were clear, albeit imaginary, and the lessons to be learned from them were obvious. Yet, nobody paid attention to the storylines illuminated during the process. By coincidence, I met a KONE colleague years later, after leaving the company. He happily told me that they were now doing all those things I explained and talked about several years ago. Sooner or later we all must face the future of our time. By denying the existence of the future we may buy time, but, at the same time, lose momentum and advantage. Therefore, building future narratives should be a lifetime, on-going path rather than a project or a process. Getting the message when the future is spelled out to you is key to leadership. Exploring IS understanding futures. Seeing the signals, reading the trends, and analyzing alternative narratives will provide a head start and allow first-mover influence when facing future challenges. It is the very difference between a forerunner and a follower. The only way to tackle change is to go with the flow—but be fully aware and prepared for where one is heading. Future speaks to us in the present—or whispers, rather—but it takes a lot of attention, skills, and experience to truly hear what it is saying.

Professor Helga Nowotny has said that one of the metanarratives of modernism is that we are able to make our future (Nowotny, 2016). It is not about good fortune or ill fortune, rather it is all about our actions. Over ten years ago, I tried to speak about the "Pebble Power" of small actions on a large scale and "a small footprint with a large impact". I also pointed out to a future vision of "Smart Cities & Smart Growth" with characteristics such as biodiversity,

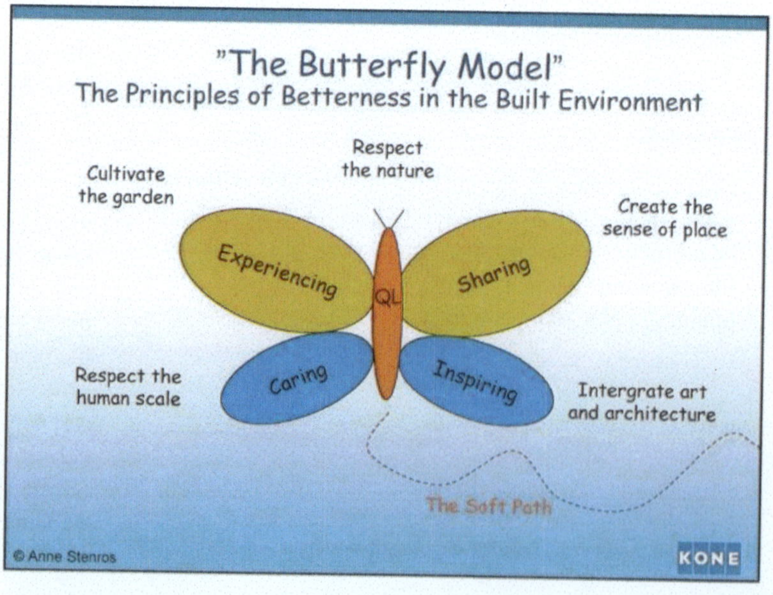

Fig. 2.1 The butterfly model of the built environment. © AS, 2009

Fig. 2.2 KONE future flow 2030. © KONE Corp., 2013

living sustainably, urban agriculture, eco-sustainability, cultivating communities, social relations, design-sensibility, hybrid typologies, and post-sustainability. Now is the moment for the true butterfly effect—the one that will change our path from crises to creativity, from disaster to dialogue, from chaos to coherence and consistency. It is up to us what the future holds for us. But first, we must SEE the future as it is unfolding in front of us. For those of us who can see… the future momentum is open but only for a short time—*carpe diem*!

> Never let a good crisis go to waste.—Winston Churchill

2 The Method: Speculative Architecture and Futuristic Activism

In 2022, a few architects, engineers, and designers formed a group called Power Animals and entered a proposal in the competition for the Finnish Pavilion at the Venice Architecture Biennale 2023. We titled our entry *NATURE WISE CITY—Rewilding Architecture and Urban Design*. The idea was that the Nature Wise City narrative happens around and in the Finnish pavilion, dressing it up as the future Finnish pavilion of the year 2059 (Stenros et al., 2022); see Fig. 2.3.

The framework of our proposal was *regenerative architecture*. Regenerative architecture focuses on revitalizing natural systems by utilizing the living systems that exist on a site. It also addresses climate change and social inequality. The method chosen was *speculative architecture and design*, which tells a story and sets up a scene, and narratives on how different drivers could shape space and culture in the future. It combines architecture fiction, design futures, architectural design thinking, and critical thinking to create alternative futures through stories, artifacts, scenarios, and environments. Our proposal is an example of creative and futuristic activism by activist-planners, architects, and designers representing the future metamorphosis of architecture in 2050 and beyond.

The future world is based on human and planetary wellbeing. Rewilding and oneness with nature are the new building blocks of holistic wellbeing, a healthy life, and a healthy planet. Restoring and rebuilding the human-nature connection is the very essence of human and planetary wellbeing. How to design a happy future in harmony with nature? In our future narrative, architects are gardeners enabling Nature to take over the built environment. Rewilding has replaced digitalization as a paradigm. Food is not processed but coproduced in co-ops by citizens in different ways and forms. The youth is involved in city design—future making is a collective effort based on a shared vision. The prevailing metanarrative, *The Narrative of Progress*, has given way to *The Narrative of Metamorphosis*, moving from globalization to micro-localism and from progress to circularity.

Through our proposal we tried to answer the crucial question: how to rebuild community spirit, the human-nature connection, and restore a sense of purpose?—that is the next big agenda in urban development. How to create a city that is coproduced by its people, especially by the youth, since they represent the future and the next generation of citizens and the city? (Stenros et al., 2022); see Fig. 2.4.

Today, future-minded and creative activists are proposing a more livable and lovable, more sustainable and inclusive, healthier and happier, and less consuming and destroying society. They are not building just imaginary

Fig. 2.3 The Nature Wise City. A proposal in the competition for the Finnish Pavilion at the Venice Architecture Biennale 2023 © Power Animals, 2022

utopias; instead, they are relying on the facts and data available.

They can plan, design, and simulate entire cities, regions, and societies in detail before finalizing their plans. They can test their visions against local data available and get local people involved in ideating, designing, and executing the plans. Different creative activist groups will be the makers of the future. They are the beacons that shine a light on causes they care about and guide the public in the direction of progress. They work through signs, symbols, stories, and spectacles. As Duncombe and Lambert say: "Our modern political terrain is a highly mediated landscape of signs and symbols, stories and spectacle. For us to stage successful battles on this cultural topography we need to observe, think, analyze, and respond creatively. We need to become artistic activists." (Duncombe & Lambert, 2021, 12). The first rule of guerilla warfare by Duncombe and Lambert is to know the terrain and use it to your advantage. By studying plausible futures, it is possible to understand alternative futures and their drivers.

To practice creative and futuristic activism is to create narratives from signals of change. The scanning of signals is the first building block, to recognize what we have NOW. Analyzing current megatrends and short-term trends is understanding the nature of the CHANGE that is underway. By building scenarios, we are creating a picture of the AFTER the change. And finally, by creating the NEXT narrative, it is possible to see what is going to be the new normal in the future; see Fig. 2.5.

Activism is also a criticism toward the prevailing metanarrative. At its best, it is an attempt to rewrite a new societal narrative and offer a better alternative to the existing

Fig. 2.4 The Nature Wise City. A proposal in the competition for the Finnish Pavilion at the Venice Architecture Biennale 2023 © Power Animals, 2022

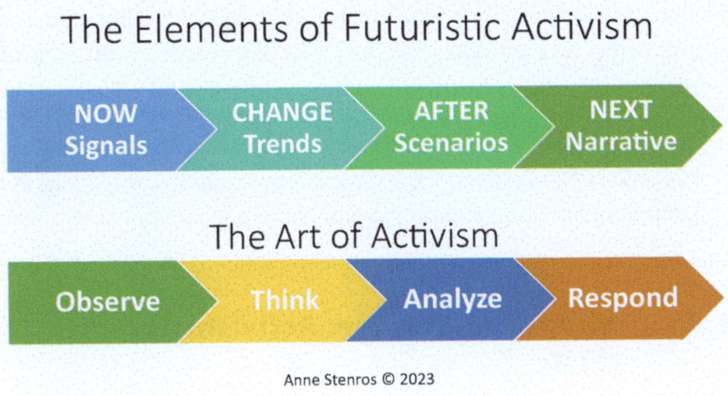

Fig. 2.5 The elements of futuristic activism © AS, 2022–23

and prevailing society. Activism is a kind of tipping point of transformation, a loud voice showing the way out of a crisis.

The work of creative activists—activist-artists, activist-designers, and activist-futurists, among others—is a positive transformative force. It is like creating a butterfly effect for betterment: step by step, they are making our society and our world better. But this "pebble power" needs all of us to join the movement. We can say that creativity needs activism and activism needs creativity, as Duncombe and Lambert have stated. Practicing creative activism as a way of life is shaping creative culture toward a nature smart society.

3 What—The Now: From Protestors' Energy to Creative Culture

The American social philosopher Eric Hoffer had this to say in his classic book on change titled *The Ordeal of Change* (1963): "The remarkable thing is that it is the crowded life that is most easily remembered. A life full of turns, achievements, disappointments, surprises, and crises is a life full of landmarks. The empty life has even its few details blurred and cannot be remembered with certainty" (Hoffer, 2006, 2).

Change is an inevitable part of human life and being a human. And yet, we mostly do not like new things, since we are not prepared for the change to come. However, change is a chance to grow. But what comes after the change has happened? How is society after transformation? What is the new normal? It's all up to us: adaptation depends on the amount of preparation that we have done.

Today, the pace of change is remarkably fast on a global scale. New ideas, new movements, new technologies, and new crises are challenging our everyday lives with a constant, ongoing flow of change. It takes a lot of energy to adapt to new situations. Resilience is a necessity to survive, and constant re-skilling and upskilling is a must. My father used to say of work that every day you learn something new is a good day. Today, one MUST learn every day new knowledge, information, and skills to be able to keep up with constant change and its demands. Change is the name of the game.

Change is also a call for action. As Hoffer has said: "A population subjected to drastic change is a population of misfits—unbalanced, explosive, and hungry for action. Action is the most obvious way by which to gain confidence and prove our worth, and it is also a reaction against loss of balance." And he continues by stating that since drastic change is one of the situations which release our energies, "there must be an abundance of opportunities, and there must be a tradition of self-reliance" to make effective use of this sudden boost of energy (Hoffer, 2006, 4).

While listening to the European Creative Industries Summit in 2022 (Prague, September 21-23), the speakers referred several times to the current challenges we meet in Europe: the climate emergency, the conflict crisis, energy crisis, aging population, drift toward populism, and refugee crisis, just to mention a few. MEP Christian Ehler summoned the current critical situation by saying that "we are in the middle of economic and social upheaval, and nothing will stay the same". He also emphasized how to help citizens with a positive message and encourage a change of behavior.

"In the case of drastic change, the uneasiness is of course deeper and more lasting. We can never be really prepared for that which is wholly new. We have to adjust ourselves, and every radical adjustment is a crisis in self-esteem: We undergo a test; we have to prove ourselves" (Hoffer, 2006, 3). This statement by Eric Hoffer shows the essence of transformation: our behavior can be unexpected and uncontrolled, also at the societal level. To cope with transformation, unleash the hidden energies.

In the very same ECIS 2022 summit (September 21-23), several statements supported moving from a sense of crisis to acts of creativity, as Kirsten Dunlop, CEO at EIT Climate-KIC, stated "from protesters' energy to makers' energy", by emphasizing:

- Creativity is key to enhancing innovation for a more resilient economy and sustainable future for the society
- In the future, it's all about creative economy and creative industries
- We need a movement by creative minds from different disciplines to create cross-innovations
- We should be looking for a mission-oriented change for the Net Zero Future
- Narratives are powerful tools for change, and
- Creative practitioners know how to "create something from nothing".

Keeping in mind what Hoffer said—that there should be an abundance of possibilities to use the energy during a drastic change—creativity, creative practitioners, and creative acts can offer a way to approach the challenge of a more resilient and sustainable net zero future. Based on EU statistics, creative industries or cultural employment represent only an average of 3.6% of total employment in Europe today (Eurostat, 2020). But creative industries have a louder voice than their numbers. Their presence is also lasting; as Dr. Johanna Leissner, Chair of EU OMC Group, reminded at the ECIS 2022 summit, cultural heritage is the creativity of the past and raised a question of how to make cultural heritage become climate neutral. The answer was to invest in the upskilling of experts. At the end of the summit, Kirsten Dunlop from EIT Climate reminded: "Crisis builds creativity". She continued by saying that small groups making narratives and stories can construct the net zero future. If we support creative culture, it will support the transformation.

Creatives are storytellers by nature. They are visionaries who "see" the future before we face it. Therefore, it is crucial and even essential to listen to their narratives. The building blocks of creative culture are (1) creativity and creative minds, (2) creative practitioners, (3) creative industries, (4) creative economy, and (5) creative actions as an outcome of the former steps. Sir Ken Robinson has said that all people are creative by nature, but it depends on how we cultivate, enable, and support the caterpillar for it to become a butterfly. A society which enables, empowers, and nurtures creativity and creative culture is better prepared against the challenges and crises to come. To overcome the great challenge of the net zero future, for example, we will need creative minds to collaborate, co-create, co-design, and co-produce alternative future narratives of positive futures; see Fig. 2.6.

Steve Duncombe and Steve Lambert say in their book *The Art of Activism* (2021) that culture is the raw material of artistic activism. "Culture is the operating system of humanity, and if we are going to change the way humanity operates, we need to be able to hack the culture." (Duncombe & Lambert, 2021, 132). Therefore, culture and creative culture are both the means and the end of transformation in society at large.

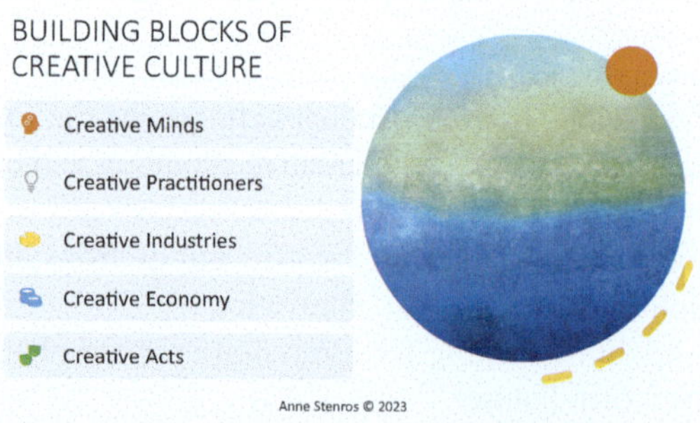

Fig. 2.6 Building blocks of creative culture © AS, 2022–23

> Learning to cope with uncertainty is one of the most precious cultural resources.—Helga Nowotny

4 Who—The Change: Activism as a Way of Life

There is a strong people-power phenomenon emerging currently: the loud voice of young people, especially the Gen Zers (born between 1995 and 2010), who are angry and unafraid to speak up. Technology has given the young the power to express their opinions and have a wide public impact. Many Gen Zers are starting a life in activism at a young age since they have experienced many great challenges during the past years: climate change, inequality and social unrest, political division, the pandemic, and economic distress, among others. Greta Thunberg was only 15 years old when she started the climate movement of Fridays for Future in 2018. Gen Z's activism looks different than the movements of the past: they are mobilizing out of fear and necessity. They have a strong sense of crisis and existential threats. With tech-enabled devices providing constant access to news, it is difficult to escape the world situation. Global data shows that 70% of Gen Zers are involved in a social or political cause. They most likely boycott a product, company, country, or state because of a political, social, or environmental point of view. Activism as a way of life is here to stay (Carnegie, 2022).

A good example of Gen Zers is the Finnish vegan activist Benjamin Pitkänen, a 25-year-old student of economics and social history. He is a founding member of Viral Vegans, a society which aims to finish all animal production. The society wants to influence both individuals and the society at large. Their major argumentation is based on animal rights. Pitkänen considers himself a global citizen, who has lived in the UK, Uzbekistan, and Tadzhikistan and resides currently in Finland. His major channel for influencing is TikTok. He is also a frequent speaker at schools (Parikka, 2022).

One way out of the current crises and drastic changes is to enable and encourage creative activism and creative collaboration. Each creative collision triggers innovative processes, and vice versa: innovation is what ignites creativity and creative collision. Ultimately, creativity is by nature questioning, challenging, and change making. The rise of citizen activism as a way of life supports creative activism in many aspects. Creative activity will carry us through times of change by offering a positive channel for the energy needed to make the better happen.

In their book on activism, Duncombe and Lambert underline the fact that artistic activism is not the preserve of the privileged; instead, cultural creativity and artistic activism flourish among communities that are marginalized within formal spheres of politics, law, and education. They also emphasize that the building blocks, the symbols, and stories that give art its content and form differ among different people and places. We should use culture to help transform culture, they say (Duncombe & Lambert, 2021, 167).

One aspect of creative culture and activism is futuristic activism. When considering the art of futuring, the visionaries are activists—and vice versa. One of the best-known examples of futuristic activism was the post-war group of young architects in London, called *Archigram*, the avant-garde neo-futurist architecture group, which existed from 1961 to 1974. During the 60s and 70s, they created several futuristic concepts for cities: The *Come and Go Project* (1963), which imagined a city as an integrated network of services and communications and led to *The Plug-in City* (1964), and *The Walking City* (1966) and *The Instant City* (1968–70), which was a concept for a nomadic, temporary city. It is said that "they acted against what they saw as a tediously conservative environment, not because of radical political sentiments, but because of the inability of art and architecture to keep pace with the products, lifestyles, and machinery that were already part of daily life". We are seeking the living city, they claimed (Andersson, 2017).

About sixty years later, in 2022, Archigram founder Sir Peter Cook exhibited his latest "*City Landscapes*" at the Louisiana Museum of Modern Art, Denmark (Barandy, 2022). The 85-year-old maestro is showcasing his visionary ideas about new cities and life forms through his drawings.

One of the most recent city scenarios, the *Inhabited Wall* (2020), presents a hill full of cave-like dwellings that nature has taken over. Only the scattered windows on the slope remind us of human presence in the green landscape. The city has finally become a living entity, like Cook forecasted and proposed already some sixty years ago. It is an image of the man-made and natural world living in harmony.

> Art needs activism and activism needs art.—Steve Duncombe and Steve Lambert

5 Why—The After: Urban Resilience

We are living in times of great transformation. After the last few years, plagued with crises, a drastic change is underway, and nothing will stay the same as what we were used to. We need to unleash our creative energies and actively seek for what's the Next and the new normal in the future. The concept of the "*Next Normal*" was coined by McKinsey in 2020 to describe the unprecedented new reality that would emerge as "a dramatic restructuring of the economic and social order in which business and society have traditionally operated." (McKinsey, 2020). This transformation is clearing the path for a new society to emerge. Only through embracing change and understanding the very essence of human capacity and the human-nature connection can we build a future of betterment for humans and the planet. It is a step-by-step journey toward a healthier planet, a healthier society, and a more resilient and sustainable man-made environment.

Now that globalization has been revealed for what it is, we are moving back to more local solutions and networks. We are currently entering the age of post-globalization and moving back toward micro-localism. Instead of the old narrative of progress and the quest for public happiness, a powerful, new metanarrative has emerged: that of Nature Smart Society, which is based on nature wisdom and the quest for human-nature wellbeing. It confronts the prevailing metanarrative of techno progress and replaces it with a narrative of harmony, health, humanity, and hope.

Nature Smart Society is a way forward based on the hierarchy of human needs: sense of trust, sense of consistency, sense of belonging, sense of being valued, sense of being heard, and sense of meaning and purpose. To support all levels of needs is creating a more resilient culture and a more humane society. Creativity as soft activism is a transformative power when coping with this change and its great challenges. Building a strong creative culture with citizen participation is a must when keeping hope, harmony, and happiness as a future goal for humanity and the planet.

Since cities are responsible for 70% of green gas emissions, it is natural to focus on cities and urban environment when rethinking how we live, work, and travel. The European Union has set as its mission to have 100 carbon-neutral cities by 2030. To support this transition and huge transformation, the EU has launched a program called CURE-Copernicus (CURE Copernicus, 2022). The focus of the program is on the nexus between climate change and digitalization. Because cities are the make-or-break points when it comes to carbon emissions, more facts, data, and science are needed to support a full-scale city transformation. "You cannot manage what you cannot measure," as Laura Hetel from EU DG Research and Innovation said at the CURE workshop (October 14, 2022). She even proposed the position of a City Science Officer in every European Union city in the future. Their task would be to create a science policy.

At the very same CURE workshop (October 14, 2022), Prof. David Ludlow said that we are going to see a revolution in urban living. That will also include the soft side, including citizens' participation. Both local experimentation and public engagement are going to be fundamentally important in the future, according to Ludlow. The mission is to deliver net zero neighborhoods. The key building blocks of resilient cities and neighborhoods are green and blue infrastructure and nature-based solutions.

As Professor Emeritus Helga Nowotny has said, future making is a collective act. We will need all stakeholders to build a more sustainable, equal, inclusive, and resilient future for a better quality of life for nature and humans. Co-design, co-creation, and co-production are the future approaches for any local initiatives. Sociologist Richard Sennet launched the concept of the *Open City*, where citizens can participate in the designing and decision-making of the built and natural environment. In a recent book—*Designing Disorder* (2020)—architect Pablo Sendra and Prof. Richard Sennett propose what they call "infrastructure for disorder" which combines architecture, politics, urban planning, and activism to develop places that nurture and support the inhabitants rather than stifle, and bring people together rather than divide (Sendra & Sennett, 2020). Their book is a manifesto for future cities where the citizen experience is at the heart of planning and implementation. The placemaking movement, which involves citizens in the designing of local communities, is spreading fast throughout Europe and transforming the traditional way of practicing city planning and urban design.

Going upwards in human needs is easier than going down. The classic pyramid of Maslow's Hierarchy of Needs shows the essentials, physiological and safety needs, as the basis of society. The higher we go, the more it is about personal needs. However, if there is a threat against the physical environment and safety in society at large, we read just our preferences and focus on adapting to the change. The bedrock of any society is the sense of trust, safety, and consistency, which support the basic needs.

If we consider the society as a whole—and especially cities and citizens—the very same needs are relevant and essential. How to respond to the spectrum of needs all the way from building a sense of trust to being heard and having a voice? The existential threats against the basic needs, physiological needs and safety needs, are always the most urgent. Therefore, climate emergency and actions against it should be high on our list of priorities. On the other hand, we have seen during the past few years how previously unheard voices have become louder and louder around Europe and the world. People feel suppressed and show their anger and frustration by demonstrating and even rebelling to support their psychological needs. The creative way to channel this energy is to support the needs of citizens at every level of hierarchy.

Especially in terms of an urban environment, there have been several new meaningful approaches to strengthen the sense of safety, of being valued and being heard through new ways of city planning, civic engagement, and placemaking. Old citizen participation is the new citizen involvement on many levels of planning, designing, producing, and using both the built and natural environment.

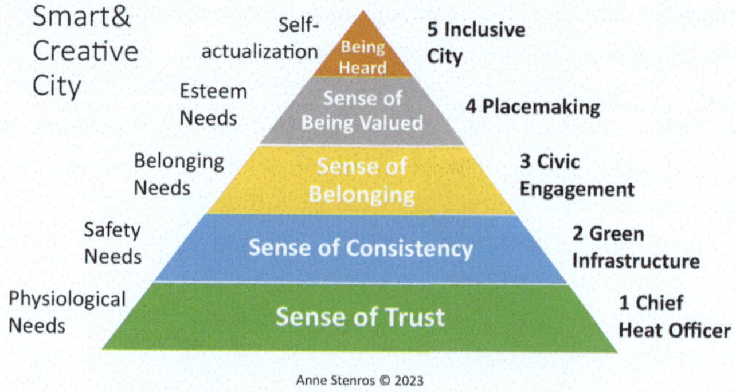

Fig. 2.7 Smart and Creative City © AS, 2022–23

In future foresight, the focus is always on the Next after the Change. The Next or New Normal is the most difficult to foresee and understand based on the present context. Currently, there are several signals and trends supporting the rise of naturalistic intelligence or nature smart throughout society. The Next—or The New Normal in society—is going to focus on the basic needs of humans and nature. The Nature Smart Society is all about healthy cities and healthy social environments; it is about living a healthy life and being safe as well as being a part of the society and having a voice.

On the city level, several new approaches and experiments are becoming mainstream during the coming decade. Among those are (1) the new and important role of the *chief climate officer*, (2) the strong development of *green and blue infrastructure*, (3) a new focus on *civic engagement* through (4) local *placemaking* and (5) as a beacon, the idea of the *inclusive city* where no one is left behind; see Fig. 2.7.

> Study nature, love nature, stay close to nature. It will never fail you.—Frank Lloyd Wright

6 How—The Next: The Nature Smart Society

6.1 Sense of Trust: Managing Resilience in Cities

The summer of 2022 was unusually hot in all of Europe. River water levels fell historically low during a long, ferociously hot summer and heatwaves were unbearable in major cities throughout Europe. There is a sense of urgency in cities to rebuild and sustain resilience to cope with extreme weather conditions and their implications. Therefore, cities are looking for new strategic ways to achieve this.

In 2021, the City of Athens appointed Eleni Myrivili as its first *Chief Heat Officer*—the only person in Europe to hold such a position. Her job is to find ways how to cope with ever hotter heatwaves in Athens. The role of Chief Heat Officer varies per city, but generally involves working with different departments and groups as well as raising awareness among the general public about the threats of heatwaves. Their purpose is to raise awareness of extreme heat risks to protect the most vulnerable people within the city (Salako & Frost, 2022).

In the future, we will see more and more—not only chief heat officers—but *Chief Climate Officers* within city organizations conducting and guiding city development toward a more resilient and safe future. They will be responsible for the overall resilience strategy of a city in a permanent way. Together with local data providers and new technologies, it is possible to simulate the impact of planning and design with great accuracy to understand the impact on microclimate.

6.2 Sense of Consistency: Rebuilding Green and Blue Infrastructure

In recent times, in the aftermath of pandemics, many European cities have recognized the importance of urban greenery. Cities are putting a lot of effort into improving the quality of life for their residents by adding more greenery. As heatwaves become more frequent and harder to handle, European cities will need to prioritize green spaces as a powerful tool against scorching temperatures. Paris has addressed the problem by planning to plant 170,000 new trees by 2026 to keep the city cool. In addition to green spaces, also blue spaces, coastal environments are growing in importance since they have been shown to improve our health and overall wellbeing. Based on recent studies, water—just like green spaces—is positively related to health, wellbeing, and happiness (Hunt, 2019).

A structural and strategic response to the challenge is to pay attention and plan in detail a *green and blue infrastructure*. The European Commission has developed a *Green Infrastructure Strategy based* on ecosystem services. The strategy aims to ensure that the protection, regeneration, restoration, and creation of green infrastructure becomes an integral part of overall spatial planning and territorial development whenever it is an alternative or complementary to standard gray choices. "Green infrastructure refers to a strategically planned network of natural and semi-natural areas with other environmental features designed and managed to deliver a wide range of ecosystem services such as water purification, air quality, space for recreation and climate mitigation and adaptation. This network of green (land) and blue (water) spaces can improve environmental conditions and therefore citizens' health and quality of life." (EU, 2022).

There is also a new way of thinking in architecture: *regenerative design*. It is more than sustainable design, as it gives back by actively contributing to the ecosystem. Said by one of the leaders of the movement Jennifer Wehling: "Regenerative design isn't just about saving the planet; it's saving the humans too. We have to ask more from building design. That means implementing architectural strategies that address air pollution, clean water, repair forests and decrease the heat island effect." (Morton, 2022).

Focusing on green infrastructure—nature as a critical element on the side of gray infrastructure (roads, bridges, rails)—is a fast-growing trend in urban planning. Many big cities, like Glasgow, Milan, Madrid, Amsterdam, and Vancouver, have started to plant trees and forests within the city. Also, many data-based platform services, such as geodesign, have already been launched to help develop green infrastructure locally. This development is the basis for the Nature Smart City and Society.

6.3 Sense of Belonging: Openness for Civic Engagement

The concept of *The Open City* developed by sociologist Richard Sennett and put in practice by architect Pablo Sendra proposes to design spaces and systems purposely open to diversity and adaptable to change. Sennett says: "In the past, thinking about democracy focused on issues of formal governance; today, it focuses on citizenship and issues of participation. Participation is an issue which has everything to do with the physical city and its design". Citizen engagement has become one of the key tools of urban design, its ongoing change, and future transformation. In the heart of this development is the citizen experience, placemaking, and co-production of activities. To demonstrate openness by practicing urban design and architecture, Sendra says: "We intend architecture as an open framework that dynamically responds and adapts to the constantly changing needs of users and the context. Buildings that become demonstrative devices of sustainable principles, and catalysts, enablers or facilitators of urban change and activation." (Sendra & Sennett, 2021, 35).

In their book, Sendra and Sennett also discuss the politics of planning and the role of activism in city planning and urban design. Pablo Sendra said that the alliance between planners and activists was something very positive,

Fig. 2.8 Dreamhamar, plan for city park. © Ecosistema Urbano, 2011–2013

and Richard Sennett emphasized that what the activist-planner does is mediate between civil society and the state. Said by him: "But what it *isn't*, is that the planner confuses his or her role with that of being a political activist. We have a different set of skills that we bring to bear in these situations." And Sendra summarized by saying that the architect-planner must show what the experiment could open up for people, what possibilities there are for them to choose from (Sendra & Sennett, 2021, 130).

Ecosistema Urbano, an architectural office based in Madrid, is a forerunner in this field. They state that they "bring a flexible way of thinking about how design and creativity can help to cope with urban vulnerabilities and threats. Combining technical interventions with a social agenda, our work helps to build resilient communities around the principles of participation and environmental education." (Ecosistema Urbano, 2022). In the pyramid of needs, the sense of belonging is essential for psychological and social wellbeing. Civic engagement is all about the people, for the people; see Fig. 2.8.

6.4 Sense of Being Valued: Placemaking with People

One example of how to foster civic engagement and community participation is placemaking with locals. *Placemaking* has grown into a strong movement during the past years in Europe. The Project for Public Spaces (PPS), an organization that promotes placemaking initiatives in different countries around the world, started using the term in the 1990s. Ethan Kent, the Executive Director of Placemaking X, says: "Placemaking is a form of community organizing, facilitating, bringing people together and challenging communities to get involved. We see it as a new environmental movement that focuses on place as a way to bring together many different issues and causes in a city". Placemaking means creating places and focuses on transforming public spaces to strengthen the connections between people and these places. Placemaking is a process centered on people and their needs, aspirations, desires, and visions, which relies strongly on community participation (Moreira, 2021).

A good example of the placemaking movement is the *Edible Cities Network*, which is an EU-funded project designing and implementing urban food innovations and sustainable urban planning for greener, more edible, and, above all, more livable cities. The EdiCitNet project brings together diverse people from across the world to find ways of tackling urgent urban challenges—from climate change to social exclusion. To do that, they're co-creating and testing edible, nature-based solutions in living labs, as well as co-developing master plans for Edible Cities of the future (Edible Cities, 2022). In the heart of their network and actions are the locals—as Helen Gallis, an environmentalist and a placemaker from Nabolagshager, said in the Edible Cities network: "The community is the expert".

6.5 Sense of Being Heard: Fighting for Inclusive City

In 2022, it is no surprise that one of the emerging issues in urban design and society is the *Inclusive City* and city for all. Already by the end of the 90s, the City of Vienna started to collaborate with female citizens to improve the quality of the urban environment. The latest development on the road is the HerCity toolbox, which is a collaborative urban development initiative by UN-Habitat and the Swedish think tank Global Utmaning (Global Challenge). The HerCity platform involves girls in urban development to make the city better for everyone. Said by HerCity: "Girls plan and design with diversity and different needs in mind. Participatory processes are key for planning a city that works for everyone. If we let citizens that are rarely heard be the experts, our cities and communities will become more inclusive, equal, and sustainable. The purpose of this initiative is to make methods and tools available to urban actors and cities globally." (HerCity, 2022).

Starting with Jane Jacobs in the 1960s and continued by Kevin Lynch in the 1970s, the voice of citizens, especially that of youth and women, has grown to become almost mainstream. On the other hand, we must remember not to take the inclusive city for granted. French author Annie Ernaux, the voice of women and the winner of the Nobel Prize in literature in 2022, has said: "You have to fight for women's rights constantly" (Kanerva, 2022). In the same way, the inclusive city is achieved every single day, over and over again; see Fig. 2.9.

Fig. 2.9 W "We don't need another hero". Statement on the sports fields of the young. © AS, Helsinki 2022

6.6 Sense of Meaning and Purpose: The Butterfly Effect

Long after creating the original Hierarchy of Needs, Maslow added the 6th and final need to his model in the 1970s. He called it *Self-Transcendence* or the sense of meaning and purpose. Self-transcendence is about transcending the self to focus on something "bigger". It is a shift in value from the self to something greater—another person, society as a whole, the universe, etc. Self-Transcendence is the realization that you are a small part of a greater whole, and that you must take responsibility for the world and other people (Davis, 2019). In my mind, this final need is the crucial element and driver in the future to survive as a society and a planet. Nature smart is a state of mind to see beyond oneself and focus on what is essential to survive as a whole, as a planet: doing and living well, and supporting the wellbeing of nature and humans together.

The building blocks of Nature Smart Society are (1) *Managing Climate Emergency* through a strategic approach, (2) *Reinforcing health and wellbeing* through blue and green infrastructure, (3) *Advocating citizen engagement* in urban design, (4) *Supporting community building* through placemaking, and (5) *Creating Inclusive City* with equal voices. In addition to these "soft power" elements, we will also need hard power, i.e., (6) *Purpose-Driven Economy* and net zero economy.

A good example of the latter is the recent news in the clothing company Patagonia. The owners, Mr. Chouinard, his wife, and their two children, have transferred their ownership of Patagonia, valued at about $3 billion, to a specially designed trust and nonprofit organization called the Patagonia Purpose Trust. The trust will oversee the commitment to run a socially responsible business and give away its profits. This kind of business activism we will hopefully see more of. Already a few years ago, Patagonia launched a new brand slogan and promise: "Making the best product matters for saving the planet". Said by Mr. Chouinard, the co-founder of the company: "Hopefully this will influence a new form of capitalism that doesn't end up with a few rich people and a bunch of poor people. We are going to give away the maximum amount of money to people who are actively working on saving this planet." Patagonia has already donated $50 million to the Holdfast Collective, and will contribute another $100 million this year, making the new organization a major player in climate philanthropy (Gelles, 2022).

When looking for net zero economy in general, radical change is the only way forward, as stated in the latest Net Zero Economy Index (NZEI) report by PWC. The fact is global carbon intensity fell only by 0.5% in 2021. Said in the report: "Achieving a 15.2% annual rate of decarbonization is now needed to bridge the gap between the climate emergency and a strong and sustainable economy." (PWC, 2022).

The new economic models circular economy, doughnut economy, purpose-driven economy, and net zero economy are all steps toward the new metanarrative of the future, replacing the old one of endless progress. The new metanarrative is about caring, sharing, creating, enabling, and protecting—it is living and doing based on lessons from nature. Nature Smart Society is not a single entity to follow, rather it is a mindset and an evolutionary, open path to take: "*I took the one less traveled by, And that has made all the difference*", as the poet Robert Frost wrote in his famous poem *"The Road Not Taken"* (1916).

The wisdom of nature is based on continuous change with sometimes drastic transformation. Evolution takes time—and like Charles Darwin famously said: "It is not the strongest of the species that survive, nor the most intelligent, but the one more responsive to change." In our current state of global crises, the only way out is a creative response to challenges. It calls for creative activism, and that's the butterfly that no one can predict and foresee, as creativity will always surprise us. Only through nature smart society can we achieve resilience, which is the basis for a healthy planet, healthy cities and social environment, and healthy citizens. That is the true butterfly effect.

> We delight in the beauty of butterfly but rarely admit the changes it has gone through to achieve that beauty—Maya Angelou

7 When—The Future: Embracing a New Global Order

Being nature smart, i.e., having naturalistic intelligence, as Howard Gardner defined it in his model of multiple intelligences, is following the natural flow of life and respecting it in its varied forms. It is about human rights as well as animal rights, and also understanding the role of diversity in nature. It is about embracing change, too—as a natural order of life—and embracing creativity—as a way out of chaos to create new order. But above all, it is embracing diversity—the way of the natural world to support the soft power of resilience through collaboration, co-existing, and co-creation culture.

The more we support creativity and creative thinking and problem solving, the more diverse solutions we will get as an answer. Experimenting, creating scenarios, and unleashing our creative energy and potential IS embracing diversity—and an inevitable change in life. And that is the true nature of creative culture: make everyone shine, in their own way and to the fullest. Nature Smart Society is about embracing change and diversity through a collaborative, creative culture.

Nature Smart Society is about unlearning, and the relearning of future urban narratives. It also means reframing the old problems to design as more meaningful problems and finding novel solutions. The question is, how to support a responsible creative culture? One answer is life-centered design, as legendary Bruce Mau has said: "Focus on designing for the welfare of all of life in order to sustain human life". The more we learn about nature, the more we learn about ourselves.

This is the question we should all ask ourselves: what is my role in building a more sustainable, resilient, human, and natural society that is inclusive, supports overall wellbeing and health, with a sense of community and hopefulness? And, in the end, achieves a sense of meaning and purpose as fulfillment?

In nature, everything is based on wellbeing and health on a microlocal scale. If that works, then the entire system is in balance. Everything in nature is also connected, but not at a cost to others. In nature, efficiency is not self-purposed; instead, it serves resilience and is based on local conditions. And, finally, a natural place as a human habitat is the nucleus of a healthy life well-lived; see Fig. 2.10.

The Ancient Greek idea of happiness, *Eudaimonia*, was based on doing and living well through virtues. It was an active mode of taking responsibility for one's own happiness by doing good to others. That is the future way of a nature smart society: doing and living well and in harmony with humans and nature. It is a life of *happiness, harmony, health, healing, humanity, and hope.*

A Nature Smart Society is a healthy societal system which is inclusive, diverse, and locally driven with place-based prosperity; it is also a well-networked, flexible, and resilient system with a citizen focus. It is made by many with a shared vision of the future; and it has continuity and consistency, but with constant change. It is an ever-changing, living system sustaining life in all its forms by adapting to the times. A Nature Smart Society is a safe and healthy place for all to flourish in all aspects of our lives: to live, love, learn, work, and play well. It is a natural human habitat of wellbeing, wealth, health, happiness, harmony, hope, and humanity. It is truly our home.

> A hallmark of a healthy creative culture is that its people feel free to share ideas, opinions, and criticisms.—Ed Catmull

Fig. 2.10 Nature Smart Society and the Elements of Life © Anne Stenros, 2022–23. Graphics by AS & Minna Takala

8 Conclusion: The Nature Wise World Order

We need to shape the future right now, in the present.— Olafur Eliasson

According to Richard Falk, world order "refers to a preferred arrangement of power and authority that is associated with the realization of such values as peace, economic growth and equity, human rights, and environmental quality and sustainability." (Falk, 2022). When looking at the future world order, we can see that the basic needs, and physiological and safety issues growing in importance due to the several current crises we are facing today: that of climate, conflicts, food, health, and energy. The next ones, psychological needs, related to respect, freedom, and human rights are also high on demand and actively discussed in public as DEI: diversity, equity, and inclusion. However, we cannot forget the importance of self-fulfillment needs: self-actualization and sense of meaning, because they can help the future transformation and the way out of crises. Growth needs act as a guiding light lending a cause, purpose, meaning, and vision of tomorrow. The emerging, new *Nature Wise World Order* has six key characteristics (Fig. 2.11).

(A) **Basic Needs:**

1. **Place:** Local sustainability and quality of the physical environment, both natural and man-made, are becoming more and more important. Food, water, shelter—the basic elements of human life—are in danger and therefore they are receiving growing attention both in cities and rural areas. A healthy environment equals a healthy life. Tackling the climate crisis both globally and locally is the biggest challenge and opportunity of the future. *This starts by supporting sustainable place-based development all around the world for the common good.*
2. **SAFETY:** Peace, personal security, employment, and overall health are seen as essential values for a happy life and holistic wellbeing. All actions should accelerate zero carbon transition in every aspect of society: lifestyles, buildings, cities, and economies. *Combining profit and purpose in value creation is critical for the future development of society.*

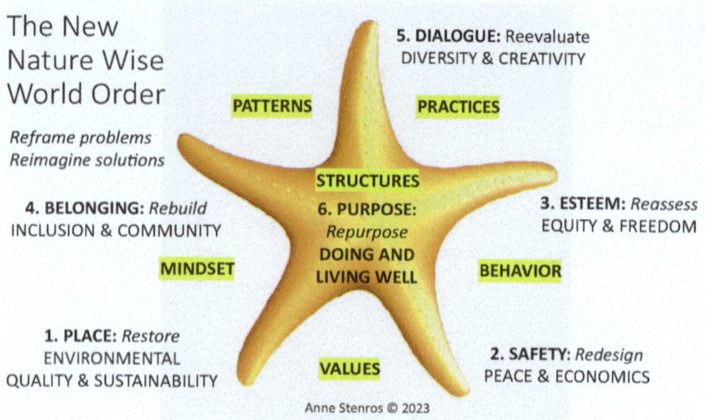

Fig. 2.11 Elements of Nature Wise World Order © AS, 2022–23. In Asia, the starfish is seen as the palm of a human hand. It is also a most resilient creature; when a part of its leg breaks off, it regenerates it. The starfish symbolizes success and prosperity

(B) **Psychological Needs:**

3. **Esteem:** Freedom, respect, human rights, and equity are getting a louder voice together with the rights of nature and animals. The ESG factors—Environmental, Social and Governance—are a must in the future, encouraging companies to act responsibly, be pro-social, environmentally-friendly, and have good corporate governance. *This means being responsible and sustainable and above all acting equitably, both locally and globally.*
4. **Belonging:** Inclusion, being part of a community, small and large, is giving meaning and supporting social health and overall wellbeing. The focus should be on building ecosystem-like local communities with wider networks for social innovations. *Leaving no one behind is one of the sustainable development goals.*

(C) **Self-fulfillment Needs:**

5. **Dialogue:** Diversity and creativity are supporting self-actualization and having a voice. Innovation will need talent from more diverse backgrounds than today. Diversity should be managed and supported step-by-step as much as creativity and innovation itself. *That is the true nature of encouraging people-powered innovation.*
6. **Purpose:** Aiming toward a socially, economically, and environmentally more sustainable worldview is a framework for happiness in life, for doing and living well. *Being mission-driven is an imperative for long-term sustainable development in the world.*

These six characteristics together create a path to follow toward the future: Tackling the climate crisis both locally and globally, combining profit and purpose, acting in a sustainable and equitable way, creating a collective movement toward the betterment of society without leaving anyone behind, supporting people-powered innovation through diversity and creativity, and, above all, being mission-driven in every action, in the long run. These elements create a resilient enough world to face the inevitable future challenges, changes, even crises, and on-going transformation.

This new worldview demands a reframing of our current problems to reimagine more meaningful solutions for the future. We have all the tools and the wisdom needed—now is the time to restart, reimagine, reframe, relearn, re-evaluate, reassess, redesign, reuse, rebuild, regenerate, restore, and repurpose our world order. But most of all, we must re-evaluate our set of values and reset our mindset to change our patterns of behavior and practices for the better. Only this can lead to new strategies and structures that are needed in the future. And that will eventually change global culture to become more sustainable, creative, equal, and responsible—the only way forward.

9 Epilog: A Lifelong Journey

The core spirit of the new nature wise world order is continuous change and improvement. It means a reframing of problems, constant unlearning, and relearning to get a better understanding of context. It is the continuous betterment of society. This aspect of lifelong willingness to learn, unlearn, and relearn, and to see creativity and curiosity as driving forces in life, is beautifully encapsulated in a small drawing by Francisco Goya when he was 82 years old and almost blind. "*Aún apprendo* – I am still learning", he titled the black-and-white chalk drawing of himself, just before his death. This humble statement shows the power of a creative and curious mind: the journey is not finished yet.

In Japanese, there is a beautiful concept of *hitoritabi*, traveling alone. Studying is a *hitoritabi*, as well as a lifetime of creative work. They both demand a lot of self-discipline, curiosity, and, above all, willingness to work and walk alone—*hitoritabi*. The force to move forward endlessly and tirelessly to get a glimpse of the future is the lifelong journey of free spirits on their way toward understanding the hidden nature of life. The Nature Wise World Order is never finished; it is a living thing, changing during times, yet, at the same time, carrying on what is the best in humanity: *Hope*.

Appendix

The Alphabets of Nature Smart Society

A ***Activism as Art*** combines the power of creativity to the movement for social change

B ***Biophilia*** is our bond with nature and ***Biophilic Design*** is rebuilding this connection

C ***Circular Economy*** is a systems solution framework tackling global challenges

D ***Diversity*** is the very essence of nature; ***Biodiversity*** supports healthy life on the planet

E ***Eco-friendly*** are goods, services, laws, and policies reducing environmental harm

F ***Framing and Reframing*** mental models, metaphors, and future narratives is essential

G ***Green Economy*** needs upskilling and reskilling in systems thinking and complexity

H ***Human-nature*** connections are an essential, vital part of life and holistic wellbeing

I ***Innovation*** powered by people and driven by technology will create new solutions

J ***JEDI***—justice, equality, diversity, inclusion—means equal access to opportunities

K ***Knowledge of Natural World*** or science is the ability to understand how nature works

L ***Learning, Unlearning,*** *and active* ***Relearning*** are essential skills throughout the life

M ***Mission-driven*** life is based on sustainable development and sustainable growth

N ***Nature Smart Society*** is environmentally, socially, and economically sustainable

O ***Originality***, initiative, creativity, and critical thinking are the new skills of the future

P ***Profit with Purpose*** is the new mantra of the future business, innovation, and capital

Q ***Quest for gender balance*** and gender equality are at the heart of holistic sustainability

R ***Regenerative Design*** supports positive forces that repair natural and human systems

S ***Sustainable Development Goals*** (SDGs) are an urgent call for action by all countries

T ***Triple Bottom Line***: Planet-People-Profit is the cornerstone of sustainable economy

U ***Utopia versus dystopia;*** to see the future scenarios through a positive lens is an imperative

V ***Vegans and vegetarian*** food and products support the reduction of carbon footprint

W ***World Environment Day*** was started by the UN in 1972 at the Stockholm Conference

X ***The X factor*** for the sustainable development is the global climate solidarity

Y ***The Youth*** should have voice on making future; they should not be left behind

Z ***Zero carbon target*** by 2050 needs a reduction of emission by 45% by 2030

***

A livable and sustainable future for all (IPCC, 2022).

References

Andersson, D. (2017, November 15). *The prophetic side of archigram*. Bloomberg. Retrieved from https://www.bloomberg.com/news/articles/2017-11-15/archigram-and-the-modern-city

Barandy, K. (2022, January 19). *Peter cook—City landscapes as architectural drawings*. Designboom. Retrieved from https://www.designboom.com/art/louisiana-museum-modern-art-peter-cook-city-landscapes-denmark-01-19-2022/

Carnegie, M. (2022, August 8). *Gen Z: How young people are changing* activism. BBC Worklife. Retrieved from https://www.bbc.com/worklife/article/20220803-gen-z-how-young-people-are-changing-activism

CURE Copernicus. (2022, October). CURE Copernicus. Retrieved from http://cure-copernicus.eu/

Davis, M. (2019, August 9). *Maslow's forgotten pinnacle: Self-transcendence*. BigThink. Retrieved from https://bigthink.com/neuropsych/maslow-self-transcendence/

Duncombe, S., & Lambert, S. (2021). *The art of activism*. OR Books.

Ecosistema Urbano. (2022, October). Ecosistema Urbano. Retrieved from https://ecosistemaurbano.com/category/by-topic/civic-engagement/

Edible Cities. (2022, October). Edible cities network. Retrieved from https://www.edicitnet.com/

EU. (2022, October). *Ecosystem services and green infrastructure*. European Commission. Retrieved from https://ec.europa.eu/environment/nature/ecosystems/index_en.htm

Eurostat. (2022, October) *Culture statistics—Cultural employment* 2020. Eurostat. Retrieved from https://ec.europa.eu/eurostat/statistics-explained/index.php?title=Culture_statistics_-_cultural_employment

Falk, R. (2022, October). *World order*. Encyclopedia Princetoniensis, Princeton University. Retrieved from https://pesd.princeton.edu/node/696

Gelles, D. (2022, September 14). *Patagonia founder gives away the company to fight climate change*. New York Times. Retrieved from https://www.nytimes.com/2022/09/14/climate/patagonia-climate-philanthropy-chouinard.html

HerCity. (2022, October). HerCity. Retrieved from https://hercity.unhabitat.org/

Hoffer, 2006 Hoffer, E. (2006, 1963). *The ordeal of change*. Hopewell Publications

Hunt, E. (2019, November 3). Blue spaces: Why time spent near water is the secret of happiness. *The guardian*. Retrieved from https://www.theguardian.com/lifeandstyle/2019/nov/03/blue-space-living-near-water-good-secret-of- happiness

Kanerva, A. (2022, October 6). Naisen elämästä ja yhteiskuntaluokista kirjoittava Annie Ernaux sai kirjallisuuden Nobel-palkinnon. *Helsingin Sanomat*. Retrieved from https://www.hs.fi/kulttuuri/art-2000009111557.html

McKinsey. (2020, March 23). *The next normal*. McKinsey. Retrieved from https://www.mckinsey.com/industries/healthcare-systems-and-services/our-insights/beyond-coronavirus-the-path-to-the-next-normal

Moreira, S. (2021, May 20). *What is placemaking?* ArchDaily. Retrieved from https://www.archdaily.com/961333/what-is-placemaking

Morton, J. (2022, March 18). *4 principles of regenerative design*. Buildings. Retrieved from https://www.buildings.com/green-design-construction/article/10196146/4-principles-of-regenerative-design

Nowotny, H. (2016). *The cunning of uncertainty* (pp. 26–27). Polity Press.

Parikka, V. (2022). Viraali vegaani. *Helsingin Sanomat*. Retrieved from https://www.hs.fi/kaupunki/helsinki/art-2000009039732.html

PWC. (2022, October). *Net zero economy index 2022*. PWC. Retrieved from https://www.pwc.co.uk/services/sustainability-climate-change/insights/net-zero-economy-index.html

Salako, T. & Frost, R. (2022, June 24). *Europe's first 'chief heat officer' explains how Athens is preparing for severe heatwaves*. EuroNews. Retrieved from https://www.euronews.com/2022/06/24/europes-first-chief-heat-officer-explains-how-athens-is-preparing-for-severe-heatwaves

Sendra, P. & Sennett, R. (2020). *Designing disorder—Experiments and disruptions in the city*. Verso.

Stenros, A. (2010). *The butterfly effect—Future trends in urban transformation*. KONE Corporation.

Stenros, A. (2022). *THE NATURE SMART CITY—Defining the next urban vision*. http://www.press.ierek.com

Stenros, A. et al. (2022). NATURE WISE CITY—Rewilding architecture and urban design. In *A competition entry by Power Animals. Open call for exhibition proposal for Pavilion of Finland; 18th International Architecture Exhibition, La Biennale di Venezia*.

Spatial Relationship Between Diet Diversity, the Food Environment and Transport: A Case Study in South Africa

Michelle Eichinger and Julian May

Abstract

Food and nutrition insecurity are prevalent issues in South Africa with health consequences. The country is seeing a paradox of diet quality resulting in the double-burden of malnutrition. While affordability and food preferences contribute to diet behaviour, the food environment influences diet patterns and food purchasing behaviour. This study examined the spatial relationship between nutrient security, purchasing behaviours and accessing the food environment. This study focused on Worcester, South Africa, a secondary city about 112 km from Cape Town. Through household surveys (n = 657), demographic, diet log within the last 24 h, food purchasing and transport behaviour data were collected. All surveys included GPS coordinates of the residence. Diet diversity score, measured by the diet log, was used as a proxy to measure nutrient security, with scores less than 4 indicating nutrition insecurity. In addition, all food retailers in the area were codified by type (supermarket, spaza, produce market, etc.) and geocoded. All minibus taxi ranks and routes were mapped. Regression and spatial analyses were performed using R Statistical Software and ArcGIS Pro. Diet Diversity scores averaged at 4 with little variation by income type. Results showed neighbourhood-level clustering of food and diet patterns with an overlapping density of spaza shops. Further, there was a spatial relationship of supermarket and spaza proximity whereas diet is more diverse near supermarkets while spazas provided a protection against food insecurity. While many residents walked to shops, taxi ranks were commonly used for larger purchases. Results from this study can inform local economic and planning officials on the relationship between the transport, food environment and nutrient security, thus providing evidence to support strategies aimed at improving healthy food availability in areas with higher rates of nutrient insecurity. These findings can help planners in prioritizing areas for healthy food accessibility.

Keywords

Food environment · Spatial analysis · Food accessibility · Food security · Diet diversity

1 Introduction

Although generally arid, South Africa's agri-food sector contributes to 5% of Gross Domestic Product (GDP) and employment, 14% of merchandise exports and a positive food balance. Yet, the country faces a food system paradox with a high prevalence of malnutrition and child stunting. During 2019–2021, nearly 20% of households experienced moderate or severe food insecurity (FAO et al., 2022a, 2022b). Nutrition security differs from food security such that nutrition security focuses on the diversity of foods available providing essential nutrient needs, whereas food security focuses on the availability and accessibility of food to people (Venugopal, 1999). In South Africa, about 6.9% of the population is undernourished while 23.2% of children under 5 years of age are stunted (FAO et al., 2022a, 2022b).

The country has also been seeing a paradox of diet quality resulting in the double-burden of malnutrition with the "coexistence of both undernutrition and overnutrition in

M. Eichinger (✉)
Vrije Universiteit Amsterdam, Amsterdam, The Netherlands
e-mail: meichinger@live.com

J. May
Director of DSI-NRF Centre of Excellence in Food Security, University of the Western Cape, South Africa, Cape Town, South Africa
e-mail: jmay@uwc.ac.za

I. Pigliautile et al. (eds.), *Urban Resilience, Livability, and Climate Adaptation*,
Advances in Science, Technology & Innovation, https://doi.org/10.1007/978-3-031-54911-3_3

the same population across the life course" (Shrimpton & Rokx, 2012). As an example, overweight/obese women likely experienced undernutrition in childhood, resulting in stunting. Since 1994, food consumption behaviours have changed, altering diet quality with increased consumption of sugar-sweetened beverages and highly processed packaged foods (Ronquest-Ross et al., 2015). As a result, this change in diet quality to high consumption of fats, sugars, salts and calories is associated with increasing rates of obesity from 19.7% in 2000 to 28.3% in 2018 (Ronquest-Ross et al.,. 2015, FAO et al., 2022a, 2022b). In addition, there has been an increasing prevalence of non-communicable, diet-related diseases, such as cardiovascular disease and diabetes (FAO et al., 2022a, 2022b).

While affordability and food preferences contribute to diet behaviour (Eichinger, 2016), the food retail environment often influences diet patterns and consumer food purchasing behaviour, and geographic proximity is an important factor in accessing healthy, affordable foods (Rose, 2010). About 65% of the population in South Africa state they are unable to afford a healthy diet (FAO et al., 2022a, 2022b). In previous research, proximity to supermarkets has been a proxy for healthy food access (Ver Ploeg et al., 2012). However, it may not be economically feasible for supermarkets to be located in low-income or low-resource neighbourhoods, nor an appropriate, adequate solution for communities experiencing higher rates of food insecurity (Horst et al., 2016). Supermarkets may also increase access to highly processed food and sugary beverages, while displacing smaller, more labour-intensive food enterprises. Instead, there has been advocacy in planning and policy to support the role of informal and small food retailers in urban areas to combat food insecurity and increase the availability of healthy, affordable food (Battersby, 2011; Battersby-Lennard et al., 2009).

In response to increasing the availability and affordability of healthy food and fresh produce, South Africa had seen an increase in "Big Food", or large commercial food and beverage retailers (Igumbor et al., 2012). While supermarkets have started to be located in low-resource communities, they do not always meet the diet patterns and strategies of low-resource residents (Peyton et al., 2015). Instead, this reveals the important role the informal food sector has in meeting the diet patterns of local residents, especially those in low-resource communities, including informal settlements. These residents often purchase their food from street vendors or spaza shops. Spaza stores are home-based shops that provide a variety of affordable foods, including some fresh fruits and vegetables, in low-income communities (Ligthelm, 2005). The informal food sector has played an important role in improving food security by increasing the affordability and accessibility of foods while also providing cash employment (Petersen & Charman, 2018). However, street vendors and spazas often sell highly processed, poor quality food and beverages (Battersby-Lennard et al., 2009).

Most households would mention food expenses as part of their regular, monthly household budgets, but transportation costs are a common consideration in household budgets (Battersby & Haysom, 2018). In a study of older, South African women, walking was the common mode of transport for food at spazas and street vendors while public transport was the more common mode to access supermarkets (Odunitan-Wayas et al., 2021). Households with ownership of a vehicle had better access to markets and higher diet diversity (Nandi et al., 2021). While there have been limited studies that explored modes of transport, there is a gap in the literature that examines food types and transport modes.

With the increasing number of supermarkets and the typical use of street vendors and spaza stores for local, affordable food in low-resource neighbourhoods, there have been limited studies examining the geographic relationship between food retailers, food insecurity, and diet diversity. This study extends the previous work that has mapped the informal food economy in South Africa by linking enterprises to a household diet recall survey (Davis et al., 2022). However, this study aims to not only examine this spatial relationship, but also the relationship between food purchasing behaviours and mode of transport to get different types of food. Finally, the study focuses on a city located in a highly productive agricultural region of South Africa with strong links to export markets that illustrate the country's food system paradox.

2 Methodology

This case study focused on the city of Worcester using a household survey to collect household demographic and socioeconomic data, home characteristics and food purchasing behaviours. The household sites were geocoded along with food retailers and minibus taxi stops in the study area.

2.1 Study Area

Worcester, a city of nearly 128,000 residents in 2020, is about 120 km from Cape Town in the Breede Valley Municipality in the Western Cape Province of South Africa. The city has about 72% coloured African, 17% White and 9% Black Africans (Stats South Africa, 2011).

2.2 Household Survey

The Food4Cities South Africa research team formed an advisory group that included experts in complex systems and nutrition held a workshop to design the household survey. The survey was then pilot tested, and changes were made accordingly to address concerns for length and redundancy. The Humanities and Social Science Research Ethics Committee of the University of the Western Cape approved the survey.

The household survey included household socioeconomic and demographic data, home energy and cooking characteristics. In addition, the survey collected a 24-h diet log. Further, each survey response included the GPS coordinates of the home location where the responses were collected. This allows for a spatial dimension to the survey data that would provide attributes for spatial analysis.

Data sampling was randomly selected based on the geographically defined Enumeration Areas (EAs) from the 2011 South Africa Census. Satellite maps for each EA guided the fieldwork which allowed them to select a starting point within the EA and then continue the survey targeting every *nth* house, in which *n* was a random number based on the day of the week. Survey data collection included all residential EAs in Worcester and Zwelethemba (126 residential EAs—111 formal residential, 7 informal residential and 8 collective living).[1]

Participants were adult men and women interviewed at home in the town of Worcester, Western Cape, South Africa. Fieldworkers requested face-to-face interviews with the participants at their homes. Surveys were administered in English, Afrikaans or isiXhosa languages, and responses were digitally recorded using ODK software on tablet computers (Davis et al., 2022).

2.2.1 Food Security

The survey included a module on food security which asked eight questions that were based on the Food Insecurity Experience Scale Survey Module (Cafiero et al., 2018). A "yes" response to any of the food security questions indicated the household experiencing or have experienced food insecurity in the last 12 months.

2.2.2 Diet Diversity and Food Purchasing Behaviour

The survey collected a 24-h diet log of the respondents. The diet was broken down to 49 categories in the survey identifying food security and nutritional assistance. After data collection, these categories were aggregated into 9 food groups. With these data, a diet diversity score was used as a proxy to measure nutrient security (Kennedy et al., 2011). Scores of less than 4 were considered as having low diet diversity, or being nutrition insecure (Steyn et al., 2006). In addition to the 24-h diet log collected in the survey, respondents were asked where they purchased each food item.

2.2.3 Home Kitchen Score

The household survey included questions on household food technology. Specifically, the survey asked respondents if their home included any of the following:

- None
- Built-in kitchen sink
- Refrigerator or combined fridge freezer
- Stove (electric or gas)
- Oven
- Deep freezer (free standing)
- Microwave.

The sum of these responses was the Home Kitchen Score. Home food technology was associated with diet diversity. For example, refrigerators allow for the storage of perishable foods, such as fruits, vegetables and meats (Faber et al., 2009). In addition, lack of a fridge has been associated with child underweight and stunting (Labadarios et al., 2001).

2.3 Local Food Retail Environment

Formal food retailers were mapped using Google Street View and then verified via a drive through the city. Informal food retailers were mapped through Google Maps. The GPS coordinates for all mapped food retailers were transferred to ESRI ArcGIS Pro for geocoding. The distance from the household to the nearest formal and informal retailer was determined using the Closest Facility Tool in ArcGIS Pro.

2.4 Data Analysis

Descriptive statistics and inferential statistics were performed using R Studio Statistical Software version 2022.12.0 (Team, 2013). Chi-square analysis was conducted to compare food security and diet diversity between racial and other socioeconomic groups. Given that the data included specific GPS locations of households and food retailers, the study was able to examine diet patterns and the food retail environment neighbourhoods. Using ArcGIS Pro version 3.0.3, spatial analyses, including spatial

[1] EAs were taken from the 2011 Census, the most recent available. However, Worcester has grown since 2011, and some residential areas were therefore not included in our sample. In particular, an informal housing area outside of Zwelethemba was not included.

autocorrelation, hot spot analysis, and kernel density, were performed. Specifically, an optimized hot spot analysis was performed to analyse patterns of food insecurity. Measuring the household distance to the nearest food retailer was done using the Closest Facility tool in Network Analysis.

3 Results

Of the 1577 attempted surveys, a total of 825 surveys were completed. After geocoding the household survey sites, 657 responses were within the Worcester Study area (Fig. 3.1) and included in this analysis. The remaining surveys fell outside the study site boundary.

Figure 3.1 details the study area with neighbourhood features, including supermarkets, spazas, density clusters by race, and taxi ranks/stops and routes. The density clusters of races show distinct segregation of neighbourhoods, which is a reflection of the apartheid past despite almost 30 years since the abolition of segregation. Black Africans are largely populated in Zwelethemba, a township east of Worcester's Central Business District (CBD). Coloureds reside south and southeast of the CBS, while White Africans reside to the north of the CBD. There is one supermarket in the Zwelethemba area and one supermarket in the Coloured area. However, there are numerous spazas located in these areas. Minibus taxi ranks are in the residential areas with routes and stops in the CBD.

Table 3.1 describes the survey respondents. The majority of respondents were female (69.3%). The mean age of respondents were 48.8 years. About two-thirds of the respondents spoke Afrikaans as their primary language, and just over half of the respondents were of Coloured African race. Over half of the respondents experienced food insecurity in the last 12 months and the mean Diet Diversity Score was 3.9. The survey respondents seem to include an oversampling of Black Africans compared to the racial demographic representation of the city. However, it is important to note that the most recent reported census had taken place in 2011 and in line with the Western Cape Province, this area has experienced significant in-migration from other parts of South Africa.

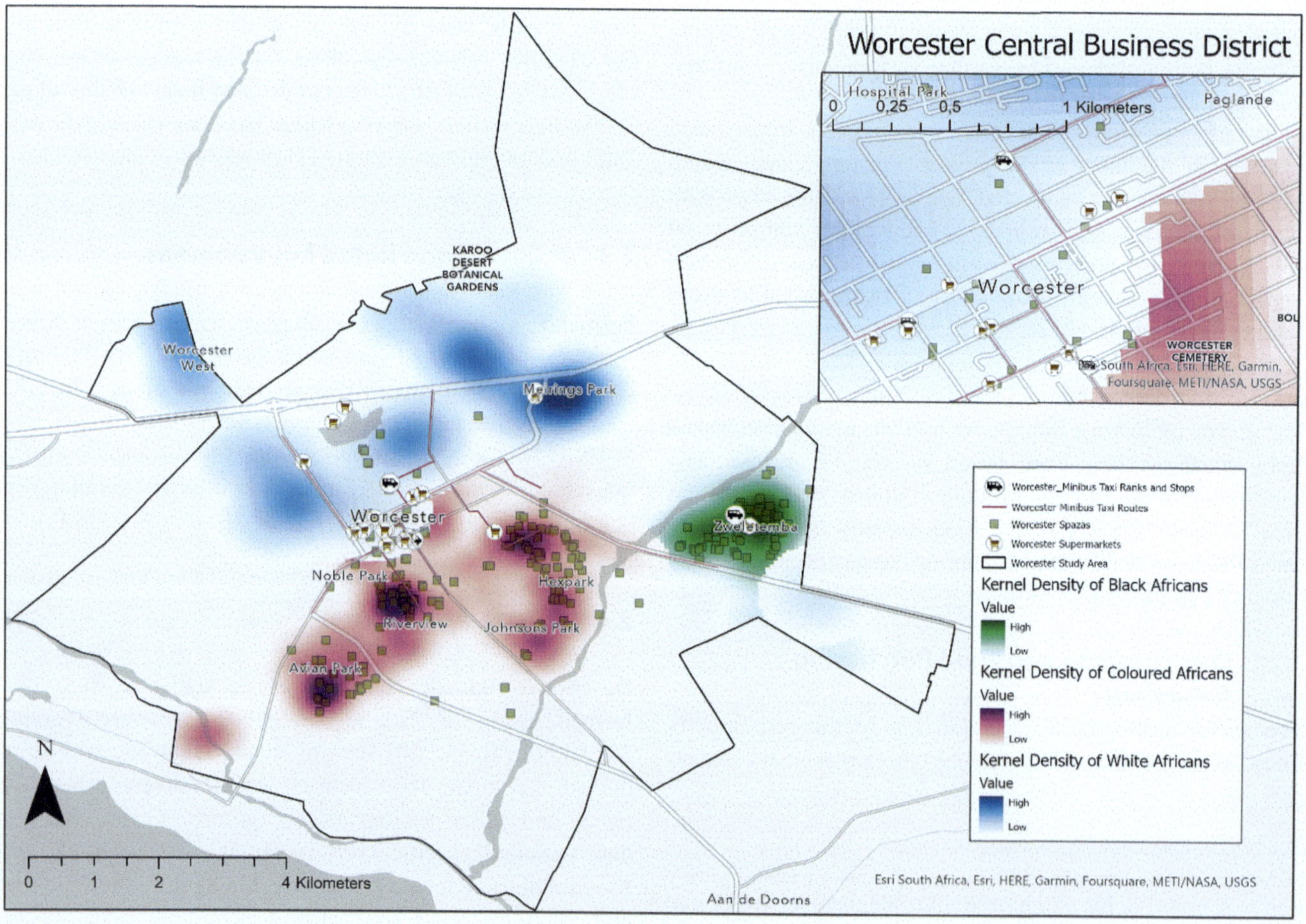

Fig. 3.1 Study area including food retail environment, minibus taxi routes and residential clusters (Author)

Table 3.1 Descriptive statistics

Characteristics	Value n=657
Gender	
Male	30.4%
Female	69.3%
Mean age in years (Standard Deviation)	48.8 (20.16)
Language	
Afrikaans	68.8%
IsiXhosa	23.9%
English	4.9%
Other	2.4%
Race	
Black African	27.5%
Coloured African	54.8%
White African	14.9%
Other	2.7%
Per cent food insecure	56.3%
Mean diet diversity score Median diet diversity score (Standard deviation)	3.96 4.0 (1.85)
Mean daily food expenses Median daily food expenses (Standard deviation)	134.96 Rand 83 Rand (221.60 Rand)

Most households earned their income through salaries (n = 322) followed by social grants (n = 148). Income was asked based on their monthly range. The majority of respondents reported their monthly income in the R1001–R2500 range (n = 193), followed by R2501–R5000 (n = 146). In 2019, the Food Poverty Line was R561 per person per month or about R1600 per household, while Lower-Bound Poverty Line (LBPL) was R810 per person per month or just less than R2500 per household. Therefore, most households report being in the LBPL. One of the limitations of this study was the survey question on income level. The income ranges used in the survey were difficult for adequate analysis.

3.1 Food Security and Diet Diversity

About 56% of the respondents reported some level of food insecurity in the last 12 months. Hot spots of food insecurity are identified in Fig. 3.2, which are located in Black and Coloured African neighbourhood areas of Worcester. In this study, Black Africans and Coloured Africans had higher rates of food insecurity than White Africans ($\chi^2 = 60.63$, $p < 0.001$). These results are consistent with national statistics on poverty and food security (Stats, 2017). In addition, the township of Zwelethemba includes a large informal settlement, and food insecurity is highly associated with informal settlements (Jonah & May, 2020).

We applied a forward Logit stepwise regression to examine the predictive relationship of food security status. Food security was measured as a binary variable (0 = Food Secure and 1 = Food Insecure). Explanatory variables were grouped as indicated in Table 3.2. Model 1 focused on home socioeconomic and racial demographics. Food insecurity was more likely reported with increasing number of children in the household, but less likely with higher education, owning a private vehicle and of White race.

Model 2 added the home food environment characteristics to the home characteristics. Food insecurity was significantly more likely among those who reported receiving nutrition assistance, but less likely with higher education and private vehicle ownership. Model 3 added food retail environment variables to the characteristics from the previous models. In addition to the significant relationships seen in Model 2, food insecurity was significantly more likely with increasing distance to the supermarket and using a taxi to get food, but less likely with increasing distance to spazas. Interestingly, those receiving nutritional assistance were more likely to experience food insecurity.

The mean and median diet diversity score was 4, which is on the threshold of low diet diversity. The top 10 most consumed foods in the 24-h diet log are listed in Table 3.3. It is important to note that fruits and vegetables were not listed, indicative of the poor diet quality that characterizes South Africa's urban areas. Commercial breads include store-bought bread and baked goods such as croissants. The food groups, Cereals, Grains and Tubers, and Fats, Oils and Ultra-Processed foods dominated the top 10 foods consumed.

Chi-square analyses were performed to examine the differences between diet and food security status and race. As seen in the frequency tables in Table 3.4, there are significant differences in diet between both groups. Food insecure households consume less Vitamin A-rich fruits and vegetables and more meats and fats. Among race, Black Africans ate more other fruits, but less legumes and dairy. White Africans ate more Vitamin A-rich fruits and vegetables and dairy, but less cereals. In addition, food security status was significantly different between races. Food insecurity was higher among Black and Coloured Africans than among White Africans ($\chi^2 = 60.63$, $p < 0.001$). These findings align with previous research on racial disparities in food security and diet patterns (Okop et al., 2019; Stats, 2019).

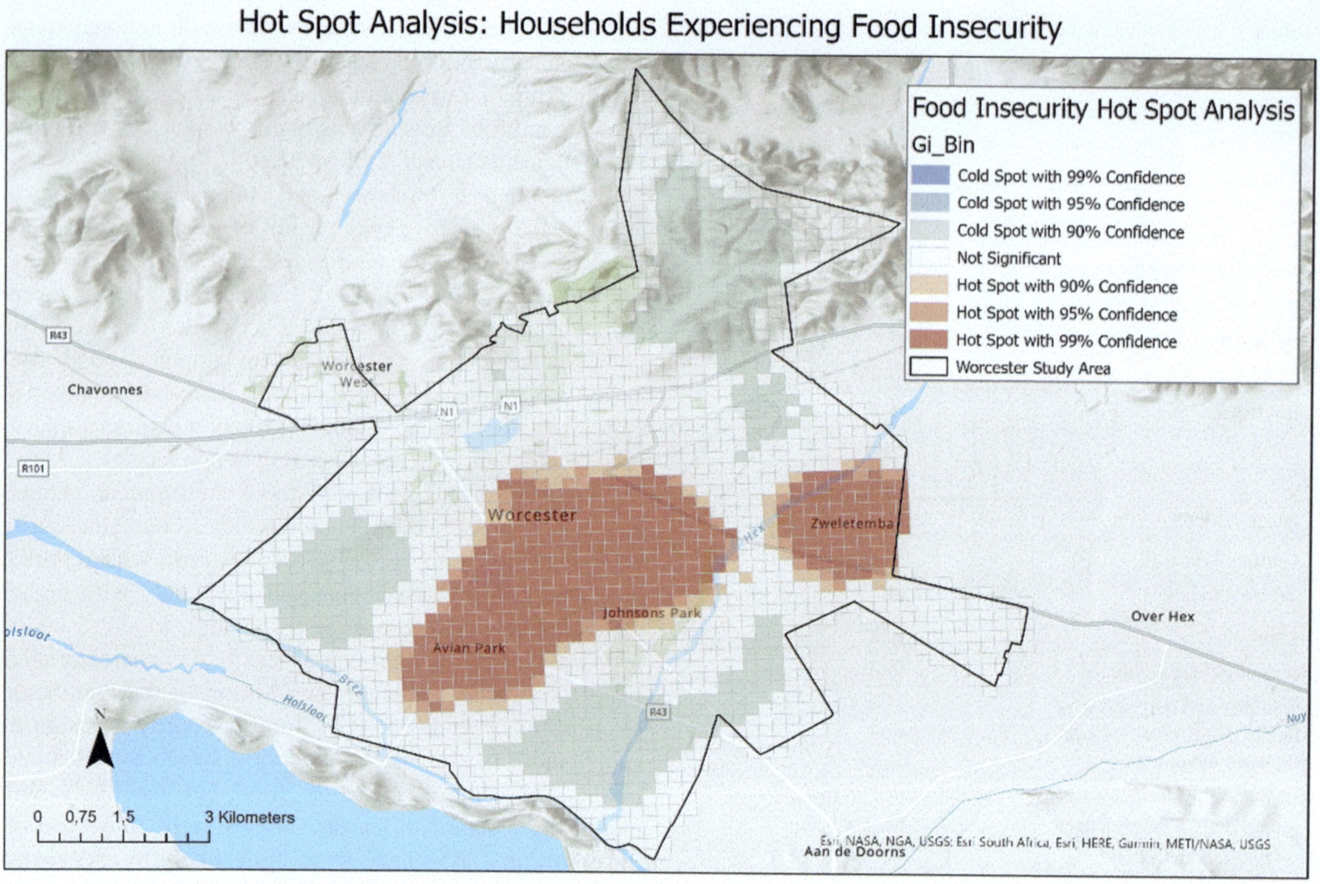

Fig. 3.2 Hot spot analysis of food insecurity (Author)

3.2 Mode of Transport to Major Food Groups

About 50.3% of the households had access to a private vehicle, and using a private car was the most common form of transport to get food. In addition, 79% of respondents reported going to the supermarket for their food, followed by spazas and street vendors at 13%. These findings differed from previous work in which supermarkets were used least frequently, while small and informal shops were more frequently used (Battersby-Lennard et al., 2009). It is nonetheless in line with the trend of supermarkets that have been reported in middle-income countries in Africa.

When examining the diet diversity score (DDS), the mean DDS was higher among those with a vehicle ($\bar{x} = 4.18$) than those with no private vehicle ($\bar{x} = 3.74$). Aside from private vehicles, residents commonly used minibus taxis and walked to get food. Using a minibus taxi was more common than walking to get rice, breakfast cereal and chicken. Walking was more common than taking the minibus taxi to get fruits, vegetables and dairy products. When determining the odds ratio, a private vehicle was the reference level for a mode of transport. Table 3.5 lists the odds ratio with the most common modes of transport to obtain major food items. According to the chi-square analysis, taking a minibus taxi for commercial bread was 8 times larger than taking a private car for those who experience food insecurity and 6.45 times larger by walking. Similarly, taking the minibus taxi for soft drinks was 8.29 times larger than taking a personal car for those who experience food insecurity, and 7.4 times larger by walking for soft drinks. Comparing those experiencing food insecurity, the odds of walking and taking a minibus taxi were significantly higher than taking a personal car for all major food groups except for Vitamin A fruits. Further, the odds of using a minibus taxi to get dark green leafy vegetables and Vitamin A-rich fruits and vegetables were nearly doubled compared to the odds of walking for households experiencing food insecurity.

4 Discussion

The agriculture sector is the largest economic sector in the Breede Valley Municipality in which this study was located. Yet, over half of the residents in the study area experienced food insecurity in the last 12 months and diet diversity was

Table 3.2 Logit regression models of food insecurity

Category		Model 1 coefficients (standard error)	Model 2 coefficients (standard error)	Model 3 coefficients (standard error)
Home environment	Number of children in household	0.141* (0.073)	-0.006 (0.082)	−0.012 (0.084)
	Home ownership	−0.130 (0.227)	−0.154 (0.250)	0.001 (0.255)
	Own private vehicle	−1.252*** (0.201)	−0.844*** (0.237)	−0.787** (0.243)
	Education level—<national certificate	−0.747** (0.246)	−0.747** (0.271)	−0.711** (0.275)
	Education level—<University degree	−1.490** (0.537)	−1.502* (0.591)	−1.350* (0.604)
	Race—coloured	0.178 (0.217)	0.470 (0.262)	0.305 (0.271)
	Race—other	0.049 (0.573)	0.265 (0.633)	0.302 (0.648)
	Race—white	−0.677* 0.345	−0.095 (0.397)	0.316 (0.441)
Home food environment	Home kitchen score		−0.104 (0.081)	−0.109 (0.082)
	Daily food expenses		−0.002 (0.002)	−0.003 (0.002)
	Receive nutrition assistance		1.942*** (0.218)	1.861*** (0.220)
Food retail environment	Distance to Supermarket			0.269* (0.126)
	Distance to Spaza			−0.326* (0.165)
	Taxi to Food			0.427* (0.218)
	Walk to Food			0.179 (0.315)
	AIC	725.81	629.74	627.80

***p<0.001
**p<0.01
*p<0.05

Table 3.3 Leading food items consumed from 24-h diet log

Top 10 foods	Number of responses
Coffee/tea	551
Sugar	469
Oils/fats	439
Rice	424
Commercial bread	414
Spices/condiments	396
Milk products	358
Chicken (fresh, raw, cooked)	352
Regular soft drinks	341
Breakfast cereals	307

relatively low. Factors influencing household food security have often focused on individual or household characteristics, such as income and nutrition assistance. This study explored the spatial relationship between food security, diet patterns and food environment. It was evident from this study that the neighbourhood environment influences diet patterns, food security and mode of transport to get food. This is consistent with the Chicago study that proximity and mode of transport influence diet patterns (Block & Kouba, 2006).

Nearly half of the residents have access to a private vehicle, and having a vehicle is associated with improving food security and diet diversity, which is consistent in the literature (Faber et al., 2009). Consistent with this study,

Table 3.4 Chi-square frequency tables of diet diversity

	Chi-square frequency table 1		Chi-square frequency table 2			Chi-square frequency table 3	
	Food secured (n=285)	Food insecure (n=370)	Black (n=181)	Coloured (n=360)	White (n=98)	No vehicle (n=324)	Own vehicle (n=331)
Cereals, grains, tubers	276	360	176	351	32	316	320
Vitamin A-rich fruits and vegetables	114	97	44	110	27	80	131
Other vegetables	124	166	82	152	20	132	158
Other fruits	46	47	38	41	7	48	45
Meats	218	283	136	283	31	243	258
Legumes	12	16	1	18	5	71	98
Eggs	80	89	42	92	17	12	16
Dairy	142	154	52	178	58	119	177
Fats	134	221	99	211	34	184	171
DF	8		16			8	
Chi-Square	18.13		85.36			20.97	
p-value	0.02		0.000			0.007	

household assets, such as vehicle ownership and home kitchen technology, seem to be protective against food insecurity and support diet diversity (Faber et al., 2009; Fitzpatrick & Ver Ploeg, 2010; Guo, 2011; Walsh & van Rooyen, 2015). Yet, receiving nutrition assistance was not enough to protect against food insecurity, which may be attributed to increasing food prices (Chakona & Shackleton, 2019).

The resident's mode of transport to get to food retailers was related to the local food environment and diet quality. Without a private vehicle, an individual's carrying capacity influences what is purchased. Residents were more likely to take a minibus taxi instead of walking to get heavier, bulk items such as rice and chicken. Commercial breads and soft drinks were the most often purchased items in spazas, and walking was the most common mode of getting there. Consistent with other studies, small retailers, such as spazas and street vendors, are more likely to sell energy-dense foods (i.e. soft drinks), and residents were less likely to consume fruits and vegetables (Okop et al., 2019; Reardon et al., 2021). While the informal sector can help protect against food insecurity, it has an influencing role in poor diet, contributing to the double-burden of obesity (Reardon et al., 2021).

However, food retail type and siting influence food access. Residents living closer to supermarkets are less likely to experience food insecurity, while the opposite was seen with distance to spazas. This reinforces the important role the informal food sector has on food security (Petersen, et al., 2018). Spazas and other informal food retail enterprises are typically clustered in socioeconomically disadvantaged areas, including areas with high concentrations of food insecurity, as demonstrated in the Black African and Coloured African neighbourhoods of this study.

4.1 Planning and Policy Implications

The impact of apartheid still permeates the urban landscape, and this study demonstrated as such with the residential segregation by race and corresponding food insecurity and food retail environment. With urbanization, South Africa has seen a rise in the number of supermarkets. While modern supermarkets have been locating in urban, low-income communities, the increase in the number of supermarkets have not alleviated food insecurity, and has not always been suitable for local consumption behaviours (Peyton et al., 2015). This study examined the relationship between where people live, to where people get their food and how this relationship affects diet patterns. In addition, the mode of transport affects diet patterns and food security. Therefore, urban planning and public policies can influence the supply and demand in the accessibility, affordability and availability of healthy food, by focusing on both people and place. Such public policies often focus on strategies targeting people or places to address poverty-related issues, yet both should be complementary (Spencer, 2004). Table 3.6 offers a framework to address the complementary supply–demand and people–place approach to public policies.

Table 3.5 Odds ratio table of food insecurity and transport type for select food items

Transport type	Food secure versus food insecure	
	Odds ratio	95% confidence interval
Rice		
Private vehicle	1.00	NA
Minibus taxi	6.14***	2.93–13.69
Walking	2.93***	1.48–5.94
Commercial bread		
Private vehicle	1.00	NA
Minibus taxi	8.00***	3.13–22.90
Walking	6.45***	3.61–12.00
Breakfast cereal		
Private vehicle	1.00	NA
Minibus taxi	7.80***	3.31–20.03
Walking	2.71***	1.20–6.21
Vitamin A vegetables		
Private vehicle	1.00	NA
Minibus taxi	7.99***	2.59–28.89
Walking	3.43*	1.31–9.27
Dark green leafy veg		
Private vehicle	1.00	NA
Minibus taxi	11.03**	1.93–100.16
Walking	6.40*	1.19–41.79
Vitamin a fruit		
Private vehicle	1.00	NA
Minibus taxi	5.64*	1.41–30.02
Walking	3.37	0.86–15.24
Dairy		
Private vehicle	1.00	NA
Minibus taxi	9.71***	3.78–28.92
Walking	4.45***	2.32–8.76
Chicken		
Private vehicle	1.00	NA
Minibus taxi	5.09***	2.34–11.70
Walking	2.59*	1.15–5.99
Soft drinks		
Private vehicle	1.00	NA
Minibus taxi	8.29***	2.57–33.39
Walking	7.40***	3.70–15.37

***$p<0.001$
**$p<0.001$
*$p<0.05$

Food systems planning has not traditionally been integrated with urban planning. Rather, food systems have largely been seen as rural and agricultural issues. Similarly, the health implications of food insecurity and inadequate diet, such as severe acute malnutrition, stunting and the emerging double-burden of malnutrition as seen with increasing rates of overnutrition, have garnered the attention of the public health sector. As a result, many efforts have focused on individual-based nutrition education and supplemental nutrition assistance programmes. This study provides evidence that food security and diet diversity have influences on the local economy, transportation and community development. Therefore, addressing food system issues, such as food security and diet diversity, is not limited to agriculture and public health sectors, but sectors targeting economic development, land use planning, transportation, housing and community development.

4.2 Limitations

While this study provided much insight into the spatial relationship of food security and food environment, it was not without some limitations. The household survey did not adequately capture household income. As a result, the study used education level and daily food expenses as proxy to this household socioeconomic characteristic. In addition, the study does not factor the input of store owners, especially from the informal food sector, as it relates to their food offerings, enterprise siting and perception of consumer purchasing and consumption behaviours. In addition, the study does not delve into the transportation sector's role in food access.

5 Conclusions

Diet diversity and food security status vary by race and proximity to retail type. The informal food retail sector, such as spazas and street vendors, cluster in socioeconomically disadvantaged neighbourhoods, and have a protective role against food insecurity. However, these small enterprises contribute to the accessibility and affordability of energy-dense foods resulting in the emergence of the double-burden of malnutrition. While many households had access to a private vehicle, many households rely on minibus taxis for bulk and heavier food items or walking to

Table 3.6 People- and place-based policy strategies

	Supply	Demand
People	Assist consumers in affording healthy food	Move consumers to affordable, healthy food
Place	Increase affordable, healthy food in retailers	Move affordable, healthy food to consumers

local spazas or street vendors. Public policies focusing on economic, transportation, housing and community development strategies offer opportunities to address place-based interventions to increase the availability, affordability and accessibility of healthy foods.

Acknowledgements This study was part of the programme of the ERA-NET Cofund FOSC that has received funding from the European Union's Horizon 2020 research and innovation programme under grant agreement No. 862555. The authors declare that they have no competing interests.

References

Battersby, J., & Haysom, G. (2018). Linking urban food security, urban food systems, poverty, and urbanisation. In J. Battersby, & V. Watson (Eds.) *Urban food systems governance and poverty in African cities* (pp. 56–67).

Battersby-Lennard, J., Fincham, R., Frayne, B., & Haysom, G. (2009). Urban food security in South Africa: case study of Cape Town, Msunduzi and Johannesburg. In *Development planning division working paper series* (Vol. 15).

Battersby, J. (2011). Urban food insecurity in Cape Town, South Africa: An alternative approach to food access. *Development Southern Africa, 28*(4), 545–561.

Block, D., & Kouba, J. (2006). A comparison of the availability and affordability of a market basket in two communities in the Chicago area. *Public Health Nutrition, 9*(7), 837–845.

Cafiero, C., Viviani, S., & Nord, M. (2018). Food security measurement in a global context: The food insecurity experience scale. *Measurement, 116*, 146–152.

Chakona, G., & Shackleton, C. (2019). Food insecurity in South Africa: To what extent can social grants and consumption of wild foods eradicate hunger? *World Development Perspectives, 13*, 87–94.

Davis, J., Magadzire, N., Hemerijckx, L. M., Maes, T., Durno, D., Kenyana, N., Lwasa, S., et al. (2022). Precision approaches to food insecurity: A spatial analysis of urban hunger and its contextual correlates in an African city. *World Development, 149*, 1–15. [105694]. https://doi.org/10.1016/j.worlddev.2021.105694

Eichinger, M. (2016). Available, affordable, accessible, appealing: The A's of influencing healthy eating and active living. *Delaware Journal of Public Health, 2*(5), 16.

Faber, M., Schwabe, C., & Drimie, S. (2009). Dietary diversity in relation to other household food security indicators. *International Journal of Food Safety, Nutrition and Public Health, 2*(1), 1–15.

FAO, European Union, CIRAD and DSI-NRF Centre of Excellence in Food Security (CoE-FS). (2022a). Food systems profile—South Africa. Catalysing the sustainable and inclusive transformation of food systems. Rome, Brussels, Montpellier, France and Bellville, South Africa. https://doi.org/10.4060/cc0071en

FAO, IFAD, UNICEF, WFP and WHO. (2022b). The State of Food Security and Nutrition in the World: 2022: Repurposing Food and Agricultural Policies to Make Healthy Diets More Affordable, FAO. Retrieved from https://doi.org/10.4060/cc0639en

Fitzpatrick, K., & Ver Ploeg, M. (2010, September). On the road to food security? Vehicle ownership and access to food. In *The Panel Study of Income Dynamics's Conference on SES and Health Across Generations and Over the Life Course*.

Guo, B. (2011). Household assets and food security: Evidence from the survey of program dynamics. *Journal of Family and Economic Issues, 32*, 98–110.

Horst, M., Raj, S., & Brinkley, C. (2016). getting outside the supermarket box: Alternatives to "Food Deserts". *Progressive Planning* 207.

Igumbor, E. U., Sanders, D., Puoane, T. R., Tsolekile, L., Schwarz, C., Purdy, C., & Hawkes, C. (2012). "Big food", the consumer food environment, health, and the policy response in South Africa. *PLoS Medicine, 9*(7), e1001253.

Jonah, C. M., & May, J. D. (2020). The nexus between urbanization and food insecurity in South Africa: Does the type of dwelling matter? *International Journal of Urban Sustainable Development, 12*(1), 1–13.

Kennedy, G., Razes, M., Ballard, T., & Dop, M. C. (2010, December). Measurement of dietary diversity for monitoring the impact of food based approaches. In *International Symposium on Food and Nutrition Security*, Rome.

Labadarios, D., Steyn, N., Maunder, E., MacIntyre, U., Swart, R., Gericke, G., & Nesamvuni, A. E. (2001). The National Food Consumption Survey (NFCS)-children aged 1–9 years, South Africa, 1999. *South African Medical Journal-Cape Town-Medical Association of South Africa, 91*(5), 62–65.

Ligthelm, A. (2005). Informal retailing through home-based micro-enterprises: The role of spaza shops. *Development Southern Africa, 22*(2), 199–214.

Nandi, R., Nedumaran, S., & Ravula, P. (2021). The interplay between food market access and farm household dietary diversity in low and middle income countries: A systematic review of literature. *Global Food Security, 28*, 100484.

Odunitan-Wayas, F. A., Faber, M., Mendham, A. E., Goedecke, J. H., Micklesfield, L. K., Brooks, N. E., & Lambert, E. V. (2021). Food security, dietary intake, and foodways of urban low-income older South African women: An exploratory study. *International Journal of Environmental Research and Public Health, 18*(8), 3973.

Okop, K. J., Ndayi, K., Tsolekile, L., Sanders, D., & Puoane, T. (2019). Low intake of commonly available fruits and vegetables in socio-economically disadvantaged communities of South Africa: Influence of affordability and sugary drinks intake. *BMC Public Health, 19*(1), 1–14.

Petersen, L., & Charman, A. (2018). The scope and scale of the informal food economy of South African urban residential townships: Results of a small-area micro-enterprise census. *Development Southern Africa, 35*(1), 1–23.

Peyton, S., Moseley, W., & Battersby, J. (2015). Implications of supermarket expansion on urban food security in Cape Town, South Africa. *African Geographical Review, 34*(1), 36–54.

Reardon, T., Tschirley, D., Liverpool-Tasie, L. S. O., Awokuse, T., Fanzo, J., Minten, B., & Popkin, B. M. (2021). The processed food revolution in African food systems and the double burden of malnutrition. *Global Food Security, 28*, 100466.

Ronquest-Ross, L. C., Vink, N., & Sigge, G. O. (2015). Food consumption changes in South Africa since 1994. *South African Journal of Science, 111*(9–10), 01–12.

Rose, D. (2010). Access to healthy food: A key focus for research on domestic food insecurity. *The Journal of Nutrition, 140*(6), 1167–1169.

Shrimpton, R. & Rokx, C. (2012). The double burden of malnutrition: A review of global evidence.

Spencer, J. H. (2004). People, places, and policy: A politically relevant framework for efforts to reduce concentrated poverty. *Policy Studies Journal, 32*(4), 545–568.

Stats South Africa. (2011). *Main place statistics: Worcester*. Retrieved June 2023, from https://www.statssa.gov.za/?page_id=4286&id=130

Steyn, N. P., Nel, J. H., Nantel, G., Kennedy, G., & Labadarios, D. (2006). Food variety and dietary diversity scores in children: Are they good indicators of dietary adequacy? *Public Health Nutrition, 9*(5), 644–650.

Team, R. C. (2013). R: *A language and environment for statistical computing*. R Foundation for Statistical Computing, Vienna, Austria.

Venugopal, K. R. (1999). Food security vs. Nutrition security. *Health Millions, 25*(2), 18–19.

Ver Ploeg, M., Breneman, V., Dutko, P., Williams, R., Snyder, S., Dicken, C., & Kaufman, P. (2012). Access to affordable and nutritious food: Updated estimates of distance to supermarkets using 2010 data (No. 1477-2017-3993).

Walsh, C., & van Rooyen, F. (2015). Household food security and hunger in rural and urban communities in the Free State Province, South Africa. *Ecology of Food and Nutrition, 54*(2), 118–137.

Spatial Analysis of Noise Contour Maps Based on Traffic Speed Using Predictor-LimA Software

Nardine El-Bardisy, Abeer Elshater, Samy Afifi, and Abdulmoneim Alfiky

Abstract

This research examines the correlations and effects of street networks on city calmness. The scope of this work covers traffic speed as one of the traffic parameters. The linkage of noise combined with traffic parameters in modern cities was explored in the relevant literature. The Aviation Design Environmental Tool (AEDT) is for airport noise mapping and simulation. Through analyzing the airport as a source of noise, this tool facilitates the selection of a case study. Predictor-LimA software was used in the research case study in Heliopolis, Cairo to create a model that included eight receiver locations. These receivers have values at six building heights: 1.5, 4, 8, 12, 15, and 18 m. The three scenarios investigated were 100, 70, and 50 kms per hour traffic speeds. Our analysis revealed a significant correlation between noise and traffic speed through the Pearson correlation coefficient. Our results also indicated that noise prediction should be considered as part of the design process for landscape designers and planners. The study ends with lessons on various scales, from the regional to design scale and community awareness. As such, we suggest a toolkit to explain the use of specific techniques to reduce speed. This toolkit includes an urban planning scale that provides for governmental decisions during the design process. Our concluding remarks confirmed the essential role of designers and community members in designing based on the soundscape for creating livable public spaces.

Keywords

Urban morphology · Traffic parameters · Noise level · Noise mitigation

1 Introduction

Egypt has faced an excessive change in urban fabric (Abusaada & Elshater, 2018). The process of upgrading the city roads in Cairo, Egypt, is a crucial and massive national project, as it is a method for reducing traffic congestion and adapting to population growth. Bridges are being built in Cairo eastern neighbourhoods as part of the Cairo East development plan (Farouk, 2021; Almoghazy, 2020; ElMoghazi, 2019; Shafik et al., 2021). Urban transformation has environmental, economic, visual, and social impacts (Hölscher and Frantzeskaki, 2021; ÖNDER, 2018, Vélez, et al., 2021). Noise pollution studies are essential in Cairo, Egypt, as it is ranked as one of the world's noisiest cities (El Samra, 2016; IQair, 2023). Through literature, some scholars spotlight the economic perspective and road access challenges, while others focus on air pollution and noise pollution as a consequence of road transformation (ElMoghazi, 2019; Puvanachandra et al., 2012; Farouk, 2021; Almoghazy, 2020; Abas & Tamura, 2003; Habib and Ouki, 2021; Ariza-Villaverde et al., 2014). Noise pollution occurs due to diverse factors including urban transformation in street networks and human activities (Abusaada & Elshater, 2021; Elshater et al., 2022a, 2022b).

Noise pollution is a critical problem that causes severe physical and mental health problems (Rossi et al., 2020; World Health Organization, 2018). The World Health Organization (2018) considers traffic noise a health

N. El-Bardisy (✉)
Department of Urban Design and Planning, Faculty of Engineering, Ain Shams University, Cairo, Egypt
e-mail: nardine.elbardisy@eng.asu.edu.eg

A. Elshater · S. Afifi · A. Alfiky
Department of Urban Design and Planning, Urban Morphology, Ain Shams University, Cairo, Egypt
e-mail: abeer.elshater@eng.asu.edu.eg

S. Afifi
e-mail: samy.afifi@eng.asu.edu.eg

A. Alfiky
e-mail: abdo.alfiky@eng.asu.edu.eg

I. Pigliautile et al. (eds.), *Urban Resilience, Livability, and Climate Adaptation*,
Advances in Science, Technology & Innovation, https://doi.org/10.1007/978-3-031-54911-3_4

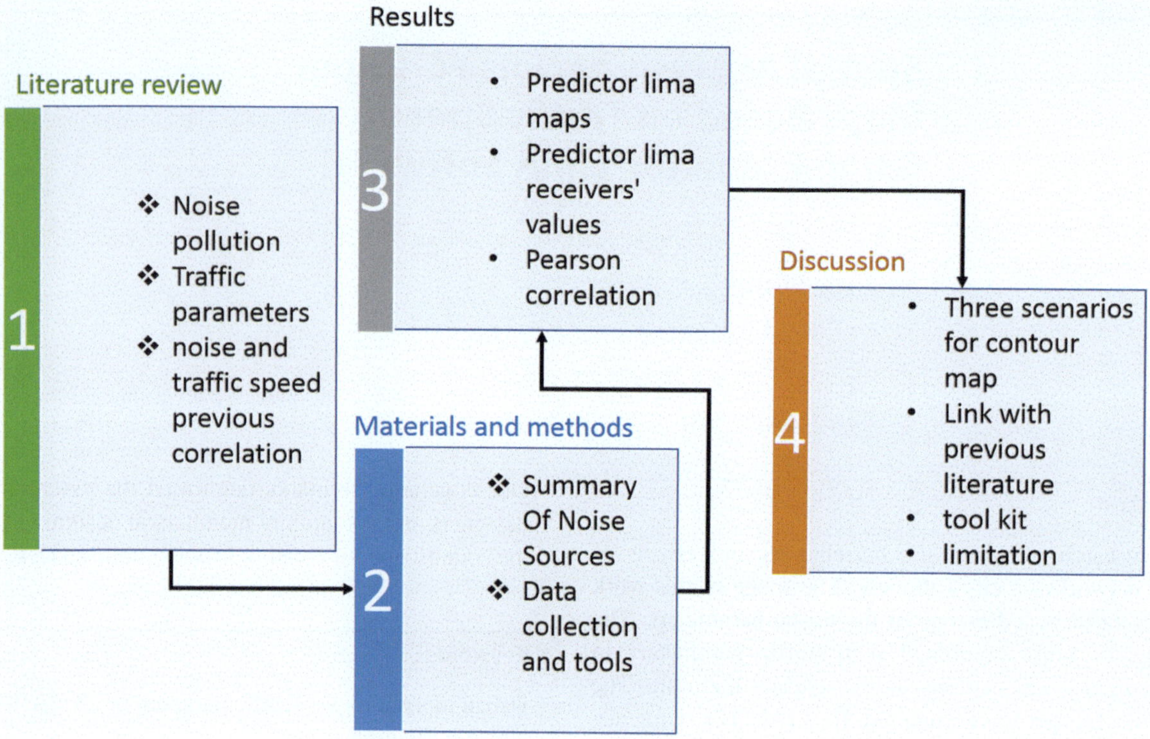

Fig. 1 Research structure

problem. This research attempts to solve this issue through noise mitigation techniques. This leads to the research's main problem: noise pollution increases due to traffic volume increases because of road width modification (Lu et al., 2019). Accordingly, this research examines the change in noise level in correlation with traffic speed (El-Bardisy et al., 2023; El-Bardisy et al, 2022).

Egypt has not done many studies on how noise and traffic speed are related. Furthermore, there is a gap in investigating case studies in the Egyptian context using prediction and contour mapping although world wide it is used for instant scholars use it to analyze wind turines studing it mechanical and environmental sound (Karadag & Kuruçay, 2021). Furthermore, an insufficient number of studies focus on the consequences of increased speed due to street network expansion. The street network is the primary source of background noise in the city, so it is essential to deeply investigate and eliminate its effect to improve the quality of noise that affects people in their everyday lives.

The toolkit introduced in the research aims to eliminate this problem from a holistic perspective, it integrates architecture and environmental health. This toolkit guides decision-makers authorities in upgrading streets combined with building configuration and land use without increasing noise levels in outdoor areas. The toolkit answers the main research question of "how can city planners and designers reduce the noise pollution consequence of urban morphological pattern changes in Cairo, Egypt?"

This research aims to predict sound pressure level values for three different scenarios of traffic speed (low, medium, and high). These scenarios envision toolkit recommendations that may ease decisions on various levels. The toolkit includes a strategic level of city planning and design at the community level as a primary contributor to the city. The central core of the economy and social development is people. People's mental and physical health is one of the main goals of city development.

This research starts with a literature review of noise pollution due to street pattern changes in Cairo, the traffic parameters, and the correlation between noise pollution and traffic speed. Afterwards, the connection between noise and traffic speed was explored in a case study. Sheraton, Heliopolis was selected as a case study in Cairo which faces massive street network development. Three scenarios were understudied to assume the impact of traffic speed on sound pressure level values. These scenarios include the study of traffic speeds of 100, 70 and 50 km/h. We assumed that the current traffic speed was 100 km/h. The primary focus of the research is the numerical statical values of traffic speed and the Equivalent Continuous Sound Pressure Level (LAeq), as well as noise contour maps. Primarily, Aviation Design Environmental Tool (AEDT) utilizes street selection. Subsequently, the current noise level was determined using the Type 1 sound level meter 2245 from Bruel and Kjaer. In addition, Predictor-LimA provides an accurate simulation of noise levels. Furthermore, the Pearson correlation analysis offers a detailed examination of the data. Figure 1 illustrates the overall structure.

The contribution here is to revisit the current morphological changes in street structures in Cairo regarding the different environmental aspects, such as noise. The present research contribution can also be recognized in providing suggestions for the mitigation measures that can be used to reduce the risk of noise pollution levels in the selected case study. The novelty is in exploring the case from Cairo, which is rarely discussed in the literature from that perspective. This research seeks to add value to the academic community as it tries to impact various scales of professional practice positively.

2 Background

The Greater Cairo Region has suffered from noise pollution over the years. Egyptian Environmental Affairs Agency reports indicate that the noise limit is exceeded in some of its 29 stations. Noise is evaluated based on the allowable limit of environmental law 4/1994 and Law 9/2009 stated by Egyptian legislation (Egyptian Environmental Affairs Agency, 2022a, 2022b; Eldakdoky and Elkhateeb, 2021). A national plan for noise pollution mitigation was placed with the cooperation of different Egyptian entities. One of these entities is the Ministry of Housing, Utilities, and Urban Development, which has seven agreed strategies, which include traffic control (Egyptian Environmental Affairs Agency, 2022a, 2022b).

Cooperation with the General Authority for Roads and Bridges attempts to eliminate noise through barriers and noise mitigation techniques (Egyptian Environmental Affairs Agency, 2022a, 2022b). The research focus of this strategy is in correlation with urban development projects that increase noise pollution levels due to street widening, followed by traffic flow growth (Faulkner & Murphy, 2022; Cole-Hunter et al., 2022; Yang et al., 2020; Lan & Cai, 2021; Lu et al., 2019). As such, this research gives a review of traffic parameters.

The traffic parameters are flow, speed, and density. Traffic flow is the quantified value of cars approaching a particular street during a specific period measured by vehicle/hr. Traffic speed is the actual speed of cars passing a clear road or the time taken for a vehicle to pass this road, it is unit is km/h. Traffic density is representative of the quality of traffic that facilities the number of vehicles per unit of road length, which is represented by cars per day per kilometre. These three variables are titled the traffic stream model (Sherief et al., 2016; TRB, 2000). This research investigates traffic speed and noise pollution. Noise pollution level and traffic speed are highly correlated because speed affects the response of sound waves in the air (Eurasto, 1995; Lan & Cai, 2021; Makarewicz & Kokowski, 2007). In this context, this research focuses on finding the statistical correlation between noise pollution level and speed, which is clarified in the next section.

Traffic speed and noise were examined from numerous perspectives. A study explained the traffic noise map method based on the speed clusters (Xue et al., 2021). After they obtain an initial traffic noise map, it is linked with road clustering, and lastly, the noise map is revised after the calculation (Xue et al. 2021). Another study explores the creation of dynamic traffic noise maps by utilizing both noise monitoring data and traffic speed information (El-Bardisy et al., 2023). Additionally, a study employed previous noise monitoring data to predict updated values, contributing to the development of a current dynamic traffic noise map (Lan & Cai, 2021).

3 Methods

3.1 The Case Study Settings

This research considered noise pollution due to the severity of the problem. This challenge is evident as Cairo, Egypt, is one of the world's noisiest cities, as mentioned in the Brüel and Kjær report, in addition to other words (Tawfeek, 2018; Martinelli, 2022, Brüel & Kjær, 2022b). Moreover, the noise level monitored by the Egyptian Environmental Affairs Agency shows that the values exceed their guideline limit (Egyptian Environmental Affairs Agency, 2022a, 2022b). Noise limit values were issued in the environment law 4/1994. Some modifications were added to environmental law 9/2009. these limits are divided based on zones as presented in their official website link: https://www.eeaa.gov.eg/Topics/77/28/Details. Power plants, industries, residential activities, traffic, and airports are noise sources (Ismail, 2013; Abas & Tamura, 2003). After examining these sources, traffic noise is considered the focal research approach, as industries and power plants are far from residential areas and within the allowable limit (Attia and Rahman, 1999, The Arab Republic of Egypt Ministry of Electricity & Energy, 2006).

Moreover, airports have a 4 km range of effects, which was considered in the study (Faiyetole & Toyin, 2019). Sheraton, Heliopolis in Cairo, was picked as a sample, given that it is near a residential area and commercial activities and is close to the airport, as simulated by the AEDT program. Shahied Said Zakariya analyzed a sample as shown in Fig. 2; it is an approximately 80 m wide two-way street with a green middle island. Recently, the road witnessed an urban transformation to ease urban mobility; the green centre island was reduced to approximately 20 m in width, and the asphalt was widened to about 30 m, including parking on each side. There was no speed radar, and speed limit signs were not placed yet.

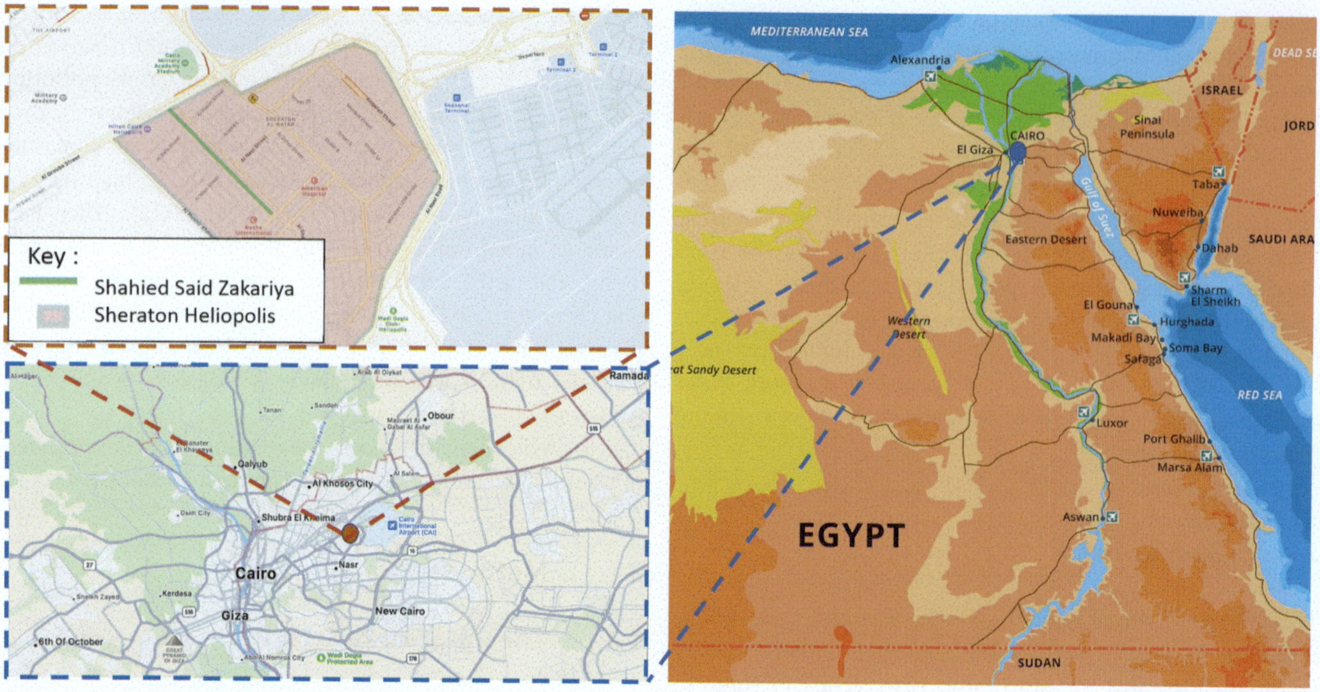

Fig. 2 Selected case study. *Source* The authors

3.2 Data Collection and Processing

The study employed a four-tool approach: Aviation Design Environmental Tool (AEDT), a sound metre (B&K 2245), Predictor-LimA, and Pearson correlations.

The first method is AEDT software (Federal Aviation Administration, 2022), which maps the noise surrounding the airports, following international standards. Professionals, authorities, researchers use the tool, and any entity related to aircraft management. The primary purpose of employing AEDT is to facilitate case study choice. The software can develop various studies, including one flight too many airport flights. Additionally, AEDT follows the International Civil Aviation Organization's (ICAO) Committee on Aviation Environmental Protection (CAEP).

The second method is the B&K 2245 device. The device measures sound levels, including all noise parameters; for instance, the average noise level, min, max, and percentages over time (Brüel & Kjær, 2022b). The device was positioned in the street, where the microphone was directed towards the traffic for around two consequential days. The measurement started on 3 September 2022 at 10 am till 5 September 2022 at 8 pm. Predictor-LimA applied these data as an input. The results are indicated in the results section.

The third method utilized was the Predictor-LimA, which produces contour maps as a noise mapping tool that predicts noise. Commonly, it is used in several projects, for instance, transportation and renewable energy industries. Numerous acoustic entities use this software, including environmental authorities, heavy industries, and educational institutes (Brüel & Kjær, 2022a). At the beginning of June 2021, a location visit was done to gather data to create a model to collect data. The model contains a vertical grid on each façade and a horizontal grid—moreover, eight receivers for exact values with control points at six heights. Some data were collected during the visit such as the height of the buildings, the material of the façade, and the type of façade involved. Additionally, the measurement from the second method was employed. Moreover, the validation of a model of the model was done through the measurements to guarantee the results.

The Pearson correlation was the final method. It was applied to analyze the outcomes of the sound level along with building height and traffic speed (Faulkner & Murphy, 2022; Cole-Hunter et al., 2022; WeijunYang et al., 2020; Lan & Cai, 2021, Lu et al., 2019). Pearson correlation provides an in-depth assessment of the linkage between those variables. The absolute values were recommended in the next step, afterwards the test statistic t-value was calculated with Eq. 1.

$$t = r\sqrt{n-2}/\sqrt{1-r^2} \tag{1}$$

where r represents the Pearson correlation, while n is the values count. Afterwards, the P-value is determined with Eq. (2).

$$P = TDIST(t, df, Tails) \tag{2}$$

where TDIST is the probability distribution continuously, while df is the freedom degree, its value equivalent to the count of values—2. Tails reflect the count of variables implemented.

4 Model of the Area

The selected street was divided into two segments, as shown in Fig. 3, and they are categorized into four types as indicated in Table 1. Figure 3 shows a base model for the fixed receivers that assist in giving specific noise values in different scenarios. The three scenarios were evaluated. The first scenario was traffic flow with a 50 km/h speed as a post intervention. The second scenario was a traffic flow of 70 km/h speed as a post intervention. In the third scenario, traffic flow was at 100 km/h as the current design.

The research identified eight receivers, placing in correspondence to each of them six receiver points within a height range from 1.5 to 18 m. Three receivers were placed in street segment 1 called Receiver A, Receiver B, and Receiver C. In street segment 2, five receivers were positioned and labeled as Receiver 1, Receiver 2, Receiver 3, Receiver 4, and Receiver 5. The results underwent statistical analysis using Pearson calculation to assess the correlation between traffic speed and noise levels numerically.

Table 1 Building configuration types

	Form	The length of façade	The height of the building
Type-1	Rectangle form	20 m	24 and 33 m
Type-2	Rectangle form	50 m	18 m
Type-3	Rectangle form	15 m	12–33 m
Type-4	Dynamic linear form	153 m	18 m

5 Results

Firstly, the results discuss the measurements in Shahied Said Zakariya at Sheraton Heliopolis. LAeq values were 56.9 and 54.2 dB daily and night, respectively, as presented in Fig. 4 and Table 2. The values exceeded the EEAA allowable limit values of 55 dB through daytime and 45 dB through nighttime. These values according to environmental law 9/2009. Moreover, the LAeq peak results were 68.8 dB during the daytime and 69 dB at nighttime.

The next part discusses the segments of the street in the case study. The first segment of the road reveals the receiver's (A, B, and C) values. However, the second segment of the street shows Receiver 1, Receiver 2, Receiver 3, Receiver 4, and Receiver 5. Each receiver has values at each specified height, along with traffic speed percentages. The software called Predictor-LimA calculates these values. The

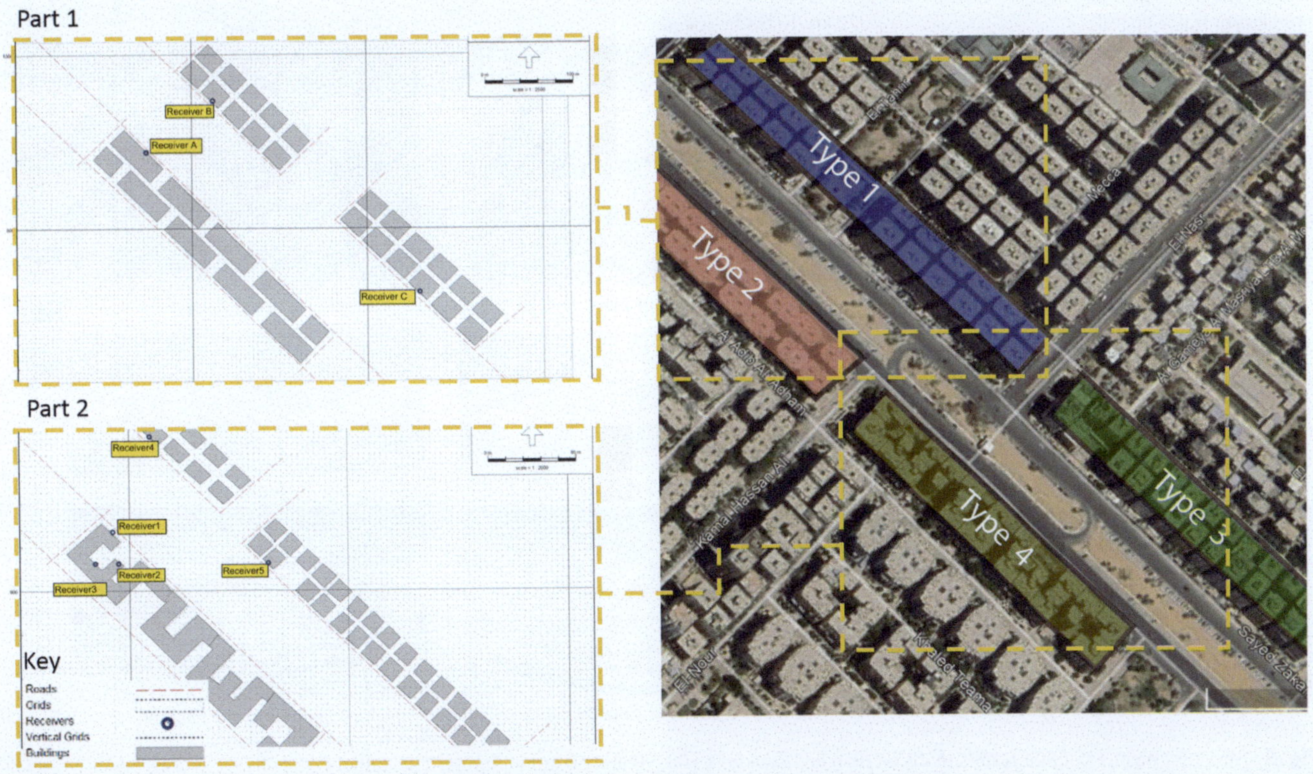

Fig. 3 A brief of the types of building and placement of receivers in the case study. *Source* Authors

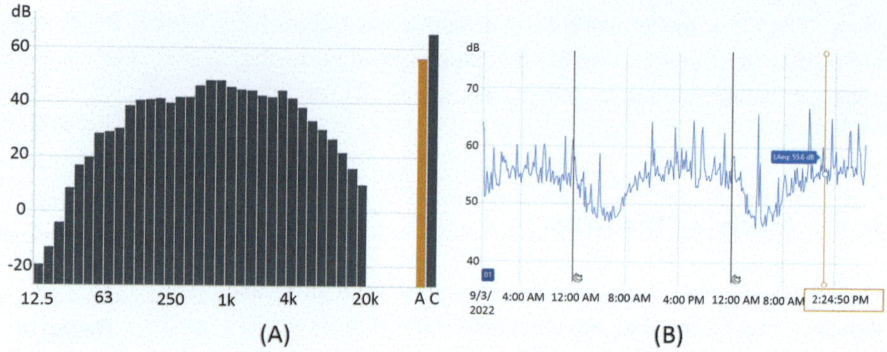

Fig. 4 **a** the spectrum of the 1/3 octave analysis bands, showing in the x-axis the Leaq (dB) value and Hz frequency in the y-axis, **b** The history LAeq (dB) during the measurement period, the dB value in the y-axis and the time in the x-axis

Table 2 LAeq (dB) measured versus allowable limit

	LAeq dB	Limit dB
Day 1	56.5	55
Day 2	56.8	55
Day 3	57.4	55
Average day	56.9	55
Max day	68.8	55
Night 1	53.6	45
Night 2	54.8	45
Average night	54.2	45
Max night	69	45

following subsections present receivers LAeq value in dB used in the Pearson correlational analysis.

5.1 Results from Street Segment 1

The three scenarios indicated a decrease in the noise level because of traffic speed reductions, as displayed in the layout. Reducing the value of the source diminishes the façade received noise. Figure 5 shows the variation between the scenarios on the plan and façade. Using the contour plan, the sound pressure level due to the road emission is 76 dB in the first scenario, between 73 and 77 dB in the second scenario, and 72–76 dB in the third scenario.

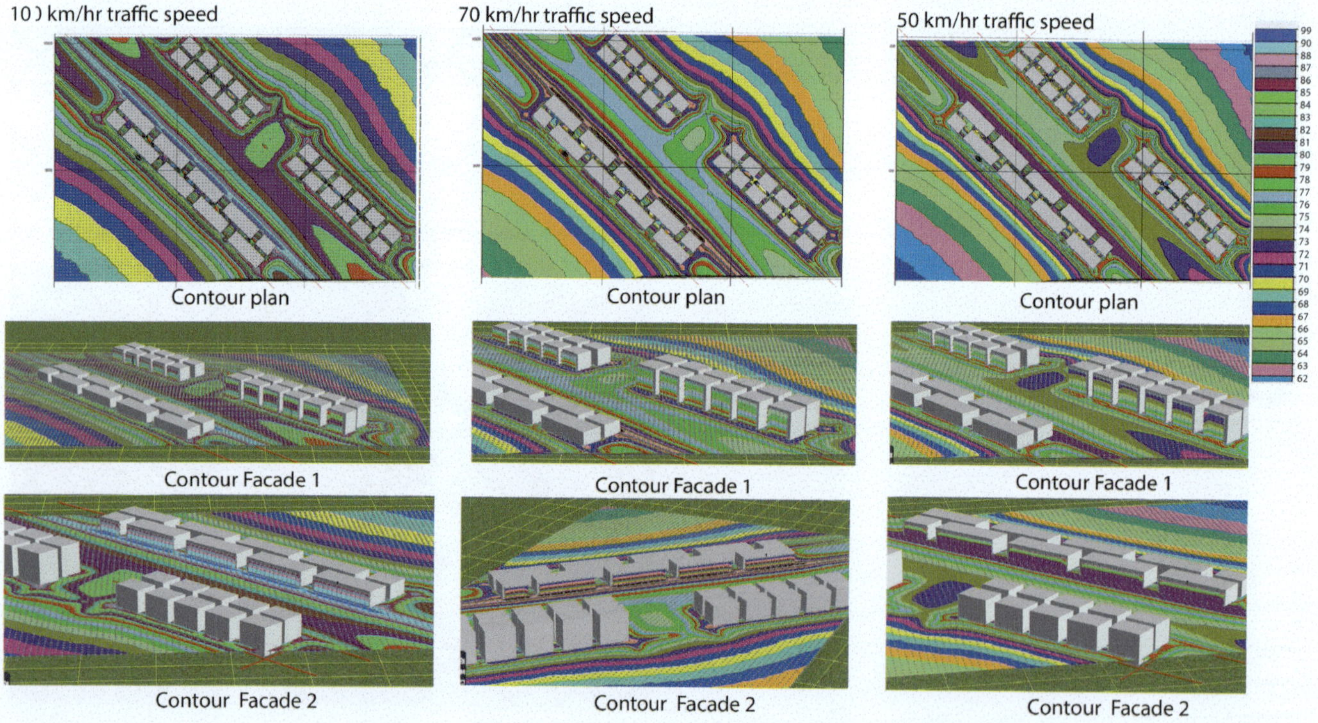

Fig. 5 Street segment 1 contour results in dB on plan and façade at 100, 70 and 50 km/h. *Source* The authors

For receiver A, the numerical results indicate the accurate noise level values with various heights. These heights are 1.5, 4, 8, 12, 15, and 18 m at traffic speeds of 100, 70 and 50 km/h. Similarly, receivers B and C produced different outcomes. The results show that the noise levels decrease as the height increases since it has become an additional source. For instance, the Receiver A (100 km/h) day value at 1.5 m was 78.6 dB, while the 18-m-high receiver recorded 71.9 dB. Additionally, the noise pollution level decreased as the speed dropped, as was evident in the value of receiver A in the daytime at the height of 1.5 m for 100 km/h was 78.6 dB. In that context, the noise pollution levels were 74.3 dB for 70 km/h and 71.9 dB for 50 km/h traffic speeds. All of the result values are indicated in Table 3.

Table 4 analyses the results by calculating the Pearson coefficient. In addition, traffic speed and noises were analyzed during the day and the night. Moreover, height and noise were analyzed. Where all results are smaller than 0.01 which shows strong evidence of correlation.

5.2 Results from Street Segment 2

Figure 6 illustrates the difference between three traffic speed cases (100 km/h, 70 km/h, and 50 km/h) on plans and façades. The contour map of three scenarios showed noise pollution decline as traffic speed dropped, which was noticeable within the contour plan. The elimination of the source noise value is directly proportional to the façade noise received. The contour plan clarifies that the noise emitted from the first scenario falls between 78 and 81 dB. In the second scenario, common values fell in the range of 74–76 dB; in the third scenario, 71–73 dB was most of the values.

The results of receivers indicated the noise levels at heights of 1.5, 4, 8, 12, 15 and 18 m, within traffic speeds of 100, 70 and 50 km/h, as well as receivers 2, 3, 4, and 5. As the height increases, the noise level decreases as the source is far from the receiver. Receiver 1 results clarify that at 1.5 m height for 100 km/h was 77.2 dB, while at 18 m height was 72.4 dB. In addition, the noise level reduced as the speed dropped, which was apparent in the Receiver 2. The values at height 1.5 m were 70.9 dB at 100 km/h, 66.6, at 70 km/h, and 64.1 dB at 50 km/h. Detailed results are provided in Table 5.

As shown in Table 6, the results were analyzed using Pearson coefficients for daytime and nighttime. Where all results were smaler than 0.01 which showed strong evidence of corelation.

It is essential to note that the smaller P-value, the more statistical meaning. The outcomes of segments 1 and 2 strongly correlated height with noise levels. Besides, noise along with traffic speed correlation is summarized in Table 7. Where all results are smaller than 0.01 which shows strong evidence of correlation.

Table 3 Street segment 1 values of receivers with unique traffic speeds

	Height values in metre	100 km/h		70 km/h		50 km/h	
		Day (dB value)	Night (dB value)	Day (dB value)	Night (dB value)	Day (dB value)	Night (dB value)
Receiver A	18	71.9	77	67.6	72.7	65.1	70.3
Receiver A	15	72.6	77.9	68.2	73.5	65.8	71.1
Receiver A	12	73.5	78.9	69.1	74.5	66.7	72.1
Receiver A	8	75.1	80.6	70.7	76.3	68.3	73.8
Receiver A	4	77.3	83	73	78.7	70.6	76.2
Receiver A	1.5	78.6	84.3	74.3	80	71.9	77.6
Receiver B	18	71.9	73.3	67.5	68.9	65.1	66.5
Receiver B	15	72.6	73.7	68.2	69.4	65.8	67
Receiver B	12	73.5	74.3	69.1	70	66.7	67.6
Receiver B	8	75	75.5	70.7	71.2	68.2	68.7
Receiver B	4	77.1	77.2	72.8	72.9	70.4	70.5
Receiver B	1.5	78.3	78.4	74	74.1	71.5	71.7
Receiver C	18	71.5	72.4	67.2	68.1	64.8	65.7
Receiver C	15	72.3	73	67.9	68.6	65.5	66.2
Receiver C	12	73.1	73.6	68.8	69.3	66.4	66.9
Receiver C	8	74.6	74.8	70.3	70.5	67.8	68.1
Receiver C	4	76.3	76.3	72	72	69.6	69.6
Receiver C	1.5	77.2	77.2	72.8	72.9	70.4	70.5

Source Authors

Table 4 The statical analysis of Pearson for segment 1

	Name	Values	Absolute values
The coefficient of Pearson correlation	Height and night noise values	−0.47893404	0.478934042
	Height and day noise values	−0.62523299	0.62523299
	Traffic speed percent with values of night noise	0.644943653	
	Traffic speed percent with values of day noise	0.769836531	
	N	54	
T-value	T-1	3.934201738	
	T-2	5.777035892	
	T-3	6.08554239	
	T-4	8.697920087	
	Df	52	
P-Value	P-Value 1	0.000248671	
	P-Value 2	4.32163E-07	
	P-Value 3	1.41034E-07	
	P-Value 4	1.02163E-11	

Source The authors

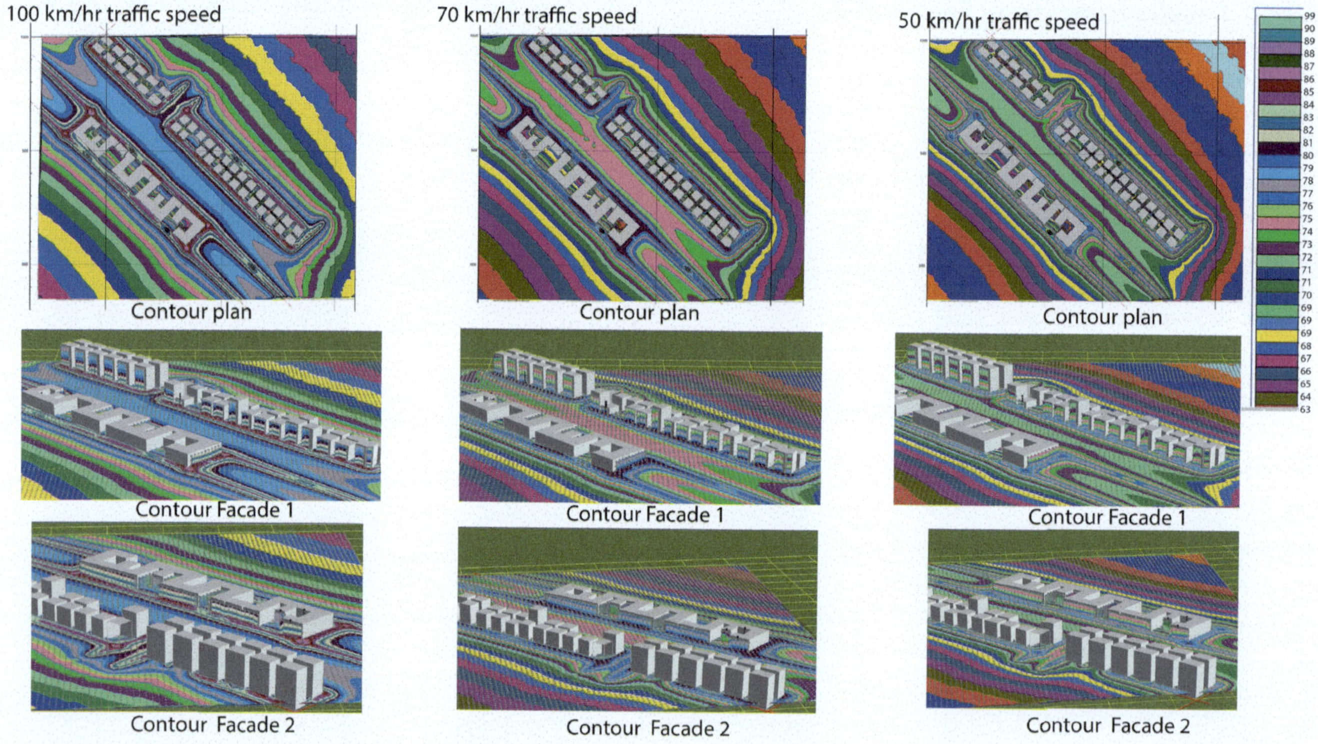

Fig. 6 Street segment 2 contour results in dB on plan and façade at 100, 70 and 50 km/h. *Source* The authors

6 Discussion and Conclusion

This research introduces the correlation between noise pollution and traffic speed as traffic parameters. Traffic is considered the primary source of noise in today's cities. The morphological pattern changes in Cairo, Egypt, cause an increase in noise pollution levels due to road widening and consequent traffic flow growth. This research outlines the shift and how it could mitigate limited car speed. The change in the pattern of urban morphology in the present case study and the street networks have a positive and

Table 5 Street segment 2 receivers values at different traffic speed percentages

		100 km/h		70 km/h		50 km/h	
	Height values in metre	Day (dB value)	Night (dB value)	Day (dB value)	Night (dB value)	Day (dB value)	Night (dB value)
Receiver 1	18	72.4	71.8	68	67.4	65.6	65
Receiver 1	15	73	72.4	68.6	68.1	66.2	65.6
Receiver 1	12	73.7	73.1	69.4	68.8	66.9	66.4
Receiver 1	8	75	74.4	70.7	70.1	68.2	67.6
Receiver 1	4	76.5	75.9	72.2	71.6	69.6	69
Receiver 1	1.5	77.2	76.6	72.9	72.3	70.3	69.7
Receiver 2	18	69.5	68.9	65.1	64.5	62.7	62.1
Receiver 2	15	69.6	69	65.3	64.7	62.8	62.2
Receiver 2	12	69.9	69.3	65.6	65	63.1	62.5
Receiver 2	8	70.3	69.7	65.9	65.4	63.4	62.9
Receiver 2	4	70.6	70	66.2	65.6	63.7	63.2
Receiver 2	1.5	70.9	70.3	66.6	66	64.1	63.5
Receiver 3	18	63.9	63.3	59.5	59	57.1	56.5
Receiver 3	15	62.9	62.3	58.5	57.9	56.1	55.5
Receiver 3	12	62.9	62.3	58.6	58	56.1	55.5
Receiver 3	8	63	62.4	58.7	58.1	56.2	55.6
Receiver 3	4	63.2	62.6	58.8	58.2	56.4	55.8
Receiver 3	1.5	63.7	63.1	59.3	58.7	56.8	56.3
Receiver 4	18	71.9	71.3	67.5	66.9	65.1	64.5
Receiver 4	15	72.5	71.9	68.2	67.6	65.8	65.2
Receiver 4	12	73.4	72.8	69	68.5	66.6	66
Receiver 4	8	74.8	74.2	70.4	69.9	68	67.4
Receiver 4	4	76.5	75.9	72.2	71.6	69.7	69.2
Receiver 4	1.5	77.3	76.7	73	72.4	70.6	70
Receiver 5	18	72	71.4	67.7	67.1	65.3	64.7
Receiver 5	15	72.6	71.9	68.2	67.6	65.8	65.2
Receiver 5	12	73.4	72.8	69.1	68.5	66.6	66.1
Receiver 5	8	75	74.4	70.7	70.1	68.3	67.7
Receiver 5	4	77.5	76.9	73.2	72.6	70.8	70.2
Receiver 5	1.5	79.2	78.6	74.9	74.3	72.4	71.9

Source The authors

negative impact. The research prefers to solve one negative consequence to achieve the optimum benefits of this update. Urban morphology changes in Cairo, Egypt, are a two-sided coin. From a practical perspective, they reduce the time to transfer between two locations in the city, create a more organized traffic system, and modernize the city to better adapt to population growth. From a disadvantageous perspective, they cause some environmental pollution, such as noise pollution, which is a vital problem based on monitoring reports and evaluation. There are lots of methods of noise mitigation. Accordingly, the research focuses on the sources and one specific parameter to find a solution. Noise control has been achieved at the source, at the receiver, and along the path, and research is now focusing on our resource (traffic roads) as the element that has been changed.

We analyzed speed because the flow had already changed, and the same for traffic volume. Noise mitigation and various techniques give the applicability of different solutions. Thus, the research will recommend other solutions if changing car speeds does not achieve the requirements. In a highly interlinked world of noise and traffic,

Table 6 The statical analysis of pearson for segment 1

	Name	Values	Absolute values
Pearson correlation coefficient	Height and night noise values	−0.24666	0.246655855
	Height and day noise values	−0.24581	0.245813049
	Traffic speed percent with values of night noise	0.504257	
	Traffic speed percent with values of day noise	0.505824	
	N	90	
T-value	T-1	2.387607	
	T-2	2.378923	
	T-3	5.477769	
	T-4	5.500634	
	Df	88	
P-Value	P-Value 1	0.019098	
	P-Value 2	0.019525	
	P-Value 3	4.04E-07	
	P-Value 4	0.011414	

Source The authors

Table 7 P-value summary

P-values		
	Segment 1	Segment 2
P-Value 1	0.000249	0.019098
P-Value 2	4.32E-07	0.019525
P-Value 3	1.41E-07	4.04E-07
P-Value 4	1.02E-11	0.011414

Source The authors

the challenge is to eliminate noise in parallel with traffic development. That was clear when the research expressed this idea by reviewing the literature, finding the connections between them, and then profoundly analyzing the statistical correlations between them.

On the one hand, the purpose of this article was to examine the relationship between traffic and noise. Current situations showed high noise exposure to façades in many buildings. On the other hand, controlling the speed led to acceptable noise exposure levels on the façades. According to the linkage, the exposure to noise increases along with block length, although façade noise exposure drops when height increases as the p-value ranges 0.01 which means there is a strong correlation. Besides, the façade length allocated near the source increases resulting in an increase in noise exposure, so Type-4 is beneficial as length is not equidistance from the source, then Type-3 with maximum length, afterward Type-1, and ending by Type-2 with mimum length.

When comparing results with the previous literature, traffic and noise were analyzed from different perspectives. The present research discusses the traffic noise map method based on speed clusters (Xue et al., 2021; Elshater et al., 2022a). After obtaining an initial traffic noise map, it is pre-treated with road clustering, and finally, the noise map is updated after the calculation (Xue et al., 2021). Another article delves into the creation of dynamic traffic noise maps by integrating noise monitoring data with information on traffic speed (Elshater et al. 2022b). Similarly, a study emphasized traffic flow and speed as primary factors influencing traffic noise levels. It was predicted by using the historical tracking of noise data for noise and speed, and then updating the values and combining the shapes to make a dynamic traffic noise map (Lan & Cai, 2021).

In street canyons, cars going over the speed limit or other problems with traffic can cause a lot of noise. A study discusses three traffic speed and flow prediction techniques: naive, parametric, and non-parametric. A certain amount of noise is investigated togaed during the sampling process, so the authors manually introduce different levels of noise based on the Beijing data set (Yu, 2021).

This paper demonstrates the benefits of a dynamic noise monitoring system that displays noise level variations with speed. Consequently, adding noise and speed correlational analysis in the design process has benefits and drawbacks. Using the analytical correlation, we could understand the current situation, and also examine design scenarios based on our findings. Secondly, the advantage allowed designers and urban planners to pick designs with lower negative impacts and improve the quality of their plans. Authorities and urban planners can develop strategies that support speed limits by fine-tuning public spaces accordingly, based on the results of our study. However, the analysis's disadvantage was the elevated cost of resolving the current noise issue since it was not planned to adapt to the existing traffic speed. In addition, it increases the design process time as the simulation is part of it.

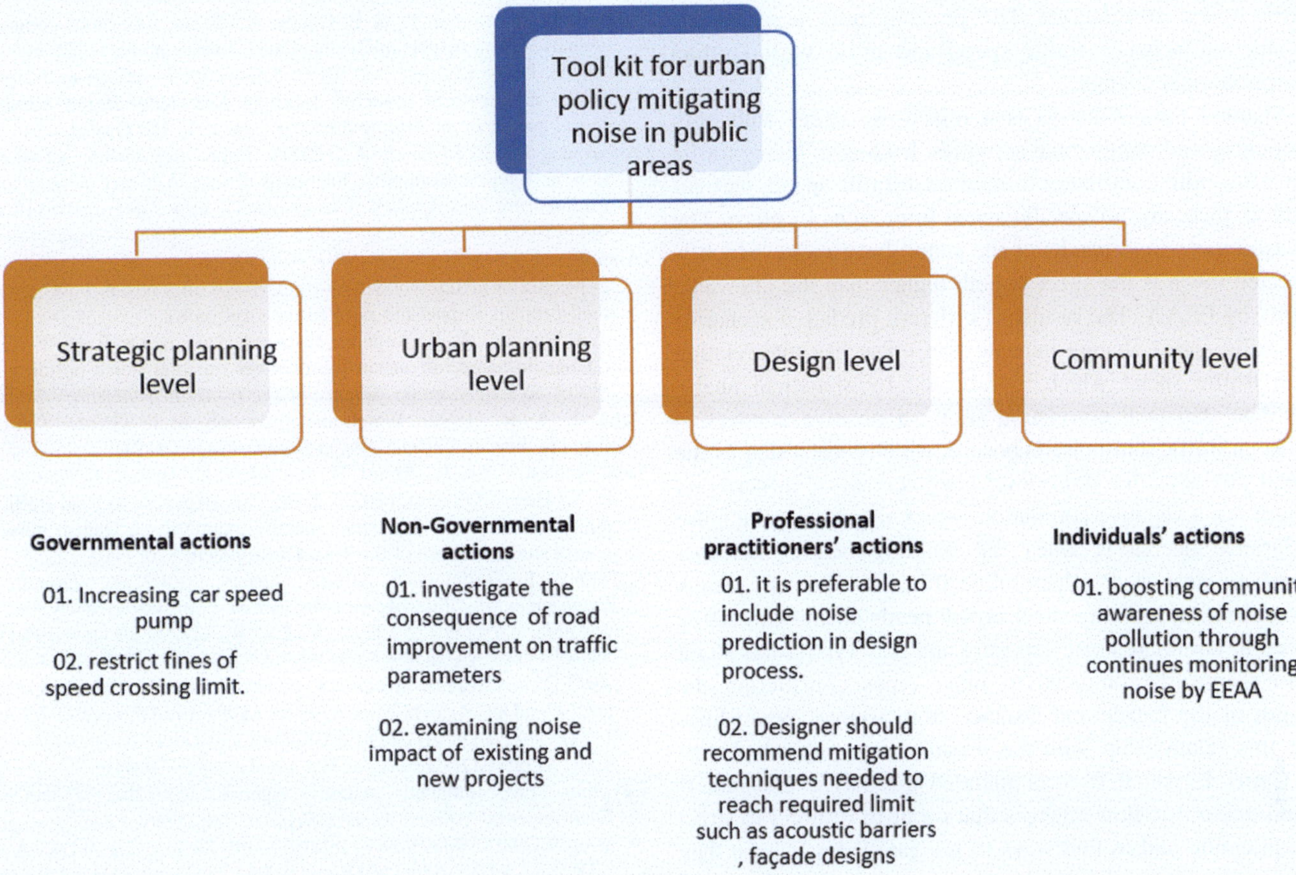

Fig. 7 Suggested toolkit intended for noise mitigation techniques as an urban policy

As a result of the case study investigation in Egypt, five lessons have been identified on which noise levels impact the outdoor environment. Firstly, authorities should study the road effects improvements on traffic parameters and then analyze the influence on noise. The second learned lesson is in placing speed bumps to limit speed. The third lesson learned is that other mitigation techniques were needed to reach the required limit, which is 50 dB at night and 60 dB during the day. The fourth lesson is that landscape designers and urban planners must involve noise prediction. By keeping track of noise, people can become more aware of how noise changes in response to environmental changes. The fifth lesson is that monitoring noise can increase citizens' awareness of noise change in correlation to ecological change.

The concluding observations from examining the case study demonstrated a toolkit for soundscape design on various levels. On the urban planning level, in the first phase, the government should increase the number of car speed bumps. Moreover, the government should restrict fines for speed crossing limits. In the following step, the government must appoint specialists to investigate the consequences of road enhancement on traffic parameters and examine the noise impacts of existing and new projects. The designer must recommend mitigation techniques to reach the required limit, such as acoustic barriers and façade designs. On the community level, community awareness of noise pollution is being boosted through the continuous monitoring of noise by EEAA. Figure 7 shows a brief about the toolkit.

One of the limitations is that there are only two parts to the case study. This study was restricted to measuring traffic speed with noise on these parts. This measurement was achieved by assuming different speed scenarios. The measurements were made at 50, 70 and 100 km/h. Additionally, this research was limited in comparing the sizes of the present study with the data available from the General Authority for Roads and Bridges and the Ministry of Transportation. It is preferable to link noise, speed, and flow together simultaneously in case data is available.

The Greater Cairo Region faces many noise pollution challenges. Although an enormous effort has been made using mitigation techniques, it is crucial to correlate traffic parameters with noise and to deeply research noise and traffic speed correlations. The present study assesses the connection between traffic speed and noise within urban morphological change.

Shahied Said Zakariya is a mixed-use street with both residential and commercial activities, leading to heavy traffic flow. Its width contributes to increased traffic speed, particularly as there are no specific speed limit signs in place. The measurements and results of the sound level metre which is directed towards the street is high higher than the allowable limits by EEAA. The results of different predicted scenarios of traffic speed change indicate that noise has interrelation with traffic speed which is validated by statistical analysis. The statistical correlation shows a positive connection between traffic and noise exposure, which means that as the speed increases, the noise level also increases. Furthermore, height has a negative correlation, which means that the noise difference decreases when the building height increases which is clear as P-values range 0.01. The present results reveal that the existing street layout needs changes in traffic speed to eliminate noise exposure and that Type-4 has slight noise exposure compared to other types considering the length of the façade and distance from source. When linking this relationship with the urban morphological change in Cairo, Egypt, it is recommended to analyze and understand any traffic flow changes that occur to respond to street modification and to find ways to mitigate them in case they exceed the required limit. Reducing car speed is one technique that can limit noise from traffic sources, as has been prescribed in the research. Different approaches can be used in some cases, including barriers, façade isolation, and many other ways. It is preferable to use the toolkit for existing and new projects, which include various scales, from urban planning to urban design and the building itself, in addition to the engagement of the community.

Building on the present research limitations, future research may utilize the same research methods for other case studies to be contextalized based on geographical, environmental conditions and background noise sources. A future recommendation should manage measurements to cope with the yield process from the General Authority for Roads and Bridges and the Ministry of Transportation to achieve similar and accurate data measurements.

References

Builes-Vélez, A. E., Suárez Vásquez, L. M., Correa Velásquez, L., & Gutiérrez Aristizábal, D. C. (2021). Analysis of urban transformation through the use of expanded research tools. *SAGE Open, 11*(3).

Abas, S., & Tamura, A. (2003).Analysis of road traffic noise level and control in Greater Cairo, Egypt. *Acoustical Science and Technology*, 358–364.

Abbass, R. A., Kumar, P., & El-Gendy, A. (2021). Emissions control scenarios for transport in Greater Cairo. *Toxics, 9*(11), 1–22.

Abusaada, H., & Elshater, A. (2018). Knowledge-based urban design in the architectural academic field. In *Knowledge-Based Urban Development in the Middle East* (pp. 204–227). IGI Global.

Abusaada, H. & Elshater, A. (2021). Improving visitor satisfaction in Egypt's Heliopolis historical district Abstract *Journal of Engineering and Applied Science, 68*(1). https://doi.org/10.1186/s44147-021-00022-y

Almoghazy, A. (2020). *Alternative Policy Solutions*. https://aps.aucegypt.edu/en/articles/608/new-roads-bad-connections-how-new-road-projects-impact-the-right-to-the-city#_edn1.

Ariza-Villaverde, A. B., Jiménez-Hornero, F. J., & De Ravé, E. G. (2014). Influence of urban morphology on total noise pollution: Multifractal description. *Science of The Total Environment*, 1–8.

Attia, F. A. R. (1999). Water and Development in Greater Cairo (Egypt). *Revista CIDOB d Afers Internacionals*, 4–23.

Cole-Hunter, T., So, R., Amini, H., Backalarz, C., Brandt, J., Brauner, E. V., Hertel, O., et al. (2022). Long term exposure to road traffic noise and all-cause and cause-specific moratality: a Danish nurse cohort study. Science of the Total Environment 820, 1-10.

Egyptian Environmental Affairs Agency. (2022a). *Egyptian Environmental Affairs Agency Mitigation Plan*. https://www.eeaa.gov.eg/ar-eg/%D9%85%D9%88%D8%B6%D9%88%D8%B9%D8%A7%D8%AA%D8%A8%D9%8A%D8%A6%D9%8A%D8%A9/%D8%A7%D9%84%D9%87%D9%88%D8%A7%D8%A1/%D8%A7%D9%84%D8%B6%D9%88%D8%B6%D8%A7%D8%A1/%D8%AE%D8%B7%D8%A9%D9%85%D9%83%D8%A7%D9%81%D8%AD%D8%A9%D8%A7%D9%84%D8%B6.

Egyptian Environmental Affairs Agency. (2022b). *Egyptian Environmental Affairs Agency Reports*. https://www.eeaa.gov.eg/ar-eg/%D9%85%D9%88%D8%B6%D9%88%D8%B9%D8%A7%D8%AA%D8%A8%D9%8A%D8%A6%D9%8A%D8%A9/%D8%A7%D9%84%D9%87%D9%88%D8%A7%D8%A1/%D8%A7%D9%84%D8%B6%D9%88%D8%B6%D8%A7%D8%A1/%D8%A7%D9%84%D8%AA%D9%82%D8%A7%D8%B1%D9%8A%D8%B1%D9%88%D8%A7%D9%84%D8%AF.

El-Bardisy, N., Elshater, A., Afifi, S., & Alfiky, A. (2023). Predicting traffic sound levels in Cairo before during and after the COVID-19 lockdown using Predictor-LimA software. *Ain Shams Engineering Journal, 14*(9). 102088. https://doi.org/10.1016/j.asej.2022.102088

El-Bardisy, N., Rashed, A. N., Elmeligy, E. A., & Aboulnaga, H. H. (2022). Ergonomic design for Pedestrians' safety in regenerative cities: Heliopolis, Cairo, Egypt. In *Remapping Urban Heat Island Atlases in Regenerative Cities* (pp. 169–186). IGI Global.

Eldakdoky, S., & Elkhateeb, A. (2021). Noise levels inside the coaches of Greater Cairo tunnel metro, line 3 tunnel. *Frontiers of Architectural Research*, 410–419.

ElMoghazi, Y. (2019). *Road safety challenges in Egypt: A discussion of policy alternatives*. Theses and Dissertations. the American University in Cairo. Accessed 2022. https://fount.aucegypt.edu/etds/790/.

Elshater, A., Abusaada, H., Alfiky, A., Elbardisy, N., Elmarakby, E., & Grant, S. (2022a) Workers' satisfaction vis-à-vis environmental and socio-morphological aspects for sustainability and decent work *Sustainability, 14*(3). 1699. https://doi.org/10.3390/su14031699

Elshater, A., Abusaada, H., Tarek, M., & Afifi, S. (2022b). Designing the socio-spatial context urban infill liveability and convivialty *Built Environment, 48*(3), 341–363. https://doi.org/10.2148/benv.48.3.341

Eurasto, R. (1995). *Road traffic noise in the speed range of 30-50 km/h*. VTT Technical Research Centre of Finland.

Faiyetole, A. A., & Toyin, I. T. (2019). *Predicting the aircraft take-off noise level. Vibroengineering PROCEDIA*, 129–134.

Farouk, M. A. (2021). *Majalla*. https://eng.majalla.com/node/177876/economyegypt%E2%80%99s-infrastructure-projects-shaping-new-future.

Faulkner, J. P., & Murphy, E. (2022). Road traffic noise modelling and population exposure estimation using CNOSSOS-EU: Insights from Ireland. Applied Acoustics 192, 1–11.

Federal Aviation Administration. (2022). *Aviation environmental tools suite*. https://www.faa.gov/about/office_org/headquarters_offices/apl/aee/research/models.

Ibili, F., Owolabi, A. O., Ackaah, W., & Massaquoi, A. B. (2022). Statistical modelling for urban roads traffic noise levels. Scientific African 15, 1-9.

Habib, A., & Ouki, M. A. (2021). *Egypt's low carbon hydrogen development prospects*. https://www.oxfordenergy.org/publications/egypts-low-carbon-hydrogen-development-prospects/.

Hölscher, K., & Frantzeskaki, N. (2021).Perspectives on urban transformation research: transformations in, of, and by cities. *Urban Transformations*, 1–14.

Howard, C. Q. (2011). Recent developments in submarine vibration isolation and noise control. In: 1st *Submarine Science Technology and Engineering Conference, Adelaide,. SA, Australia*

IQair. (2023). *Air quality in Egypt*. Accessed june 2023. https://www.iqair.com/egypt.

Ismail, M. R. (2014). Sound preferences of the dense urban environment: Soundscape of Cairo. *Frontiers of Architectural Research, 3*(1), 55–68.

Karadag, I., & Kuruçay, E. (2021). Noise emission from building integrated wind turbines: a case study of a tall building. *Sakarya University Journal of Science, 25*, 563–570. https://doi.org/10.16984/saufenbilder.884517

Brüel & Kjær. (2022a). *BSKV*. https://www.bksv.com/en/services/downloads/environmental-software/predictor-7810.

Brüel & Kjær. (2022b). *BKSV*. https://www.bksv.com/en/instruments/handheld/sound-level-meters/2245-series.

Lan, Z., & Cai, M. (2021). Dynamic traffic noise maps based on noise monitoring and traffic speed data. *Transportation Research Part D, 94*, 1–20.

Lu, X., Kang, J., Zhu, P., Cai, J., Guo, F., & Zhang, Y. (2019). Influence of urban road characteristics on traffic noise. *Transportation Research Part D: Transport and Environment, 75, 136-155*

Makarewicz, R., & Kokowski, P. (2007). Prediction of noise changes due to traffic speed control. *The Journal of the Acoustical Society of America, 122*.

Martinelli, L. (2022). *Global noise pollution map*. http://lukasmartinelli.ch/gis/2016/04/03/openstreetmap-noise-pollution-map.html.

Onder, M. S. (2016). Social impact of the urban transformation (Diyarbakir Sampe). *Recent Ideas and Research in Social Sciences*, 147–157.

Puvanachandra, P., Hoe, C., El-Sayed, H. F., Saad, R., Al-Gasseer, N., Bakr, M., & Hyder, A. A. (2012). Road traffic injuries and data systems in Egypt: Addressing the challenges. *Traffic Injury Prevention, 13*(1), 44–56. https://doi.org/10.1080/15389588.2011.639417

Rossi, I. A., Vienneau, D., Ragettli, M. S., Flückiger, B., & Roosli, M. (2020). Estimating the health benefits associated with a speed limit reduction to thirty kilometres per hour: A health impact assessment of noise and road traffic crashes for the Swiss city of Lausanne. *Environmental International, 145*, 1–8.

El Samra, G. (2016). Community noise in greater cairo. *Egyptian Journal of Occupational Medicine*, 267–285.

Shafik, N., Mansour, Y., Kamel, S., & Morcos, R. (2021). The impact of the Cairo streets development project on the independent mobility of children: A field study on the streetsof heliopolis, Egypt. *Infrastructures, 6*(7), 1–20.

El Sherief, M. M., Ramadan, I. M., & Ibrahim, A. M. (2016). Development of traffic stream characteristics models for intercity roads in Egypt. *Alexandria Engineering Journal, 55*(3), 2765-2770.

Tawfeek, F. (2018). *Egypt independant*. https://egyptindependent.com/cairo-ranked-second-noisiest-city-world/.

The Arab Republic of Egypt Ministry of Electricity & Energy. (2006). *15OMWe kuraymat integrated solar combined cycle power plant project*. Energy & Environment Consultant (E&E).

TRB. (2000). *Highway capacity manual*. National Research Council.

Yang, W., He, J., He, C., & Cai, M. (2020). Evaluation of urban traffic noise pollution based on noise maps. *Transportation Research Part D 87*, 1–14.

World Health Organization. (2018). *Environmental noise guidelines for the european region*. World Health Organization. https://www.euro.who.int/__data/assets/pdf_file/0008/383921/noise-guidelines-eng.pdf?ua=1

Xue, W., Huang, Z., Zhao, B., Yang, W., Lan, Z., & Cai, M. (2021). Updated traffic noise map method based on speed cluster. *Applied Acoustics*.

Yu, J. J. Q. (2021). Citywide traffic speed prediction: A geometric deep learning approach. *Knowledge-Based Systems, 212*, 1–13.

Critical Factors Affecting the Design and Use of Elevated Urban Spaces: The Sky Garden, London

Ahmed Ehab and Tim Heath

Abstract

The Covid-19 pandemic and subsequent pandemics have and will continue to have a significant impact on public spaces in cities around the World. One of the major challenges and debates for governments and professionals alike is therefore how to create enjoyable and usable environments in high-density cities, which also feel safe. Elevated urban gardens have become increasingly popular public and social spaces in large cities with their unique qualities in terms of location, accessibility, scale, intensity of experience, etc. Indeed, the past two decades have seen the re-emergence of exciting social and public places 'in the sky' which justifies the need to research these spaces in more detail. Due to their location, searching questions can be raised with regard to the design and management of safe vibrant spaces within or on contemporary buildings. This paper focuses on analysing the Sky Garden at 20 Fenchurch Street, London using direct observation and walk-along interviews to provide qualitative and quantitative data before, during, and after the Covid pandemic. The direct observation explores how the space changed in terms of accessibility, circulation, and activities during these periods of time. The study also uses semi-structured interviews to investigate the qualitative relationship between visitors' behaviour and the design of the pace. Participants (n = 23) were interviewed by the researcher while walking in the Sky Garden. This paper examines critical issues such as accessibility, circulation, activities, design obstacles, security and safety, etc., in exploring the possibilities and opportunities for the future design of elevated urban gardens. The research findings include a study of human activities, features that need improvements, design strategies, and an analysis of the potential need for new rules and regulations relating to the use of such spaces.

Keywords

Sky garden · Vertical public realm · Covid-19 · Social resilience · London

1 Introduction

A new and major challenge in creating a safe environment during the recent or any future pandemics will be spatial density (Afrin et al., 2021; Megahed &Ghoneim, 2020). When we think about the future of creating public spaces, there are questions to be asked, such as who really owns the space, the number of people able to be in that space at a time, and how they should group (Geary et al., 2021; Martínez & Short, 2021). Traditional urban public spaces and their timeless values remain immensely important, however, it is becoming important to examine new ways of attaining and sustaining such values in the multi-layered cities of the future. Indeed, it is important to investigate possible new values and modes of publicness, that are crucial for vibrant cities in new high-density conditions (Lak et al., 2020; Zhongming et al., 2021).

One of the main essential components of enjoying London's cityscape is being in its public spaces. These public spaces play a major role in the quality of life, environment, economy, and tourism for both residents and visitors (Carmona, 2019; Holy-Hasted & Burchell, 2022). As cities

A. Ehab (✉) · T. Heath
Department of Architecture and Built Environment, University of Nottingham, University Park, Nottingham NG7 2RD, UK
e-mail: A.Abdelsalam@lboro.ac.uk

T. Heath
e-mail: Tim.Heath@nottingham.ac.uk

A. Ehab
School of Design and Creative Arts, Loughborough University, Loughborough, Leicestershire LE11 3TU, UK

I. Pigliautile et al. (eds.), *Urban Resilience, Livability, and Climate Adaptation*,
Advances in Science, Technology & Innovation, https://doi.org/10.1007/978-3-031-54911-3_5

become denser with new business and residential developments, tall building design needs to be increasingly considered in relation to its intricate urban contexts (Al-Kodmany, 2020; Oldfield, 2019). Indeed, hybrid buildings with vertical access for the public are now seen as having the potential to be regarded as a vertical extension of the public realm (Cho et al., 2015; Mualam et al., 2019).

Vertical green social spaces are types of spaces, elevated above, or even below the ambient ground plane, but privately owned and managed are increasingly being offered as public spaces (Hadi et al., 2018; Osmundson, 1999; Pomeroy, 2013). These hybrid spaces are a new typology of public space, with characteristics and regulations quite different from the familiar models of public open space, such as city squares and parks (Abdelsalam, 2018; Cho et al., 2015; Oldfield, 2019). Many types of hybrid urban spaces are emerging, such as elevated spaces and multi-level spaces that are enhanced with nature. In our cities, we are now seeing pedestrian bridges, shopping malls, and transport interchanges, which contain green spaces and may lead people up to sky courts, sky gardens, and even sky parks (Samant, 2019; Viñoly et al., 2015).

Historically, many developers have failed to understand the importance of such spaces being used to improve amenities, well-being, good health, productivity, and social interaction, or earning hospitality revenue; as a result, they were often omitted from design proposals for cost-saving economic reasons (Triguero-Mas et al., 2020; Lee & Park, 2018; Taib et al., 2010). It has been pointed out that a city government can require such spaces as a condition of planning approval for a major building, through using incentives such as greater permitted height, more parking spaces, and reduced property taxes (Pomeroy, 2012, 2013). This is a requirement in Singapore, where they use the 'LUSH Index', as a way of calculating how much green private and public space, and green facade has been provided by a developer (Timm et al., 2018). Singapore, perhaps more than anywhere else in the world, has embraced the concept of vertical urbanism (Feng, 2021; Yuen & Hien, 2005). It's now uncommon for any new Singaporean high-rise not to consider the provision of sky gardens or vertical social communal spaces. Indeed, sky gardens are widely spread across both public and private high-rise housing developments (Bay, 2004; Samant & Hsi-En, 2017). Also, mixed-use high-rise design that can include shopping malls, residential development, offices, entertainment, and public sky gardens all within one building are proliferating internationally in cities across Asia, Europe, and the US (Generalova & Generalov, 2020; Oldfield, 2019).

Since the world's first underground railway network, in 1863, London has been a multi-layer city (Aleta et al., 2017; Madanipour, 2019), but it is now experiencing an evolution as a 'skyscraper city'. London in the twenty-first Century has witnessed an increasing trend for towers taller than 100 m with some of these buildings including elevated urban spaces (Morato, 2022; Viñoly et al., 2015). This new phenomenon of elevated gardens has been designed and tested in a number of completed projects, such as the Sky Garden (indoor garden, skyscraper) at 20 Fenchurch Street, The Garden (outdoor garden, medium tall) at 120 Fenchurch Street, and Crossrail Place (semi-indoor-outdoor garden, transport interchange) at Canary Wharf. As these spaces become part of the City's public space offer, the debate continues about whether these innovative places can become vibrant and inclusive urban spaces (Wood & Safarik, 2019; Hadi et al., 2018).

The Covid-19 pandemic caused significant social and economic uncertainty and caught the development industry off-guard (Dwijendra et al., 2021). The temporary closure of construction sites because of the lockdown reduced the number of completions and reduced the revenue in projects that had reached completion (Zamani et al., 2021). Developers also hesitated to consider or invest in new projects. The confidence-shattering effect of the pandemic further emphasises the critical need for resilience and adaptability in the built environment (Lak et al., 2020). The pandemic has created numerous complex obstacles, so it is essential to take a comprehensive approach that incorporates multiple pragmatic solutions (Mills et al., 2020; Navaratnam et al, 2022).

In considering pandemic adaptability, public spaces should be able to accommodate expanded physical distancing—three to four times what is normal—between occupants during a health crisis (Afrin et al., 2021; Shroff, 2020). As the world continues to adapt post-Covid-19, it is essential for building designers and developers alike to become proactive in their design responses. This means creating strategies that reduce risk, promote hygiene habits, and also design social spaces that are flexible enough for high levels of interaction while providing a safe environment (Bereitschaft & Scheller, 2020). The design and development community, therefore, needs to emerge from the defensive stage of the pandemic response to a more forward-looking approach by creating new, innovative responses (Navaratnam et al, 2022). In addition, the repurposing of existing design tools and incorporation of new thinking around user experience design and technologies is required (He et al, 2021).

The purpose of this research study is to explore how people use elevated (vertical) green social spaces before, during, and after the Covid-19 pandemic. The specific objectives are to analyse the real-life cognitive experience of visitors to London's Sky Garden and to examine critical issues such as accessibility, circulation, activities, limitations of visitors, and social distancing. The research aims to outline solutions for designing safe and adaptable vertical

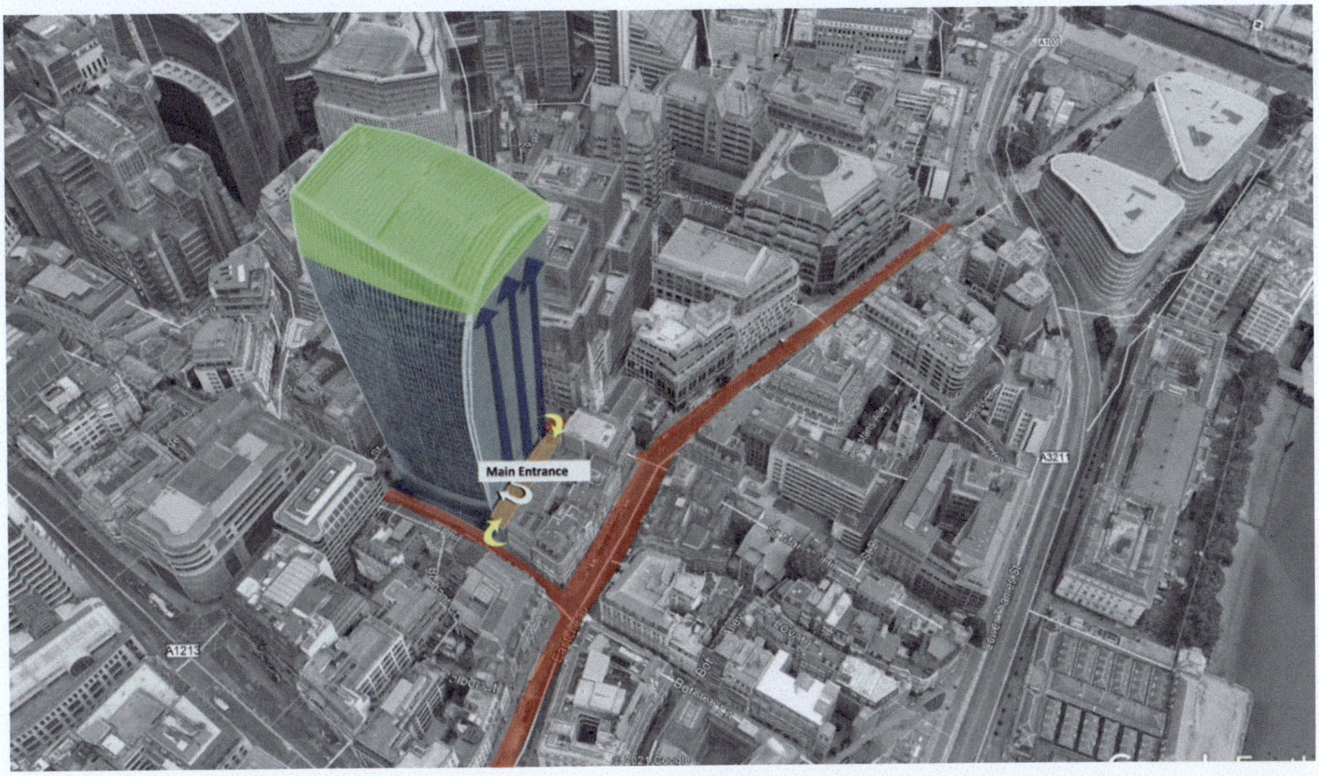

Fig. 1 Accessibility from the street level. *Source* Author

social spaces that can operate effectively, and add value for developers and operators, thereby allowing premises to resume operations and adapt to a 'new normal' situation.

2 Case Study—Sky Garden, London

There are a great variety of vertical social spaces in London, but this research is focused on the Sky Garden at 20 Fenchurch Street, in London. The Sky Garden was selected due to its unique typology, location, size, and management approach (no entrance charge), and that it remained open, when legally permitted, during the Covid pandemic.

The Sky Garden is on the top three floors of the 'Walkie Talkie' skyscraper in the heart of London's financial district (Figs. 1 and 4). The Sky Garden refers to the space on levels 35–37 as well as its dedicated entrance on the ground floor. The Sky Garden is accessible by only one access point from the ground level on Philpot Lane, on the south-west corner of the building (Figs. 2 and 3). Visitors must have a valid free ticket or restaurant reservation to pass through airport-style security checks and take the elevator up to Level 35 (Sky-Garden, 2015).

Fig. 2 Access from Philpot Lane. *Source* Author

Fig. 3 Entrance security checks, Sky Garden. *Source* Author

The primary method for booking visits to the Sky Garden is through its website, with tickets available up to three weeks in advance on a weekly basis. However, access is not guaranteed if the maximum capacity of the space has been reached, and visitors who miss their designated entry time may be denied entry (Viñoly et al., 2015). Once maximum capacity is reached, visitors who have stayed for over one hour will be asked to leave. Re-entry to the Sky Garden is not permitted for pass holders unless granted permission, and visitors must undergo security screening again (Sky-Garden, 2015). These restrictions also applied to the researcher.

The Covid-19 pandemic has had a significant impact on the accessibility of the Sky Garden. The pandemic has resulted in a reduction of the number of visitors allowed from a maximum of 600 to a safer limit of 200, and the elevator capacity was limited to a maximum of six guests from the same household (Fig. 5). Due to the limited access points and enclosed indoor environment, the Sky Garden was temporarily closed during the government lockdown restrictions in 2020.

The pandemic-induced closure of the Sky Garden also presented a significant challenge to the maintenance of its ventilation and indoor environment. The effectiveness of the ventilation and air filtration systems is crucial in providing a comfortable and healthy indoor environment for visitors.

Fig. 4 Three-D plan of the different activities and functions, Sky Garden. *Source* Author

Fig. 5 London Sky Garden, 20 Fenchurch Street, London. *Source* Author

Therefore, when the space reopened to the public, it was imperative to ensure that the ventilation and air filtration systems were fully operational and capable of providing a safe and healthy indoor environment.

The ventilation system at the Sky Garden employs a combination of natural and mechanical ventilation to achieve a comfortable and healthy indoor environment. Natural ventilation is achieved through the use of vents and louvres, while mechanical ventilation includes air handling and heat recovery units to circulate fresh air and extract stale air. Additionally, the Sky Garden uses a sophisticated air filtration system to eliminate pollutants and airborne particles, ensuring a clean indoor atmosphere (Schoenefeldt, 2016; Sky-Garden, 2015).

3 Methodology

There were two main research methods utilised in this research: (i) direct observation and (ii) semi-structured walk-along interviews.

3.1 Direct Observation

The first method used in this study was direct observations at different times over three years (2019–2022). The research team visited the Sky Garden before, during, and post the Covid-19 pandemic for data collection, pertaining to the movement and legibility of the space, and to observe people's activities within the space. The manual methods used for measuring the public life interaction in the space involved counting, mapping (collecting subject locations), tracing (drawing people's movement), and tracking (shadowing or following people) (Gehl & Svarre, 2013). The original research intention was to observe human behaviour and design factors in elevated public garden spaces. The research proposal found a whole new direction at the start of 2020 with the effect of a pandemic, limiting freedom of movement and social mixing.

There are various tools for analysing and registering people's behaviours in public spaces. The observation tools described are primarily manual tools that can be replaced by automated registration. The crucial difference is that human registration presents more than 'cold facts' as details from the site area, which might have a significant influence on the study analysis (Hanzl & Ledwon, 2017).

The fieldwork of the 'before and after' was essential for data collection to gather an understanding of how the space was experienced and how it then changed in terms of accessibility, circulation, and activities. This method required observation and data collection during 2019, and then to apply the method during and after the conditions caused by the pandemic. There were also many subsequent visits during the gradual relaxation of regulations throughout 2022. The author visited the Sky Garden to observe the layout and the users visiting the spaces (see dates and times in Table 1).

The author of this study developed a detailed, digital computer model of the Sky Garden using Building Information Modelling (BIM) software, Autodesk *Revit*

Table 1 Data collection date and time

Sky Garden (direct observation field study)		
Date	Time	Regulations
Monday, 23/12/2019	15:00–17:30	Pre-pandemic
Friday, 27/12/2019	9:30–12:00	Pre-pandemic
Saturday, 11/07/2020	14:00–15:30	New rules with Covid-19
Monday, 19/10/2020	12:30–14:00	New rules with Covid-19
Friday, 21/05/2021	23:00–00:30	New rules with Covid-19
Monday, 24/05/2021	13:15–15:00	New rules with Covid-19
Thursday, 11/11/2021	11:00–13:00	Post-pandemic
Thursday, 31/12/2021	18:00–20:30	Post-pandemic
Saturday, 03/07/2022	15:00–17:00	Post-pandemic
Monday, 05/07/2022	9:30–12:00	Post-pandemic

2022, to map and analyse the data collected during site observation visits (Ehab & Heath, 2023). This model was utilised to illustrate and simulate the changes in users' activities and movements over time. To analyse the controlling 'pinch points' within the space, the research employed *depthmapX*, which enabled the assessment of visual accessibility through the production of 'point isovists' and 'isovist paths'.

Point isovists are polygons that represent the visually accessible area from a specific location, while isovist paths depict how the view changes when moving through the space. Moreover, *depthmapX* facilitated the simulation of different types of pedestrian behaviour by providing various ways for an agent to choose where to walk, such as towards larger spaces, along lines of sight, or occluded areas of their view. The development team of *depthmapX* have highlighted the versatility of the software in enabling the analysis of pedestrian behaviour in complex environments (depthmapX development team, 2017). Overall, the use of BIM software and *depthmapX* served to enable the author to conduct a comprehensive analysis of the Sky Garden's spatial qualities and user behaviour.

The observational studies were conducted during weekends and weekdays to allow direct and equal comparison with three different conditions: 'pre-pandemic'; 'during pandemic'; and 'post-pandemic'. The observations were held in hourly-based intervals both on weekends and weekdays. During the official 'lockdown' period, the Sky Garden was totally closed and there was, therefore, no observation or data collection. The period of time referred to as 'during pandemic' in this study is post-lockdown, when not everybody had been vaccinated, but limited degrees of relaxation for exercise, social meetings, and travel were permitted. 'Post-pandemic' is after the date that the UK government announced the relaxation of all regulations. The author, in conducting the observations, faced the same problems as the visitors of the public. These included being confined to set times that had to be booked in advance, limiting the population within a timeslot (to optimise social distancing), not being permitted to stay more than two hours, and not being permitted to visit twice in one day.

3.2 Semi-Structured Walk-Along Interviews

Walk-along interviews in the Sky Garden were used to explore the physical experience of the space users and how this can impact the types of social interaction and activities (Carpiano, 2009). The semi-structured interviews were conducted with people from different age groups with all of the participants aged 18 and above. Twenty-three (n = 23) interviews were conducted and analysed (n = 23) with visitors to Sky Garden. The duration of interviews took around 20 min each. The participants were recruited on the site, with an explanation of the purpose and duration of their contribution, plus their signature on an ethics consent form.

The author conducted a theme-based analysis using various qualitative data sets. Data were analysed using content analysis, which is an effective method for descriptive aims (Schreier, 2014), guided by a summative approach (Hsieh & Shannon, 2005). This strategy was used to examine the ideas that make up the theme and sub-theme and how they interact with each other. The final stage was to investigate the evidence of relationships between the overarching themes, identifying the quotes that were initially hard to classify and fit into the themes and sub-themes. Sub-themes fit under the major themes in the write-up (Table 2).

3.2.1 Qualitative Data Analysis

A total of 23 interviews were completed in June 2021; 39.1% (9) were males and 61.9% (14) were females. The average age of participants was 28.2 years. Around 34.7% of the interviewees lived in London, 39.1% lived elsewhere in the UK, and 26.08% were overseas tourists visiting London. The analysis of the interviews highlights five overarching themes: the purpose of the visit; activities; accessibility; design concerns; suggestions for design features and activities.

Table 2 Displaying the main themes and sub-themes

Themes	Sub-themes
The choice of the space and the purpose of the visit	Free to visit
	Exploring the city
	Meeting a friend
	Good attraction to show a visitor
	Social media
Activities	Checking the views
	Taking pictures
	Relaxation
	Food and beverage
Accessibility	Online booking
	Accessibility from the ground level
	One-way circulation system
	Security
	Publicness
Design concerns	Seating spaces
	Stairs
	Plants
	Open space (terrace)
	Restaurant design
Suggestions for design features and activities	Features that improve the design quality of the space
	Features that encourage physical activity
	Features that encourage social interaction

4 Results

4.1 Observation Study Results

According to the Sky Garden management team, the theoretical maximum occupancy of the space is 600 visitors. The pre-pandemic analysis showed that the average occupancy per hour during the hourly intervals did not exceed an average of 310 visitors (51.6% of the maximum) on weekends and 260 visitors (43.3%) on weekdays (Fig. 15). Many people of different age groups were visiting the space, such as tourists, families, elderly people, and youth. The observation study reported six different optional and social activities taking place in the roof garden (Fig. 6). Walking was the major activity taking place; around 190 visitors were moving around the Sky Garden levels taking pictures of the city views and enjoying the green environment. The study showed an average number of 82 visitors were on stationary activities such as sitting, eating, relaxing, reading, watching, and listening to others. The most dominant activity that stimulated social interaction in the garden was on the outdoor observation platform where visitors were taking pictures (Figs. 6 and 7).

The data collected during the pandemic showed a massive difference in terms of accessibility, circulation, activities, and the number of visitors. The analysis showed a significant decrease in the number of people visiting the Sky Garden at one time. The average occupancy per hour reached about 110 visitors (18.3% of the maximum) on weekends and 90 visitors (15%) on weekdays (Fig. 15).

The connectivity map produced by entering the researcher's data into the application of *depthmapx* indicates that the weakest connection zones that need further management and circulation control are the entrance, the stairs, and the outdoor terrace (Fig. 9). Most of the space syntax analysis aligned with the field observations, but the practical use of spaces is not fully reflected in space syntax visual graphic analysis. The circulation of the space was changed from free movement to a one-way system to manage and control the pinch points (social distancing hazard) (Fig. 8). The management team reviewed all areas where a queue might occur and added markings to the ground to clearly show safe social distancing for guests to observe. The floor signage outside the toilets also helped to ensure that guests maintained their distance when waiting to use the facilities. The floor markers indicated the direction of movement permitted for all guests around the venue and indicated the one-way circulation system.

The floor markers outlined a safe distance for guests; however, they hugely affected the movement patterns and limited the sense of publicness. As another safety measure, face coverings were mandatory in the Sky Garden unless visitors were seated at a restaurant table or in the bar areas.

Fig. 6 The visitor's activities pre-pandemic, Sky Garden, London. *Source* Author

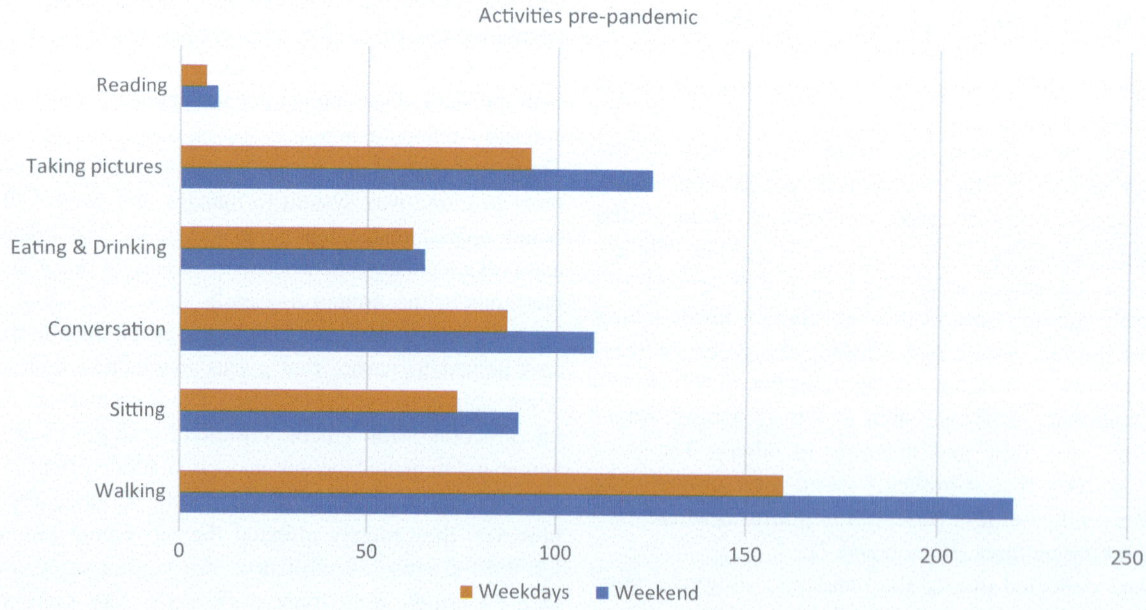

Fig. 7 The average number of visitor's activities pre-pandemic, Sky Garden, London. *Source* Author

Fig. 8 One-way circulation system to maintain social distancing and control the pinch points, Sky Garden. *Source of 3D model and image*: Author

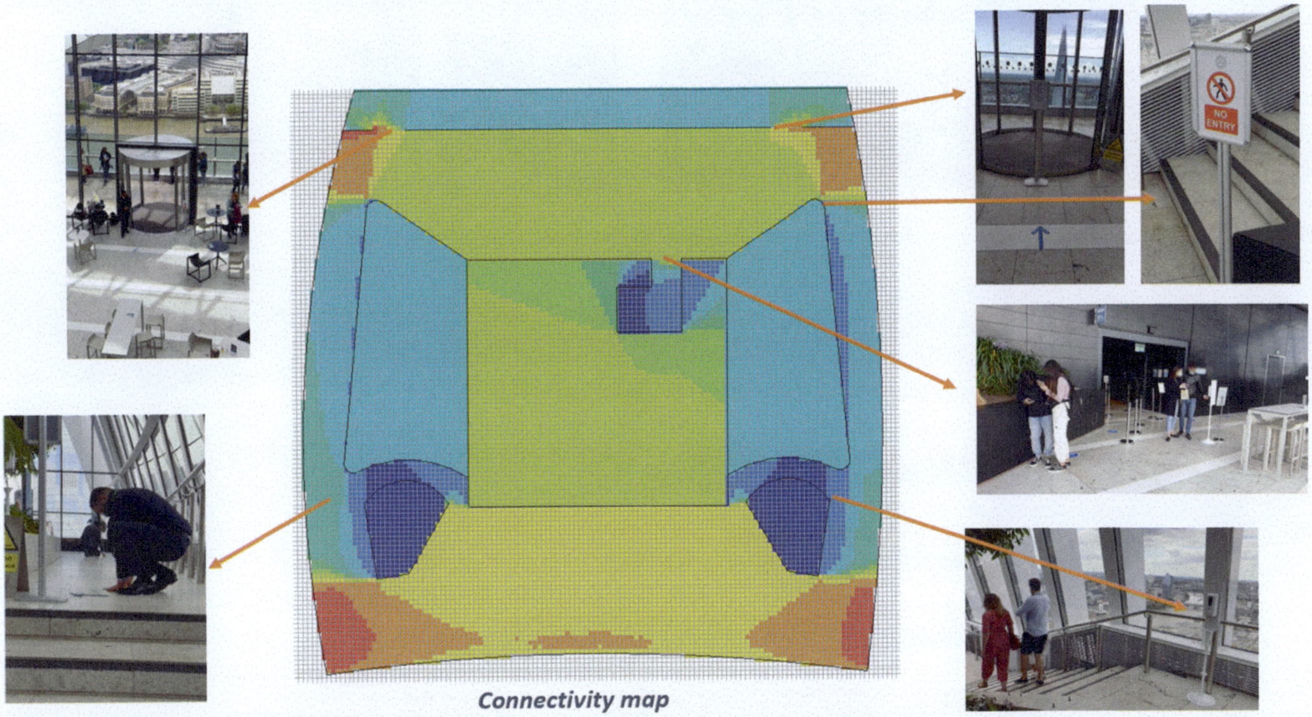

Fig. 9 Connectivity map (using depthmapX). Each location is coloured according to how many other locations are visible from it. The range runs from blue, for low, through green, yellow to red for many visible locations. *Source* Author

Further analysis indicates that the one-way circulation system during the pandemic had a major impact on activities in the Sky Garden (Figs. 10 and 11). The majority of the visitors (n = 70 visitors) preferred to spend more time in stationary activities. The number of people involved in stationary activities, such as eating, drinking, and relaxing, increased compared to the number of people engaged in movement activities. There was a significant decline reported in the average number of people sauntering in Sky Garden at one time, which was around 42 visitors. Most of the movement activities were taking place on the outdoor terrace where people felt safer outdoors taking pictures and enjoying the view. The study reports the emergence of new activities within the Sky Garden, such as visitors using the space for working on their laptops, as well as using the newly installed chaise longues near the plants, for leisure

Fig. 10 The average number of visitor's activities during the pandemic, Sky Garden, London. *Source* Author

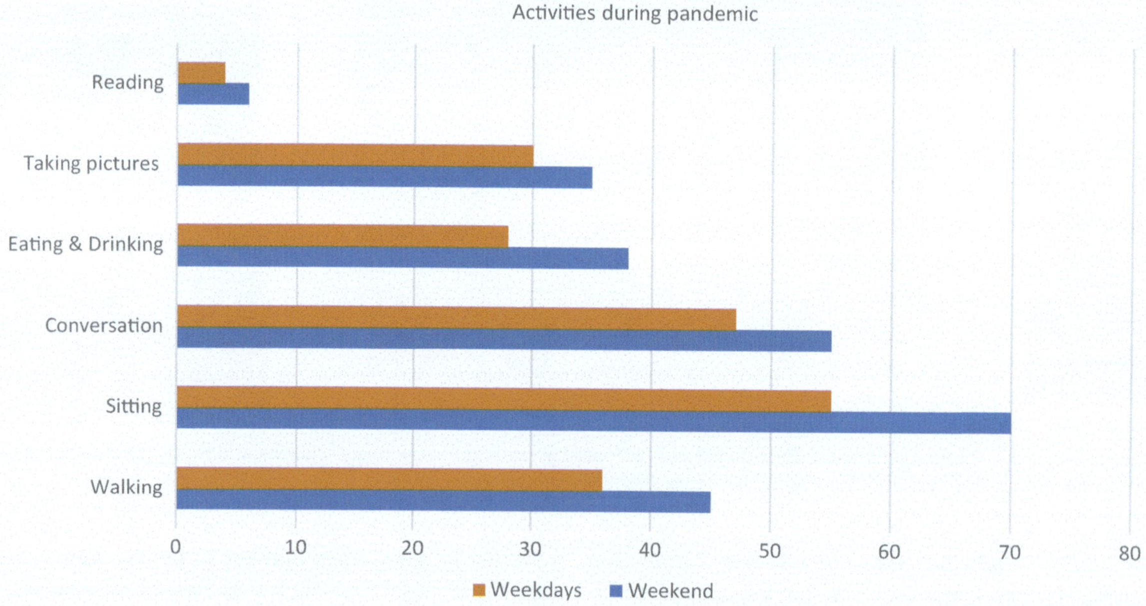

Fig. 11 The visitor's activities during the pandemic, Sky Garden, London. *Source* Author

and relaxation. These chairs lounges were introduced by the Sky Garden management during the pandemic and have proven to be popular among visitors.

The data analysed during the post-pandemic time showed that the roof garden activities and use changed to become more active than during the pandemic time. The average occupancy per hour reached 220 visitors (36.6% of the maximum) on weekends and 185 visitors (30.8%) on weekdays (Fig. 15). Although the Covid regulations were more relaxed, visitors preferred to spend more time in stationary activities such as eating, drinking, chatting, and relaxing (Fig. 12). The Sky Garden hosts a variety of special events throughout the year, catering to different interests and age groups. These events include live music performances, cultural exhibitions, fitness classes, DJ nights, Halloween parties, and New Year's Eve parties. These new themes and events of the Sky Garden were inviting and interactive for most visitors, encouraging them to spend more time in the venue. However, most of these events were not free for the public, and visitors had to pay for valid tickets to enter the Sky Garden during the organised events (Figs. 13 and 14).

Prior to the pandemic, these special events were a regular occurrence in the Sky Garden. However, during the pandemic, these events were temporarily halted to comply with the restrictions imposed by the government. As restrictions have lifted, the Sky Garden has resumed these events, which have increased in frequency to attract visitors. The events offer visitors a unique and engaging experience that ranges from sunrise yoga classes to cocktail-making workshops. The frequency of the events varies according to the season and the availability of the venue. Nonetheless, they usually occur several times per week, with an increase during peak times such as holidays and special occasions. These special events are designed to provide visitors with an unforgettable and enjoyable experience, enhancing the Sky Garden's status as a popular destination in London.

4.2 Interview Results

4.2.1 The Choice and the Purpose of the Visit

The most common reasons for visiting the Sky Garden were: 'for the panoramic view'; 'exploring the city of London'; 'meeting a friend'; and that 'it's free to visit'. A significant number of participants referred to the fact that booking a free ticket encouraged them to visit the Sky Garden. Some participants mentioned that they knew about the Sky Garden from social media posts. Most of the participants mentioned that the views and advertisements stating that the Sky Garden is London's highest public park were their main reasons for visiting and exploring the space.

> I have seen some pictures of it on social media, and it looks very beautiful. I am visiting London by myself, so I have just been doing random sightseeing to explore the city (Female, aged 25 years, tourist from the USA).
>
> It is free to visit, and the views are very good (Male, aged 27 years, living in London).

4.2.2 Activities

The main activities that participants undertook at the Sky Garden were: 'enjoying the views'; 'taking pictures'; 'drinking cocktails'; 'meeting a friend'; 'reading a book'; and 'eating'. The most popular two activities mentioned by the participants were 'checking the views of London', and 'taking pictures' (11 respondents; 47.8% of the interview sample). Thirty per cent of respondents (n=7) considered the outdoor terrace as the best vantage point for taking

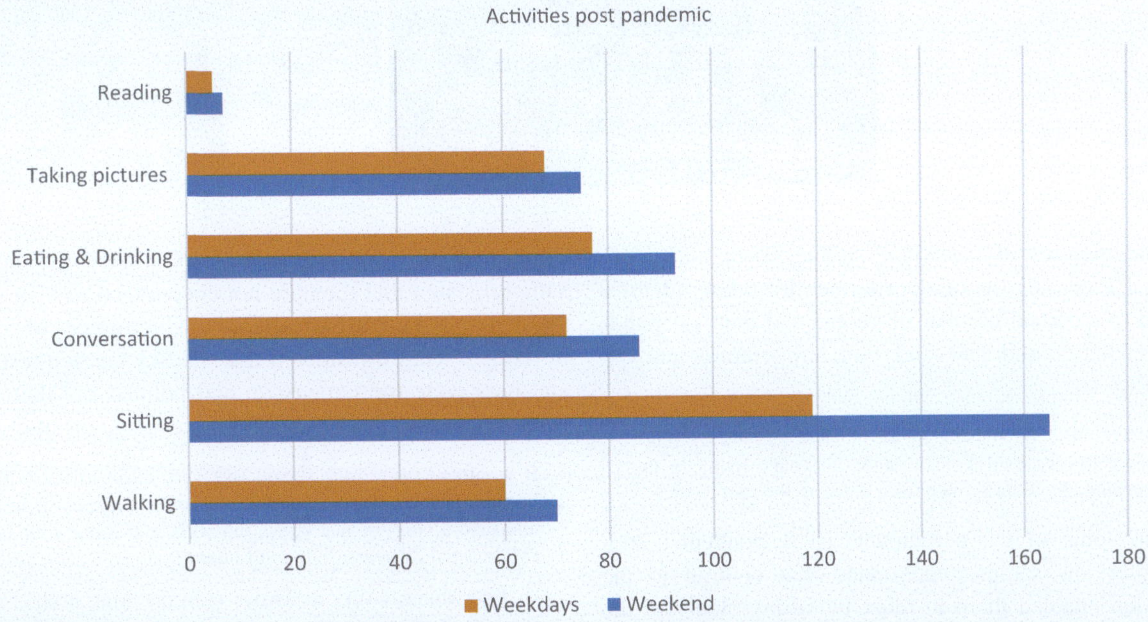

Fig. 12 The average number of visitors activities during the post-pandemic, DJ night, Sky Garden, London. *Source* Author

Fig. 13 The visitor's activities during the post-pandemic, DJ night & morning Yoga, Sky Garden, London. *Source* Author

panoramic pictures of London's skyline. They also mentioned that accessing the outdoor terrace for taking pictures is limited to a certain number of people and that it depends on the weather conditions (wind, rain, temperature, and visibility) (Fig. 16).

> I wish we could go outside to the terrace but, we couldn't because of the weather (Female, aged 24 years, living in the UK, restrained by security men from entering the roof terrace).

Moreover, some of the participants (6 respondents; 26% of the interview sample) mentioned that coming to the Sky Garden enabled them to relax by sitting closer to the plants and reading a book. However, a significant number of participants (11 respondents; 47.8% of the interview sample) stated that they are not encouraged to visit the Sky Garden on a regular basis due to the high prices of food and beverages. They considered that the Sky Garden restaurants and bars were too expensive, and pointed out that visitors cannot consume their own food or beverage on site.

> It would be quite nice if you live very close to the building because it's free to come up here so you can just come to relax and read a book or just come and work and bring your laptop (Male, aged 28 years, living in London).
>
> Everything is very expensive even the drink itself, I can't see myself coming here on a regular basis" (female, aged 27 years, living in the UK).

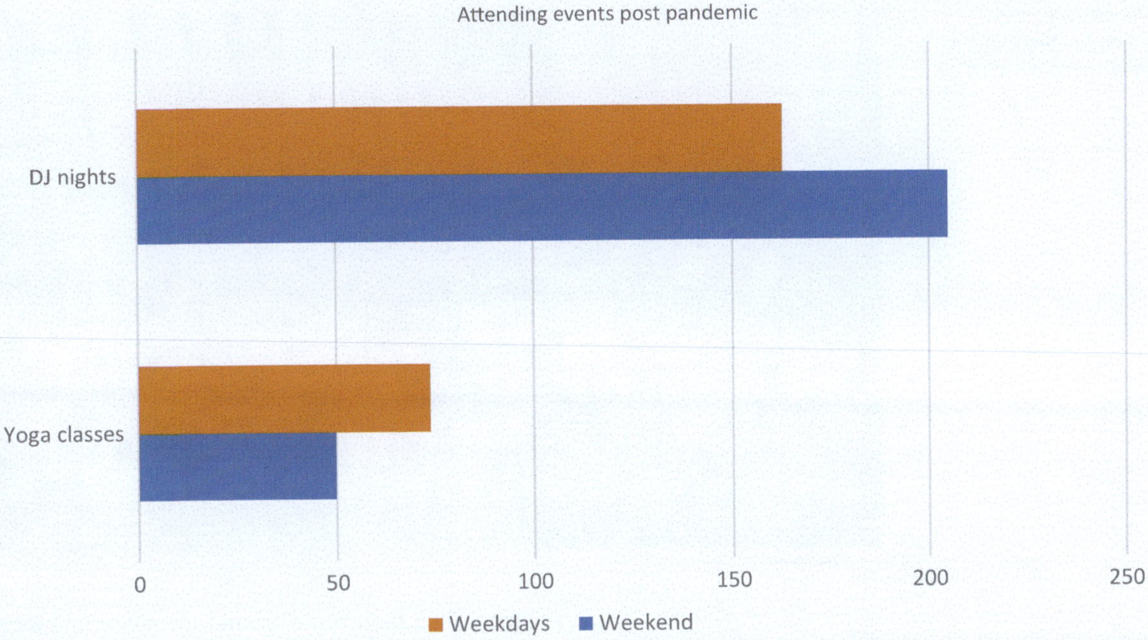

Fig. 14 The average number of visitors during the post-pandemic events, DJ night, Sky Garden, London. *Source* Author

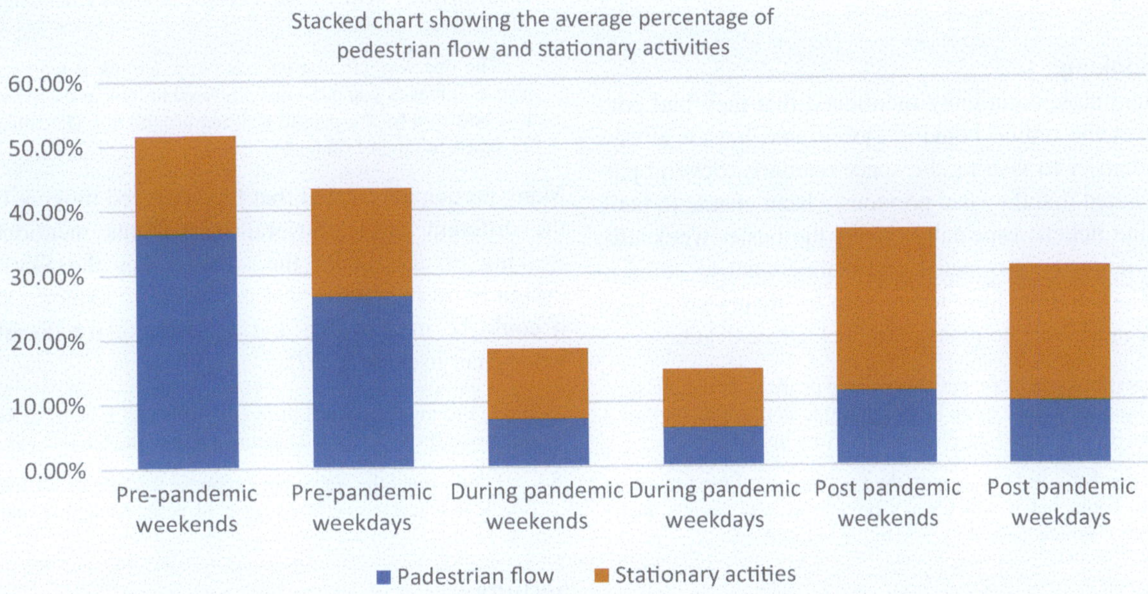

Fig. 15 Stacked chart showing the average percentage of pedestrian flow and stationary activities at the Sky Garden before, during, and after the Covid-19 pandemic for both weekdays and weekends, as counted by the author's observation. *Source* Author

4.2.3 Accessibility

The accessibility of the Sky Garden is an important factor that influences people's behaviour, activities, and interaction in the space. When the participants were asked how they found the accessibility in the light of the Covid-19 risk, four main themes emerged: 'online booking', 'circulation', 'security', and 'publicness'. Nearly half of the participants (n = 11) mentioned that the extra safety level was understandable, and they found it good that the Sky Garden is accessible directly from the ground level.

> The accessibility was pretty easy—it took me around 10 min to reach the top level. The garden wasn't busy during my visit, so I was feeling safe and not worried about social distancing (Male, aged 25 years, living in London).

Fig. 16 Visitors activities, 34th floor, London Sky Garden, London. *Photo source* Author

The accessibility and the security level are understandable, especially with the current situation of the Covid-19 pandemic. I think the security guards and the management team are doing their best to make the place safe (Male, aged 33 years, tourist from the Philippines).

Online Booking

Participants most commonly mentioned that they had concerns about the online booking system and how it always acts as a barrier to visiting the space regularly. Several participants noted that they had previously been unable to book their online tickets, especially during the busier weekends. Some participants also stated that if they want to come to the Sky Garden, they have to pre-plan for their visit at least one week in advance to manage to book a free ticket.

> The accessibility is okay, but the booking phase is annoying. It would be nice to have more flexible visiting hours slots on the websites (Female, aged 26 years, tourist from Spain).

> I remember a little issue with the booking, that I tried many times to book, but it was always busy especially during the weekends (Female, aged 45 years, living in London).

Circulation

Participants had mixed responses when asked about the Sky Garden's circulation system. Two main sub-themes were discussed: the one-way circulation system; and the different levels in the Sky Garden. The most mentioned responses regarding the one-way circulation system were that it was recognised as necessary during the current situation, but participants also stated that they were annoyed at having to go around again if they missed a particular view of the city. Some other participants reported favourably that the one-way circulation system helped them to be more purposeful on their route around the space and their enjoyment of the city views.

> if you pass something on the left and you want to see it again you must go all the way back. I guess that was a little bit annoying, so I went around a couple of times (Male, aged 27 years, living in London).

> The one-way circulation was okay. I think it works really nice as it makes you more concentrated to look at the views and less bothered by the people walking around you (Female, aged 33 years, living in London).

Many participants stated that they enjoyed moving between the different levels. Several participants mentioned that walking up and down the steps makes the Sky Garden design more interactive and enjoyable to explore, although it could be inaccessible and challenging for disabled and wheelchair users (Fig. 17).

> It's fun to move between different levels and it helps to keep it organised (Female, aged 23 years, living in the UK).

> I don't mind the stairs, but i suppose it might be difficult for people in wheelchairs (Male, aged 34 years, living in London).

Security

Participants also spoke about how they felt safe within the Sky Garden and were asked what they think about the security system. Participants commonly reported that the security level in the building makes them feel safe during their visit. However, some of the participants (n = 5) mentioned that they were disturbed at being 'watched' by the security guards in the space.

> The high-security level is necessary as it makes me feel safe inside the building (Female, aged 26 years, living in London).

> The security level was a bit annoying, to have so many security guards in the space (Female, aged 25 years, living in the UK).

Fig. 17 One-way circulation system, London Sky Garden, London. *Source* Author

Publicness

When asked to describe the level of publicness in the Sky Garden compared to a public park, most of the participants struggled to consider the Sky Garden as a truly public space or as a private space. Participants had mixed responses describing the garden as a 'private-public space' (n = 13). The group of participants who considered the Sky Garden as a private space (n = 8) mentioned specific critical factors such as the *pre-booking*, the *security system*, the *Covid rules*, and the use *regulations*. Some participants raised an issue that most of the seating spaces in the garden felt private, as most (but not all) of these seating spaces are connected to the bars, café, and restaurant, implying an obligation to purchase.

> It's more like a private space, as the booking phase and the one hour permitted visiting slot puts a pressure on me although, no one asked me to leave the space before (Female, aged 26 years, living in London).

> They offer you free entry but then there are bars around here, you can spend your money in. I guess this is how they make their profit. I found it fairly similar to most of the public spaces these days; there are always some food stalls around (Male, aged 26 years, living in London).

> Well, the only thing that was interesting was that most of the seats were connected to the bar. There are a few spots that you could sit in that are not part of the bars and the restaurant. It felt a little bit less public than another public park, as it seems that if you want to sit you have to buy something (Female, aged 25 years, tourist from the USA).

4.2.4 Design Concerns

Participants were asked what they liked and disliked about the design of the Sky Garden. Twenty-six per cent of the participants (n = 6) reported that they were satisfied with the design of the garden and they didn't have any concerns. The presence of natural elements (plants, and trees) were enjoyed by almost all of the participants. Seventy-four per cent of the participants (n = 17) reported that they had concerns about the design of the space and that improvements were needed for a better visiting experience. The most frequently mentioned themes discussed by the participants included design concerns related to the 'privacy and positioning of seating spaces', 'lack of different tropical plants and flowers', 'stair accessibility for disabled people', 'outdoor terrace design', and the 'restaurant design'.

> I think with the name itself Sky Garden, I would like to see different kinds of flowers and roses. If they can add more green features, it will differently encourage more people to come and visit, not only for the view (Male, aged 33 years, tourist from the Philippines).

> I don't know how it would work if you were handicapped in any way, or if you have a disability. A lot of the garden experience is on the stairs so you wouldn't be able to see any of that if you are taking the enclosed elevator (Female, aged 25 years, tourist from the USA).

4.2.5 Design Features and Suggested Activities

When asking the participants what design features and activities they would like to see in the Sky Garden, three main themes emerged: 'features that improve the design quality of the space', 'features that encourage more physical activity', and 'features that encourage more social interaction'. The most frequently suggested specific features to improve the design quality of the space were: 'adding electricity sockets', 'providing a drinking fountain'. 'playing soft background music', and 'placing new relaxing seats and a quiet area near the plants'.

> The place was a bit noisy so I would like to listen to soft music in the background (Female, aged 45 years, living in London).

> A drinking fountain just to refresh yourself a bit (Female, aged 27 years, living in London).

> More relaxing and quiet spaces to sit, away from the bar and the café, where you can read a book or work on your laptop (Female, aged 26 years, living in London).

The most frequently mentioned features that would encourage participants to be active in the Sky Garden were the plants and the views. Many of the participants mentioned that they would like to see more diversity and variations of plants and flowers, with name tags and information for each group of plants. Participants also stated that a *guided tour* that gives information about the city's views and identifiable landmarks would be an interesting activity that would encourage walking in the space.

> If it was more like a garden space where you can walk around and see different plants, maybe like having tags of what each plant is, I think that would be interesting (Female, aged 26 years, living in the UK).

> I think it might be worth considering having a guided tour that explains what you are seeing, and the different views and landmarks of the city (Female, aged 26 years, tourist from Spain).

Participants had mixed responses when asked about what activities and design features would encourage them to interact with others. The most commonly mentioned responses include 'live music events' and 'placing some musical instruments such as a piano'. Some participants also suggested photograph opportunities such as 'a photo booth' and '360 viewing platform'. A few participants also mentioned other activities as events like 'table tennis area', 'kids' zone', and 'night club' during weekends.

> I think life soft music and a piano would be nice, so people could go randomly, sit down and play music (Female, aged 35 years, tourist from France).

> Maybe something like a photo opportunity, it might be a little corner design for a photo booth or a photo frame (Female, aged 25 years, living in the UK).

> Having music events will be quite good and it will attract more people to come (Male, aged 27 years, living in London).

5 Discussion

The study analysed and discussed the design problems and principles that need to be considered when designing vertical urban spaces. As discussed in the Sky Garden case study there are many design aspects that could be improved in terms of accessibility, circulation, design, and management. There are nine key qualities that the study's participants most frequently referred to when describing their ideal roof garden: accessibility; provision of activities; comfort; sociability; quality of management; publicness; security; green nature; and natural ventilation.

Results from both the observation study and the interviews indicate that good pedestrian access is one of the main factors for an urban space to function well (Carmona et al, 2008; Whyte, 1980). Accessibility refers to the capability of all users to access space, regardless of any existing or potential restrictions (Persson et al, 2015). Reframing pedestrian access could encourage logical movement and desired behaviours, while improving the level of inclusiveness (Pineo, 2020; Aelbrecht & Stevens, 2019). Carr et al. (1992) noted that three forms of physical access exist: visual, physical, and symbolic. Visual access lets people view the space before they enter it; physical access controls who can or cannot use a public area; while symbolic access indicates if the space is safe and welcoming through spatial clues (Mehta, 2014; Németh, 2009).

Although the access to the public facilities and visitor areas in the Sky Garden is free of charge and open to the public, the physical and visual accessibility remain the most significant limitations that define the publicness of the space. The Covid-19 pandemic had a major impact on Sky-Garden's accessibility. For a time, it was closed altogether, and then when access was permitted, there was the requirement to bring in social distancing rules. This reduced the permitted number of guests and isolation of elevator occupants into family groups. During the pandemic, there was widely accepted advice to have good natural ventilation, but the environment at the top of a skyscraper, with a large south-facing glass wall, is mostly controlled by artificial environmental systems.

The pandemic has forced the managers of elevated urban spaces to rethink sightlines and wayfinding to provide good visibility in terms of horizontal and vertical directions. To ensure safety, these strategies should emphasise visibility from both inside and outside the points of the entrance and exit. A larger and more visible area allows for increased safety measures such as social distancing to be upheld while allowing pedestrians greater freedom of movement (Cho et al., 2015; Honey-Rosés et al., 2020). An environment with a heightened sense of connectivity will not only reduce the risk of exposure, but also promote accessible routes within public spaces (Afrin et al., 2021; Pineo, 2020).

A well-connected space is integrated with local movement and pattern systems; the movement is itself an activity that often generates other activities (Carmona et al., 2008; Moore, 2021). Legible spaces are essential for the identification of prominent activity nodes. To be effective, elevated roof gardens should provide a balanced mix and visibility to all activities; it should encourage users to engage with the environment and also have moments to pause (Ehab & Heath, 2024; Pomeroy, 2013).

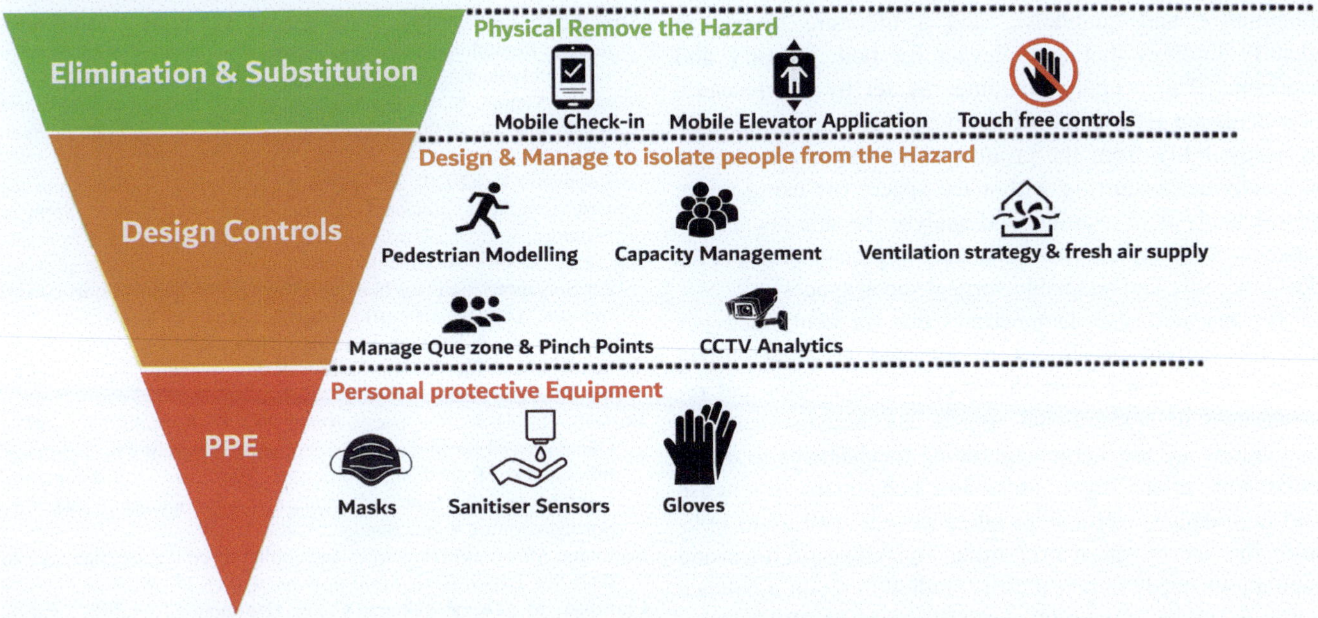

Fig. 18 The Hierarchy of controls is a commonly accepted system for reducing risk. The icons to the right represent some of the specific solutions that can be implemented in the vertical social space to effectuate each control action. *Source* Author

Seating organisation is one of the fundamental tactics to enable social activities in a vertical green space. Providing various kinds of seating in the roof gardens is therefore highly advisable. Movable chairs and benches with different orientations can improve the variety and choice of seating amenities, in terms of comfort and user experience (Huang & Franck, 2018). Seating arrangement and orientation can improve social interaction and passive activities such as people watching (Cho et al., 2015; Neto & Munakata, 2015). Interactive elements such as fountains, plants, installations, game facilities, interactive displays, pianos, swings, and public art sculptures are highly recommended to activate the roof garden design. These interactive elements bring a unique character to the roof garden while also serving as an attraction that encourages visitors to stay longer in the roof garden, come back another time, or recommend it to friends (Milne & Pojani, 2022).

Public roof garden management should encourage inclusion through skilfully 'light-handed' regulations, rather than the hard exclusion of undesirable behaviours through restrictions on people's activities (Hadi et al., 2018; Oldfield, 2019). These new guidelines should limit the number of restrictions to essential and reasonable ones, employ plans for users' participation in the roof garden management, and regulate spatial uses and actions, rather than just confining or prohibiting them (Ehab et al., 2023).

The building industry needs new strategies to manage risk and hygiene to combat the transformation of Covid-19 and future diseases, but also new thinking on how to create resilient and vertical social spaces that allow for high-value interaction and increased health and wellness (Morawska et al., 2021). A number of possible solutions include a combination of operational, technological, and architectural inventions (Mills et al., 2020). These innovative solutions are firmly rooted in science and built upon the hierarchy of controls, a widely recognised industry standard for risk management. Safety organisations such as OSHA (U.S. Occupational Safety and Health Administration) and HSE (UK Health and Safety Executive) advocate this solution to manage risks effectively (Shroff, 2020). To reduce the probability of risks in any built environment, this hierarchical system categorises design solutions or operational measures according to their effectiveness.

The inverted pyramid graphic creates a hierarchy of solutions, with those at the top considered to be more effective than those at the bottom (Sehgal & Milton, 2021; Morawska et al., 2021) (Fig. 18). The use of personal protective equipment (PPE) was the initial response to the pandemic and will continue to serve an important purpose in the short to medium term but these are considered less effective over the long term (Gandhi & Marr, 2021). Design and engineering solutions at the top of the pyramid are, however, more suitable long-term solutions and are considered to be more effective (Mills et al., 2020).

6 Conclusions

This paper presents a research study on the current design and management of the Sky Garden, London before, during, and post the Covid-19 pandemic. The research leveraged different methods for collecting data, including quantitative and qualitative, and a literature review to identify findings that are relevant for both industry and academia. The primary outcomes as set by the research objectives are (i) analysis of the design critical issues such as accessibility from the ground level, visitor circulation, and activities occurring within the space; (ii) using space syntax analysis to examine and analyse the safe social distance for visitors; and (iii) to outline suggested solutions for designing safe and adaptable vertical social spaces.

The research also demonstrates that the evolving types and the current conditions of elevated (vertical) urban spaces in London require the re-conceptualisation and re-assessment of conventional approaches. Management and regulations are inevitable negotiating mechanisms to determine appropriate users, uses, and behaviours in intense and high-density vertical social spaces, as well as to optimise the use of space over time. The range of rules and regulations needs to be carefully studied to retain a balance between safety, adaptability, and inclusiveness. If this range does not consider the public perception of urban spaces, anticipating the whole experience of users, not just a few images or renders, then it will not be complete.

In response to such challenges, strategic design guidance is needed, that would work better at all levels, enabling communication among the various actors, including the public, to facilitate the decision-making process, and ultimately deliver improved outcomes (Bussell et al., 2023; Ehab et al., 2023). This requires future investigation of new efficient and flexible approaches, to understand, assess, and guide the design and management of emerging vertical urban spaces that would signify their qualities and performance. Public engagement, consultation, alternative design options, and ideas gathering are all essential before real money is spent on real buildings (Ehab et al., 2023b).

Future research being undertaken by the author and colleagues will evaluate and assess the relative merits of different types of vertical and enclosed social spaces, including innovative methods of public consultation using smart technology. Virtual Reality (VR) and augmentation of VR modelling for environments such as indoor sky parks and outdoor open rooftop gardens is the framework for a form of consultation that can harvest public approval and design ideas, generate enthusiasm and motivation, stimulate funding and initiatives, and create better city spaces.

References

Abdelsalam, A. E. (2018). Sustainable vertical urbanism as a design approach to change the future of hyper density cities. *Journal of Advance Research in Mechanical & Civil Engineering, 5*(7), 2208–2379.

Aelbrecht, P., & Stevens, Q. (Eds.). (2019). *Public space design and social cohesion: An international comparison*. Routledge.

Afrin, S., Chowdhury, F. J., & Rahman, M. M. (2021). Covid-19 pandemic: Rethinking strategies for resilient urban design, perceptions, and planning. *Frontiers in Sustainable Cities, 3*, 668263.

Aleta, A., Meloni, S., & Moreno, Y. (2017). A multilayer perspective for the analysis of urban transportation systems. *Scientific Reports, 7*(1), 1–9.

Al-Kodmany, K. (2020). *Tall buildings and the city: Improving the understanding of placemaking, imageability, and tourism*. Springer Nature.

Bay, J. H. (2004). Sustainable community and environment in tropical Singapore high-rise housing: The case of Bedok Court condominium. *ARQ: Architectural Research Quarterly, 8*(3–4), 333–343.

Bereitschaft, B., & Scheller, D. (2020). How might the Covid-19 pandemic affect 21st century urban design, planning, and development? *Urban Science, 4*(4), 56.

Bussell, C., Ehab, A., Hartle-Ryan, D., & Kapsalis, T. (2023). Generative AI for immersive experiences: Integrating text-to-image models in VR-mediated co-design workflows. In *International Conference on Human-Computer Interaction* (pp. 380–388). Springer Nature Switzerland.

Carmona, M. (2019). Principles for public space design, planning to do better. *Urban Design International, 24*(1), 47–59.

Carmona, M., De Magalhaes, C., & Hammond, L. (2008). *Public space: The management dimension*. Routledge.

Carpiano, R. M. (2009). Come take a walk with me: The "Go-Along" interview as a novel method for studying the implications of place for health and well-being. *Health & Place, 15*(1), 263–272.

Carr, S., Francis, M., Rivlin, L. G., & Stone, A. M. (1992). *Public space*. Cambridge University Press.

Cho, I. S., Heng, C. K., & Trivic, Z. (2015). *Re-framing urban space: Urban design for emerging hybrid and high-density conditions*. Routledge.

depthmapX development team. (2017). depthmapX (Version 0.6.0) [Computer software]. https://github.com/SpaceGroupUCL/depthmapX/

Dwijendra, N. K. A., Akhmadeev, R., Tumanov, D., Kosov, M., Shoar, S., & Banaitis, A. (2021). Modeling social impacts of high-rise residential buildings during the post-occupancy phase using dematel method: A case study. *Buildings, 11*(11), 504.

Ehab, A., & Heath, T. (2023). Exploring immersive co-design: Comparing human interaction in real and virtual elevated urban spaces in London. *Sustainability, 15*(12), 9184.

Ehab, A., Burnett, G., & Heath, T. (2023a). Enhancing public engagement in architectural design: A comparative analysis of advanced virtual reality approaches in building information modeling and gamification techniques. *Buildings, 13*(5), 1262.

Ehab, A., Heath, T., Burnett, G. (2023b). Virtual reality and the interactive design of elevated public spaces: Cognitive experience vs VR Experience. In: Stephanidis, C., Antona, M., Ntoa, S., Salvendy, G. (Eds.). *HCI International 2023 Posters*. HCII 2023. Communications in Computer and Information Science, vol 1836. Springer, Cham. https://doi.org/10.1007/978-3-031-36004-6_24

Ehab, A., Heath, T. (2024). Rethinking the design of vertical green spaces in the post-pandemic era: Visitor behaviour and real-life cognitive experience at crossrail place, London. In: Cheshmehzangi, A., Sedrez, M., Zhao, H., Li, T., Heath, T., Dawodu, A. (Eds.). *Resilience vs Pandemics. Urban Sustainability*. Springer, Singapore. https://doi.org/10.1007/978-981-99-8672-9_2

Feng, Q. (2021). Reviewed interpretations and inspirations on the development and strategies of Garden City Theory in Singapore. *Journal of Architectural Research and Development, 5*(3).

Gandhi, M., & Marr, L. C. (2021). Uniting infectious disease and physical science principles on the importance of face masks for Covid-19. *Med, 2*(1), 29–32.

Geary, R. S., Wheeler, B., Lovell, R., Jepson, R., Hunter, R., & Rodgers, S. (2021). A call to action: Improving urban green spaces to reduce health inequalities exacerbated by Covid-19. *Preventive Medicine, 145*, 106425.

Gehl, J., & Svarre, B. (2013). Public space, public life: An interaction. In *How to study public life* (pp. 1–8). Island Press.

Generalova, E., & Generalov, V. (2020). Mixed-use high-rise buildings: a typology of the future. In *IOP conference series: Materials science and engineering* (Vol. 753, No. 2, p. 022062). IOP Publishing.

Hadi, Y., Heath, T., & Oldfield, P. (2018). Gardens in the sky: Emotional experiences in the communal spaces at height in the Pinnacle@ Duxton, Singapore. *Emotion, Space and Society, 28*, 104–111.

Hanzl, M., & Ledwon, S. (2017). *Analyses of human behaviour in public spaces*. ISOCARP, Portland, Oregon, USA.

Huang, T. S., & Franck, K. A. (2018). Let's meet at Citicorp: Can privately owned public spaces be inclusive? *Journal of Urban Design, 23*(4), 499–517.

He, W., Zhang, Z. J., & Li, W. (2021). Information technology solutions, challenges, and suggestions for tackling the Covid-19 pandemic. *International Journal of Information Management, 57*, 102287.

Holy-Hasted, W., & Burchell, B. (2022). Does public space have to be green to improve well-being? An analysis of public space across Greater London and its association to subjective well-being. *Cities, 125*, 103569.

Honey-Rosés, J., Anguelovski, I., Chireh, V. K., Daher, C., Konijnendijk van den Bosch, C., Litt, J. S., Mawani, V., McCall, M. K., Orellana, A., Oscilowicz, E., Sánchez, U., Senbel, M., Tan, X., Villagomez, E., Zapata, O., & Nieuwenhuijsen, M. J. (2020). The impact of Covid-19 on public space: An early review of the emerging questions—Design, perceptions and inequities. *Cities & Health, 5*(sup1), S263–S279. https://doi.org/10.1080/23748834.2020.1780074

Hsieh, H. F., & Shannon, S. E. (2005). Three approaches to qualitative content analysis. *Qualitative Health Research, 15*(9), 1277–1288.

Lak, A., Shakouri Asl, S., & Maher, A. (2020). Resilient urban form to pandemics: Lessons from Covid-19. *Medical Journal of the Islamic Republic of Iran, 34*, 71. https://doi.org/10.34171/mjiri.34.71

Lee, D., & Park, B. (2018). Effects of elevated public skygarden on employees' psychological restoration-A case study of Seoullo 7017; 12(2), 51–60.

Madanipour, A. (2019). Rethinking public space: Between rhetoric and reality. *Urban Design International, 24*, 38–46.

Mehta, V. (2014). Evaluating public space. *Journal of Urban Design, 19*(1), 53–88.

Martínez, L., & Short, J. R. (2021). The pandemic city: Urban issues in the time of Covid-19. *Sustainability, 13*(6), 3295.

Megahed, N. A., & Ghoneim, E. M. (2020). Antivirus-built environment: Lessons learned from Covid-19 pandemic. *Sustainable Cities and Society, 61*, 102350.

Milne, C., & Pojani, D. (2022). Public art in cities: what makes it engaging and interactive?. *Journal of Urban Design*, 1–20.

Mills, S., Brahney, A., & Mackay, S. (2020). New Solutions for the New Normal. *CTBUH Journal*, (4).

Moore, G. M. (2021). Connectivity of public open space: Its meaning for different functions. *Journal of Urban Design, 26*(3), 279–295.

Morato, L. C. (2022). Opportunities and challenges of municipal planning in shaping vertical neighbourhoods in greater London. *Urban Planning, 7*(4), 267–283.

Morawska, L., Allen, J., Bahnfleth, W., Bluyssen, P. M., Boerstra, A., Buonanno, G., Cao, J., Dancer, S. J., Floto, A., Franchimon, F., Greenhalgh, T., Haworth, C., Hogeling, J., Isaxon, C., Jimenez, J. L., Kurnitski, J., Li, Y., Loomans, M., Marks, G., & Yao, M. (2021). A paradigm shift to combat indoor respiratory infection. *Science, 372*(6543), 689–691. https://doi.org/10.1126/science.abg2025

Mualam, N., Salinger, E., & Max, D. (2019). Increasing the urban mix through vertical allocations: Public floorspace in mixed use development. *Cities, 87*, 131–141.

Navaratnam, S., Nguyen, K., Selvaranjan, K., Zhang, G., Mendis, P., & Aye, L. (2022). Designing post Covid-19 buildings: Approaches for achieving healthy buildings. *Buildings, 12*(1), 74.

Németh, J. (2009). Defining a public: The management of privately owned public space. *Urban Studies, 46*(11), 2463–2490.

Neto, O. A., & Munakata, J. (2015). Seat choice and distance judgment in public spaces. *Perceptual and Motor Skills, 121*(2), 548–567.

Oldfield, P. (2019). *The sustainable tall building: A design primer*. Routledge.

Osmundson, T. (1999). *Roof gardens: History, design, and construction*. WW Norton & Company.

Persson, H., Åhman, H., Yngling, A. A., & Gulliksen, J. (2015). Universal design, inclusive design, accessible design, design for all: Different concepts—one goal? On the concept of accessibility -historical, methodological and philosophical aspects. *Universal Access in the Information Society, 14*(4), 505–526.

Pineo, H. (2020). Towards healthy urbanism: Inclusive, equitable and sustainable (THRIVES)–an urban design and planning framework from theory to praxis. *Cities & Health*, 1–19.

Pomeroy, J. (2013). *The Skycourt and Skygarden: Greening the urban habitat*. Routledge.

Pomeroy, J. (2012). Room at the top—the roof as an alternative habitable/social space in the Singapore context. *Journal of Urban Design, 17*(3), 413–424.

Samant, S. (2019). Cities in the Sky: Elevating Singapore's Urban Spaces. *International Journal of High-Rise Buildings, 8*(2), 137–154.

Samant, S., & Hsi-En, N. (2017). A tale of two Singapore sky gardens. *CTBUH Journal, 3*, 26–31.

Schoenefeldt, H. (2016). Question time. *CIBSE Journal, 2016*(9), 70–74.

Schreier, M. (2014). Varianten qualitativer Inhaltsanalyse: ein wegweiser im dickicht der Begrifflichkeiten. In *Forum Qualitative Sozialforschung/Forum: Qualitative Social Research* (Vol. 15, No. 1).

Sehgal, N. J., & Milton, D. K. (2021). Applying the hierarchy of controls: What occupational safety can teach us about safely navigating the next phase of the global Covid-19 pandemic. *Frontiers in Public Health, 1688*.

Shroff, R. (2020). The Tall Building Strategically Reconsidered—Seattle 2030: The Post-Crisis Tower. *CTBUH Journal*, (4).

Sky-Garden. (2015). Sky Garden Visitor Rules & Regulations. https://skygarden.london/terms-conditions/. Accessed 4 May 2020.

Taib, N., Abdullah, A., Fadzil, S. F. S., & Yeok, F. S. (2010). An assessment of thermal comfort and users' perceptions of landscape gardens in a high-rise office building. *Journal of Sustainable Development, 3*(4), 153.

Timm, S., Dearborn, L., & Pomeroy, J. (2018). Nature and the city: Measuring the attention restoration benefits of Singapore's urban vertical greenery. *Technology| Architecture+ Design, 2*(2), 240–249.

Triguero-Mas, M., Anguelovski, I., Cirac-Claveras, J., Connolly, J., Vazquez, A., Urgell-Plaza, F., Cardona-Giralt, N., Sanyé-Mengual, E., Alonso, J., & Cole, H. (2020). Quality of life benefits of urban rooftop gardening for people with intellectual disabilities or mental health disorders. *Preventing Chronic Disease, 17*, E126. https://doi.org/10.5888/pcd17.200087

Viñoly, R., Blomberg, C., & Blanes, M. (2015). Challenges and benefits of integrating public space into tall buildings. *CTBUH Research Paper*, 282–287.

Whyte, W. H. (1980). *The social life of small urban spaces*.

Wood, A., & Safarik, D. (2019). Skybridges: A history and a view to the near future. *International Journal of High-Rise Buildings, 8*(1), 1–18.

Yuen, B., & Hien, W. N. (2005). Resident perceptions and expectations of rooftop gardens in Singapore. *Landscape and Urban Planning, 73*(4), 263–276.

Zamani, S. H., Rahman, R. A., Fauzi, M. A., & Yusof, L. M. (2021). Effect of Covid-19 on building construction projects: Impact and response mechanisms. In *IOP conference series: Earth and environmental science* (Vol. 682, No. 1, p. 012049). IOP Publishing.

Zhongming, Z., Linong, L., Xiaona, Y., Wangqiang, Z., & Wei, L. (2021). *Cities and pandemics: Towards a more just, green and healthy future.*

Human Experience and Well-Being in Urban Environments

Assessment of the Impact of the Office Environment on Productivity Based on Employees' Satisfaction

Kousar Abedini and Hector Altamirano

Abstract

Office employees spend the most amount of time in an indoor environment. The indoor environment has a substantial influence on the office occupants' comfort, satisfaction, work performance, and overall productivity. This research aims to investigate the impact of various environmental factors on the productivity of office workers and their overall satisfaction with the workplace environment. Specifically, the study seeks to answer the following research questions: Which environmental factors most significantly affect office workers' satisfaction and productivity? How does satisfaction with the workplace environment differ between two different offices located in different firms? And what is the impact of office satisfaction on employee productivity? For this purpose, a qualitative methodology was adopted, and a set of questions was compiled in a questionnaire and sent to two different offices at the Podium Building Surveying and Kendall Kingscott firms, respectively. The results obtained were analysed using simple statistical tools. The results indicated that employee satisfaction levels were mostly affected by temperature, noise distractions, and personal preferences. Employees at office 2 located at Kendall Kingscott firm were more dissatisfied with their office environment as compared to office 1 at Podium Building Surveying. The reason for this was found to be the lack of an air conditioning system at office 2. The results indicated that productivity is affected by office satisfaction levels as many of the employees at office 2 were found to be dissatisfied with some of the environmental factors (such as temperature, noise, and air quality). Due to this dissatisfaction, office 2 employees were found to be less productive than employees in Office 1. Finally, it was recommended that an air conditioning system is to be installed at office 2 to further enhance employee satisfaction.

Keywords

Satisfaction · Employee · Office · Productivity · Environment

1 Introduction

The workplace environment is often considered to have a considerable impact on the performance of an employee which can either be negative or positive (Chandrasekar, 2011). It is further believed that most individuals spend close to half their lives indoors, and ergo, the indoor environment can have a substantial influence on an individual's mental state, workplace behaviours, skills, and general productivity (Singh, 1996). A better workplace atmosphere is believed to result in improved outcomes and improved productivity. Dissatisfaction, congested workspaces, and the physical environment, as per numerous publications relating to the research of numerous workplaces and buildings, play a significant role in the reduction of workers' performance. Various publications in several companies and office buildings have shown that variables such as dissatisfaction, crowded offices, and the environment of the workplace have an essential part in the loss of employee productivity (Candido et al., 2019; De Been & Beijer, 2014; Fassoulis & Alexopoulos, 2015).

Due to changes to many variables including the social environment, the introduction of information technology,

K. Abedini (✉) · H. Altamirano
The Bartlett School of Environment, Energy and Resources, University College London (UCL), London, UK
e-mail: kousar.abedini.ucl@gmail.com

H. Altamirano
e-mail: h.altamirano-medina@ucl.ac.uk

I. Pigliautile et al. (eds.), *Urban Resilience, Livability, and Climate Adaptation*, Advances in Science, Technology & Innovation, https://doi.org/10.1007/978-3-031-54911-3_6

and the flexible method to organizes office work processes, the office working environment has evolved since the 1990s (Hasun, 2005). The motivation to work and work performance is significantly improved if the employees are physically and emotionally fit. In addition, a suitable working environment contributes to a reduction in absenteeism. It can, therefore, improve employees' performance and can lead to higher workplace productivity (Boles et al., 2004).

The current paper adheres to a commonly used structure for a research study. It commences with an introduction that offers background and context into work satisfaction and set study objectives, followed by a literature review that establishes key concepts and summarizes past research on office satisfaction and productivity. The methodology section explains the research design and techniques employed. The results section reports the study's findings, while the discussion section scrutinizes and interprets them. Lastly, the conclusion summarizes the key outcomes and presents recommendations for future research and practice.

1.1 Defining Work Satisfaction

Work satisfaction, as per Woodworth (1964), relates to the orientation of the feelings workers have at work and work satisfaction is the key component to motivating and encouraging employees to achieve greater results. Over the years, many people have determined what workplace satisfaction entails. As per Clark (1997), if the staff is unhappy with the task assigned or they do not have proper rights or if the conditions at work are not satisfactory or if colleagues are unhelpful, or if the officers/managers are not respectful to them and if the employees are not involved in the decision-making process, it can lead to employees not trusting their organization.

Moreover, the author stated that in the present world, companies cannot sustain unsatisfied employees because such employees won't be able to meet the demands due to which they may be fired from work leading to companies sustaining additional expenses in hiring new personnel. As a consequence, companies benefit from a flexible working environment in which employees feel valued and part of the company. The morale of the workers must be kept high which will be seen in their work productivity; otherwise, they will make fewer efforts to improve their productivity if their morale is poor.

1.2 Office Work Environment

Two (2) major elements form the basis of the workplace environment, which includes work and context. Work includes every part of a job, involving how the work is accomplished and finished, and also includes duties such as training of the tasks, monitoring of own employment activities, a work performance sense, as well as an intrinsic value for a task. The study also centred on the essential component of job happiness. The findings suggest a positive link between working conditions and work satisfaction, as reported in studies by Sousa-Poza (2000) and Skalli et al., (2008). In addition, these studies described the second component known as the context of job satisfaction including physical circumstances of work and social conditions of work (Skalli et al., 2008; Sousa-Poza, 2000).

In one other study, Spector (1997) noted that companies often overlook the importance of creating conducive working environments, which can have detrimental effects on employees' performance. Such environments should provide a sense of safety, promote positive interactions among colleagues, recognize and reward achievements, incentivize successful operations, involve employees in decision-making processes, and communicate their value to the company. When employees perceive themselves as integral to the organization's success, they tend to be highly committed and develop a sense of ownership.

It was argued by Petterson and Arnetz (1998) that the communication of workers in offices is essential for the accomplishment of organizational objectives. Furthermore, the author explains the need for information to be communicated correctly and in a timely fashion to make the company processes work effectively. The author further argues that it is difficult to fulfil organizational objectives if there is a dispute among co-workers.

1.3 Environmental Factors

Many factors of the office environment including light, temperature, Indoor Air Quality (IAQ), humidity, and noise, play a significant role in organizational work satisfaction. A prior study conducted by Moss (2007) revealed the perceived impact of employment on health in the working environment. This research employed a national sample of 2048 employees, whose physical and mental health was evaluated by the effects of their professions. Analyses of regression have demonstrated a substantial correlation between workers' responses and their health results. Shikdar and Sawaqed (2003) also found a significant link between the performance metrics and the qualities of health, infrastructure, and environmental factors.

The link between the working environment and productivity has been given greater emphasis. Laboratory and field research demonstrated that there might be considerable impacts on the well-being and performance and thus productivity of office employees on the physical and chemical variables in the workplace. Workplace factors such as humidity, IAQ, and noise have important connections with the workplace satisfaction and productivity of workers (Fassoulis & Alexopoulos, 2015; Moss, 2007). Similarly, IAQ might directly affect health concerns and contribute to discomfort on the job (Palacios et al., 2021; Shek & Chan, 2008). Thus, environmental factors directly affect the health of the workers and lead to dissatisfaction and ultimately low productivity in the offices.

1.4 Research Statement

Noble (2009) argues that greater emphasis should be made on recognizing and addressing the work environment since employees may suffer from chronic stress when they experience negative perceptions of their workplace. The current study is designed to assess the effect of the office design, layout, and environmental factors (lighting quality, humidity, temperature, light, and noise) on the performance of office workers based on their overall satisfaction in the Podium Building Surveying and Kendall Kingscott offices. Podium is a building surveying company in London, while Kendall Kingscott is a construction consultancy in London. The study aims to investigate the relationship between office design, layout, and environmental factors (such as lighting quality, humidity, temperature, and noise) and the performance of office workers.

1.5 Research Questions

The specific research questions that the study aims to answer are:

- What is the relationship between the office working environment and the level of performance of employees?
- Which environmental factors most affect the satisfaction of office employees?
- What is the impact of employee satisfaction on office productivity?

By addressing these research questions, the study seeks to provide insights into how the work environment impacts employee performance and satisfaction, and how improvements in office design and environmental factors can enhance employee productivity.

1.6 Research Objectives

The general objective of the current research is to assess the effect of the office environment on office productivity based on the occupant's satisfaction. While the specific aims of the current research are described as:

- To evaluate the impact of the office environment on the office employee's satisfaction.
- To develop a tool for the assessment of office productivity based on occupants' satisfaction.
- To investigate which environmental elements have the greatest influence on workforce satisfaction and productivity.

1.7 Significance of the Study

The results of the current study are anticipated to open the way for the authorities to acknowledge the variables affecting office working conditions and the performance of workers in the companies. Such approval might be used to enhance employees' working circumstances. The results of this research will also help the companies to understand how to deal with employee concerns and their working environment and to see office layout and the environment as essential elements in boosting employee satisfaction and productivity.

It may serve as a basis for further studies in the current research field. The findings will also identify environmental elements that most impact the satisfaction of employees in their surroundings. In particular, at the construction and consultancy-related firms, the research will give information and actions to enhance employees' working environment and boost productivity.

1.8 Scope of the Study

The current research emphasizes office employees who are working in two different office environments of establishments namely Podium Building Surveying and Kendall Kingscott. This research will examine how office satisfaction and productivity are affected by the environmental factors of the two respective offices.

2 Literature Review

Most office workers spend half their time indoors, and research has shown that the indoor environment has a significant impact on their mental health, behaviour, skills, and performance (Sundstrom et al., 1994). Therefore, improving the physical atmosphere of the office is crucial to increasing productivity and employee satisfaction. However, defining and measuring employee productivity has proven to be a challenging task in the literature, particularly in service environments (Brown et al., 2009). The traditional definition of productivity, based on the ratio of input cost to output cost, may not accurately reflect the actual impacts on company structure. Consequently, the elements that impact employee productivity remain unclear in terms of conceptualization, measurement, and testing.

Enhancing staff productivity has become a critical issue for most organizations as it directly influences their income (Gummesson, 1998; Sels et al., 2006). Employee productivity can be assessed by measuring the efficiency of an employee or the number of employees. Productivity is usually measured by the number of products or services an employee manages to produce in a specific period (Piana, 2011). An organization's success depends significantly on its employees' productivity (Cato & Gordon, 2012; Gummesson, 1998; Sharma & Sharma, 2014). Although the methods of evaluating office productivity do not seem to be widely accepted, it seems that an assessment of productivity on a self-basis is still better than no measure of productivity (Clements-Croome, 2002; Haynes, 2008a).

Haynes (2008a) conducted a study to assess the Indoor Environmental Quality (IEQ) elements that impact employee productivity. The study involved a questionnaire survey of 996 employees in 26 offices and an internet poll of 422 employees. The physical components of comfort, including ventilation, heating, natural lighting, artificial lighting, décor, cleanliness, overall comfort, and physical security, were assessed, and compared to other factors such as workplace design, employees' communications, and disturbances. The author found that while the physical aspects of comfort showed no improvement in tenants' perceptions of productivity, interaction and distraction had a significant impact on perceived productivity. Hameed and Amjad (2009) also conducted a study of 107 employees in 12 Pakistani banks and found a clear link between office layout and employee performance. They conducted a questionnaire-based survey and discovered that illumination was the most important element affecting productivity. The impacts of noise, lighting, and temperature on productivity have also been discovered. Lee and Brand (2010) collected 384 questionnaires from office workers in three manufacturing businesses in Michigan (USA) and analysed them. The study found that the perception of control over physical environmental factors moderated the relationship between reported interruptions and perceived work productivity. Specifically, the study demonstrated that a sense of control over the physical conditions of the offices moderated the attitudes and outcomes of the participants. The study by Lee (2007) examines the relationship between indoor environmental quality and worker satisfaction and performance in Leadership in Energy and Environmental Design (LEED) certified buildings. The study involved surveying workers in four LEED-certified buildings in the United States and found that better indoor environmental quality was associated with higher worker satisfaction and performance. The author suggests that creating healthier and more comfortable indoor environments in buildings can lead to better outcomes for workers and organizations.

Office layout and design also have an essential role in employee satisfaction and productivity. There are two types of attempts to relate the office environment to the performance of its people. These include office layout and office comfort (Haynes, 2008a). Haynes (2008a) investigated the impact of office comfort on productivity. The authors use a mixed methods approach, including a literature review and a survey questionnaire of office workers in the United Kingdom. The study found that office comfort, as measured by factors such as temperature, lighting, noise, and air quality, was significantly related to productivity. In particular, workers who reported higher levels of comfort in their offices also reported higher levels of productivity. The study concludes that improving office comfort can have a positive impact on worker productivity. The study by Charles and Veitch (2002) investigated the effects of workstation size, partition height, and window provision on environmental satisfaction in open-plan offices. The results indicated that employees had a greater sense of environmental satisfaction when they had larger workstations, higher partitions, and access to windows with a view of the outdoors. These findings suggest that improving the physical environment of the workplace can contribute to higher levels of employee satisfaction.

Participants in one study which was conducted by (Mark et al., 2008), reported considerably higher stress, frustration, workload, effort, and pressure after only 20 min of interrupted performance. A two-second interruption can be enough to interrupt someone's stream of thinking (Haynes et al., 2019). Returning to a task after being distracted can take up to 25 min on average (Mark et al., 2008). In addition, it may take up to 15 min to return to the same flow conditions as before the interruption once the train of thought has broken (Lister & DeMarco, 1987). The occupants may manage their privacy and other factors of working in a private office, for example, air quality, noise, lighting, and temperature. Occupants of open-plan office spaces typically lack the same level of control over such issues, resulting in increased levels of unhappiness with the workplace (Kim & de Dear, 2013; Samani, 2015).

Just like privacy is important in the workplace, interaction in the office is also essential to share ideas and data, which is the primary purpose for the presence of offices. Formal gatherings at specified meeting venues that let the occupants regulate their levels of time and focus can be utilized to facilitate interactions. Interactions can also be spontaneous and serendipitous, occurring in a variety of settings and making it difficult for an occupier to maintain control. The layout of an office can be designed to encourage contact and communication among employees.

2.1 Environmental Factors and Their Overall Impact on Employee Satisfaction and Productivity

2.1.1 Thermal Comfort and Humidity

Temperature is a crucial factor that affects the productivity of employees in both large and small workplaces. Studies have shown that when the temperature exceeds 25 °C, the performance of call centre employees declines (Niemela et al., 2002); Federspiel et al., 2002). However, changing the normal temperature of a room within workers' comfort zone has no substantial influence on their productivity, as observed by Federspiel et al. (2002). On the other hand, Pilcher et al., (2002) observed that high temperatures of greater than 25.4 °C have a detrimental effect on office workers' job performance. Extreme heat and cold also impact workers' performance, with cognitive tasks being affected by temperatures of 10 °C and 32.22 °C (Pilcher et al., 2002).

In a study by Roulet et al. (2006), involving 66 workplaces across 9 EU nations, questionnaires were used to measure tenant satisfaction with comfort, and it was discovered that, in the summer, residents' reported productivity was related to temperature, while this relationship was less evident in the winter. The study found that high summer temperatures had an adverse effect on work productivity. Tanabe et al., (2015) investigated the relationship between thermal comfort and workplace productivity. The authors surveyed to measure thermal sensation and workplace productivity 480 participants working in six different office buildings in Tokyo. They found that individual thermal satisfaction was significantly associated with workplace productivity and that there was a positive correlation between the two variables. The study suggested that improving thermal comfort in the workplace can lead to increased productivity. Kekäläinen et al. (2010) investigated the impact of lower summer interior temperature on perceived productivity following the refurbishment of a Heating, Ventilation, and Air Conditioning (HVAC) system in workplace buildings in Helsinki, Finland. The researchers used two questionnaire assessments on job effectiveness and psychosocial work environment to assess the tenants' productivity. The study found that work efficiency had decreased before the remodelling and had significantly improved following the renovation. Prakash (2005) compared the impact of IEQ in Leadership in Energy and Environmental Design (LEED) certified buildings to that of non-LEED certified buildings, using a web-based Centre for Built Environment (CBE) questionnaire. The author found that sunlight and temperature had a positive impact on the inhabitants' sense of productivity in LEED-certified buildings, whereas in non-LEED-certified buildings, the building's overall ambience and ability to accommodate multiple activities had a greater impact on productivity.

Toftum et al., (1998) found that excessive humidity and high air temperatures can cause discomfort and stuffy air. Additionally, another study (Toftum, 2002) found that a relative humidity of over 65% was considered uncomfortable. When the relative humidity rises to between 80 and 95%, mould can begin to grow, which can pose significant health risks. Lenoir et al., (2012) used a questionnaire and an interview to assess the productivity of employees at the ENERPOS building in La Reunion (France) in the Indian Ocean. The personnel and students at this building were given a BUS questionnaire. While average productivity increases were recorded by students of 11.25%, staff reported a 17.5% rise in average output. All tenants interviewed indicated increasing productivity in these air-conditioned mixed-mode buildings during the interview sessions, as a result of the enhanced IEQ.

2.1.2 Light

Several studies have shown that natural lighting in offices is associated with improved employee health, which is also something that employees like (Sakuraya et al., 2020). Workplace illumination can impact employee performance in a variety of ways, including eye strain and visual comfort, as well as cognitive performance and problem-solving abilities, by interfering with physiological factors such as circadian cycles (Boyce et al., 2006; Juslén & Tenner, 2005).

Kim and Kim (2007) conducted a study to assess the acceptable degree of tolerance to variations in light levels and discovered that such adjustments had no significant influence on reading task productivity or letter recognition performance. Similarly, Baird and Thompson (2012) discovered that illumination had a substantial influence on productivity and conducted research on how occupants evaluated IEQ in their interior environment using the BUS questionnaire. The authors experimented with several forms of lighting in the office, including natural light, artificial light, sunshine, sun glare, and light glow. Newsham (2005) investigated the impact of task lighting on worker satisfaction, productivity, and energy efficiency in an office environment. The authors conducted a field study in which they compared the effects of different types of lighting on worker satisfaction and performance. The study found that task lighting had a positive impact on worker satisfaction and productivity and that energy-efficient lighting was not necessarily associated with decreased worker satisfaction or productivity. The study concluded that task lighting can improve worker satisfaction and productivity while also reducing energy consumption in the workplace.

2.1.3 Noise

Noise is one of the major issues identified in IEQ and can come from various sources, such as outdoor noise (e.g., road noise), interior mechanical noise (e.g., printer noise), and indoor human activity noise (e.g., conversations and phone calls), which is the most significant source of noise impacting employee performance in the office environment (Banbury & Berry, 2005). Noise in the work environment can have several effects, including physiological effects (e.g., headaches and weariness), psychological effects, cognitive effects, and social effects (Rasila & Jylhä, 2015). However, excessive noise levels can negatively impact productivity, resulting in a drop in organizational productivity and the quality of services and goods. It is recommended that workplace noise levels be controlled and reduced to a standard level of less than 85 dB to enhance comfort and human productivity (Akbari et al., 2013). Nonetheless, the impact of office noise is subjective and can depend on various factors, such as the occupants' capacity to regulate noise, the type of job activity, gender, and personality type.

Kaarlela-Tuomaala et al. (2009) studied 31 employees before and after shifting from a private office room to an open-plan workplace using questionnaires. The findings revealed that during a working day, the average noise level increased significantly, affecting the occupants' perceived productivity. The authors concluded that professional employees should not work in an open-plan workplace. Mak and Lui (2012) conducted a questionnaire study of 369 office employees in 57 air-conditioned offices in Hong Kong. They discovered a clear and substantial link between noise and temperature, as well as occupant productivity. Noise in the background, shutting doors, workers' activity, and distractions from both within and out of the office environment were all recognized as sources of low productivity. In another study conducted by Leaman and Bordass (2001), the authors used a BUS questionnaire to examine the effect of Indoor Environmental Quality (IEQ) on perceived employee performance in United Kingdom office buildings. They found that noise levels, particularly random noises, and irrelevant talk, were most closely associated with perceived productivity. Additionally, tenants who felt comfortable with their building's IEQ were more productive than their peers.

In summary, noise is a crucial element that can affect office building comfort and productivity. While it can foster a sense of belonging among office workers and improve productivity by offering timely feedback, excessive noise levels can negatively impact productivity. It is recommended that workplace noise levels be controlled and reduced to a standard level of less than 85 dB.

2.1.4 Indoor Air Quality (IAQ)

The air inside the workplaces is commonly polluted with several different substances such as volatile organic compounds that humans have not evolved to deal with, as well as concentrations of body smells and microorganisms that are considerably greater than outside. Sick Building Syndrome (SBS) has long been a source of worry among office designers and occupiers. In a post-occupancy study of the recently built Council House, 2 in Melbourne, Australia, Paevere and Brown (2008) found a 5.8% development in employee productivity as a result of improved IAQ, with 77% of the building's inhabitants believing the building had a positive or neutral impact on their productivity. While the authors acknowledged the difficulty of differentiating between building-related influences on productivity and other factors such as work reorganization, the study's focus group interviews were conducted simultaneously with the BUS questionnaire. Kamaruzzaman and Sabrani (2011) investigated how occupants perceived the impact of IAQ on their performance by distributing questionnaires to tenants in four office buildings. The findings indicated that the increase in work productivity resulting from the influence of IAQ was marginal.

2.1.5 Relationship Between Satisfaction and Productivity

Three methods are commonly used to measure the productivity of office employees: simulated task measurement, real productivity output measurement, and self-assessed productivity evaluation (Haynes, 2008a). While simulated task measurement can be carried out over a short period, its reliability over a longer duration is questionable. Real productivity metrics are highly desired, but not all office jobs can be easily classified as productivity output. As a result, some researchers have used workplace satisfaction as a substitute for productivity measurement (Haynes, 2008a; Leaman, 1995; Oseland & Bartlett, 1999). However, while productivity and satisfaction are correlated, using satisfaction as a proxy for productivity can pose problems of validity in research.

Leaman (1995) found a significant correlation ($r=0.93$, $p=0.0033$) between employees who reported dissatisfaction with their interior environment and those who reported a decrease in productivity due to their office environment. Similarly, Oseland and Bartlett (1999) found a strong correlation ($0.92<r<0.98$) between productivity and satisfaction among occupants of 10 office buildings. However, the authors noted that the strong correlation may be somewhat due to the way questions were asked, such as "Considering the effect on your performance, how satisfied are you with the office facilities and services?" In addition, Leaman (1995) identified employees' perceptions of their ability to influence their working environment as a crucial factor in their productivity. Oseland and Bartlett (1999) also found a correlation between perceived control over environmental circumstances and productivity (r = 0.48). Therefore, measuring genuine factors under examination may provide more accurate insights into the productivity of office workers, and self-assessed measurements may be considered when measurable productivity measures are not available (r = 0.48).

3 Methodology

The above literature suggests that the office layout and design, office privacy, and environmental factors (humidity, temperature, light, IAQ, and noise) affect employees' satisfaction in many different ways, and each of these factors has a different impact on employees' general productivity (Fig. 1). It was also found that there is not much research on the impact of the office environment on workers in the engineering and construction-related office environment. Therefore, the current study seeks to investigate how these factors impact employee productivity based on their overall satisfaction.

The current portion covers many methodological techniques employed by the researcher. The current chapter covered the themes of research design, the target population, sampling and sampling processes, the data recovery tools, data collecting procedures, the validity and reliability of the instruments utilized for the study, and ethical concerns including data analysis methodologies.

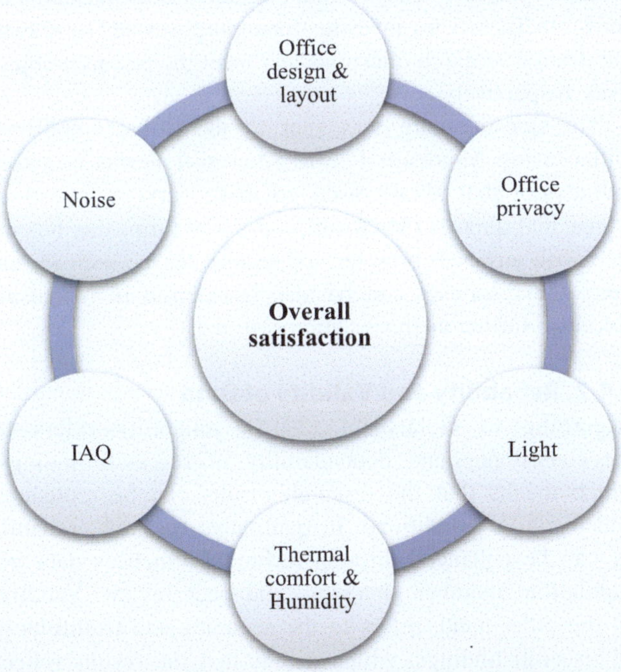

Fig. 1 Variables that affect the overall employee satisfaction

3.1 Research Design

"Research designs are types of inquiry within qualitative, quantitative, and mixed methods approaches that provide specific guidance for research procedures within a research design" (Cresswell, 2014). As a result, when academics choose between qualitative, quantitative, or mixed methodologies research, they must also determine what sort of study they will undertake within these three methodologies. Researchers typically use quantitative techniques to address research questions that require numerical data, qualitative methodology to questions needing textual data, and mixed methodologies techniques to address questions that require both numerical and textual data (Naoum, 2019).

The current study utilizes the qualitative method for data collection as employed by many researchers before to investigate employee satisfaction and overall productivity (Hameed & Amjad, 2009; Haynes, 2008a; Lee & Brand, 2010). The qualitative research technique is a technique for examining and interpreting the meaning and significance that people or groups attribute to a social or human problem. Qualitative research differs from quantitative research in that it focuses on exploring and interpreting complex phenomena that are difficult to measure quantitatively and are often perceived and expressed through language and meaning, whereas quantitative research is focused on data investigation and experiments (McLeod, 2019).

3.2 Target Population and Offices

The population that was targeted in the research was office employees from two different construction and engineering-related companies. The study focused on employees and administrative assistants consisting of about 50 people in two separate offices. The reason for this selection is that these office employees take part in daily operations in construction and architecture-related activities. This study, therefore, hopes to receive more valuable data to help construction-related company managers overcome productivity-related issues in offices and to improve workers' performance through the improvement of their working environments and satisfaction levels as discussed in the literature review, satisfaction is correlated with office productivity.

The study selected the target offices for data collection in London's Podium Building Surveying and

Kendall Kingscott buildings using convenience sampling. Convenience sampling is a non-probability sampling technique that involves selecting participants who are conveniently available to the researcher, rather than through a random selection process (Babbie, 2015). Furthermore, these offices are also the representatives of typical engineering and construction-related offices in London. Both offices were open-plan designs. An open-plan office is one with a floor plan designed in such a way that encourages people to collaborate face-to-face (Hedge, 1982).

3.3 Sample Size and Sampling Strategy

The current research uses convenient and random sampling techniques for qualitative data collection. Convenient sampling is perhaps the easiest technique to sample as respondents are selected based on willingness to attend/participate. Valuable findings can be made using this sampling strategy, but the findings are likely to outcome in considerable bias, since persons who engage in the research of their own free will may not be the representative of the sample as people who refuse to participate may differ from those who choose to participate (Sedgwick, 2013). The current survey was conducted based on the availability and readiness of office employees and was distributed to 50 individuals in the two separate offices.

In the first office, the total number of employees asked was 23 while the number of employees asked in the second office was 27. Only eight employees responded to the questionnaire in the first office while the number of respondents that responded in the second office was seven. This shows the total sample size of the target population which was 15. It would have been better to add some interviews in different circumstances to increase the sample size. However, due to the COVID-19 restrictions, the survey interviews could not have been performed.

3.4 Data Collection Method

Cresswell (2014) defines primary data as the one involving the researcher's development of a tool for the respondents' administration, and secondary data is defined as data that is already accessible in published or unpublished existing materials. The current study only utilized the primary data and employed a questionnaire methodology to gather data for the study.

3.4.1 Questionnaire

A questionnaire is a common way to collect data of various kinds in a research problem (Rattray & Jones, 2007). It is frequently employed in research to gather data on specific circumstances and behaviours to investigate people's or groups' opinions and attitudes (Rattray & Jones, 2007). According to a study by Sutrisno et al. (2016), questionnaire-based surveys are widely used in office satisfaction research because they are efficient, cost-effective, and easy to administer. Additionally, this method can capture a broad range of factors that contribute to office satisfaction, such as workplace design, environmental factors, and social aspects of the workplace. A study by Sundstrom et al., (1982) found that employees' self-reported satisfaction with their physical work environment was significantly related to their productivity and job satisfaction indicating that questionnaires can be a reliable indicator of office satisfaction. In another study, Harter et al. (2002) found that employees who were engaged and satisfied with their jobs had higher levels of productivity and customer loyalty. The researchers used questionnaires to measure job satisfaction and engagement and found that higher levels of both were associated with better work outcomes. Overall, questionnaires are a widely used method for measuring office satisfaction due to their ease of administration, ability to collect a broad range of data, and evidence supporting their validity and reliability.

In response to these study aims and research issues, the current study created the questionnaire with direction from the University Supervisor and the help of the previous literature. Table 1 shows the literature that was used to create the current survey questionnaire for each variable.

The questionnaire also contained the background part of the respondents for the purpose of identifying the sample demographics and the suitable portions of questions as shown in Table 1 for investigational purposes and to answer the desired research questions and meet the research objectives, respectively.

The questionnaire was sent to the office employees online using Microsoft Forms. Microsoft Forms is a free online tool that allows users to make surveys, questionnaires, and quizzes (Microsoft, n.d.). The forms can be created with any web browser and results can be seen in real time; the responses can be also transferred to Microsoft Excel for further analysis (Microsoft, n.d.).

3.4.2 Reliability and Validity of Data

According to Golafshani (2015), reliability refers to the consistency and dependability of research findings, which means that the research results can be replicated under similar conditions. In qualitative research, reliability can be enhanced through techniques such as data triangulation, member checking, and peer review. Validity, on the other hand, refers to the accuracy and truthfulness of research findings, which means that the results reflect what they are intended to measure. In qualitative research, validity can be enhanced through techniques such as

Table 1 Creation of questions based on the literature

Independent variables	Questions	Literature source
Office layout	Degree of satisfaction with the office space for the office work and storage facilities	Lee (2007)
	Degree of satisfaction with the office privacy	Lee (2007)
	Degree of satisfaction with the office furniture	Lee (2007)
Air quality	Satisfaction with the indoor air quality	Oseland and Bartlett (1999)
Noise	Degree of satisfaction with the noise levels in the workspace	Lee (2007)
	Office satisfaction with the internal noise distractions (ringing cell phones, conversations, etc.)	Haynes (2008b)
	Office satisfaction with external noise distractions (traffic noise)	Haynes (2008b)
Lighting	Degree of satisfaction with the lighting at the workplace	Lee (2007)
	Degree of control of lighting at the workplace	Newsham (2005)
Temperature	Degree of satisfaction with the temperature at the workplace in the summer and winter seasons	Tanabe et al., (2015)
	Degree of control of temperature at the workplace	Tanabe et al., (2015)
Windows	Degree of office satisfaction influenced by windows in the workspace	Charles and Veitch (2002)
Productivity	Degree of the impact of environment on office productivity	Leaman (1995)

credibility, transferability, dependability, and confirmability (Golafshani, 2015). These techniques aim to ensure that the research findings accurately represent the perspectives and experiences of the participants and that they are not influenced by the researcher's biases or assumptions (Golafshani, 2015) While Bashir et al., (2008) define reliability as the consistency and stability of the results obtained from a particular research instrument or technique, while validity refers to the accuracy and truthfulness of the results obtained from a research study. In the current study, the respondents' consistent findings were used to attain dependability. The data's reliability was ensured by gathering information from relevant respondents, with a focus on essential factors such as the working environment and employee satisfaction, variables that hampered employee productivity, correct wording of directions, and logical arrangement of the questions asked.

3.4.3 Data Analysis

The questionnaire data were analysed using percentages in this study. The obtained data was verified for consistency, and the distribution's answers were represented using frequencies and percentages. Tables and graphs were utilized to present the findings. Microsoft Excel was used to analyse the results of the study.

3.5 The Originality of the Methodology

The methodology presented here is not particularly novel, as it relies on commonly used qualitative research methods such as convenient and random sampling techniques and questionnaire-based surveys. However, the study's strength lies in its focus on a specific population of office employees in the construction and engineering-related industry, which can lead to valuable insights for managers in this field. Additionally, the study's use of a literature review to guide the creation of the questionnaire helps to ensure that the questions are relevant and grounded in existing research. Overall, while the methodology is not novel, it is well-suited to address the research question at hand and can provide useful insights into office satisfaction and productivity in the construction and engineering-related industry.

4 Results and Discussions

This chapter discusses the analysis and interpretation of the collected data obtained via questionnaires to meet the current study's objectives. A total of fifty people work at the two offices of Podium Building Surveying and Kendall Kingscott, respectively. For the current study data collection purpose, a sample of 15 employees was selected (based on employee participation), and fifteen (15) questionaries were attached and sent to respondents/employees via Microsoft Forms. The total number of questions asked was twenty-nine (29), and the results of the overall analysis are listed below.

4.1 Data Analysis

4.1.1 Respondents Gender

The gender distribution of employees was used in the current research for the purpose of filling out the questionnaires, as indicated in Table 2.

The table above indicates that 53.4% of male respondents and 46.6% of female respondents correspondingly responded to the questionnaires sent to them.

4.2 Office Noise Distractions

The presence of all types of noise disturbances (such as the ringing of cell phones, personal and telephone conversations, a loud printer or a noisy heating and air conditioning system, and outside traffic) in the office somehow affects employee satisfaction, as shown in Table 3.

Table 3 indicates that the majority of employees in both offices (50% in Office 1 and 42.8% in Office 2) reported being affected by noise distractions, which caused disruptions to their work and negatively impacted their satisfaction levels. Specifically, 37.5% in Office 1 and 42.8% in Office 2 reported that noise distractions often led to an unsatisfactory atmosphere in their workspace. In contrast, only a small proportion of respondents in Office 1 (12.5%) and Office 2 (14.4%) reported being unaffected by noise distractions and satisfied with their office environment. These findings suggest that a significant number of employees were dissatisfied with the noise levels and that noise had a negative impact on the office environment in both offices.

Table 2 What is your gender?

Gender	Frequency	Percentage
Female	7	46.6
Male	8	53.4
Total	15	100

Table 3 Do you get disturbed by noise sources like the ringing of cell phones, personal and telephone conversations, a loud printer or a noisy heating and air conditioning system, and/or outside noise?

Responses	Office 1		Office 2	
	Frequency	% age	Frequency	% age
Not at all	1	12.5	1	14.4
To some extent	4	50	3	42.8
Often	3	37.5	3	42.8
Mostly	0	–	0	–
Always	0	–	0	–
Total	8	100	7	100

Table 4 Do employees agree or disagree with the lighting system in the office?

Responses	Office 1		Office 2	
	Frequency	% age	Frequency	% age
Agree	6	75	2	28.6
To some extent	2	25	4	57.2
Neutral	0	–	1	14.2
Total	8	100	7	100

4.3 Office Lighting

The employees were asked about the significance of lighting in their offices and whether their work is provided with a good amount of lighting so that they can perform their tasks easily. The majority of the respondents (75%) in Office 1 responded that they have a good amount of light in their offices, which means they were satisfied with the lighting conditions in their offices. While in office 2, only 29% of the respondents agreed and were satisfied with the lighting, and more than half (57%) were not satisfied with the lighting in their office. This can be seen in Table 4.

When asked about the employee's control of light on their working desks, the vast majority of the respondents (88% in office 1 and 86% in office 2) responded that they have no control of lighting on their desks as indicated in Fig. 2. This shows that even though, in Office 1, employees had no control over the lighting on their desks, the lighting provided by the management in the office was good enough most of the time and the majority of the employees were generally satisfied with it. But for office 2, most employees neither had any control over the lighting on their work desks, nor the majority of them were satisfied with their office lighting.

4.4 Office Temperature

Employees were also asked about the thermal comfort and temperature in their offices and whether they were satisfied with the temperature of their offices in both the winter season and summer seasons. The figures below imply the effect

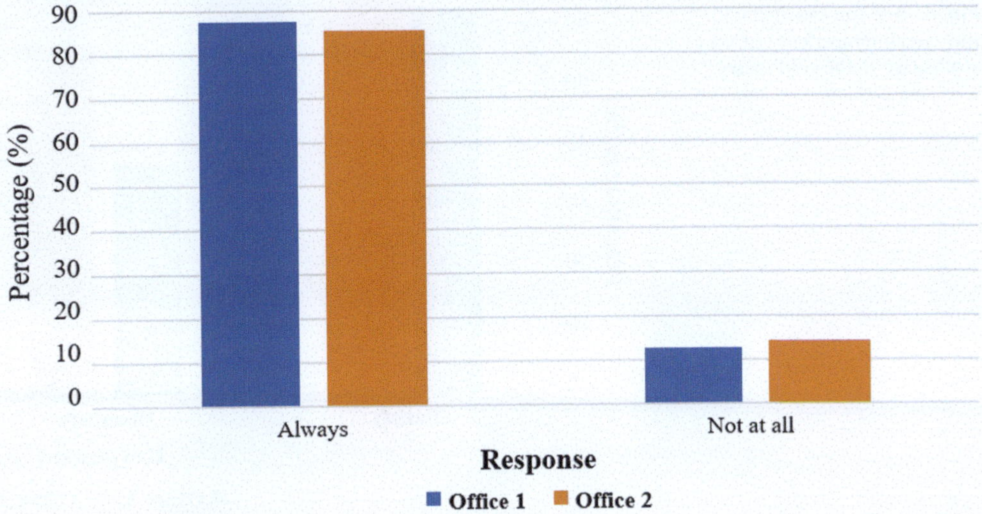

Fig. 2 Do you have control over the lighting on your desk (i.e., adjustable desk light on the desk)?

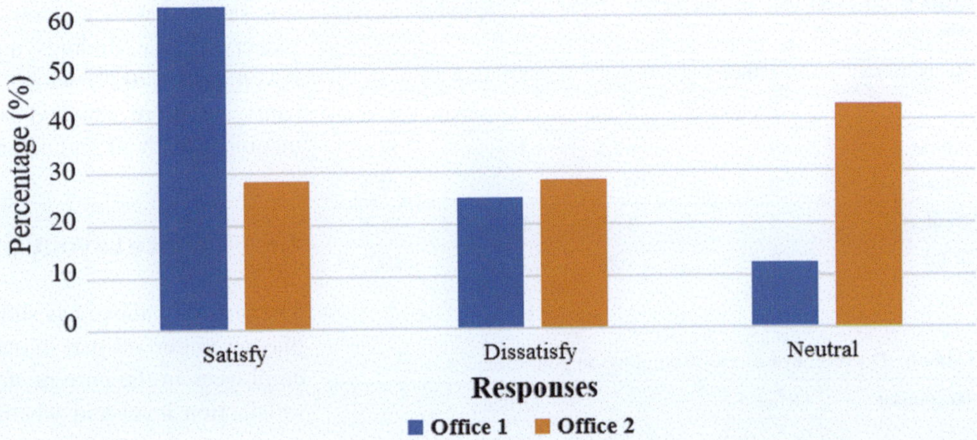

Fig. 3 Are you satisfied with the overall temperature of your workspace in the summer season?

of temperature on the satisfaction level of employees in the summer and winter seasons.

Figure 3 shows that the vast majority (62.5%) of the employees in Office 1 were satisfied with the temperature in their offices in the summer season while only 28.5% of the employees in Office 2 were satisfied with the office temperature. The reason for this could be that Office 1 had an air conditioning system installed while there was no air conditioning system in Office 2 as mentioned by the employees in their responses.

The trend of employee satisfaction in the winter season was similar to that observed in the summer season for both offices 1 and 2. However, in the winter season, more than half (57.1%) of the employees in office 2 were dissatisfied with their office room temperature (as shown in Fig. 4), likely due to the absence of an air conditioning system. In contrast, in the summer season, only 28.5% of employees were not satisfied with the temperature while 43% were neutral. Table 5 shows that in Office 1, 62.5% of employees responded that they could control the temperature of their office, while in Office 2, 57.2% of employees could not control their office temperature. The presence of an air conditioning system in Office 1 allows employees to control the temperature according to their satisfaction, as reported by Tse and So (2007), whereas the absence of an air conditioning system in Office 2 means that most employees were unable to control their room temperature. According to Abdou et al. (2006), employees who can adjust or control the temperature of their workplaces can reduce absenteeism by 43%, which in turn can impact individual productivity. The inability to control the room temperature can make employees drowsy or disoriented, further reducing their productivity (Abdou et al., 2006).

4.5 Office Windows

Employees were also asked about their access to windows in their offices as windows affect the workplace environment in many ways. It has been shown that 75% of

Fig. 4 Are you satisfied with the overall temperature of your workspace in the cold season?

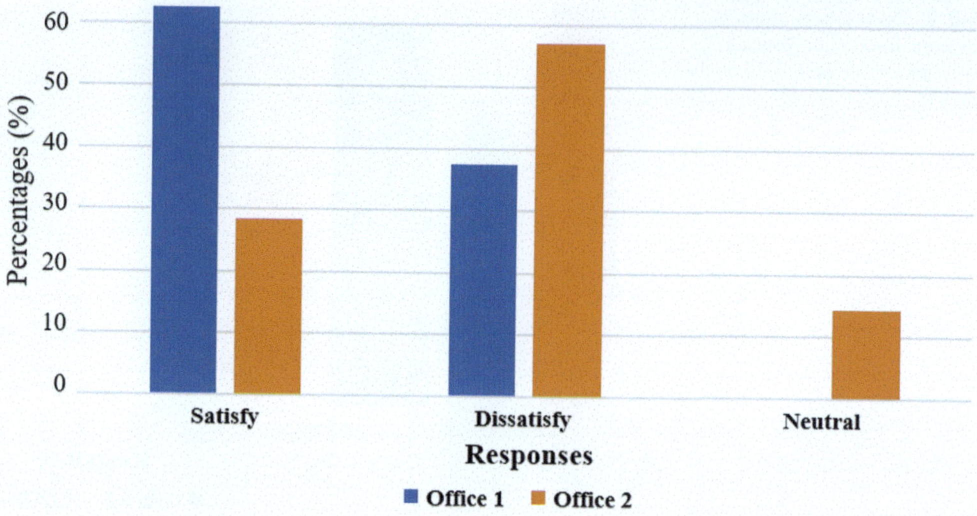

Table 5 Are you able to control the airflow or temperature in your offices?

Responses	Office 1		Office 2	
	Frequency	% age	Frequency	% age
Agree	5	62.5	2	28.6
To some extent	0	0	4	57.2
Neutral	3	37.5	1	14.2
Total	8	100	7	100

Table 6 Do you have access to windows in your office?

Responses	Office 1		Office 2	
	Frequency	% age	Frequency	% age
Yes	6	75	7	100
No	2	25	0	0
Total	8	100	7	100

employees in Office 1 and 100% of employees in Office 2 have access to Windows (as seen in Table 6). Windows are useful for the flow of fresh air in the office environment (Browning, 2015). Windows also are a good source of natural light (Browning, 2015). The employees were questioned about their satisfaction level with the access to windows. All of the employees that had access to Windows were satisfied with it. Figure 5 shows the reason behind this satisfaction of employees.

Figure 5 shows that 50% of employees in Office 1 and 60% of employees in Office 2 responded that they are satisfied with the windows due to the incoming natural sunlight from these windows. Similarly, 41.6% and 40% of employees in Office 1 and Office 2 were satisfied because of the incoming fresh air through these windows into the indoor office environment. While only 1 (8.4%) of the employee was satisfied with windows due to the view of the outside scenery. These findings are in line with the analysis of Kwon et al. (2019) which demonstrated that office occupants are more satisfied with access to windows due to incoming fresh air and natural light.

4.6 Office Layout and Design

The literature above has shown that office layout and design play an important part in overall employee satisfaction. The employees in the current study were also asked about their satisfaction level and whether they were satisfied with the space size available for their daily work, personal storage facilities, and available office furniture. Table 7 shows that the vast majority (75%) of the office employees in Office 1 were satisfied with the space available for work with 12.5% very satisfied. In the case of office 2, 43% of respondents said that they were very satisfied with the available space in the office, 28.5% said that they were satisfied and 28.5% said that they were neutral (neither satisfied nor dissatisfied). This shows that even in both offices, the majority of the employees were satisfied with the office available space for daily operation but the level of satisfaction of office 2 employees is more than that of office 1.

Table 7 also shows that half of the employees (50%) in Office 1 and 42.8% in Office 2 were found to be satisfied with the available furniture layout at the office. While there 25% of employees in office 1 and 14.3% in office 2 were not satisfied with the furniture layout at the office. Similarly, 25% of employees in Office 1 and 14.3% in Office 2 were neither satisfied nor dissatisfied. There was also 1 very satisfied employee and 1 employee who was very dissatisfied with the furniture layout in office 2. This shows that there are many differences among employees when it comes to furniture layout and the satisfaction level

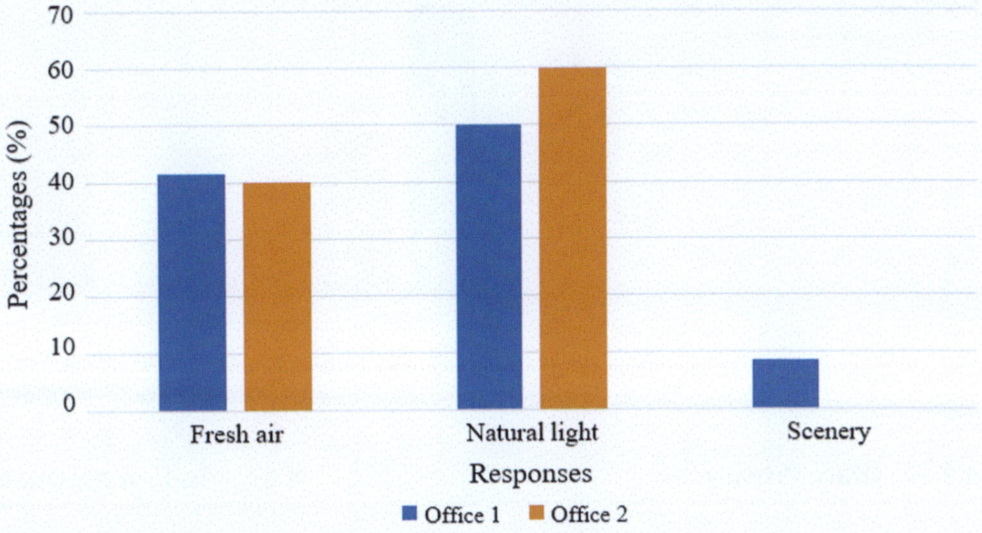

Fig. 5 If you have access to the window, then why are you satisfied with it?

Table 7 Employees' level of satisfaction with the office layout (office space, furniture, and storage facilities)

		Satisfaction with office space		Satisfaction with office furniture		Satisfaction with Storage facilities	
	Responses	Frequency	% age	Frequency	% age	Frequency	% age
Office 1	Very satisfy	1	12.5	0	0	0	0
	Satisfy	6	75	4	50	7	87.5
	Neutral	1	12.5	2	25	0	0
	Dissatisfy	0	0	2	25	1	12.5
	Very dissatisfy	0	0	0	0	0	0
	Total	8	100	8	100	8	100
Office 2		*Satisfaction with office space*		*Satisfaction with furniture comfort*		*Satisfaction with Storage facilities*	
	Very satisfy	3	43	1	14.3	2	28.5
	Satisfy	2	28.5	3	42.8	1	14.3
	Neutral	2	28.5	1	14.3	2	28.5
	Dissatisfy	0	0	1	14.3	2	28.5
	Very dissatisfy	0	0	1	14.3	0	–
	Total	7	100	7	100	7	100

of each employee is dependent on personal preferences. Similarly, Table 7 also shows the satisfaction level of employees based on storage facilities in the office. Storage facilities are needed in offices for the storage of files, equipment, and personal belongings of employees. 87.5% of the employees in Office 1 were satisfied, while 12.5% of employees were not satisfied with the storage facilities. While in the case of Office 2, 14.3% responded that they were satisfied, 28.5% were very satisfied, 28.5% were neutral, and 28.5% were dissatisfied. This shows that Office 1 has adequate storage facilities for its employees which is why the majority of the employees are satisfied while in Office 2, more than half said that either they are dissatisfied or neutral, which means the storage facilities in Office 2 are not as good as Office 1. Finally, Fig. 6 shows the satisfaction level of employees based on the overall design of the offices. It can be seen that half (50%) of the employees in Office 1 were satisfied with the overall office design compared to 28.5% in Office 2. Similarly, 28.5% of the employees in Office 2 were found to be not satisfied with the office design as compared to Office 1 where there was not a single employee was not satisfied with the office design. This shows that Office 1 has a good office design compared to Office 2, which keeps its occupants satisfied.

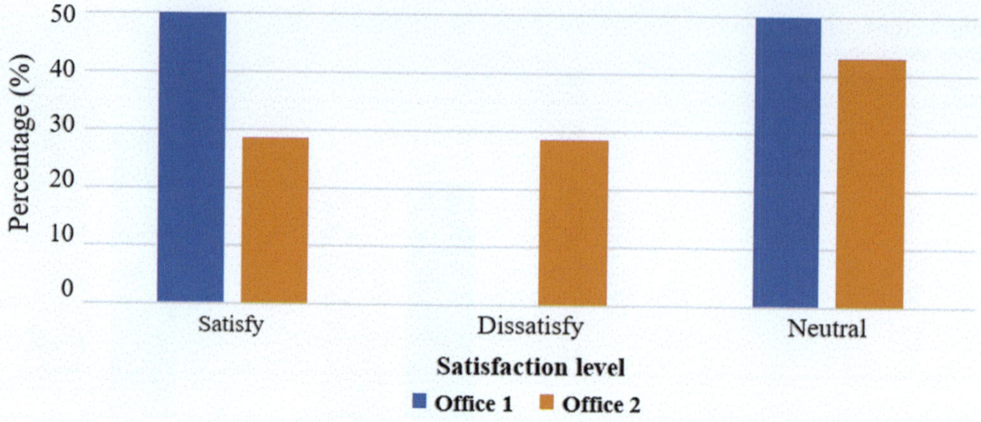

Fig. 6 Are you satisfied with the overall design of the offices?

4.7 Office Privacy

Employees were also asked about office privacy as the literature above shows that it also affects employee satisfaction and comfort in offices (Kim & de Dear, 2013); therefore respondents were asked whether their office gives them enough privacy or not.

Figure 7 shows that even the number of employees that agree that their offices provide them with enough privacy is almost the same (25% in office 1 and 28% in office 2) but more than half (57.1%) of the employees in office 2 also responded that their office environment does not provide them with enough privacy as compared to 25% in office 1. This shows that Office 1 has more privacy than Office 2. An ample amount of privacy is considered good for office work as indicated by Haynes et al., (2019) that lack of privacy and just two seconds of interruption can be enough to disrupt someone's stream of thinking.

4.8 Indoor Air Quality (IAQ)

IAQ is considered to be very important for occupant's health and literature indicated that there was a 4.9% improvement in employee productivity as a result of a better IAQ in an office environment (Paevere and Brown, 2008). In the current research, the employees were also questioned about the IAQ in their offices. They were asked if there was any issue with air quality in their offices. Figure 8 shows that 87.5% of the employees in Office 1 think that there is no issue with the air quality as compared to 57.1% in Office 2. This means that 43% of the respondents were not satisfied with the air quality in Office 2 compared to only 1 employee (12.5%) in Office 1.

The employees who responded that there was some issue with the air quality were then asked about the specific issue regarding the air quality. One employee in Office 1 responded that the air was too warm. In the case of office

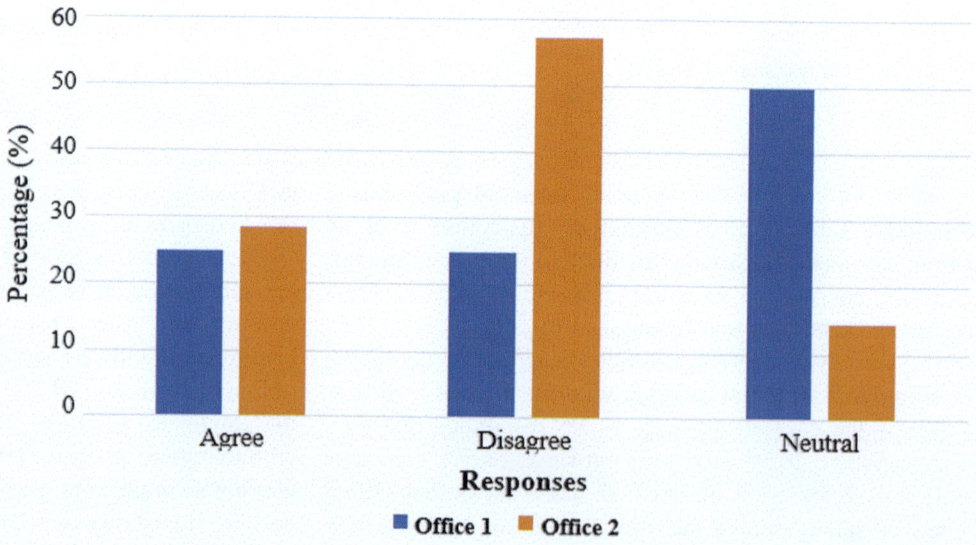

Fig. 7 Do you think that your office layout gives you enough privacy?

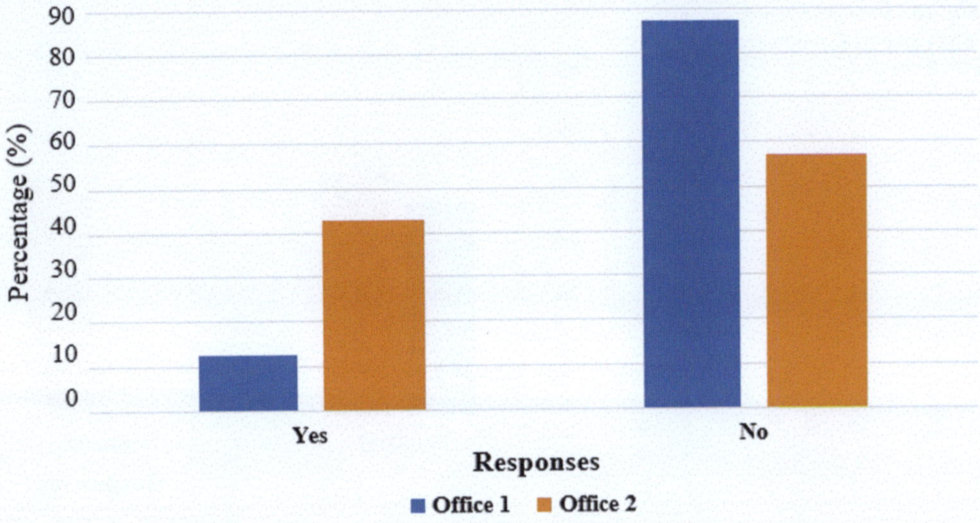

Fig. 8 Do employees think there is an issue with air quality?

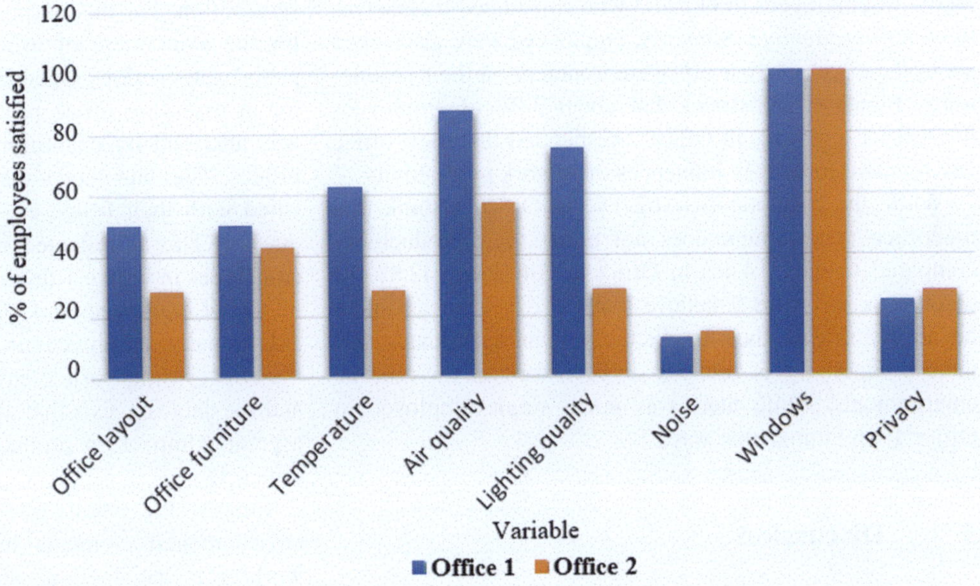

Fig. 9 Percentage of employees satisfied with office design, layout, and the environment

Table 8 How productive are you in general?

Responses	Office 1		Office 2	
	Frequency	% age	Frequency	% age
Productive	6	75	3	42.8
Not Productive	0	0	2	28.6
Neutral	2	25	2	28.6
Total	8	100	7	100

2, one employee responded that the air was too dry and the other two responded that there was low airflow in the office. These results indicate that inside office 2 there is an issue with ventilation and proper ventilation is needed to enhance the airflow and hence the satisfaction level of employees.

4.9 Office Productivity and Satisfaction

Figure 9 shows the percentage of employees in both offices 1 and 2 who are satisfied with the various variables of their office environment. The remaining percentage of employees were either dissatisfied or neutral about the office environment's impact on their satisfaction. It can be seen that Office 1 employees are more satisfied with their environment as compared to the employees of Office 2. Finally, the employees of both offices were asked about their productivity in the workplace.

Table 8 shows that the majority of the respondents in Office 1 (75%) think that they are productive in their office work while in the case of Office 2, only 42.8% of the employees think they are productive. There was not a single non-productive employee in Office 1 while there were

Fig. 10 How office environment affect employee productivity?

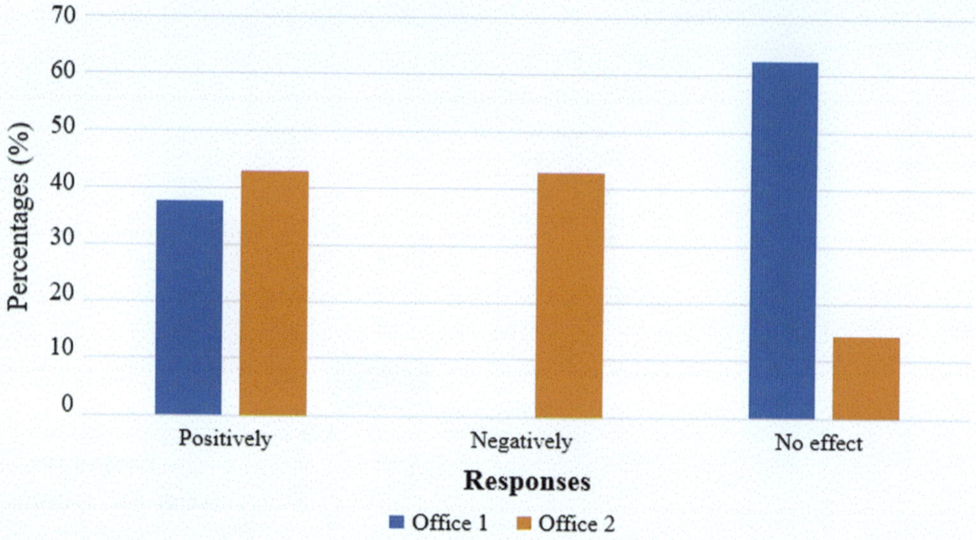

28.6% of employees in Office 2 who considered themselves to be non-productive. Similarly, employees were also asked about the effect of their office environment on their productivity. Figure 10 illustrates that about 38% of employees in Office 1 and 43% in Office 2 responded that their office environment positively influences their work productivity.

While the vast majority (62.5%) in Office 1 think that the office environment does not affect their productivity compared to only 14.4% in Office 2. Similarly, 42.8% of employees in Office 2 believe that the office environment negatively affects their productivity (which indicates that their productivity is reduced or decreased in their office environment), while there was not a single employee in Office 1 who thinks that way.

5 Discussion

In the discussion section, the study's objectives and their corresponding findings were presented and analysed. The results are analysed to answer the research question of how workplace environment affects employee satisfaction. The results show a link between office environmental characteristics and productivity, which is similar to prior studies by Moss (2007) and Fassoulis and Alexopoulos (2015). This can be seen in Figs. 9 and 10 which show that employees in Office 1 are more satisfied with their working environment, which led to higher productivity. In comparison, employees in Office 2 are less satisfied with their environment, which leads to lower productivity when compared to their counterparts in Office 1.

The results are also analysed to determine how can office productivity be assessed based on occupants' satisfaction. For this purpose, a simple tool was developed for the assessment of office productivity based on occupants' satisfaction. Occupants' satisfaction is a key factor in assessing office productivity (Leaman, 1995). This was analysed by examining data from employees in two offices. The data shows that employees who are more satisfied with their office environment tend to be more productive. This is reflected in the fact that the majority of employees in Office 1 think they are productive, while only 42.8% of employees in Office 2 think they are productive. Additionally, employees in both offices were asked about the effect of their office environment on their productivity, and the data suggests that a positive office environment has a greater impact on productivity in Office 2 compared to Office 1.

Finally, the results are analysed to determine which environmental elements have the greatest influence on workforce satisfaction and productivity. The results showed that temperature, lighting, air quality, and noise are the environmental elements that have the greatest impact on employee satisfaction and productivity. Although noise was determined to be the most impactful environmental element influencing satisfaction in both offices, it affected both offices equally and cannot be regarded as the primary contributor to Office 2's decreased productivity. However, lighting, temperature, and air quality were identified as the key environmental elements influencing satisfaction and productivity in Office 2. Furthermore, the study found that thermal comfort and the absence of an air conditioning system substantially impacted employee satisfaction and productivity in one of the offices, emphasizing the significance of these environmental elements in the workplace.

6 Conclusions

This portion covers the most important findings that came out of the current research and the conclusions that were drawn. The working environment is critical in encouraging employees to complete their tasks effectively. In today's competitive organizational contexts, the capacity to recruit, retain, and inspire high-performance employees is becoming increasingly crucial. This study aims to evaluate the impact of the office environment on office employees' satisfaction and to assess office productivity based on occupants' satisfaction. To meet the study aims and objectives, the current study investigated the workplace environment and its impact on workplace satisfaction and productivity. For the determination of employee satisfaction and productivity in the workplace, qualitative data was used that was obtained from the questionnaire survey. The questionnaire survey included different variables of the office environment and their impact on employee satisfaction. Environmental variables were selected based on a literature review. Since the literature suggests a correlation between satisfaction and productivity therefore to draw the relationship between employee satisfaction and productivity, the data was analysed. Some of the conclusions that can be drawn from the analysis of the data are:

- Noise distractions caused workplace dissatisfaction in almost half of the employees in both of the offices. Lighting conditions play a substantial role in office satisfaction, and an office with a good amount of light was found to be satisfactory for employees.
- Temperature and thermal comfort were important environmental factors affecting employees' satisfaction levels. It was found that employees in office 2 were not satisfied (57.1%) with indoor temperature. In comparison, most Office 1 employees were found to be satisfied (62.5%).
- Access to windows in both offices resulted in high satisfaction among employees due to incoming fresh air and natural light
- No employee in Office 1 was dissatisfied with the office design, while in Office 2, it was found that about 30% of the employees were not satisfied with the general office design and layout. Thus, indicating that office layout and design are important in employee satisfaction.
- Indoor air quality was not found to be a big issue in Office 1, while due to poor ventilation in Office 2, employees were not satisfied with the airflow in the office.
- The office environment affected the satisfaction level of employees in the office which then affected the office productivity. This indicates that employees with high satisfaction levels are more productive compared to employees who are less satisfied with the environment.
- Environmental factors that most affected the satisfaction levels were noise, lighting, temperature, and air quality, while lighting, temperature, and indoor air quality were the ones most affecting office productivity.

6.1 Limitations of the Study

The current study has some limitations which include:

- The current study only analysed qualitative data which might not truly indicate the levels of some environmental factors in offices.
- The current study considered only a few variables of the environment and their impact on the employees' satisfaction, while many other factors such as social environment, health and safety, psychological factors, etc., impact employee satisfaction.
- The current study assumes that highly satisfied employees are more productive, while the employees who are not satisfied with the environment are not very productive.
- The sample size of the target population is small which might not be a good indicator of overall office satisfaction and thus productivity.
- The information about the target offices and buildings, such as design, orientation, and layout is limited.

6.2 Further Work

Based on the current study limitations, the following are the important considerations for further work in the future:

- There is a need for a quantitative study to measure and record the environmental parameters and their impact on workplace satisfaction.
- There is a need to enhance the questionnaire survey by adding interviews.
- There is a need for a large sample size of the target population to get a better representation of office satisfaction and productivity.
- More information about the offices and buildings' orientation, layout, and design is needed.
- The questionnaire questions need to be enhanced and modified to get clear responses from the respondents.

References

Abdou, O. A., Kholy, G. M., & Abdou, A. A. (2006). Correlation between indoor environmental quality and productivity in buildings. In *International Association for People—environment Studies conference proceedings on CD-Rom* (pp. 11–16).

Akbari, J. et al. (2013). Relationship between lighting and noise levels and productivity of the occupants in automotive assembly industry. *Journal of Environmental and Public Health.*

Babbie, E. R. (2015). *The practice of social research* (14th ed.). CENGAGE Learning Custom Publishing.

Baird, G., & Thompson, J. (2012). Lighting conditions in sustainable buildings: Results of a survey of users' perceptions. *Architectural Science Review, 55*(2), 102–109. https://doi.org/10.1080/00038628.2012.667941

Banbury, S. P., & Berry, D. C. (2005). Office noise and employee concentration: Identifying causes of disruption and potential improvements. *Ergonomics, 48*(1), 25–37. https://doi.org/10.1080/00140130412331311390

Bashir, M., Afzal, M. T., & Azeem, M. (2008). Reliability and validity of qualitative and operational research paradigm. *Pakistan Journal of Statistics and Operation Research, 4*(1), 35. https://doi.org/10.18187/pjsor.v4i1.59

Boles, M., Pelletier, B., & Lynch, W. (2004). The relationship between health risks and work productivity. *Journal of Occupational and Environmental Medicine, 46*(7), 737–745. https://doi.org/10.1097/01.jom.0000131830.45744.97

Boyce, P. R. et al. (2006). Lighting quality and office work: two field simulation experiments. *Lighting Research & Technology (London, England: 2001), 38*(3), 191–223. https://doi.org/10.1191/1365782806lrt161oa

Brown, J. et al. (2009). Using human resource management (HRM) practices to improve productivity in the Canadian tourism sector (pp. 1–15).

Browning, B. (2015). Healthier workplaces, happier employees. *People and Strategy, 38*(3).

Candido, C., et al. (2019). Designing activity-based workspaces: Satisfaction, productivity and physical activity. *Building Research and Information, 47*(3), 275–289. https://doi.org/10.1080/09613218.2018.1476372

Cato, S. T., & Gordon, J. (2012). Relationship of the strategic vision alignment to employee productivity and student enrolment. *Research in Higher Education Journal, 15.*

Chandrasekar, K. (2011). Workplace environment and its impact on organisational performance in public sector organisations. *International Journal of Enterprise Computing and Business Systems, 1*(1), 1–19.

Charles, K. E., & Veitch, J. A. (2002). Environmental satisfaction in open-plan environments: 2. Effects of workstation size, partition height and windows. Institute for Research in Construction. National Research Council Canadá. Internal Report.

Clark, A. E. (1997). Job satisfaction and gender: Why are women so happy at work? *Labour Economics, 4*(4), 341–372. https://doi.org/10.1016/s0927-5371(97)00010-9

Clements-Croome, D. J. (2002). Specifying indoor climate data. In *Naturally Ventilated Buildings: Buildings for the Senses, the Economy and Society.*

Creswell, J. W. (2014). *Research design: Qualitative, quantitative and mixed methods approaches* (4th ed.). Sage.

De Been, I., & Beijer, M. (2014). The influence of office type on satisfaction and perceived productivity support. *Journal of Facilities Management.*

Fassoulis, K., & Alexopoulos, N. (2015). The workplace as a factor of job satisfaction and productivity: A case study of administrative personnel at the University of Athens. *Journal of Facilities Management.*

Federspiel, C. C. et al. (2002). Worker performance and ventilation: Analyses of individual data for call-center workers (No. LBNL-50124). Berkeley, CA (United States).

Golafshani, N. (2015). Understanding reliability and validity in qualitative research. *The Qualitative Report.* https://doi.org/10.46743/2160-3715/2003.1870

Gummesson, E. (1998). Productivity, quality and relationship marketing in service operations. *Handbuch dienstleistungsmanagement* (pp. 843–864). Gabler Verlag.

Hameed, A., & Amjad, S. (2009). Impact of office design on employees productivity: A case study of banking organizations of Abbottabad, Pakistan. https://www.semanticscholar.org/paper/d1bc089f2e32dcd1762ab63dc75631186fe5b2ff. Accessed 27 Apr 2023.

Harter, J. K., Schmidt, F. L., & Hayes, T. L. (2002). Business-unit-level relationship between employee satisfaction, employee engagement, and business outcomes: A meta-analysis. *The Journal of Applied Psychology, 87*(2), 268–279. https://doi.org/10.1037/0021-9010.87.2.268

Hasun, F. M. (2005). An overview of workplace environment and selected demographic factors towards individual's health and performance enhancement. *Synergizing OSH for Business Competitive*, 45–53.

Haynes, B. P. (2008a). Impact of workplace connectivity on office productivity. *Journal of Corporate Real Estate, 10*(4), 286–302. https://doi.org/10.1108/14630010810925145

Haynes, B. P. (2008b). The impact of office comfort on productivity. *Journal of Facilities Management, 6*(1), 37–51. https://doi.org/10.1108/14725960810847459

Haynes, B. P., Suckley, L., & Nunnington, N. (2019). Workplace alignment: An evaluation of office worker flexibility and workplace provision.

Hedge, A. (1982). The open-plan office: A systematic investigation of employee reactions to their work environment. *Environment and Behavior, 14*(5), 519–542.

Juslén, H., & Tenner, A. (2005). Mechanisms involved in enhancing human performance by changing the lighting in the industrial workplace. *International Journal of Industrial Ergonomics, 35*(9), 843–855. https://doi.org/10.1016/j.ergon.2005.03.002

Kaarlela-Tuomaala, A., et al. (2009). Effects of acoustic environment on work in private office rooms and open-plan offices—Longitudinal study during relocation. *Ergonomics, 52*(11), 1423–1444. https://doi.org/10.1080/00140130903154579

Kamaruzzaman, S. N., & Sabrani, N. A. (2011). The effect of indoor air quality (IAQ) towards occupants' psychological performance in office buildings. *Journal Design+ Built, 4*(1), 49–61.

Kekäläinen, P., et al. (2010). Effect of reduced summer indoor temperature on symptoms, perceived work environment and productivity in office work: An intervention study. *Intelligent Buildings International, 2*(4), 251–266.

Kim, S.-Y., & Kim, J.-J. (2007). Influence of light fluctuation on occupant visual perception. *Building and Environment, 42*(8), 2888–2899. https://doi.org/10.1016/j.buildenv.2006.10.033

Kim, J., & de Dear, R. (2013). Workspace satisfaction: The privacy-communication trade-off in open-plan offices. *Journal of Environmental Psychology, 36*, 18–26. https://doi.org/10.1016/j.jenvp.2013.06.007

Kwon, M., Remøy, H., & van den Bogaard, M. (2019). Influential design factors on occupant satisfaction with indoor environment in workplaces. *Building and Environment, 157*, 356–365. https://doi.org/10.1016/j.buildenv.2019.05.002

Leaman, A. (1995). Dissatisfaction and office productivity. *Facilities, 13*(2), 13–19. https://doi.org/10.1108/02632779510078120

Leaman, A., & Bordass, B. (2001). Assessing building performance in use 4: The Probe occupant surveys and their implications. *Building Research and Information, 29*(2), 129–143. https://doi.org/10.1080/09613210010008045

Lee, Y. S. (2007). The relationship between indoor environmental quality and worker satisfaction and performance in leadership in energy and environmental design (LEED®) certified buildings.

Lee, S. Y., & Brand, J. L. (2010). Can personal control over the physical environment ease distractions in office workplaces? *Ergonomics, 53*(3), 324–335. https://doi.org/10.1080/00140130903389019

Lenoir, A., Baird, G., & Garde, F. (2012). Post-occupancy evaluation and experimental feedback of a net zero-energy building in a tropical climate. *Architectural Science Review, 55*(3), 156–168. https://doi.org/10.1080/00038628.2012.702449.Lister. T.R., & DeMarco, T. (1987). *Peopleware: Productive projects and teams*. Dorset House.

Mak, C. M., & Lui, Y. P. (2012). The effect of sound on office productivity. *Building Services Engineering Research & Technology: BSER & T, 33*(3), 339–345. https://doi.org/10.1177/0143624411412253

Mark, G., Gudith, D., & Klocke, U. (2008). The cost of interrupted work: more speed and stress. In *Proceedings of the SIGCHI conference on human factors in computing systems* (pp. 107–110).

Mcleod, S. (2019). *Qualitative vs Quantitative Research: Methods & Data Analysis*.

Microsoft. (n.d.). Introduction to microsoft forms. https://support.microsoft.com/en-us/office/introduction-to-microsoft-forms-bb1dd261-260f-49aa-9af0-d3dddcea6d69. Accessed 11 Sep 2021.

Moss, K. J. (2007). *Heat and mass transfer in buildings* (1st ed.). Routledge.

Naoum, S. G. (2019). *Dissertation research and writing for built environment students* (4th ed.). Fourth edition Routledge.

Newsham, G. (2005). Task lighting effects on office worker satisfaction and performance, and energy efficiency. *LEUKOS the Journal of the Illuminating Engineering Society of North America, 01*(04), 7–26. https://doi.org/10.1582/leukos.2004.01.04.001

Niemela, R., et al. (2002). The effect of air temperaturę on labour productivity in cali centres-a case study. *Energy and Buildings, 34*, 759–764.

Noble, A. (2009). Building health promotional work setting identifying the relationship work characteristics and occupational stress. *Promotional International Journal, 18*(4), 351–359.

Oseland, N., & Bartlett, P. (1999). *Improving Office Productivity: A guide for business and facilities managers*. Longman.

Paevere, P., & Brown, S. (2008) *Indoor environment quality and occupant productivity in the CH2 building: Post-occupancy summary*.

Palacios, J., Steele, K., & Zheng, S. (2021). Human health and productivity outcomes associated with indoor air quality: A systematic review. *SSRN Electronic Journal*. https://doi.org/10.2139/ssrn.3881998

Petterson, I. L., & Arnetz, B. B. (1998). Psychosocial stressors and well-being in health care workers. The impact of an intervention program. *Social Science & Medicine (1982), 47*(11), 1763–1772. https://doi.org/10.1016/s0277-9536(98)00245-7

Pilcher, J. J., Nadler, E., & Busch, C. (2002). Effects of hot and cold temperature exposure on performance: A meta-analytic review. *Ergonomics, 45*(10), 682–698. https://doi.org/10.1080/00140130210158419

Piana, V. (2011). Productivity. http://www.economicswebinstitute.org/glossary/prdctvt.htm. Accessed 21 Sep 2021.

Prakash, P. (2005). *Effect of indoor environmental quality on occupant's perception of performance: A comparative study*.

Rasila, H., & Jylhä, T. (2015). The many faces of office noise—Case contact center. *Facilities, 33*(7/8), 454–464. https://doi.org/10.1108/f-11-2013-0085

Rattray, J., & Jones, M. C. (2007). Essential elements of questionnaire design and development. *Journal of Clinical Nursing, 16*(2), 234–243. https://doi.org/10.1111/j.1365-2702.2006.01573.x

Roulet, C.-A., et al. (2006). Perceived health and comfort in relation to energy use and building characteristics. *Building Research and Information, 34*(5), 467–474. https://doi.org/10.1080/09613210600822279

Sakuraya, A., et al. (2020). What kind of intervention is effective for improving subjective well-being among workers? A systematic review and meta-analysis of randomized controlled trials. *Frontiers in Psychology, 11*, 528656. https://doi.org/10.3389/fpsyg.2020.528656

Samani, S. A. (2015). The impact of personal control over office workspace on environmental satisfaction and performance. *Journal of Social Sciences and Humanities, 1*(3), 163–175.

Sedgwick, P. (2013). Convenience sampling. *BMJ (Clinical research ed.), 347*(2), f6304–f6304. https://doi.org/10.1136/bmj.f6304

Sels, L., et al. (2006). Linking HRM and small business performance: An examination of the impact of HRM intensity on the productivity and financial performance of small businesses. *Small Business Economics, 26*(1), 83–101. https://doi.org/10.1007/s11187-004-6488-6

Sharma, M. S., & Sharma, M. V. (2014). Employee engagement to enhance productivity in current scenario. *International Journal of Commerce, Business and Management, 3*(4), 595–604.

Shek, K. W., & Chan, W. T. (2008). Combined comfort model of thermal comfort and air quality on buses in Hong Kong. *The Science of the Total Environment, 389*(2–3), 277–282. https://doi.org/10.1016/j.scitotenv.2007.08.063

Shikdar, A. A., & Sawaqed, N. M. (2003). Worker productivity, and occupational health and safety issues in selected industries. *Computers & Industrial Engineering, 45*(4), 563–572. https://doi.org/10.1016/s0360-8352(03)00074-3

Singh, J. (1996). Review : Health, comfort and productivity in the indoor environment. *Indoor + Built Environment: The Journal of the International Society of the Built Environment, 5*(1), 22–33. https://doi.org/10.1177/1420326x9600500105

Skalli, A., Theodossiou, I., & Vasileiou, E. (2008). Jobs as Lancaster goods: Facets of job satisfaction and overall job satisfaction. *The Journal of Socioeconomics, 37*(5), 1906–1920. https://doi.org/10.1016/j.socec.2008.04.003

Sousa-Poza, A., & Sousa-Poza, A. A. (2000). Taking another look at the gender/job-satisfaction paradox. *Kyklos: International Review for Social Sciences, 53*(2), 135–152. https://doi.org/10.1111/1467-6435.00114

Spector, P. E. (1997). *Job Satisfaction: Application, assessment, causes, and consequences*. SAGE Publications.

Sundstrom, E., Herbert, R. K., & Brown, D. W. (1982). Privacy and communication in an open-plan office: A case study. *Environment and Behavior, 14*(3), 379–392. https://doi.org/10.1177/0013916582143007

Sundstrom, E., et al. (1994). Office noise, satisfaction, and performance. *Environment and Behavior, 26*(2), 195–222. https://doi.org/10.1177/001391659402600204

Sutrisno, A., Yulianti, D., & Setiawan, M. (2016). Measuring office employee satisfaction: A study on the factors that determine satisfaction. *Procedia-Social and Behavioral Sciences, 224*, 193–200.

Tanabe, S.-I., Haneda, M., & Nishihara, N. (2015). Workplace productivity and individual thermal satisfaction. *Building and Environment, 91*, 42–50. https://doi.org/10.1016/j.buildenv.2015.02.032

Toftum, J., Jørgensen, A. S., & Fanger, P. O. (1998). Upper limits of air humidity for preventing warm respiratory discomfort. *Energy and Buildings, 28*(1), 15–23. https://doi.org/10.1016/s0378-7788(97)00018-2

Toftum, J. (2002). Human response to combined indoor environment exposures. *Energy and Buildings, 34*(6), 601–606. https://doi.org/10.1016/s0378-7788(02)00010-5

Tse, W. L., & So, A. (2007). The importance of human productivity to air-conditioning control in office environments. *HVAC&R Research, 13*(1), 3–21. https://doi.org/10.1080/10789669.2007.10390941

Woodworth, D.G. (1964). *Job-satisfaction and personality: A study of research scientists*. University of California, Berkeley. https://www.proquest.com/openview/c8b0b082a965567fb7c9481ef37fd433/1?pq-origsite=gscholar&cbl=18750&diss=y. Accessed 1 Sep 2021

Impact of Streetscapes on Anxiety: A Physiological Evidence

Sayna Anand and Tina Pujara

Abstract

Urbanization has shown a significant impact on the mental health of people in current times. Growing research evidence shows that the street as an urban space may have an impact on people's mental health, either positively or negatively. Anxiety is a common mental health issue affecting people psychologically and physiologically. Few studies have highlighted the psychological impact of specific streetscape elements, but there is a significant research gap in exploring the associated physiological impact of such urban settings on people. This exploratory study involves nine mentally and physically healthy participants walking at stable speed for 5–10 min on the four selected commercial neighborhood streets in Delhi, India, with baseline measurements done before the walk to evaluate the psychological and physiological effects of the streetscapes (identified for study) on anxiety.

During the walk, heart rate and heart rate variability were continuously measured to assess the physiological impact of anxiety, and STAIS-5, I-PANAS-SF, and anxiety-based likert scale questionnaires were used to measure the psychological responses. Each walk ended with a post-walk interview of the participants. The results were compared among the streets. The analysis indicates relatively higher positive psychological responses, heart rate variability, and lower heart rate on street with wider sidewalk, comparatively better green quality, façade permeability, and traffic buffer. However, the heart rate showed a minimal change across the streets. Negative responses to the psychological tests and lower heart rate variability were observed for two streets with low green quality and less walking space. The study analyzes further to explain the experiential quality of the streets with the significant elements impacting anxiety levels. The conclusion highlights this work as scientific evidence of which streetscape elements and qualities may significantly impact anxiety.

Keywords

Mental health · Anxiety · Physiological · Psychological · Streetscapes · Heart rate variability

1 Introduction

Urbanization has exacerbated mental health problems. In comparison to rural areas, cities have greater rates of mental health issues, including a nearly 40% higher risk of depression, a 20% higher risk of anxiety, double the risk of schizophrenia, as well as higher levels of loneliness, isolation, and stress (McCay et al., 2017). Anxiety disorders affected an estimated 284 million people worldwide in 2017, making it the most common mental health or neurodevelopmental disorder (Dattani et al., 2021). Urban environments can either trigger or restore from such mental health issues. The recently emerging field of design focuses on designing cities for better mental and physical health highlighting the need to look into urban design and planning features responsible for bringing the change e.g., studies such as Navarrete-Hernandez and Laffan (2019), Knöll et al. (2018) and Hidalgo (2019) which have studied specific urban environment aspects impacting mental health factors. Urban factors such as residential density, land use, proximity to facilities and amenities, street configuration and design, housing form, greenery/green spaces, daylight, noise, and

S. Anand · T. Pujara (✉)
Research Group 'Happiness By Design', Department of Architecture and Planning, Indian Institute of Technology (IIT) Roorkee, Roorkee, India
e-mail: tina.pujara@ar.iitr.ac.in

S. Anand
e-mail: sanand@ar.iitr.ac.in

I. Pigliautile et al. (eds.), *Urban Resilience, Livability, and Climate Adaptation*, Advances in Science, Technology & Innovation, https://doi.org/10.1007/978-3-031-54911-3_7

climate are all associated with mental well-being, impacting anxiety, stress, and other mental health issues (Barton et al., 2015). In cities, streets make up more than 80% of all public places. They can foster economic growth, operate as front yards for homeowners, and provide a secure environment for those traveling by foot, bicycle, car, or public transportation. (Ink, 2021). Streets can be classified as a channel for movement, Social Artifact (gathering, shopping), Container (a habitat, ecosystem, amenities, and other features), and Health artifact (physical and mental health) (Hidalgo, 2019). The mental health aspect of the streets may have a varied impact in the form of stress, anxiety, depression, or restorative characteristics. A study by Dillon (2005) highlights that the large number of potential stressors for anxiety in today's urban context can result in clinical symptoms that occur during the everyday occupation of the city and its spaces in response to threats of violence, crime, and terrorism among a very wide range of individuals. Streets form a significant part of the daily life of the people and thus can have a prominent impact on the experience and further on the mental health of the people.

Street and mental health research has gained attention in recent studies. There are few studies that have addressed singular streetscape elements such as lighting (Fotios et al., 2018), greenery (de Vries et al., 2013; Navarrete-Hernandez & Laffan, 2019), built height (Asgarzadeh et al., 2012), while even fewer have explored the multiple aspects of streetscape design in their studies (Barros et al., 2021; Park & Garcia, 2019). These studies have covered diverse aspects of mental health, such as oppressiveness, happiness, stress, safety, restorative properties, and anxiousness. Few studies have addressed anxiety as a mental health issue, and when it has found mention, it is mostly with reference to the individual streetscape elements rather than a comprehensive overview of the streetscape, showing a significant gap in studies of the association of streetscape and anxiety.

Anxiety is one of the most common mental health issues among people around the world. According to statistics, 56 million Indians suffer from depression, with another 38 million suffering from anxiety disorders (Bhatia, 2020). Anxiety, as per Spielberger, can be trait or state anxiety. Trait anxiety is an individual's proclivity to respond, whereas state anxiety is a transitory emotion characterized by physiological arousal and consciously perceived feelings of apprehension, dread, and tension. The difference between trait and state anxiety is analogous to the difference between potential and kinetic energy (Endler & Kocovski, 2001). The relationship between streetscapes and anxiety is under researched, while few studies specifying the individual streetscape elements are roadside trees (Elsadek et al., 2019), street view (visual enclosure), walkability (Wang et al., 2019), and facades (Hollander & Anderson, 2020) and all these studies focused on other psychological factors apart from anxiety also as part of the research. A study by Hollander and Anderson (2020) on street facades and their association with agoraphobia, trait anxiety, pleasantness, and likelihood of visiting concludes that trait anxiety had no significant effect on the level of arousal due to facades while state anxiety's impact on the users has not been researched in past literature. This paper focuses on the state anxiety of the users which can affect their experience on streets of everyday use both psychologically and physiologically. When people are stressed or anxious, their autonomic nervous systems get active. As a result, they can be stimulated psychologically or physiologically (e.g., rapid heart rate, shortness of breath, anxiety, and irritability). Heart rate is particularly responsive to stress or feelings of fear/anxiety among the physical symptoms linked with Autonomic Nervous System (ANS) activation (Lee et al., 2015). Contrary to heart rate, heart rate variability (HRV) is now understood to be a significant and reliable indicator of the autonomic nervous system's health and is also a measure of resilience to stress (Reiner, 2008 cited in Lee et al., 2015). Anxiety is a response to stress (Robinson, 1990). Such physiological responses are tested along with psychological factors to understand the impact of the urban environment on the people.

1.1 Physiological and Psychological Impact of Anxiety

Physiological evidence in urban studies is very limited and mostly highlights the benefits of green on users' mental health, while research based on varied urban environments and its association with mental health is scarce. Considering the user experience, a study highlighted that the evocative power of a space influences people's emotions associated with that location, which is not just based on visual experiences but primarily due to the combination of sensory experiences (Boumezoued et al., 2020). A pedestrian's strong sense of place can plausibly increase the impact of micro-scale streetscape elements on them. The impact can be, as mentioned, physiological and psychological both. Recent studies show that biosensing is becoming increasingly prominent for testing the environmental experience, and they complement the psychometric data (Chen et al., 2018). Anxiety, particularly, has been found to be associated with heart rate and heart rate variability (HRV), as mentioned previously. HRV and heart rate are one of the known cardiovascular measures of the Autonomic Nervous System (ANS) (refer to Fig. 1).

HRV is the method widely used in academic literature for reflecting ANS. The ANS is divided into two parts: the sympathetic system (SNS), which excites, and

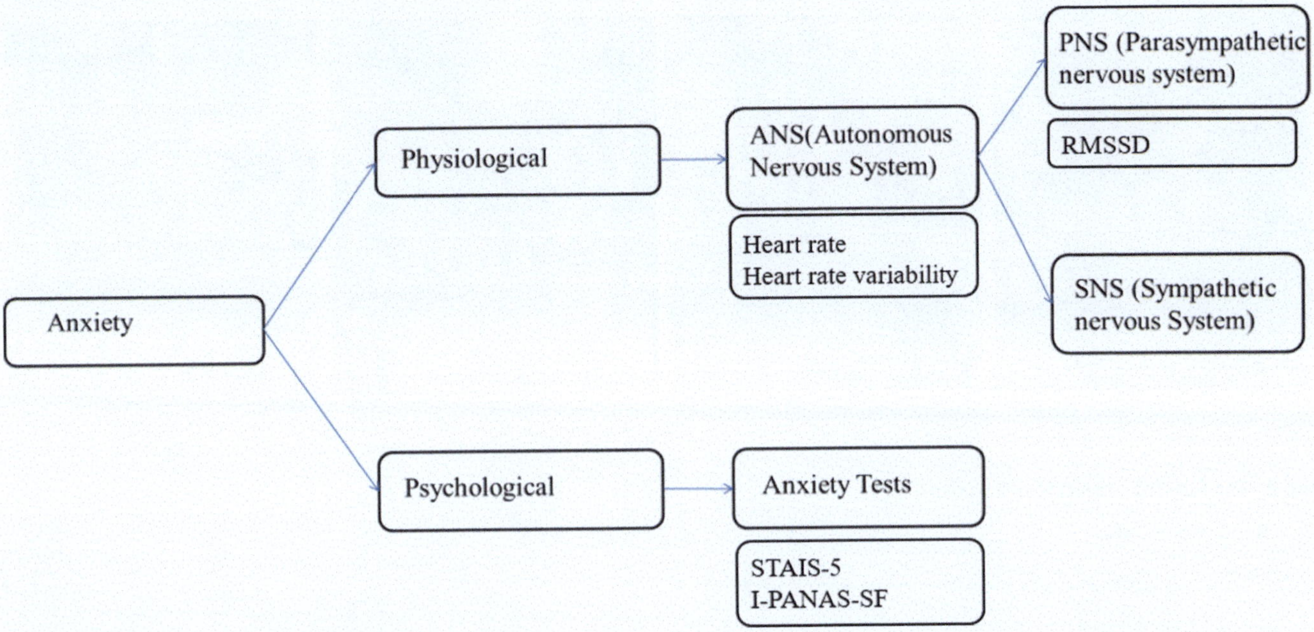

Fig. 1 Physiological and psychological parameters used for the study. *Source* Author

the parasympathetic system (PNS), which inhibits (Paül i Agustí et al., 2019). The electrodermal (sweat gland) and cardiovascular (blood circulatory system) responses are the most commonly used indices of ANS activation. The autonomic function is assessed using noninvasive heart rate variability techniques; in particular, time domain analysis assesses parasympathetic activity using root mean square of successive differences (RMSSD) (Brown et al., 2013). RMSSD (Root Mean Square of Successive Differences), or the square root of the average values of the sum of the differences to the square of all consecutive RR intervals in a recent study on perception of urban spaces was found higher at places with more positive reports at the historic center along the route in a study as compared to the other urban spaces (Paül i Agustí et al., 2019).

The physiological data collection method is a relatively newer method used to understand the impact of urban environments on people's emotions and mental health. It does have some advantages as well as limitations associated. Collecting physiological data may also aid in understanding the differences between perceived experience and bodily reactions to an environment (Dritsa & Biloria, 2021). As noted, most studies engaging physiological data collection method about the numerous obstacles exist to using human health sensors in non-laboratory settings to monitor real-time stress responses. Due to the volume of data, they collect and the time it takes to manage and interpret it, sensors that measure both physical and emotional stress may be challenging to employ in health research. Additionally, it could be challenging for participants to follow a protocol that calls for them to wear a noticeable gadget, guarantee its performance, and perform tasks that take more time and effort than typical daily chores. This may clarify why studies that use these sensors to record stress responses generally have tiny sample sizes (Kondo et al., 2018). This study posed similar challenges and had a limited number of participants involved in the study for data collection aiming to understand the association of anxiety with the streetscape while analyzing the physiological reactions of the participants.

2 Methodology

2.1 Street Selection

The neighborhood commercial streets were selected in the South Delhi area in India; Delhi is a metropolitan city that has been found to have high mental health issues, highlighted in a recent study by Agarwal (2019). More than 10 South Delhi commercial neighborhood streets were visited, after which the four were shortlisted based on their design qualities. The selected stretches are of length 500–900 m (refer to Figs. 2 and 3a, b). The reasons for including the neighborhood commercial street as the context of study are, firstly, it is easily approachable by the neighborhood population and is an essential part of the daily activities of the people. Secondly, there is an untapped potential of these streets for psychological restoration and their role in alleviating anxiety on an everyday basis. Figures 4a, b and 5a, b show the characteristics of the four streets with varying street qualities. The four streets selected were Green

Green Park Market Street, Green Park | Veer Savarkar Marg (Street), Lajpat Nagar | Guru Gobind Singh Marg (Street), Malviya Nagar | Fifth Avenue Road, Lodhi Colony

Fig. 2 The selected streets (*Source* Author)

Green Park Market Road, Green Park (GP) *Fifth Avenue Road, Lodhi colony (FA)*

a

Veer Savarkar Marg, Lajpat Nagar (VS) *Guru Gobind Singh Marg, Malviya Nagar (MN)*

b

Fig. 3 **a** Location plan also showing distance walked on Green Park Market Road, Green Park (GP) and Fifth Avenue Road (FA). Street plan source: Open Street Maps, https://www.openstreetmap.org/. **b** Location plan also showing distance walked on Veer Savarkar Marg, Lajpat Nagar (VS) and Guru Gobind Singh Marg, Malviya Nagar (MN). Street plan source: Open Street Maps, https://www.openstreetmap.org/

Park main market street (Green Park) (refer to Fig. 4a), Fifth Avenue Road (Lodhi colony) (refer to Fig. 4b), Veer Savarkar Marg (Lajpat Nagar) (refer to Fig. 5a), and Guru Gobind Singh Marg (Malviya Nagar) (refer to Fig. 5b).

The streetscape elements were shortlisted based on an extensive literature review which highlighted the various elements plausibly impacting mental health, specifically anxiety, and a few added based on the informal discussions with the users done before the study was conducted. They were rated by the researcher based on the criteria mentioned after multiple visits to the site. Lighting as a streetscape element was not considered while selecting the street as it was beyond the scope

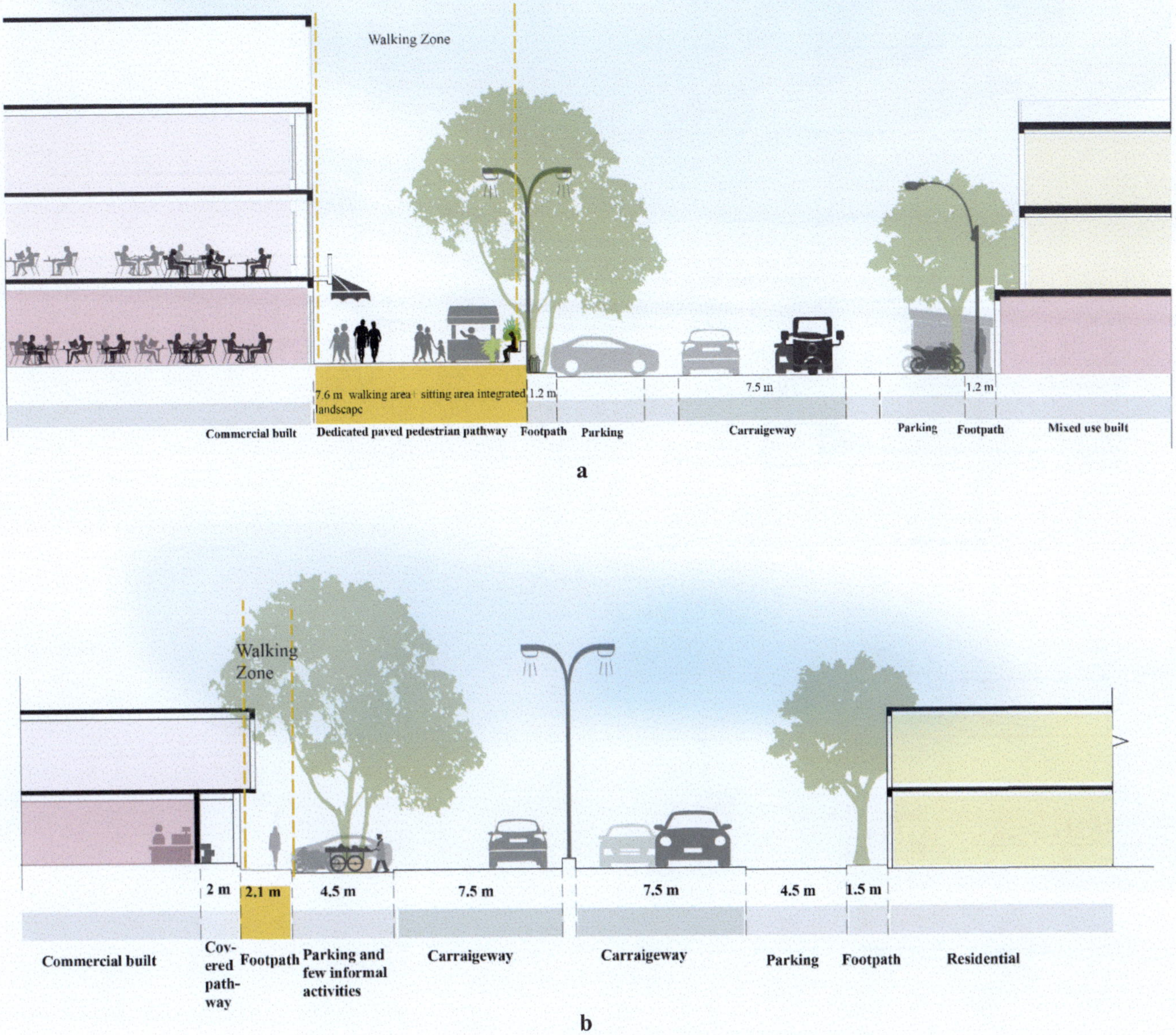

Fig. 4 **a** Green Park Market Road (GP) typical street section of the selected walking route (measured in meters). *Source* Author. **b** Fifth Avenue Road (FA) typical street section of the selected walking route (measured in meters). *Source* Author

of the study. The measurements of the streetscape feature and scoring process are shown in Table 1. The streets were qualitatively analyzed by the researcher based on the methodology used by Zhao et al. (2019), highlighting the streetscape elements variation through likert scale rating, and some elements were directly measured, such as sidewalk width.

2.1.1 Qualitative Characteristics of the Streets

Since the selected streets were commercial neighborhood streets, the built use and variety of businesses are of a similar kind, such as daily needs shops, medical stores, and restaurants, etc., and are serving the neighborhoods around. The streetscape elements varied across the streets, such as varied sidewalk width, greenery (quality and quantity), traffic buffer, and other differentiating elements, as mentioned in Table 1.

Green Park Market Road, Green Park (GP) is an upscale market with a dedicated walking path which is buffered from parking and traffic by landscaped barrier.

The market street has many upscale restaurants and local shops which cater to the neighborhood. The height of the building is, on average, 3–4 floors. Informal vendors also

Fig. 5 **a** Veer Savarkar Marg (VS) typical street section of the selected walking route (measured in meters). *Source* Author. **b** Guru Gobind Singh Marg (MN) typical street section of the selected walking route (measured in meters). *Source* Author

Table 1 Measurement of streetscape features by the researcher

Streetscape feature	Scores			
	1	2	3	4
Greenery quantity	No greenery	Low greenery /scanty vegetation	Moderate quantity	Significant amount of greenery
Greenery quality	No greenery	Some trees and limited vegetation	Moderate amount of trees with bushes/ hedges	Mixed vegetation types and variety of trees along with other landscape features
Sidewalk width	Average sidewalk width measured in meters			
Traffic buffer type	Not present	Man-made fence/separation	Man-made separation with vegetation	Vegetated separation with significant segregation (visual and physical)
Facade quality - Visual permeability	Blank wall /very low visual permeability at each business front	Some visual permeability with windows and doors present	Moderate visual permeability with larger doors and windows	High visual permeability
Facade quality -articulation	Very low articulation	Some articulation	Moderate articulation	High articulation
Facade quality-color	No/very low color variation	Some variation of building and accent colors	Moderate variation of building and accent colors	High variation and number of accent colors
Seating area/Furniture	Rated the furniture and seating area based on quantity and quality			
Built height	Average height measured			
Enclosure ratio of the street	Average enclosure ratio measured (street width vs. built height)			
Pedestrian density	None /very few	Some	High	Extremely high
Informal activities	None	A few	Some	Many
	Number of informal activities (vendors etc.) were counted at each and then rated as per the quantity			

occupied the walking space but added some vibrancy to the space (refer to Fig. 4a).

Fifth Avenue Road, Lodhi Colony (FA), is a local neighborhood market with upscale restaurants and daily shops. This street has an old Meherchand Market also, which was not part of the study area as it added to the historical context. There were many shops closed as well, which affected the facade quality and walking experience of the street. The height of the buildings was one or two floors only, and this street has one interesting element of wall art on the other side of the street, which added a little to the experience but was not very significant. The pedestrian density here was the lowest as compared to the others (refer to Fig. 4b).

Veer Savarkar Marg, Lajpat Nagar (VS) is a heavy commercial neighborhood market; it adjoins a city-level famous market which increases the pedestrian density on this road. The greenery was least on this road, and walking width was also minimum, with no buffer from parking and traffic. Traffic also remains high on this road. Some of the shops opened directly on footpaths while others had divided into the basement and ground floor to be accessed via stairs (refer to Fig. 5a).

Guru Gobind Singh Marg (Geetanjali Malviya Nagar Road), Malviya Nagar (MN), is a neighborhood market street with a larger green buffer along the sidewalk as compared to the other streets. A ledge around the green area was being used for sitting. The sidewalk was wide for half of the route and then reduced from 6 m to 3.6 m walking further. The green area had a strong landscaped buffer to the parking and traffic, which added to the positive experience. The area with shorter sidewalk width had some walking area occupied by vehicles which reduced the usage of the ledge and sitting area also while adding to the negative experience (refer to Fig. 5b).

Considering the varied quality of the streetscapes expressed in the figures, this study is able to analyze the impact of the streetscape elements while the context in terms of land use and built use remains the same. Post the existing streetscape analysis and selection, the physiological and psychological study for testing anxiety was planned.

2.2 Physiological and Psychological Tests

HRV was measured using a Polar H10 heart rate device (chest strap) linked with the Elite HRV app on an android phone. Elite HRV is found to be a reliable smartphone application for measuring HRV (Moya-Ramon et al., 2022). The Heart rate and HRV collection method, as suggested in a study by Song et al. (2015), is to be collected at each minute and average to be compared between the sites which were followed in this study also. The psychological test in the study is used, as previously discussed, to correlate the results of physiological measurements. When compared to criterion measures such as Holter ECG and Bittium FarosTM, the Polar H10 is proven as the most accurate wearable device and reports less R-R interval error detection (Hinde et al., 2021). The heart rate was measured by the Polar H10 using an electrocardiogram, or ECG. It is the most accurate option because it reads the electrical pulses your heart produces (Buscemi, 2021). For HRV parameters with a strong parasympathetic contribution (e.g., SDNN, LnRMSSD, LnHF, and SD1/SD2), the Polar H10 chest strap has been found to provide particularly valid and reliable measurements for children as compared to PPG (photoplethysmography) system for HRV measurement in a study by Speer et al. (2020).

Along with the physiological indices, psychological tests were conducted, which involved the State Anxiety test that is STAIS-5(Short version of STAI). The Spielberger State-Trait Anxiety Inventory (STAI) is the most frequently used measure of state and trait non-disorder-specific anxiety. The questionnaire is valid for usage for both healthy and clinical populations (Zsido et al., 2020). The STAI test has both state and trait questions, with 20 questions for each. This study considers the state anxiety of the users on the streets, and for practical reasons, a short version of the test is required to be tested for which STAIS-5 fulfills the requirement of a shorter survey questionnaire. Excellent internal consistency and reliability were shown for the STAI scales' five-item short form (Zsido et al., 2020). Other psychometric tests included a shorter version of PANAS (Positive and Negative Affect Schedule) that is I-PANAS-SF (International PANAS Short Form), and one item likert scale question for anxiety. This international version in the short form of PANAS (Positive and Negative Affect Schedule) was validated in a study by Thompson (2007) for cross-cultural use. The PANAS was created with the goal of providing a valid measure of two broad and mostly independent elements involved in emotional experience: positive and negative affect (PA and NA). It is one of the tools (STAI, POMS, PANAS, and VAS) which can be used for testing state-dependent anxiety (Rossi & Pourtois, 2012). One item likert scale question for anxiety for asked to the participants along with other two anxiety tests for at the moment anxiety measures of the participants. It was found in a study that participants can easily complete a one-item anxiety question with a five-point likert response scale, which accurately measures anxiety (Davey et al., 2007).

The study was conducted using quantitative and qualitative techniques. The method involved both psychological and physiological tests. A total of 9 participants were included in the experiment. The participants were either slightly or not at all familiar with the streets. All the participants were asked to wear the HR (heart rate) and HRV (heart rate variability) devices and were supposed to walk in the selected four urban streets for 6–12 min. The physiological measurement used for various research based on environmental psychology and urban design has a limited number of participants in the study. A within-subject study was performed. The subjects were directly taken to the site and performed the walk while the heart rate was continuously monitored and recorded for 6–12 min. After the walk, the participants answered the questionnaires, and an unstructured interview based on their experience was conducted. The walking speed was kept consistent (being recorded) to avoid walking as a confounding factor in increasing heart rate, but a sudden increase or decrease in speed was recorded. The study focused on the impact of the street environment in healthy adult samples with controlled confounding effects of social interaction and physical activity. The experiment enables us to understand the immediate psychological impacts even after leaving the environment. The study considers the fact that person-to-person heart rate may vary based on physical activity; this situation has been dealt with in previous studies with prior laboratory calibration (Paül i Agustí et al., 2019). Considering the feasibility of the experiment, the short-term baseline heart rate and heart rate variability were measured for each participant (ST,~5 min). The participants involved were instructed not to smoke or consume alcohol during the experiment.

2.3 Steps Followed for the Study

The participants included in the study were between the age of 23–30 years old healthy adults with four females and five male participants. The process of the experiment involved participants reaching the first street and wearing the Polar H10 chest strap (refer to Fig. 6). While in resting state the participants were asked their current mental state by using WHO-5 and STAIS-5 for testing resting state anxiety levels. After the responses to these questionnaires, the baseline heart rate measurement was taken for 5 min (Short term heart rate variability). The participants were then taken to the starting point of the street with the bluetooth connected app for Elite HRV and Polar Beat, recording the heart rate and R-R intervals of the walk (refer to Fig. 6). The Polar Beat app also recorded the geo-location of the walk and the speed of the participant while walking. The researcher walked a few meters behind the participant to observe the movement and to ensure minimal impact of external factors. After each walk, participants were asked to fill a survey form with STAIS-5, likert scale anxiety test, and I-PANAS-SF, including other experience-related questions in the survey schedule, followed by an unstructured interview to decipher their responses which reduced the possibility of any researcher biases associated with other types of interviews (structured and semi-structured). Considering the feasibility of the process in terms of time and availability of the participants, all the streets were covered on the same day, one after another. The study was conducted for a week with limited participants on each day. The time of the walk was kept consistent for all the participants past 4:30

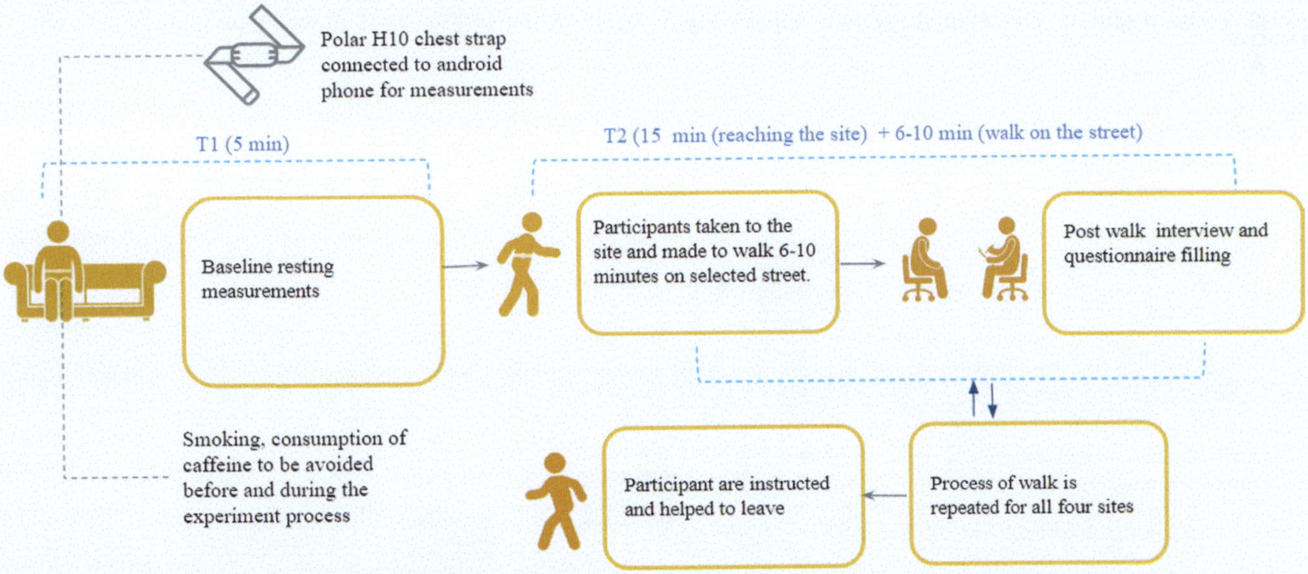

Fig. 6 Process followed for the study. *Source* Author

PM to minimize the impact of the temperature, which was quite high in the morning, considering the Indian context. The days of the study were all having similar weather conditions. The participants were provided with water bottles, and refreshments post the walk.

3 Statistical Results

Given the practicality of on-site and real-time surveying, the streets surveyed were restricted to four, and the analysis included a comparison, associational, and descriptive approach as part of the quantitative technique. The quantitative analysis was performed using SPSS (Statistical Package for the Social Sciences); the mean values for the tests conducted on each street were compared to understand the impact on the physiology and psychology of the people in four varied street environments. The association between streetscape elements and anxiety (physiological and psychological aspects) was analyzed using the Spearman correlation. Regarding the investigated streetscape variables, there may be some interrelation between them; however, the method selected for the study made sure that it did not impact the results of the study. The advantage of researching in a natural setting (not laboratory-based) is that the subject experiences a real-time environment resulting in a more reliable response. Spearman correlation was used to illustrate a relationship between the physiological and psychological aspects of anxiety with streetscape elements, as the data was non-parametric and quality based. To support the study objective, a mixed method approach was used, and qualitative methods such as interviews were also used along with quantitative techniques. Interviews have been proven to be useful in exploring the views, experiences, beliefs, and/or motivations of individuals to gain a deeper understanding of social phenomena, as explained by Gill et al. (2008). Post-walk interviews were analyzed based on the content recorded and the transcript of the conversation with each participant.

There were 36 responses received from 9 participants, each walking on all of the four selected streets. Before the walk, baseline measurements for all participants revealed that they were not anxious before walking, and the mean percentage score of WHO-5 for females was 80, while for males was 76.8, indicating a good well-being score for the participants. The data points in terms of R-R intervals were obtained from the Kubios HRV application, which was found compatible with Elite HRV applications' results. The results represent the parasympathetic nervous system (PNS) as well as the sympathetic nervous system (SNS) results. The RMSSD part of the PNS results was used to explain the results obtained; the stress index from SNS results was also noted. But for correlation, RMSSD and HR were used along with psychological parameters. When it comes to developing links with subjective sentiments, RMSSD was recognized as the most appropriate measurement (Levenson et al., 2017, cited in Paül i Agustí et al. (2019). The average RMSSD scores for each street for all the participants were obtained, and the results were compared on all four streets among the 9 participants. Higher RMSSD (in ms) was found on Green Park Market Road (GP) and lower on both Fifth Avenue Road, Lodhi colony (FA), and Veer Savarkar Marg, Lajpat Nagar (VS). A slight increase in mean RMSSD can be observed in Guru Gobind Singh Marg, Malviya Nagar (MN). The mean values of RMSSD for FA and VS are comparable with 6.9 and 6.4, respectively, indicating a minimal difference between these two streets (refer to Fig. 7). A min-to-min analysis of RMSSD

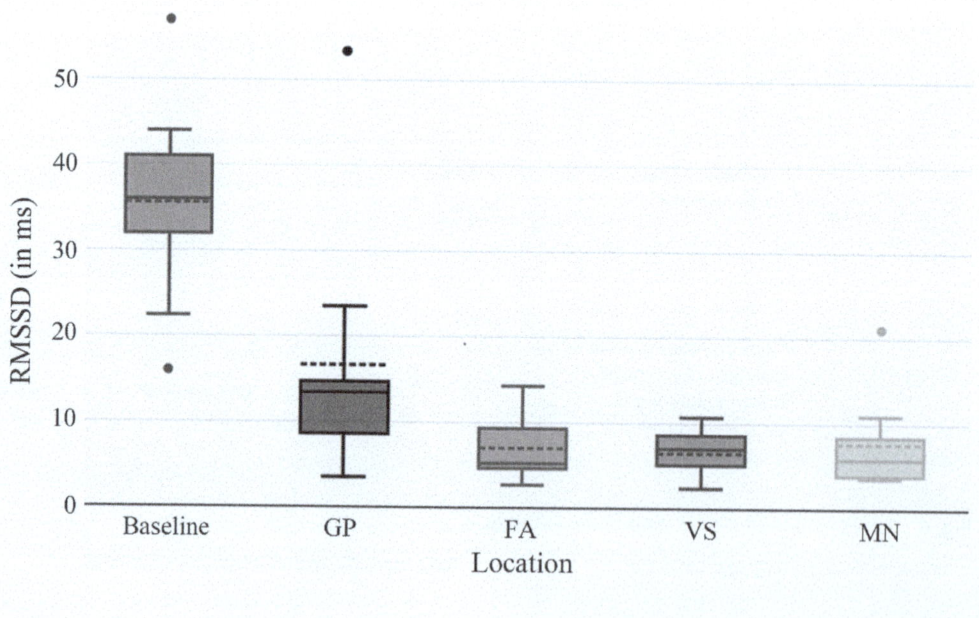

Fig. 7 Box plot showing distribution of RMSSD values as per the location for the nine participants. *Source* Author

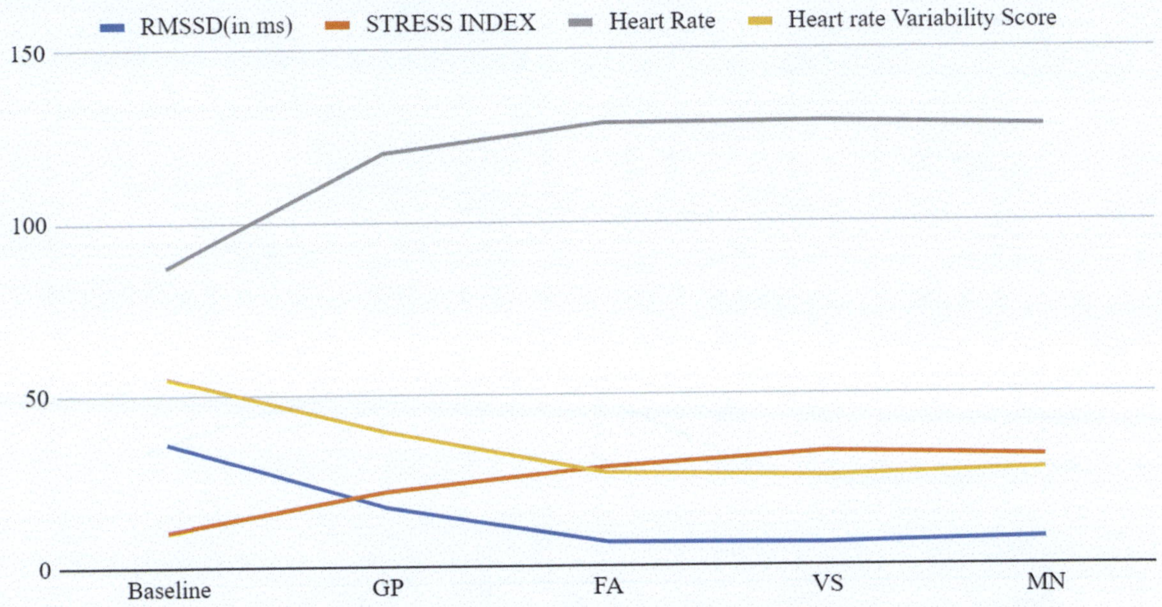

Fig. 8 Showing the mean Heart Rate and Heart Rate Variability Measurements corresponding to the location. *Source* Author

was done, which showed minimal significant change each minute along each route, so the average RMSSD for each participant was noted on each street. Lower RMSSD indicates possible anxiousness among the participants.

The average RMSSD (PNS), STRESS index (SNS), HRV Score, and HR were compared among the streets. It was observed that HRV score was lower in both FA and VS, and a slight increase was seen on street MN. Street GP has been found to have the lowest STRESS index, HR, and highest RMSSD as compared to others (refer to Fig. 8), indicating a positive impact on the participant. The physiological parameters at a certain level correspond to the psychological results obtained. The STAIS-5 test results show higher anxiety levels on FA and VS; the mean score of both the street is equal to 9.1 and 9.0, respectively, which is near the cutoff score of 9.5 indicating high state anxiety while considering anxiety results for females specifically, it was 11 which is higher than the cutoff score indicating that they could be potentially clinically anxious. STAIS-5 results show the least on the street GP, which is equal to 5.1 (mean), followed by street MN with 6.3 (mean), which indicates lower anxiety levels in these locations and no clinical anxiety. I-PANAS-SF also shows the least negative (NA = 5(Avg)) and most positive responses (PA = 21.66 (Avg)) on location GP while I-PANAS-SF (PA) does not show a significant change on the other three streets (FA, VS, MN) (refer to Fig. 9). RMSSD and heart rate as PNS indices are used for correlation analyses which are considered valid parameters to measure the impact of HRV on anxiety.

The rated streetscape elements are correlated with the physiological and psychological results of all the participants. Table 2 shows Spearman's correlation matrix highlighting the significant factors affecting the anxiety levels of the participants. RMSSD was found to be significantly correlated to sidewalk width, facade quality, green quality, and sitting quality of the streets ($p < 0.05$ level). The State anxiety test (STAIS-5), Anxiousness, and Negative affect-NA (I-PANAS-SF) show a strong and significant negative correlation with sidewalk width, green quality, green quantity, traffic buffer, seating quality and quantity ($p < 0.01$) (refer to Table 2).

4 Post Walk Interview Analysis

An unstructured interview was carried out post each walk; each participant shared their experience post the walk on every street; the observations by male and female participants differed based on the perception of the surrounding. GP (Green Park Main Market Road) was very comfortable for all the participants to move around safely and comfortably. The key aspects of the streetscape elements of this street which enhanced the positive experience were the wide walking space and sitting space. Informal eateries were the focal point for the moving crowd.

Green Park Main Market Road (GP)

Female respondents responded by saying, "Green Park market was really nice; It had all the vendors, shops, and all that. It was really nice to see hawkers and formal shops. The platform is so wide that you can walk around and not feel claustrophobic, so it was really nice." "The experience

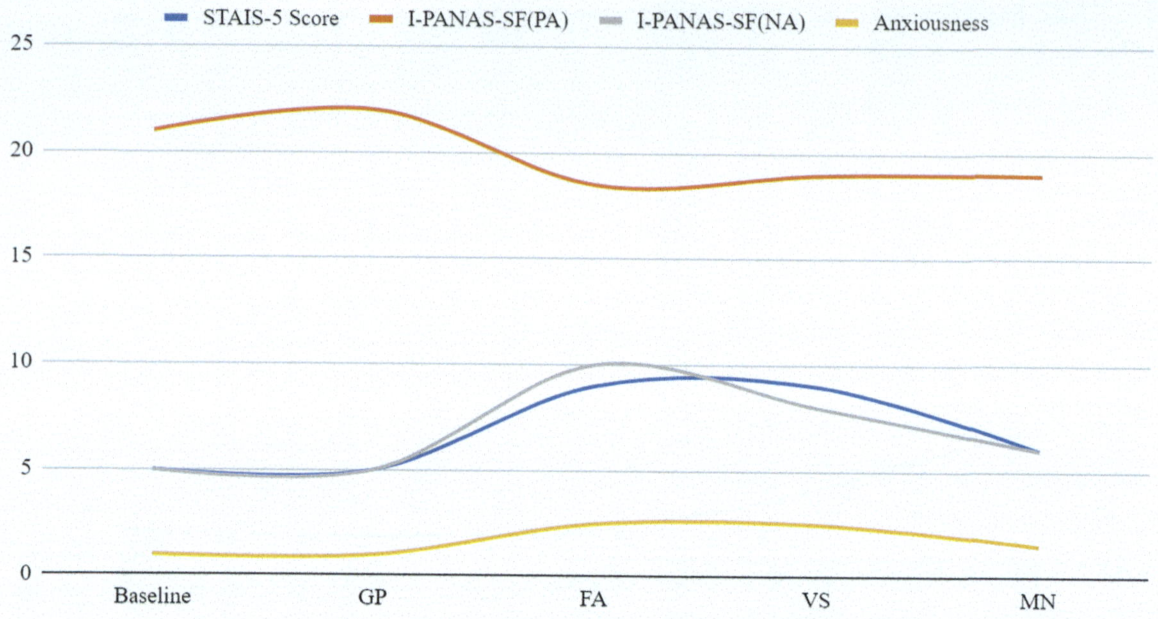

Fig. 9 Showing the mean values of psychological tests corresponding to the location. *Source* Author

was nice; the street was really calm and cozy." In response to why you feel nice and not anxious on Green Park Market Street, this female respondent answered by mentioning that "The space outside the shops and sitting spaces with trees, it gives more liveliness to the space." She continued mentioning the best part of the street to be the 'sitting area.' Some of the male respondents mentioned the green park market as the "Quietest and cleanest of the markets; I was able to walk more freely here." "It was clean and had a green belt also; it had many shops and a good sitting area also."

Fifth Avenue Road, Lodhi Colony (FA)

The female response to this street was that "when I start the walk, it's a bit posh area, but while moving ahead, it gets a bit scary". The built-use and informal vendors beyond a certain point were affected in a negative way. One of the male respondents "Overall experience was more like average", the concern on this street was because of the closed shops, which reduced the pedestrian density significantly at certain points. The public art side of the street was also a bit impactful, as mentioned by two of the participants. The issue of lack of maintenance was also highlighted.

Veer Savarkar Marg, Lajpat Nagar (vS)

Female respondents to this street answered that "The social proximity is too high; people are literally banging into each other. There could have been a proper footpath; there are footpaths, but not in good shape." Another response was that "The market is nice and vibrant because of many shops but becomes very crowded after a point." The major issue of personal space invasion could be observed on this street. Intrusion in personal space can lead to stress, anxiety, aggression, and negative mood (Engelniederhammer et al., 2019).

Guru Gobind Singh Marg, Malviya Nagar Market Road (MN)

Some of the major responses for Malviya Nagar Street were the walking width, vibrancy, and how the sidewalk was away from traffic with a wide green buffer, although some of them did not find it as clean and maintained. A female respondent said that "It's a nice market, it's my first time here, a lot of space here except for one stretch, which is also not as bad as Lajpat." The best part, as mentioned by her, was the ample walking space and how it was so a little disconnected from the road and traffic via a green buffer. This response was common with male participants, also highlighting the fact that it has a good buffer from traffic. One of the male respondents also mentioned a similar observation that the initial part of the street was good, but while walking ahead, the walking space became a bit narrow, occupied by vendors and vehicles (refer to Fig. 5b for variation along the stretch), one of them highlighted that the impact of sitting spaces was less as compared to the shops and buildings on the same side. One respondent highlighted that the user group made him feel a bit unsafe; otherwise, it was exciting to see children playing and skating, which was the best part. Cleanliness and maintenance were also pointed out as an issue.

Table 2 Correlations between Physiological and psychological parameters with streetscape elements

Parameters			Sidewalk width	Facade quality	Green quality	Green quantity	Traffic buffer	Built height	Enclosure	Informal activities	Pedestrian density	Sitting quality	Sitting quantity
Spearman's rho	RMSSD	Correlation Coefficient	0.355*	0.333*	0.355*	0.132	0.257	−0.146	0.078	0.085	−0.146	0.355*	0.132
		Sig. (2-tailed)	0.033	0.048	0.033	0.444	0.131	0.397	0.652	0.622	0.396	0.033	0.444
	HR	Correlation Coefficient	−0.227	−0.255	−0.227	−0.095	−0.170	0.023	−0.108	−0.073	0.025	−0.227	−0.095
		Sig. (2-tailed)	0.182	0.134	0.182	0.583	0.323	0.895	0.532	0.672	0.884	0.182	0.583
	STIAS	Correlation Coefficient	−0.653**	−0.679**	−0.653**	−0.477**	−0.596**	−0.047	−0.127	0.087	0.049	−0.653**	−0.477**
		Sig. (2-tailed)	0.000	0.000	0.000	0.003	0.000	0.784	0.462	0.613	0.776	0.000	0.003
	ANX	Correlation Coefficient	−0.682**	−0.654**	−0.682**	−0.455**	−0.599**	0.070	−0.071	0.079	0.155	−0.682**	−0.455**
		Sig. (2-tailed)	0.000	0.000	0.000	0.005	0.000	0.686	0.680	0.647	0.366	0.000	0.005
	PA	Correlation Coefficient	0.216	0.238	0.216	0.041	0.135	−0.073	0.120	0.127	−0.055	0.216	0.041
		Sig. (2-tailed)	0.206	0.163	0.206	0.813	0.431	0.671	0.486	0.460	0.749	0.206	0.813
	NA	Correlation Coefficient	−0.723**	−0.836**	−0.723**	−0.537**	−0.664**	−0.182	−0.262	0.044	−0.076	−0.723**	−0.537**
		Sig. (2-tailed)	0.000	0.000	0.000	0.001	0.000	0.288	0.122	0.798	0.660	0.000	0−001

* *Correlation is significant at the 0.05 level (2-tailed)*
** *Correlation is significant at the 0.01 level (2-tailed)*

The interview analysis showed which streetscape elements were focused on by the participants in their conversation post-walk. The result highlights that the maximum number of times sidewalk width/walking width was emphasized on each street. Two of the streets (FA and VS) show most of the negative responses (refer to Table 3). Although the street characteristics were very different from each other such as street FA had the least pedestrian density and a distance from the shops with a relatively more walking area than street VS, they were talked mostly with negative responses showing issues with low walking space, less greens, maintenance, and traffic buffer issues. One different aspect of public art (wall art), which was present on one of the streets (FA) on the opposite side of the walking zone, was highlighted by few as a positive aspect but was overpowered by the other streetscape elements, which added to the negative experience. Although the greens in quantity were considered more in MN Street, the impact and usage of those greens were obstructed by vehicles parked, limited easy access to greens, and low maintenance on half of the stretch decreased its quality of usage. Street GP is found to have the most positive responses, with the highest mention of its walking width as comfortable and wide, enhancing a positive walking experience (refer to Table 3). Figure 10 highlights the significant elements which were emphasized by the participants in their conversations, impacting their experience positively and negatively. The significant elements which impacted participants' anxiety levels and experiences based on conversation analysis of the interviews are sidewalk width, greenery, maintenance, a buffer from traffic, and facades (refer to Fig. 10).

Table 3 Interview conversation analysis chart showing the frequency of the streetscape elements' occurrence in conversations and also the positive response (text in italics) and negative response (text in bold) corresponding to the element mentioned

	Frequency of mentioned streetscape elements by participants				Total frequency
Streetscape elements	GP	FA	VS	MN	
Wide walking space	*9*	**2**	**5**	*4*	20
Sitting areas	*3*	NA	NA	NA	3
Greens	*5*	**1**	**2**	**2**	12
				2	
Maintenance	*2*	**4**	**2**	**1**	9
Buffer from traffic	*2*	**1**	**2**	*3*	8
Varied Land Use	*1*	**1**	NA	NA	2
Freedom to walk	*1*	NA	NA	*1*	2
Facade	*3*	**4**	*1*	*4*	12
Vendors and informal zone	*1*	**1**	*3*	NA	5
Encroachment on footpath (vehicles and vendors)	NA	**1**	**2**	**1**	4
Security	NA	**2**	NA	**1**	3
Public art	NA	*2*	NA	NA	2
Pedestrian density	NA	**2**	**4**	NA	6

5 Discussion

The results indicate that heart rate variability is affected by the surrounding street environment. The streetscape elements which were found statistically significant were sidewalk width, facade quality, green quality, and seating quality. These elements are not independent of each other. Both the streets GP and MN had significant physical traffic buffers (landscaped), and the sidewalk width in both cases was comfortable for people to walk. However, the other factors did contribute to affecting the overall experience. Although the physiological results on the street MN were not statistically significant, which could be due to the tiredness affecting the low RMSSD, the psychological responses indicated that the street was found vibrant and comfortable in walking but not as positive as street GP. In the case of streets FA and VS, both the streets have low walking space and low maintenance as well. The physiological and psychological results on these streets indicate comparable results, but there was a major difference in the pedestrian density of these streets. Street VS had very high pedestrian density, which significantly affected the personal space while walking, and street FA on the other hand, had very low pedestrian density, which also made respondents uncomfortable and enhanced their feeling of insecurity among them. Thus, pedestrian density could not be found as a significant factor as it is dynamic in nature and can affect in different ways if there is enough walking space that their personal space does not get invaded. This can be observed in street MN where the walking space and pedestrian density were high, but the density was not perceived as high due to comfortable and wide walking space. Greenness on the street GP and MN both streets was high; in fact, on street MN, the traffic buffer was a large green space that buffered the traffic noise and negated the feeling of being too close to traffic, which added to the positive responses and low levels of anxiety. This observation has been highlighted in a study mentioning that by increasing the distance between stressors and lowering their perceptual prominence, natural areas, and features can lessen exposure to challenging environmental situations (Hartig et al., 2014). The seating space could also be considered in quantity at Street MN as higher than Street GP as the ledge around the

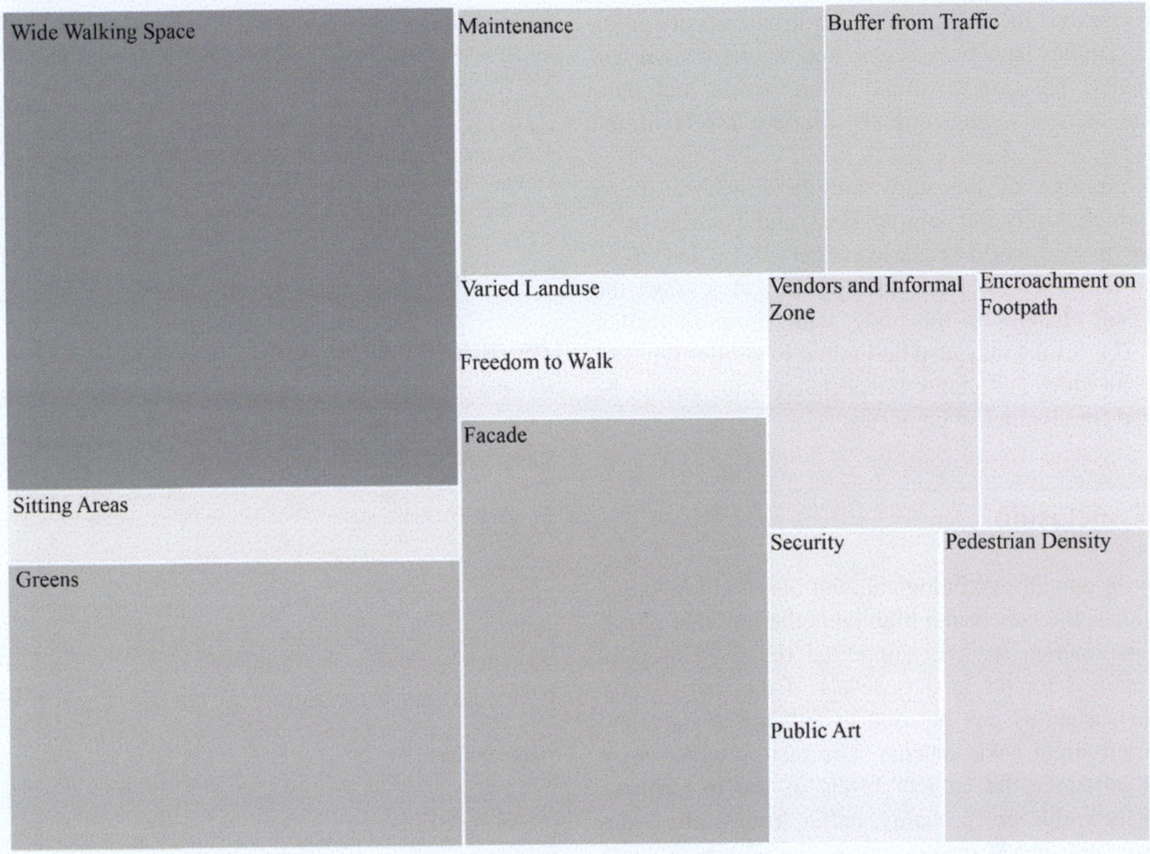

Fig. 10 Tree map representing the proportion of mentioned streetscape elements by the participants based on their significance of impact. *Source* Author

green space on Street MN was used for sitting and was continuing along the whole route. However, the quality of sitting space on the street GP was rated better as the sitting spaces were smaller (suitable for two people to sit) and personalized while being evenly distributed along the route, which provided ownership of the space for the person sitting while being close to green (hedges, planters, and trees) and enjoying the watching the passerby without personal space invasion around the sitting area enhancing the sense of control. It has also been suggested in the literature that behavior settings that allow users to territorialize and customize space convey a degree of control that not only gives users freedom and comfort but also supports social behavior (Mehta, 2013). This could highlight the plausible contribution of a sense of control on the user, further reducing anxiety.

Green Quality was found to be associated with both HRV and psychological parameters of anxiety, while the quantity was only with psychological parameters of anxiety. Streetscape green quality has been observed to be more associated with stress than quantity, as emphasized in a study by de Vries et al. (2013). The observation of this study also aligns with this literature, suggesting that green quality can impact anxiety (physiologically) more than quantity considering the fact that street MN having higher green quantity was not found as positive as street GP with lower quantity but better quality. In all the cases, the other side of the street was not found significant as per the responses of the participants, but in the case of Fifth Avenue Road (FA), the other side had public wall art which was briefly affecting the experience but was not of utmost importance as per the respondents.

6 Strengths and Limitations of the Study

The strength of the study is that, while understanding the impact of the environment on people's mental health, specifically anxiety, physiological results add significant support for deciphering the psychological reactions and add to the advantage of understanding the subjective emotions of the people. The study offers insights into not just psychological but how physical health can get impacted by anxiety in certain types of urban environments. This methodology in urban design is fairly new and highlights the various advantages of it in health and behavioral research. Till now, the research has been limited to urban versus green and scarce studies have been researched among busy urban environments. This study offers urban streets and their streetscape elements impacting the anxiety levels of the people.

The limitations of the study are the time-consuming process which limits the sample sizes and is difficult to implement in real-world scenarios. Apart from that effect of motion and tiredness from walking may also affect the physiological changes in the body, which require further research. The results may also be limited to similar types of metropolitan cities and cultural aspects and thus cannot be generalized for all cities or countries.

7 Conclusion

After physiological, psychological, and observations based on interviews, the conclusion highlights the fact that physiological parameters strongly supported the psychological results obtained for the anxiety levels of the participants. Heart rate variability can be used to indicate anxiety levels in varied street environments. The factors which were found to influence the anxiety levels of the participants are sidewalk width, green quality, buffer from traffic, seating quality, and facade quality. Other factors which could be considered important based on the results of the study are green quantity, seating quantity, and maintenance which became distinguishing elements in two positively rated environments (GP and MN). Sidewalk width was found to be the most significant element of impact based on the quantitative and qualitative analysis of the streets. The results indicate that the quantity and quality of the streetscape elements have a strong influence on people's perceptions. Looking at two similar street environments also, the quality made a difference in opinions and anxious feelings. Anxiety on the street has been observed to be higher in females than in males compared physiologically and psychologically; as noted from the study, this can be taken further for research to observe the gender-based perceptions and mental health impacts on them. This indicates the design of streetscapes keeping in mind the gender perspective to alleviate anxiety, is an important aspect for future research.

Considering the association of physiological markers (HRV) and psychological parameters (state anxiety), the results coincide at a certain level. The most positive psychological responses at one street were strongly associated with better heart rate variability, i.e., the physiological reaction. This can be taken forward, and more physiological parameters can be added to observe the overall impact of the street environment on the human body and mind.

Acknowledgements I would like to acknowledge the contribution of the participants who were residents of Delhi for being part of the study and also the Department of Architecture and Planning, IIT Roorkee, for sponsoring the devices to measure heart rate variability. Also, I would sincerely thank Prof. Dr. Rohit Verma, from the Department of Psychiatry and National Drug Dependence Treatment Centre, All India Institute of Medical Sciences (AIIMS) Delhi, India, for his responses to queries on anxiety and physiological parameters.

Annexure

Quantitative Characteristics Measured for the Streets:

Streetscape feature	Scores			
	GP	FA	VS	MN
Greenery quantity	3	2	1	4
Greenery quality	4	2	1	3
Sidewalk width	7.6 m	2.1 + 2 m	2m	6 m (Part 1) and 3.6 m (Part 2)
Traffic buffer type	4	2	1	4
Facade quality	4	1	2	3
Sitting QL	4	2	1	3
Sitting QN	3	2	1	4
Built height	3–4 floors	1–2 floors	4–5 floors	3–4 floors
Enclosure ratio of the street (Built height/Street Width)	0.40	0.21	0.70	0.38
Pedestrian density	2	1	4	3
Informal activities	3	2	4	1

References

Agarwal, K. (2019). *From the City of Djinns to the City of Sins: how Delhi has facilitated mental wellbeing through urban design*. Centre for Urban Design and Mental Health. https://www.urbandesignmentalhealth.com/edition-6.html

Asgarzadeh, M., Lusk, A., Koga, T., & Hirate, K. (2012). Measuring oppressiveness of Streetscapes. *Landscape and Urban Planning, 107*(1), 1–11. https://doi.org/10.1016/j.landurbplan.2012.04.001

Barros, P., Mehta, V., Brindley, P., & Zandieh, R. (2021). The restorative potential of commercial streets. *Landscape Research, 46*(7), 1017–1037. https://doi.org/10.1080/01426397.2021.1938983

Barton, H., Thompson, S., Grant, M., & Burgess, S. (2015). *The Routledge Handbook of Planning for Health and well-being: Shaping a sustainable and Healthy Future*. Routledge, Taylor & Francis Group.

Bhatia, A. (2020). *World Mental Health day 2020: In numbers, the burden of mental disorders in India*. NDTV. https://swachhindia.ndtv.com/world-mental-health-day-2020-in-numbers-the-burden-of-mental-disorders-in-india-51627/

Boumezoued, S., Bada, Y., & Bougdah, H. (2020). Pedestrian itinerary choice: Between multi-sensory, affective and syntactic aspects of the street pattern in the historic quarter of Bejaia, Algeria. *International Review for Spatial Planning and Sustainable Development, 8*(4), 91–108. https://doi.org/10.14246/irspsda.8.4_91

Brown, D. K., Barton, J. L., & Gladwell, V. F. (2013). Viewing nature scenes positively affects recovery of autonomic function following acute-mental stress. *Environmental Science &; Technology, 47*(11), 5562–5569. https://doi.org/10.1021/es305019p

Buscemi, J. (2021). *Polar H10 Heart Rate Sensor Review*. Verywell Fit. https://www.verywellfit.com/polar-h10-heart-rate-sensor-review-4778846

Chen, Z., Schulz, S., Qiu, M., Yang, W., He, X., Wang, Z., & Yang, L. (2018). Assessing affective experience of in-situ environmental walk via wearable biosensors for evidence-based design. *Cognitive Systems Research, 52*, 970–977. https://doi.org/10.1016/j.cogsys.2018.09.003

Dattani, S., Ritchie, H., & Roser, M. (2021). *Mental health*. Our World in Data. https://ourworldindata.org/mental-health [Online Resource].

Davey, H. M., Barratt, A. L., Butow, P. N., & Deeks, J. J. (2007). A one-item question with a Likert or visual analog scale adequately measured current anxiety. *Journal of Clinical Epidemiology, 60*(4), 356–360. https://doi.org/10.1016/j.jclinepi.2006.07.015

de Vries, S., van Dillen, S. M. E., Groenewegen, P. P., & Spreeuwenberg, P. (2013). Streetscape greenery and health: Stress, social cohesion and physical activity as mediators. *Social Science & Medicine, 94*, 26–33. https://doi.org/10.1016/j.socscimed.2013.06.030

Dillon, R. (2005). Designing urban space for psychological comfort: The kentish town road project. *Journal of Public Mental Health, 4*(4), 10–20. https://doi.org/10.1108/17465729200500027

Dritsa, D., & Biloria, N. (2021). Mapping the urban environment using real-time physiological monitoring. *Archnet-IJAR: International Journal of Architectural Research, 15*(3), 467–486. https://doi.org/10.1108/arch-02-2021-0041

Endler, N. S., & Kocovski, N. L. (2001). State and trait anxiety revisited. *Journal of Anxiety Disorders, 15*(3), 231–245. https://doi.org/10.1016/s0887-6185(01)00060-3

Engelniederhammer, A., Papastefanou, G., & Xiang, L. (2019). Crowding density in urban environment and its effects on emotional responding of pedestrians: Using wearable device technology with sensors capturing proximity and psychophysiological emotion responses while walking in the street. *Journal of Human Behavior in the Social Environment, 29*(5), 630–646. https://doi.org/10.1080/10911359.2019.1579149

Elsadek, M., Liu, B., Lian, Z., & Xie, J. (2019). The influence of urban roadside trees and their physical environment on stress relief measures: A field experiment in Shanghai. *Urban Forestry & Urban Greening, 42*, 51–60. https://doi.org/10.1016/j.ufug.2019.05.007

Fotios, S., Monteiro, A. L., & Uttley, J. (2018). Evaluation of pedestrian reassurance gained by higher illuminances in residential streets using the Day-Dark Approach. *Lighting Research & Technology, 51*(4), 557–575. https://doi.org/10.1177/1477153518775464

Gill, P., Stewart, K., Treasure, E., & Chadwick, B. (2008). Methods of data collection in qualitative research: *Interviews and focus groups. British Dental Journal, 204*(6), 291–295. https://doi.org/10.1038/bdj.2008.192

Hartig, T., Mitchell, R., de Vries, S., & Frumkin, H. (2014). Nature and health. *Annual Review of Public Health, 35*(1), 207–228. https://doi.org/10.1146/annurev-publhealth-032013-182443

Hidalgo, A. K. (2019). *Streets for Mental Health: An Interdisciplinary Approach to Restorative Urban Design* (Doctoral thesis). University of Calgary. https://doi.org/10.11575/PRISM/35717

Hinde, K., White, G., & Armstrong, N. (2021). Wearable devices suitable for monitoring twenty four hour heart rate variability in military populations. *Sensors, 21*(4), 1061. https://doi.org/10.3390/s21041061

Hollander, J. B., & Anderson, E. C. (2020). The impact of urban façade quality on Affective Feelings. *Archnet-IJAR: International Journal of Architectural Research, 14*(2), 219–232. https://doi.org/10.1108/arch-07-2019-0181

Ink, S. (2021). *Streets for pandemic response & recovery*. National Association of City Transportation Officials. https://nacto.org/publication/streets-for-pandemic-response-recovery.

Knöll, M., Neuheuser, K., Cleff, T., & Rudolph-Cleff, A. (2018). A tool to predict perceived urban stress in open public spaces. *Environment and Planning B: Urban Analytics and City Science, 45*(4), Article 4. https://doi.org/10.1177/0265813516686971

Kondo, M. C., Jacoby, S. F., & South, E. C. (2018). Does spending time outdoors reduce stress? A review of real-time stress response to outdoor environments. *Health & Place, 51*, 136–150. https://doi.org/10.1016/j.healthplace.2018.03.001

Lee, J., Kim, J. K., & Wachholtz, A. (2015). The benefit of heart rate variability biofeedback and relaxation training in reducing trait anxiety. *Han'guk Simni Hakhoe chi. Kon'gang = The Korean journal of health psychology, 20*(2), 391–408. https://doi.org/10.17315/kjhp.2015.20.2.002

Levenson, R. W., Lwi, S. J., Brown, C. L., Ford, B. Q., Otero, M. C., & Verstaen, A. (2017). Emotion. In *Handbook of psychophysiology, 4th ed* (pp. 444–464). Cambridge University Press.

McCay, L., Bremer, I., Endale, T., Jannati, M., & Yi, J. (2017). Urban Design and mental health. *Mental Health and Illness in the City*, 1–24. https://doi.org/10.1007/978-981-10-0752-1_12-1

Mehta, V. (2013). *Street: A quintessential social public space* (1st ed.). Routledge.

Moya-Ramon, M., Mateo-March, M., Peña-González, I., Zabala, M., & Javaloyes, A. (2022). Validity and reliability of different smartphones applications to measure HRV during short and ultra-short measurements in elite athletes. *Computer Methods and Programs in Biomedicine, 217*, 106696. https://doi.org/10.1016/j.cmpb.2022.106696

Navarrete-Hernandez, P., & Laffan, K. (2019). A Greener Urban Environment: Designing Green Infrastructure Interventions to promote citizens' subjective wellbeing. *Landscape and Urban Planning, 191*, 103618. https://doi.org/10.1016/j.landurbplan.2019.103618

Park, Y., & Garcia, M. (2019). Pedestrian safety perception and urban street settings. *International Journal of Sustainable Transportation, 14*(11), 860–871. https://doi.org/10.1080/15568318.2019.1641577

Paül i Agustí, D., Rutllant, J., & Lasala Fortea, J. (2019). Differences in the perception of urban space via mental maps and Heart Rate Variation (HRV). *Applied Geography, 112*, 102084. https://doi.org/10.1016/j.apgeog.2019.102084

Reiner, R. (2008). Integrating a portable biofeedback device into clinical practice for patients with anxiety disorders: Results of a pilot study. *Applied Psychophysiology and Biofeedback, 33*(1), 55–61. https://doi.org/10.1007/s10484-007-9046-6

Robinson, L. (1990). Stress and Anxiety. *Nursing Clinics of North America, 25*(4), 935–943. https://doi.org/10.1016/S0029-6465(22)02991-7

Rossi, V., & Pourtois, G. (2012). Transient state-dependent fluctuations in anxiety measured using STAI, POMS, PANAS or VAS: A comparative review. *Anxiety, Stress, & Coping, 25*(6), Article 6. https://doi.org/10.1080/10615806.2011.582948

Song, C., Ikei, H., Igarashi, M., Takagaki, M., & Miyazaki, Y. (2015). Physiological and psychological effects of a walk in urban parks in fall. *International Journal of Environmental Research and Public Health, 12*(11), Article 11. https://doi.org/10.3390/ijerph121114216

Speer, K. E., Semple, S., Naumovski, N., & McKune, A. J. (2020). Measuring heart rate variability using commercially available devices in healthy children: a validity and reliability study. *European Journal of Investigation in Health, Psychology and Education, 10*(1), Article 1. https://doi.org/10.3390/ejihpe10010029

Thompson, E. R. (2007). Development and validation of an internationally reliable short-form of the positive and negative affect schedule (PANAS). *Journal of Cross-Cultural Psychology, 38*(2), Article 2. https://doi.org/10.1177/0022022106297301

Wang, R., Lu, Y., Zhang, J., Liu, P., Yao, Y., & Liu, Y. (2019). The relationship between visual enclosure for neighbourhood street walkability and elders' mental health in China: Using street view images. *Journal of Transport & Health, 13*, 90–102. https://doi.org/10.1016/j.jth.2019.02.009

Zhao, J., Wu, J., & Wang, H. (2019). Characteristics of urban streets in relation to perceived restorativeness. *Journal of Exposure Science & Environmental Epidemiology, 30*(2), 309–319. https://doi.org/10.1038/s41370-019-0188-4

Zsido, A. N., Teleki, S. A., Csokasi, K., Rozsa, S., & Bandi, S. A. (2020). Development of the short version of the spielberger state—Trait anxiety inventory. *Psychiatry Research, 291*, 113223. https://doi.org/10.1016/j.psychres.2020.113223

Envisioning Cities of the Future: A Malaysia Youth Perspective Using Concept Mapping

Stephen T. Homer, Eliza Berezina, and Colin Mathew Hugues D. Gill

Abstract

The global urban population now exceeds that of those who live in rural areas for the first time in human history. Spurred by the advances in technology, resources are funnelled into our cities, with an estimated 75% of global resources being used by cities which only accommodate 2% of the landmass. Looking at how far societies have developed within cities in a short space of time creates an alluring area of study of what future cities will consist of. Writers have created works to describe their ideal cities and societies of the future for centuries, but also there have been works of dystopias where society collapses. Such writings highlight the wants and concerns of their contemporary society. This study sought to explore the vision of future cities based on youth residing within Malaysia, with the focus question being: How do you envision the city of the future 100 years from now? The participants' thoughts were collected and organized through a concept mapping methodology, which includes creating statements, sorting statements, multidimensional scaling, hierarchical cluster analysis and label clusters. Concept mapping was used within the present study as it provides an exploratory bottom-up research design and was undertaken in a half-day workshop. 106 unique statements were generated by 21 participants, which after being sorted, created seven clusters of Rights, Social Development, Environment, Sustainability, Welfare, Transportation and Technology. The results suggest that the participants wanted a more liberal and supportive urban environment which differs from the current conservative stance of Malaysia, with cities that are healthier and cleaner pointing towards Malaysia facing an obesity crisis and the environmental challenges of a quickly developing economy. However, much of this necessitates technology that may or may not exist in the city of the future.

Keywords

Future cities · Urbanization · Malaysia · Concept Mapping · Youth · Development

1 Introduction

For the first time in human history, more people are living in cities and urban areas than those who live in rural settlements, and the number of people transitioning to urban living is set to continue to rise (Norman, 2018; World Bank, 2020). The urban landscape has changed drastically in a relatively short space of time, partially driven by technological development. As technology becomes ever more powerful and commonplace, for example, the iPhone 6 clock speed is 32,600 times faster than the best Apollo computers, meaning the smartphone has more processing power than what was used to take a man to the moon in 1969. However, broader technological development has assisted, such as within food production. Advances in the management of crops, the breeding of high-yield varieties, and the improved use of fertilizers and minerals have enabled food production to be trebled over the past fifty years whilst the human population has only doubled

S. T. Homer (✉)
Future Cities Research Institute, Sunway University Malaysia, Kuala Lumpur, Malaysia
e-mail: stephenhomer@sunway.edu.my

E. Berezina
Psychology Department, School of Medical and Life Sciences, Sunway University, Kuala Lumpur, Malaysia
e-mail: elizab@sunway.edu.my

C. M. H. D. Gill
Chancellery, Sunway University, Kuala Lumpur, Malaysia
e-mail: hewg@sunway.edu.my

I. Pigliautile et al. (eds.), *Urban Resilience, Livability, and Climate Adaptation*,
Advances in Science, Technology & Innovation, https://doi.org/10.1007/978-3-031-54911-3_8

(Murphy, 2007). All of these technological developments have allowed cities to thrive, appearing as the apex of the anthroposphere with material and energy being provided from external sources, leading to a dynamic interaction between the city and its surrounding biosphere, hydrosphere, atmosphere and geosphere (Hannon & Zaman, 2018). As urbanization spreads and the complex dynamic interactions between city and countryside inevitably lead to change, for either the better or, the worse, it is inevitable that future cities will become the critical focal points for resolving the challenges of globalization, climate change, sustainable development and continued population growth (Hannon & Zaman, 2018).

Evaluating how far societies have already changed during urbanization and assessing the extent to which the future city may bring further change creates an interesting area of study. Where are we heading now? Many writers have speculated about the ideal city and society or warned of a potentially bleak future. These texts attempt to develop a discourse about the current societal atmosphere and then to prompt or prevent potential changes that the authors see as desirable or dangerous (Iavarone, 2022), for example, The City of the Sun by Campanella in 1602 and New Atlantis written in 1627. Thus, by evaluating the differences between an existing society and a writer's utopian dream or dystopian nightmare, it is possible to posit a narrative of contemporary desires and fears. This can also be reversed, with dystopian writings emphasizing the worries and challenges present in society at the time of writing, an example here being Nineteen Eighty-Four by George Orwell, written in 1949 in the shadow of Nazi Germany and Soviet Russia.

Although some may be interested in researching potential dystopias, the present research aims to sketch out the elements of a potential utopia by exploring how young adults residing in Malaysia envision cities developing over the next century. How will people work, live, and move around? Thinking about the limitations of current cities, participants were asked to identify what they thought were the key features of their imagined future city without the restrictions of current technologies and materials. The paper continues with a literature review looking at how cities have developed and what challenges we are facing. Following this section, the paper discusses the selected method of concept mapping within the methodology section before reporting the outcome of the study within the results section. Proceeding with this, the results are evaluated for what they actually mean within the discussion section before the paper is concluded.

2 Literature Review

The importance of cities across the globe can be seen through the 'funnelling' of resources to these locations, with 75% of planetary resources being transferred to cities which occupy only 2% of the land area (Hannon & Zaman, 2018). However, this has led to the uncontrolled growth of cities creating new issues such as overcrowding, increased crime and increases in the spread of illnesses (Sanchez-Corcuera et al., 2019). Historically looking at how cities have dealt with this creates a paradox of urban planning and non-planning paired with the interdependence of order and chaos, as planning has tended to be vague and generalized leading to hidden conflicts (Schubert, 2019). There has been a great focus on improving the form of cities in preparation for their assumed future role and function, such as adapting to the pressures of urban migration (Kassens-Noor & Hintze, 2020). This is incredibly important as cities are already facing pressures through the challenges of urban governance, social justice, climate change and urban health to the point that existing policy solutions are becoming overwhelmed (Norman, 2018). Contemporary cities have also faced the COVID-19 disaster, which refocused attention on changing perceptions of cities' resilience (Maestos et al., 2021). The epidemic altered how people inhabited and used physical areas beginning in March 2020. (Khan et al., 2021). Large urban centres and cities around the world took centre stage in the fight against the pandemic because of COVID-19's greater rates of infection in urban areas and the need for containment to stop the virus' spread (Strielkowski et al., 2022). 90% of documented COVID-19 cases came from cities and the surrounding areas, making them the pandemic epicentre.

To tackle the vast array of challenges and create resilient cities for the future, many propositions of sustainable urban spaces have been proposed, culminating in the United Nations Sustainable Development Goal (SDG) number 11: Sustainable Cities and Communities. Yet, the idea of sustainability is ambiguous, which makes it difficult to achieve without creating complicated socio-ecological structures and controversial multidimensional, normative, and philosophical features (Homer, 2022). This is future exacerbated by city planning being highly controversial as there are normative elements and attempts to control development spaces that have social repercussions (Schubert, 2019). Some studies would suggest the core of the city consists of seven critical services and infrastructure components; city administration, education, healthcare, public safety, real estate, transportation and utilities (Washburn et al., 2009). Transportation is not only a critical infrastructure needed within cities, but it also has historically demonstrated that cities have changed to accommodate the evolution of urban transport (Cugurullo et al., 2020). This has been seen from the narrow, winding roads of medieval cities, which were straightened and widened to allow for a horse-drawn carriage, to the accommodation of the steam train and the rail network through to the modern city and its central focus around the car. Thus a contextual change can lead to the cities of the future looking and operating considerably differently from their contemporary namesakes (Rogers, 2019).

Technology has no doubt played an ever-growing role in everyday life, but as cities become technologically enabled and transformed into 'smart' cities, there are several issues with this. Studies, such as Cugurullo (2018), suggest 'smart' cities have a tendency to be isolated and unconnected and isolated, leading towards 'fragmented urbanism' and away from human-centred cities that are sought (Aina et al., 2019). Thus, there has been a growing push to rethink smart cities from the ground up, where technologies collaborate with the citizenry and assist them in shaping the city around them (Joss, 2018). As modern technologies have spread, there has been a growing form of function debate within cities regarding human behaviour and needs as modern technology offers such flexibility, it may be possible to delink form and function (Kassens-Noor & Hintze, 2020). The flexibility of modern technology can be exemplified through the increased use of the Internet of Things (IoT), where citizens' homes have increased capabilities leading to an improved quality of life (Sanchez-Corcuera et al., 2019). However, whilst some technology, such as IoT, is widely accepted, another technology is not so embraced. This can be seen with the use of autonomous cars; even though a single autonomous car that is shared between users could remove up to 11 conventional cars, the attitude of the citizenry towards them has slowed their adoption (Cugurullo et al., 2020). Yet, this is not surprising as the accidents during testing and other issues have been highly published. Thus even though the technology is attempting to remove human emotion and subjectivity in some processes, the human element still exists through the adoption or rejection of a particular technology. Whilst there have been calls for more people-oriented smart cities, this involvement of the citizenry is highly complex and also considered problematic, requiring engagement at three levels; institutional arrangements, conceptual discourse and on-the-ground practices (Joss, 2018). However, there are indications of persistent marginalization processes of engagement that favour neoliberal interests at the expense of quality of life for some citizens (Kassens-Noor & Hintze, 2020). This marginalization is further exacerbated as the idea of the public as a singular whole and seeking the common good is consistently being replaced with multiple co-existing publics, referred to as heterogeneous public, which may have diverging interests (Joss, 2018).

However, the disorder can have a stimulating effect on cities, building upon the three Ds of density, diversity and disorganized, as city planning's complexity brings with it perspectives of unpredictability and uncertainty (Schubert, 2019). This becomes more apparent when trying to design, plan and build cities for an unclear future. How drastic will the effects of climate change become? Will the individual car still be the preferred method of transport? There are thousands of questions such as these, which those responsible for our cities have no answer to, and thus it may be that disorganization becomes dominant as ad-hoc fixes are adopted to attempt to cater for the uncertainties as they come into fruition. With such an amount of complexity surrounding a magnitude of interdependent challenges, the planning and design of cities tend not to be done well (Rogers, 2019). Thus, a visionary perspective is required to peer into what the future has in store for cities carrying with it the predictions and dreams of what future life will consist of (Iavarone, 2022). Needless to say, this reinforces the previous uncertainties. However, these uncertainties bring about pragmatic solutions, and thus cities have been central to creativity, productivity and social connectivity, which have been vital to technological development and human advancement (Hannon & Zaman, 2018). Yet, Kassens-Noor and Hintze (2020) caution against technological advancement, especially Artificial Intelligence (A.I.), as there is the possibility of technology turning our current urban areas into ghost cities.

3 Methodology

The present study used concept mapping as this methodology uses an exploratory bottom-up research design to develop a series of participant-generated ideas, which are then clustered and ranked in terms of importance and implementation feasibility. The concept mapping methodology consists of five procedures: first, participants are presented with a brainstorming prompt and asked to generate statements; second, participants are asked to sort statements into cognitively related groups and rate each statement; third, the sorting data is being used to conduct multidimensional scaling (MDS); fourth, a hierarchical cluster analysis (HCA) is performed to establish statistically robust clusters of the cognitively related statements; fifth and finally, the clusters are named based upon the participants' grouping (Kane & Trochim, 2007). The advantage of concept mapping is that it allows for the generation of novel ideas and then systematically organizes these ideas, guided by grounded theory, into statistically robust clusters (Rosas & Kane, 2012; Sarantakos, 2005).

3.1 Participants

In line with the recommendations of Rosas and Camphausen (2007) for concept mapping, an opportunity sample of 21 young adults residing in Malaysia aged between 18 and 30 (Mean = 23; SD = 2.83) with the workshop being conducted at a private university within Selangor, which is Malaysias' most populous state. The sample comprised nine males and eleven females, with one participant preferring not to say. The ethnicity of the sample was made of 11 Chinese Malaysians, 4 Indian Malaysians, 1 Malay and five other ethnicities. The highest educational attainment of the participants was 13 at undergraduate degree, four postgraduate degrees, one college level, one high school and one preferring not to say. The sample was recruited via social media. Participants were invited to attend a four-hour workshop with lunch and refreshments being provided. At the end of the workshop, participants were given a voucher worth approximately USD 10, which could be exchanged for food delivery or travel services.

3.2 Procedure

3.2.1 Stage 1—Create Statements

During the first part of the workshop, the following brainstorming prompt was used:

> How do you envision the city of the future, 100 years from now?" How will people work? Live? And move around? Think about the limitations of current cities and consider the key features of your imagined future city without the confines of current technologies and materials. Think about bout tangible and intangible elements that this city requires to function.

Participants recorded their ideas through the online brainstorming tool offered by The Concept System® Global MAX© browser-based software, which was accessed by scanning a Q.R. code. This session was completely free-form, and we were encouraged to speak to one another. Participants could see their own and others' entries, meaning participants could minimize duplication whilst building upon one another's ideas. To ensure all ideas were captured, there was no limitation on the number of statements generated during the brainstorming session. At the end of the brainstorming session, the participants were asked to review the entire list of statements, and through group discussion, a consensus was reached on which statements should be removed or refined, e.g. repeated statements or statements making more than one point, until a face valid set statements remained.

3.2.2 Stage 2—Group Statements

Participants were asked to drag and drop the generated statements into groups using the Concept System® Global MAX© table top sorting function. The instructions were; all statements should be placed in a group of at least two, all statements should not be placed into a single pile, and statements should not be formed into a pile for miscellaneous themes. Participants were also asked to name each group of statements they constructed. No further instructions were given to ensure that participants grouped the statements to reflect what they perceived to be the inherent cognitive relationships between the statements. Once the participants had grouped the statements, they were then asked to rate each statement on Importance (−3: Very Unimportant to 3: Very Important) using a 7-point Likert scale.

3.2.3 Stage 3—Multidimensional Scaling

The Concept System® Global MAX© multidimensional scaling (MDS) function was used to produce a point map, with each point representing the location of a statement relative to other statements in terms of their grouping, with statements in the same group being closer to each other and statements in different groups being further apart (Kane & Trochim, 2007). MDS was carried out in line with the parameters recommended by Sturrock and Rocha (2000) and sorted the statements into a two-dimensional space using the participants' statement groupings to generate a point map with a 99% likelihood that the final arrangement was not stochastic. Thus, the unit of analysis is transposed from the qualitative statements into a quantitative indicator which represents the statements. Once positioned within the point map, the distance between the points indicates how often each statement had been grouped together by the participants. Points closely grouped together indicate many of the participants grouped the statements together, whilst those which are far apart were not grouped often, if at all, together by the participants.

3.2.4 Stage 4—Hierarchical Cluster Analysis

The Concept System® Global MAX© hierarchical cluster analysis (HCA) function was used to develop robust statement clusters. Clusters were generated through an iterative maximum likelihood analysis of the second stage, using the data from the sorting task to identify the minimum number of clusters into which the statements could be organized. Given that there may be more than one optimal solution, participants were asked to review each iteration of the clustering process in order to ensure that the final clusters selected were semantically consistent and reflected the issue under investigation (Homer, 2021).

3.2.5 Stage 5—Labelling Clusters

The Global MAX software displayed the 10 group names generated by participants in stage two that were closest to the clusters developed in stage four. Participants were asked to review the content of the clusters and nominate one of the group names for each cluster. The nomination of names continued until a consensus emerged.

4 Results

4.1 Stage 1—Create Statements

The participants generated a total of 162 initial statements from their brainstorming exercise. After putting the statements back to the group for refinement, the list was reduced and refined to 106 statements which can be seen in Table 8.2.

4.2 Stage 2—Group Statements & Stage 3—Multidimensional Scaling

Once the participants had grouped and named all the statements individually, the individual results were then merged into a single matrix within The Global MAX software in preparation for MDS. The MDS returned a final Stress of 0.2879 after 12 iterations, which is within the acceptable values of below 0.32. The point map generated from the MDS can be seen in Fig. 1, where each point represents a statement, and the associated number refers to the individual statement contained within Table 2.

4.3 Stage 4—Hierarchical Cluster Analysis & Stage 5—Labelling Clusters

From the point map, the cluster playback function incorporated in The Concept System® Global MAX© was used to cycle through each hierarchical reduction in cluster numbers. In an effort to reduce the inherent subjectivity of cluster analysis, the results of each reduction were reviewed by the participants until a consensus was reached about the appropriate number of clusters. Through this process, the participants came to an agreement on a seven-cluster solution. Proceeding this, the ten most fitting names identified by the software and the contents of the cluster itself were reviewed by the participants to reach an agreement on naming each cluster. The final cluster names were as follows; Rights, Social Development, Environment, Sustainability, Welfare, Transportation and, Technology. The final agreed cluster map can be seen in Fig. 2. Once this had been confirmed by the participants, the ratings could be overlaid upon the clusters with individual statement average ratings can be seen in Table 1, and the summary statistics can be seen in Table 2.

Fig. 1 Point map (*Source* Authors)

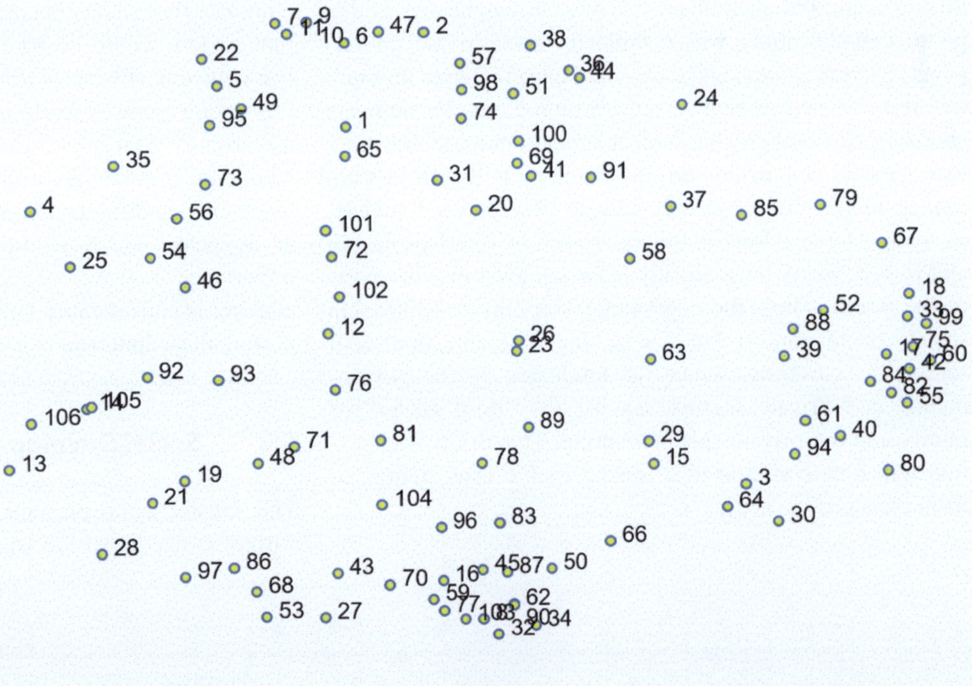

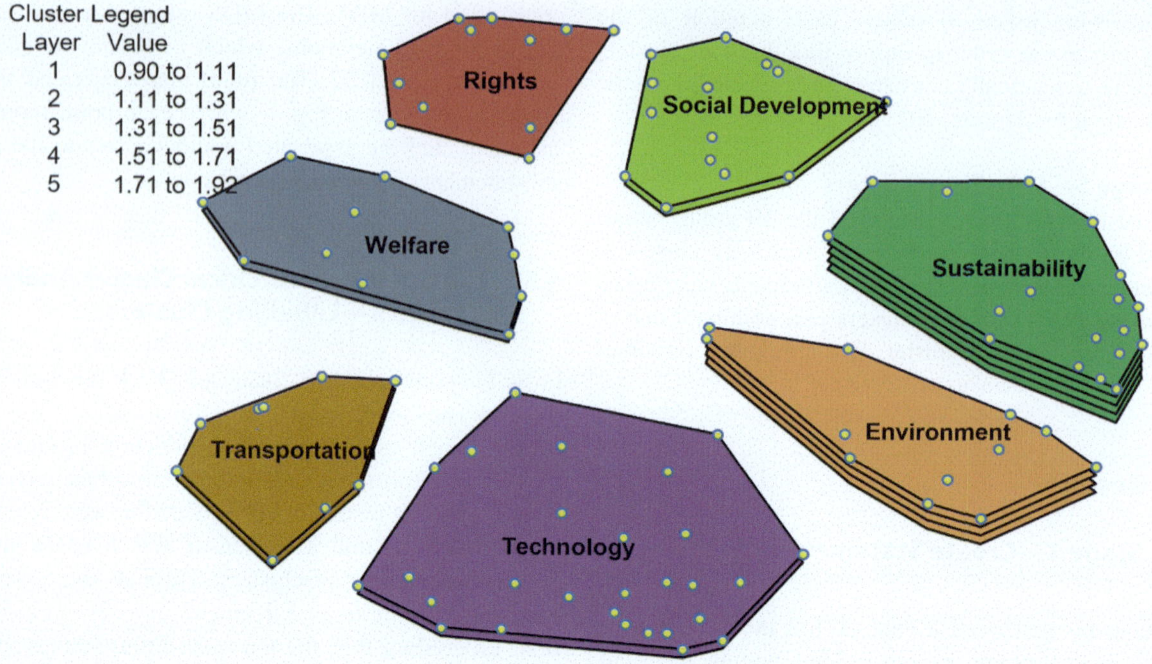

Fig. 2 Cluster rating map (*Source* Authors)

5 Discussion

The participants generated a wide and varied list of statements for the city they envisioned in 100 years' time. Most notably, there is a lack of hard infrastructure-based statements and many more towards 'softer' societal influences. Thus, although the city is constructed from the 'hard' infrastructure, the citizenry is the ones who create the society through their day-to-day interactions with not only the infrastructure but each other. It is vital to emphasize, at this point, that the study was completely open to the participants' interpretations, and they could have included any factors they wished as there were no limitations. Participants choosing to highlight the 'soft' elements may demonstrate that they do not desire drastic changes in the infrastructure of future cities, but they seek to develop civil society, as many of the points highlighted are not futuristic at all, and in fact, many have already been achieved in other parts of the world. Thus, the commentary on current Malaysian society could suggest that it is lacking the elements of civil society, as currently, society is dominated by the governmental and private sectors, leaving the "third sector" of family and the private sphere wanting. This discussion section will now continue by covering each cluster generated from the study.

5.1 Rights Cluster

Whilst the rights cluster does not specifically focus just on citizens' rights in our city in the future, it takes a broader societal viewpoint of the current environment. This can be seen in #7, Gun ban, #11, No more weapons, #9, No more police force and #10, No more soldiers, which would suggest there may be concerns over growing violence and ineffective security being offered, thus removing both would eliminate these concerns. Another poignant comment on current society would be #2 LGBTQIA+are free to roam, #5 Pro-choice is still not a thing, and #6 Drag queens can perform point towards freely are future city being more liberal, as currently, Malaysian society is relatively conservative. An additional commentary on present society can be seen in #1, You'll own nothing and you'll be happy and, #95, Capitalism at its peak, which is dichotomous yet rated closely on importance, however when regarding feasibility, capitalism peaking is considerably larger and thus participants are suggesting that capitalism is going to grow in the future.

5.2 Social Development Cluster

This cluster again presents interesting commentary on the current concerns within society. Statements #91 Affordable

Table 1 Cluster & statement importance rating table

#	Statement	Mean import	#	Statement	Mean Import
Cluster: Technology		1.27	Cluster: Sustainability		1.92
86	Advanced hospital equipment	2.33	60	Pollution free	2.48
71	Easy mobility for physically disabled individuals	2.33	75	Household wastes are separated so that recycling can be implemented	2.43
50	Free internet on all devices at all times/ locations	2.24	58	Free drinking water everywhere	2.43
53	Big data for health information	2.14	85	Better resource management	2.33
87	Full wifi coverage	2.10	99	Plants, EVERYWHERE!	2.29
97	Smarter public transport tracking system	2.05	18	Sustainable globalization	2.19
90	Better internet stream	1.95	52	Having sustainable communities or urban eco-villages	2.19
83	Cashless payments available everywhere	1.90	79	Smoking not allowed in most spaces	2.19
16	Free phones and computers	1.86	17	Clean beaches	2.00
77	Technologically advanced cities with new types of powerful minerals	1.86	39	Better sewage management	2.00
62	Robots for manual work such as cleaning	1.71	33	A very clean and controlled environment	2.00
34	Lots of technology	1.71	42	Fresh organic produce only	1.90
45	A.I. powered home kitchens	1.67	82	Completely halting the usage of fossil fuels	1.81
8	Integrated human-AI system for A.R. and V.R. world	1.57	55	Hydroponic farming system	1.76
68	Drone deliveries	1.48	88	Aesthetics that are not just eco-brutalism	1.48
59	A.I. automated surgery procedures	1.48	84	Rain gardens	1.48
96	More security such as CCTV	1.48	67	Ban paper	1.19
70	Power bank booths everywhere	1.29	37	Irregular development	0.33
81	Attractive traffic routes for pedestrians and bikes	1.29	Cluster: Rights		0.90
78	Aircon & escalator provided for canopy walk	1.19	47	No ethical issues in the government especially bribes	2.43
27	Everyone can fly on their own with technology	0.95	65	Free welfare for all	2.05
89	Living in a protective shield capsule isolating humans	0.86	7	Gun ban	1.57
66	Holographic posters	0.81	11	No more weapons	1.24

(continued)

Table 1 (continued)

#	Statement	Mean import	#	Statement	Mean Import
103	AI servants everywhere	0.19	9	No more police force	1.10
104	Hovering city	0.14	10	No more soldiers	1.05
32	Mind control technology	0.10	2	LGBTQIA + are free to roam	0.81
43	Reanimating the dead is possible	−0.52	5	Pro-choice is still not a thing	0.76
76	Humans with superpower	−0.57	22	Cities will implement the purge, to keep crimes in control	0.67
48	A.I. worshipping religions	−0.67	1	You'll own nothing and you'll be happy	0.52
Cluster: Social development		1.30	95	Capitalism at its peak	0.43
91	Affordable housing and shelters for vulnerable individuals	2.29	6	Drag queens can perform freely	0.05
51	No poverty	2.24	49	New cult religions	-0.90
74	No homelessness	2.10	Cluster: Welfare		1.17
41	Less hierarchical work environments	1.71	54	Free and really accessible healthcare	2.33
69	Free food available	1.71	46	Emergency kits booth (free)	2.10
100	Social spaces where different groups can merge together	1.71	12	Everything is within walking distance	1.81
31	People being skilled at how to live in a community	1.57	73	Reduced work hours	1.81
98	Upskilling workshops ran by the municipal gov and NGOs to tackle social issues	1.43	72	Sign language interpretation	1.76
57	No privatized entities monopolizing industries	1.19	101	More work from home options	1.62
36	Controlled economies	1.19	102	Open work spaces	1.57
24	A religion of sustainable communities	0.71	56	Everyone's fat because of the lack of movement	0.24
20	Party city	0.48	4	Poor ethical practice in medicine	0.14
44	Bring back the barter system	0.05	25	Eating is not a daily need anymore	0.14
38	No religion	−0.19	35	Immorality	−0.67
Cluster: Environment		1.57	Cluster: Transportation		1.29
64	Solar power or hydropower everything	2.29	13	There is very accessible public transport	2.24
30	Automated systematic cleaning protocols for waste on land, river, and sea	2.29	106	No traffic jam	2.14
94	Water purifying system	2.29	21	Parcel delivery reduced to minutes from placing an order	1.71
29	Self-sustaining homes	2.19	19	Bicycles to replace cars	1.62
61	Weather sensors and warnings	1.95	92	Free parking	1.52

(continued)

Table 1 (continued)

#	Statement	Mean import	#	Statement	Mean Import
80	Recycle car wastes	1.86	105	Wider roads	1.10
63	Cultural architecture in major cities	1.76	93	Cafés extending to pedestrian walks	0.60
23	More pet friendly infrastructure	1.57	28	Teleportation as a mode of transport	0.52
40	Genetically modified plants that glow (no need for street lights)	1.38	14	Ban personal transport	0.14
3	Man-made atmosphere	1.00			
15	Man-made meat	0.71			
26	Colonized on the moon and in space	−0.43			

Table 2 Cluster summary statistics table

	Count	Average	Median	Min	Max	Std	Variance
Sustainability	18	1.92	2.00	0.33	2.48	0.52	0.27
Environment	12	1.57	1.81	−0.43	2.29	0.78	0.61
Social Development	14	1.30	1.50	−0.19	2.29	0.75	0.57
Transportation	9	1.29	1.52	0.14	2.24	0.70	0.49
Technology	29	1.27	1.48	−0.67	2.33	0.88	0.77
Welfare	11	1.17	1.62	−0.67	2.33	0.96	0.92
Rights	13	0.90	0.81	−0.90	2.43	0.82	0.67

housing and shelters for vulnerable individuals, #51 No poverty, #74 No homelessness, and #69 Free food available demonstrate that participants want the cities of the future to care for the less fortunate. The idea of supporting the vulnerable alongside #36 Controlled economies could suggest a move towards a more socialist society, as participants may be concerned by what they deem a current lack of economic equality. However, this would depend upon who is in control of the economy, which the participants have not elaborated upon. There are, however, more statements which would hint at a move away from a capitalist society, particularly #57 No privatized entities monopolizing industries but also the alleviation of power distance through #41 Less hierarchical work environments. As a whole, the social development cluster appears to highlight that, at present, there is too much inequality within society, and the cities of the future need to move to address this.

5.3 Environment Cluster

Whilst participants named this cluster environment, it appears to be more around the city environment and creating a more liveable city rather than the natural environment. Whilst #80 recycles car wastes, #94 water purifying system and #30 automated systematic cleaning protocols for waste on land, river and sea all appear to be focused upon the natural environment; however, this is not explicitly stated. When looking at other statements grouped in the cluster, such as #63 cultural architecture in major cities and #29 self-sustaining homes, there is a clear shift to making the living environment more pleasant and liveable. This may be seeking to address current issues such as the standard of living and citizens' well-being in the current economic climate coming out of the COVID-19 pandemic and the cost of living increase. Some more drastic and futuristic ideas come from statements such as #40 Genetically modified plants that glow (no need for street lights), #3 Man-made atmosphere and #15 Man-made meat, may indicate that participants believe science can be used in the future to alleviate some of the problems currently being faced globally.

5.4 Sustainability Cluster

The sustainability cluster was rated significantly higher than the other clusters at average importance of 1.92, whilst the next cluster was the environment at 1.57 importance. Whilst the environment took focus upon the living environment, the sustainability cluster focused upon the natural environment with statements such as; #85 Better resource management and #99 Plants, EVERYWHERE!. However, even within this cluster, there is still an underlying focus on the liveability of the citizenry, similar to the environment cluster. This can be seen in #58, Free drinking water everywhere and #33, A very clean and controlled environment. What should be noted, though, is the desire for some basic institutional changes to improve the sustainability of Malaysian cities, such as #75 Household wastes are separated so that recycling can be implemented, which reflects the low current rating of recycling in Malaysia and the participants desire for this to be improved. A similar story can be seen in #79, Smoking not allowed in most spaces, which again is a small change but is emphasized by being displayed in this study. Considering the participants were asked to consider 100 years in the future, with both recycling and no smoking still being thought of as problematic, may demonstrate how institutionally embedded these issues are and that it is necessary to have a large societal shift to address them.

5.5 Welfare Cluster

Welfare was rated considerably lower on importance, being the second least important out of the seven clusters from this study. This in itself raises questions, as although the environment cluster and sustainability clusters focused upon the liveability of cities, the welfare elements are not so highly rated. These ratings come from the fact that the items are bi-polar, in that there are some that are largely positive and high importance but also some which are decremental to society and rated low importance or negative. This in itself presents an interesting commentary on current society in Malaysia, as participants had the opportunity to create a futuristic utopia, yet some ideas state a dystopian view suggesting that participants believe these statements are not easy to resolve. Examples of these dystopian views can be seen in #4, Poor ethical practice in medicine, with average importance of 0.14 and #35, Immorality, with average importance of –0.67. Whilst the strong positive statements include #46 Emergency kits booth (free) with average importance of 2.10 and #54 Free and really accessible healthcare with an average importance of 2.33, but again these statements are not particularly creative as many countries around the world have already achieved this and may again reflect upon the lack of accessibility to healthcare in current Malaysian society.

5.6 Transportation Cluster

The transportation cluster had two key statements; #13 There is very accessible public transport with an average importance of 2.24 and #106 No traffic jam with an average importance of 2.14; this clearly reflects upon the current situation in Malaysia with estimates suggesting that citizens spend upon to an hour a day in traffic jams with few alternatives as public transportation is still limited. The lower rated statements present more interesting and move away from short-term concerns; #93 Cafes extending to pedestrian walks rated at 0.60 with the vision of Parisian café but currently impractical with the pollution and dirty environment highlighted in earlier clusters, #28 Teleportation as mode of transport rated at 0.52 but would alleviate many of the traffic and pollution problems discussed within this study and finally #14 Ban personal transport rated at 0.14 but reflects that most citizens in Malaysia have a personal vehicle and also work vehicles, with more vehicles registered in Malaysia that the population.

5.7 Technology Cluster

The technology cluster presented another interesting cluster; not only was it the largest, but it again featured statements which were bipolar, with some being rated considerably high and others being rated the lowest for the whole study. The most important statements, again, lacked creativity but gave a running commentary of the current issues within Malaysian society; these included; #71 Easy mobility for physically disabled individuals with an importance of 2.33, #86 Advanced hospital equipment with an importance of 2.33 and, #50 Free internet on all devices at all times/locations with the importance of 2.24. Whilst the more creative statements, such as; #103 AI servants everywhere with an importance of 0.19, #104 Hovering city with an importance of 0.14 and #32 Mind control technology with an importance of 0.10, these statements may represent technological marvels but may lack feasibility as well as benefit to society in our future cities. However, the even more creative statements were rated even lower; #43 Reanimating the dead is possible with importance −0.52, #76 Humans with superpower with importance −0.57 and, #48 A.I. worshipping religions with importance −0.67; however, this may be due to the clarity around how these statements would benefit our future city and society is even more unclear.

5.8 Limitations

Limitations from this study are focused on two areas; small sample size and narrow demographic. The small sample size is appropriate for the selected methodology and is good for an in-depth exploratory study such as this one, but it means that the generalisability can be brought into question. Similarly, the narrow demographic of the study, those residing within Malaysia and aged between 18 and 30, means that a wider group of possible participants have been excluded, which again can impact the generalisability of the study.

5.9 Future Research

Future research should seek to alleviate the limitations; this can be done by statistically validating the results to a larger sample of Malaysian residents in the 18 to 30-year-old category before then proceeding to validate (or seek differences) from a broader scope of demographics such as age, geographic location or educational level.

6 Conclusion

For the first time in human history, those living in urban areas now outnumber those who live in rural areas. Looking at how far societies have developed within cities in such a short space of time creates an alluring area of study of what future cities will consist of. Writers have created works to describe their ideal cities and societies of the future for centuries, but also there have been works of dystopias where society collapses; such writings highlight the wants and concerns of their contemporary society. This study sought to explore how youth residing in Malaysia envision cities developing 100 years in the future. How will people work? Live? And move around? Thinking about the limitations of current cities, participants were asked to identify what they think are the key features of their imagined future city without the confines of current technologies and materials. Many of the ideas generated within this study and rated as highly important focus on simply improving existing services and infrastructure, something that most individuals would not expect to take 100 years, but it may be that participants believe some of these challenges are institutionally embedded and thus require a fundamental societal shift. Whilst there were a few ideas that were creative and futuristic, most statements were aimed at creating more liveable cities.

From these results, we can deduce that the participants of this study do not seek technologically advanced cities in the future but simply seek to make cities more human-centric. This human-centric emphasis can be seen through the clusters of rights, social development and welfare, which present much more 'soft' elements; whilst many of these elements require technology to enable them to be realized, technology components themselves were isolated and separated. The criticism of technologically advanced cities is reflected in the academic literature, as they maintain a focus on technology and lack the humanistic social element of attending to the needs of the community (Yigitcanlar et al., 2019) due to the trend of being focused on the business and professional classes (Hollands, 2008). Where ICT can assist citizen engagement, there hasn't been much research to explore the actual behaviours because of this ignorance of the underclasses and the marginalized (Granier & Kudo, 2016). This criticism contends that the emphasis on "smartness" does not ensure a city's sustainability and, in turn, liveability and that the intended outcomes can only be attained by developing the idea of a sustainable urban form (Almeida et al., 2018). Whilst predicting the future is unlikely to prove correct, many of the authors who wrote about their utopian dream have never seen it realized; the significance can be drawn from the narrative developed about which contemporary issues in cities our participants sought to resolve.

References

Aina, Y. A., Wafer, A., Ahmed, F., & Alshuwaikhat, H. M. (2019). Top-down sustainable urban development? Urban governance transformation in Saudi Arabia. *Cities, 90*, 272–281.

Almeida, V. A. F., Doneda, D., & da Costa, E. M. (2018). Humane smart cities: The need for governance. *IEEE Internet Computing., 22*(2), 91–95.

Cugurullo, F. (2018). Exposing smart cities and eco-cities: Frankenstein urbanism and the sustainability challenges of the experimental city. *Environment and Planning a: Economy and Space, 50*(1), 73–92.

Cugurullo, F., Acheampong, R. A., Dusparic, I., & Gueriau, M. (2020). The transition to autonomous cars, the redesign of cities and the future of urban sustainability. *Urban Geography, 42*(6), 833–859.

Granier, B., & Kudo, H. (2016). How are citizens involved in smart cities? Analyzing citizen participation in Japanese "Smart Communities." *Information Polity., 21*(1), 61–76.

Hannon, J., & Zaman, A. U. (2018). Exploring the phenomenon of zero waste and future cities. *Urban Science, 2*. https://doi.org/10.3390/urbansci2030090

Hollands, R. G. (2008). Will the real smart city please stand up? *City, 12*(3), 303–320.

Homer, S. T. (2022). Perceptions of smart sustainable cities: A scale development study. *Quality & Quantity*. https://doi.org/10.1007/s11135-022-01507-5

Homer, S. T. (2021). Perceived corporate citizenship: A scale development and validation study adopting a bottom-up approach. *Quality & Quantity*. https://doi.org/10.1007/s11135-021-01184-w

Iavarone, A. H. (2022). Seeking Cities of the Future: Techno-City Visions From the 1960s to the Present. *International Journal of*

Digital Innovation in the Built Environment, 11(3). https://doi.org/10.4018/IJDIBE.306257

Joss, S. (2018). Future cities: asserting public governance. *Palgrave Communications, 4*(36). https://doi.org/10.1057/s41599-018-0087-7

Kane, M., & Trochim, W. M. K. (2007). *Concept mapping for planning and evaluation*. Sage Publications Inc.

Kassens-Noor, E., & Hintze, A. (2020). Cities of the Future? The Potential Impact of Artificial Intelligence. *AI, 1*(192–197). https://doi.org/10.3390/ai1020012

Khan, I., Iftokhar, M. N., Ali, S. H., & Khalid, S. (2021). Cities and COVID-19: Navigating the new normal. *Global Sustainability, 4*(e12), 1–6.

Maestosi, P. C., Andreucci, M. B., & Civiero, P. (2021). Sustainable urban areas for 2030 in a post-covid-19 scenario: focus on innovative research and funding frameworks to boost transition towards 100 positive energy districts and 100 climate-neutral cities. *Energies, 14*.

Murphy, D. J. (2007). *People, plants and genes: the story of crops and humanity*. Oxford Scholarship Online.

Norman, B. (2018). Are autonomous cities our urban future? *Nature Communications, 9*. doi:https://doi.org/10.1038/s41467-018-04505-0.

Rogers, C. D. F. (2019). Realizing visions for future cities: an aspirational futures methodology. *Proceedings of the Institution of Civil Engineers—Urban Design and Planning, 174*(4), 125–140.

Rosas, S. R., & Camphausen, L. C. (2007). The use of concept mapping for scale development and validation in evaluation. *Evaluation and Program Planning, 30*(2), 125–135.

Rosas, S. R., & Kane, M. (2012). Quality and rigor of the concept mapping methodology: A pooled study analysis. *Evaluation and Program Planning, 35*, 236–245.

Sánchez-Corcuera, A., Nuñez-Marcos, A., Sesma-Solance, J., Bilbao-Jayo, A., Mulero, R., Zulaika, U., Azkune, G., & Almeida, A. (2019). Smart cities survey: Technologies, application domains and challenges for the cities of the future. *International Journal of Distributed Sensor Networks, 15*(6). https://doi.org/10.1177/1550147719853984

Sarantakos, S. (2005). *Social research* (3rd ed.). Palgrave MacMillan.

Schubert, D. (2019). Cities and plans—the past defines the future. *Planning Perspectives, 34*(1), 3–23.

Strielkowski, W., Zenchenko, S., Tarasova, A., & Radyukova, Y. (2022). Management of smart and sustainable cities in the post-COVID-19 Era: lessons and implications. *Sustainability, 14*.

Sturrock, K., & Rocha, J. (2000). A Multidimensional scaling stress evaluation table. *Field Methods, 12*(1), 49–60.

Washburn, D., Sindhu, U., Balaouras, S., Dines, R. A., Hayes, N., & Nelson, L. E. (2009). Helping CIOs understand "smart city" initiatives. *Growth, 17*(2), 1–17.

World Bank. (2020). *Urban development*. https://www.worldbank.org/en/topic/urbandevelopment/overview#1

Yigitcanlar, T., Kamruzzaman, M., Foth, M., Sabatini, J., da Costa, E., & Ioppolo, G. (2019). Can cities become smart without being sustainable? A systematic review of the literature. *Sustainable Cities & Society., 45*, 348–365.

Evaluating Daylighting Performance Within Existing Schools Expansion in Egypt

Nessrin El Helaly, Morad Abdelkader, Hanan Sabry, and Ashraf Nessim

Abstract

Daylighting is an essential resource for improving the indoor quality environment in school buildings. During the last decades, Egypt has faced some problems; the increase in annual school enrollment, high classroom density, and inadequate lands manifest in public existing schools expansion. Which may cause visual discomfort in classrooms. This study aims at evaluating the daylighting performance at public school expansion classrooms in two cities of Egypt: Cairo and Aswan. This study has considered two typical design prototypes for classrooms facing south orientation. Two situations for each classroom prototype have been analyzed and evaluated. In the first one, the two studied classrooms do not face any obstructions, while in the second, they face a neighborhood building. The daylighting analysis and simulation were based on Honeybee and Ladybug. Two aspects of daylighting analysis have been used to evaluate performance; Spatial Daylight Autonomy (sDA), and Annual Sunlight Exposure (ASE). This study showed that classrooms oriented to the south without external obstruction had a high value of ASE (58.85, 55.61%), and sDA (88.49, 88.79%) for Cairo City, while in Aswan ASE values were (66.14, 62.56%), and sDA were (89.32, 89.43%), it indicates sufficient daylight levels in classrooms. Moreover, it also showed glare problems. For the existing neighborhood buildings, the sDA values were less than (75%), whereas the ASE values did not exceed (10%). It is concluded that south orientation and obstructing buildings significantly impact the amount of daylight and visual comfort.

Keywords

Daylighting performance · Prototype classrooms · Angle of obstruction · With no obstruction · With neighborhood building

N. E. Helaly (✉)
Faculty of Engineering, Department of Architecture, Ain Shams University, Cairo, Egypt
e-mail: nesren_melhelaly@yahoo.com; g18052794@eng.asu.edu.eg

N. E. Helaly · M. Abdelkader · H. Sabry · A. Nessim
Architecture Engineering Department, Ain Shams University, Cairo, Egypt
e-mail: moradabdelkader@gmail.com

H. Sabry
e-mail: hanan_sabry@eng.asu.edu.eg

A. Nessim
e-mail: a.nessim@eng.asu.edu.eg

1 Introduction

Numerous standards and recommendations emphasize the value of daylight in classrooms (Syaheeza et al., 2018). Creating daylit spaces that encourage occupant comfort, health, and performance remains an unresolved problem for designers and building researchers (Zomorodian & Tahsildoost, 2019). During a renovation project, Salary et al. (2018) investigated how important school building characteristics significantly affect student test scores. The authors developed a building assessment tool to help designers and decision-makers with a comprehensive source of requirements for planning and renovating school buildings. Rahman and Tuhin (2019) stated that the appropriate use of daylight in educational facilities decreases operating costs, enhances students' vision and perception, and promotes their comfort and productivity. Daylight impact and artificial light on the performance of university students was examined by Alvarez (2020), who compared two groups of classrooms, the first was entirely artificially lit, and the other one had windows and mixed natural and artificial light. The author discovered clear evidence that Students perform better on exams in classrooms with windows than in basements (13.17 and 773%). In another

I. Pigliautile et al. (eds.), *Urban Resilience, Livability, and Climate Adaptation*,
Advances in Science, Technology & Innovation, https://doi.org/10.1007/978-3-031-54911-3_9

research, Authors investigated how primary school students' health and wellbeing are affected by daylight exposure. The primary dependent variables in this study were actigraphy data, which were documented. In contrast, to people who experience little or no daylight, their findings demonstrated that learning in classrooms with natural light will result in good sleep duration and quality (Boubekri et al., 2020).

Evaluating a design solution's daylighting performance is a challenging issue. Designers have reference to a few quantitative metrics for assessing this performance (Boubekri & Lee, 2017). The design and efficiency of light wells and other ways to catch daylight are developed and evaluated. Costanzo et al. (2018) encouraged the use of Climate-Based Daylight Modeling (CBDM) metrics to enhance daylight performance at elementary schools in Agira, Italy with various orientations, shapes, and functions. The authors stated that dynamic daylight simulations are essential for effectively guiding the design process. Akin et al. (2021) tackled the requirement for interactive and integrated design tools for immersive environments (IEs) towards improved enhancement of architectural solutions regarding daylighting illuminance. The authors Implemented the immersive tool (HoloArch), which promotes design advancements during the daylighting-informed design process. According to this study, (IEs) have great potential for performing design processes in conceptual architectural domains and providing immediate feedback. Mohelníková et al. (2020) examined the interior thermal and visual environments in school classrooms in a few chosen school buildings in temperate regions of central Europe.

The evaluation of the school building envelope revealed that the structures remain extremely efficient despite being renovated. Thermal stability inside buildings and daylighting efficiency are impacted by solar gain. De Luca et al. (2019) compared the accuracy of more reliable climate-based daylight simulations to the static daylight factor (DF) employing spatial daylight autonomy. Results revealed a deficiency in the DF's ability to estimate how much daylight should be allowed into interior building spaces in Tallinn, Estonia. Bakmohammadi and Noorzai (2020) examined a novel approach that maximizes visual comfort while minimizing lighting energy consumption. Results indicated that this new approach offers architects a framework for creating a suitable classroom that meets their needs. Zhang et al. (2017) presented simulation optimization tools to find the optimal solution for reducing energy demand by 28–24%, Thermal discomfort in summertime by 9–23%, and increasing UDI by 15–63% in three types of schools (buildings) in China considering; orientation, room depth, corridor depth, window- to- wall ratio WWR, and glazing material, and shading devices. Ziaee and Vakilinezhad (2022) studied how the sky cloudiness affected the ideal light shelf characteristics specified for typical classrooms in two Iranian cities. Results indicated that using a light shelf with WWR in Tehran, which had a less cloudy sky, was superior to Sari in terms of supplying daylight and thermal comfort. Moreover, a correlation between the rotation angle and the amount of daylight in each city concerning WWR.

Internal and external elements affect the quality and amount of daylight entering a building. The surface reflection of the environment, such as the ground and exterior barriers, impacts the externally reflected component. Internal surfaces, such as wall, floor, and ceiling finishes and furniture, affect the internally reflected component (Mohelnikova & Hirs, 2016). Maria and Prihatmanti (2017) investigated the daylight features in high schools in historical buildings. Due to several impediments discovered, the results demonstrated that the amount of daylight entering the classrooms was below the required amount. Tharim et al. (2022) examined daylighting levels in Malaysian classrooms using a range of parameters that affect daylighting design failure factors, such as bad orientation, limited glazing areas, double-loaded buildings with single-loaded windows, and the closeness of the existing structures. They showed that the classroom design is unsuitable for managing daylight. Santos et al. (2017) studied the availability of daylight in congested urban regions in a reference room with varied depths in Canyon. The authors observed the relationship between the sky angle and the sDA, taking into account the dimensions of the street and any obstructing neighborhood buildings. Regression analysis results confirmed that room depth was typically the most important factor in maximizing indoor daylight.

The passive design must benefit from the environment to improve illumination, particularly solar radiation. As a result, building shading design is essential as a passive strategy to maximize indoor visual comfort and decrease building energy consumption by managing the amount of solar radiation quantity (Hernández et al., 2017). The building's orientation and the lack of suitable shading are the main causes of the problems. Octarino (2022), studied the indoor daylighting quality and energy use efficiency through designing the building envelope in the Kanisius Kalasan Junior High School building in Indonesia. The study was conducted by simulating the building's design and construction drawing document. The author reported that shading devices significantly influence the building's performance. In order to improve daylight performance for classrooms with bilateral openings in an Indonesian City, Mangkuto et al. (2022) proposed design options for outside shade devices with different depths and heights. The interaction between a shade device's depth and elevation is unique as was found, depending on the direction of the structure and the presence of direct sunlight. Moreover, shading elevation has more of an impact than shading depth.

Windows in buildings are generally intended to admit daylight, appreciate vistas, and allow for cross ventilation. Numerous studies looked at the WWR ratio to determine the best window-to-façade ratio. Ayoosu et al. (2021), studied the daylighting performance of ten existing lecture halls within eight universities in Nigeria. The author revealed that increasing WWR from 13 to 30% enhanced the daylighting performance. Dewi et al. (2022) investigated the daylighting distribution due to WWR in the educational building in Depok, Indonesia. WWR was considered from 30 to 60% using glazing material of different shading coefficients (SC), and it was found that WWR (40%) with SC (0.42) and WWR (60%) with SC (0.95) obtained the desired outcomes for the north and south façade respectively to opaque double skin façade building (DSF). In Tehran, Iran, Zomorodian and Tahsildoost (2017) analyzed the impact of window design (WWR, window shading, orientation, window U value, and glazing SGC solar heat gain coefficient) on a classroom model with a north and south façade. WWR, glazing SHGC (solar heat gain coefficient), and window shading were the most effective factors. Motazedian (2019), revealed that windows with WWR from 25 to 35% on the south façade achieve the best daylighting performance in primary school classrooms.

For southern facades, various studies have recommended relatively simple and effective daylighting techniques, but addressing the problem is more challenging for less desirable orientations. Michael and Heracleous (2017) put up suggestions for enhancing visual comfort in Cyprus schools. The research was conducted using simulation models and a questionnaire. Their findings demonstrated that glare occurs in classrooms that are oriented to the east and west due to strong lighting contrast and intense visible light sources in the field of view. A parameterization approach developed by Wen et al. (2023) in Shanghai University Library, China, using four shading techniques: vertical slats (VS), a performed aluminum sheet (PAS), serrated windows with southern orientation (SN–S), and serrated windows with northern orientation (SW-N) According to their results, (SW–S) and (PAS) performed quite poorly, whereas (SW–N) offered a higher degree of illumination and was more successful at eliminating glare. Samadi et al. (2020), created a kinetic shading system with independent units that rotate in three dimensions around their centers and move in two dimensions in response to the sun. According to their results, the employment of such a shade system in ideal circumstances can significantly enhance the efficiency of interior daylight.

Much research has been done to approach Egyptian public schools in the field of daylighting performance, and the majority of this research has been carried out in the environmental analysis. Saleem et al. (2016), assessed the existing indoor thermal comfort in Egyptian public primary school buildings. Abdin and Mahmoud (2017), investigated the variables (orientation, WWR, U value, SHGC, shading, and occupancy) with adverse impacts on the energy consumption of classrooms of schools in Giza, Egypt. They established parameters for optimization for thermal comfort, preparing a checklist to follow for the design. Abdallah (2017), analyzed the children's classrooms in two government schools that were studied for thermal comfort and indoor CO_2 concentration. New Assuit City, Egypt. On the other hand, a few studies investigated visual comfort: Rakha et al. (2023), investigated the enhancement of daylighting efficiency in a classroom at a public school in Cairo, Egypt, as a result of retrofitting (shading units, number of slats rows, depth, and inclination angle).

The authors' result showed that a reduced amount of space exposed to direct sunlight by 91% and decreased ASE from 31.5 to 2.7%.

During the last decades, Egypt faced some problems. The majority of schools in metropolitan areas have overcrowding owing to the restricted number of available classrooms, according to information reviewed by the Egyptian Ministry of Education. That affected the educational process. All these lead to real problems in existing public school expansion, like increased annual school enrollment, high classroom density, and inadequate lands. It has been found that the proper orientation is not always realized, mainly due to the limited space allocated for the new building to be added to the existing school. This may lead to allocating the building in less desirable orientations (south, east, or west). Secondly, the presence of external obstruction may affect the efficiency of daylighting performance in classrooms as shown in Fig. 1 which presents four examples of a public school's existing expansions. Therefore, this study gives an insight into visual performance for existing public school expansion in Egypt through a series of daylighting simulations. Figure 1 represents four examples of existing school expansion layouts.

2 The Aim of the Research

The aim of the present study was to evaluate daylighting performance within classroom expansion in two cities of Egypt: Cairo and Aswan. This work was concerned with the visual performance of the classrooms. It investigates the effect of south orientation on indoor daylighting performance in two situations. First, without obstruction, and second with neighborhood buildings. Only the results relating to the classrooms are presented in this paper.

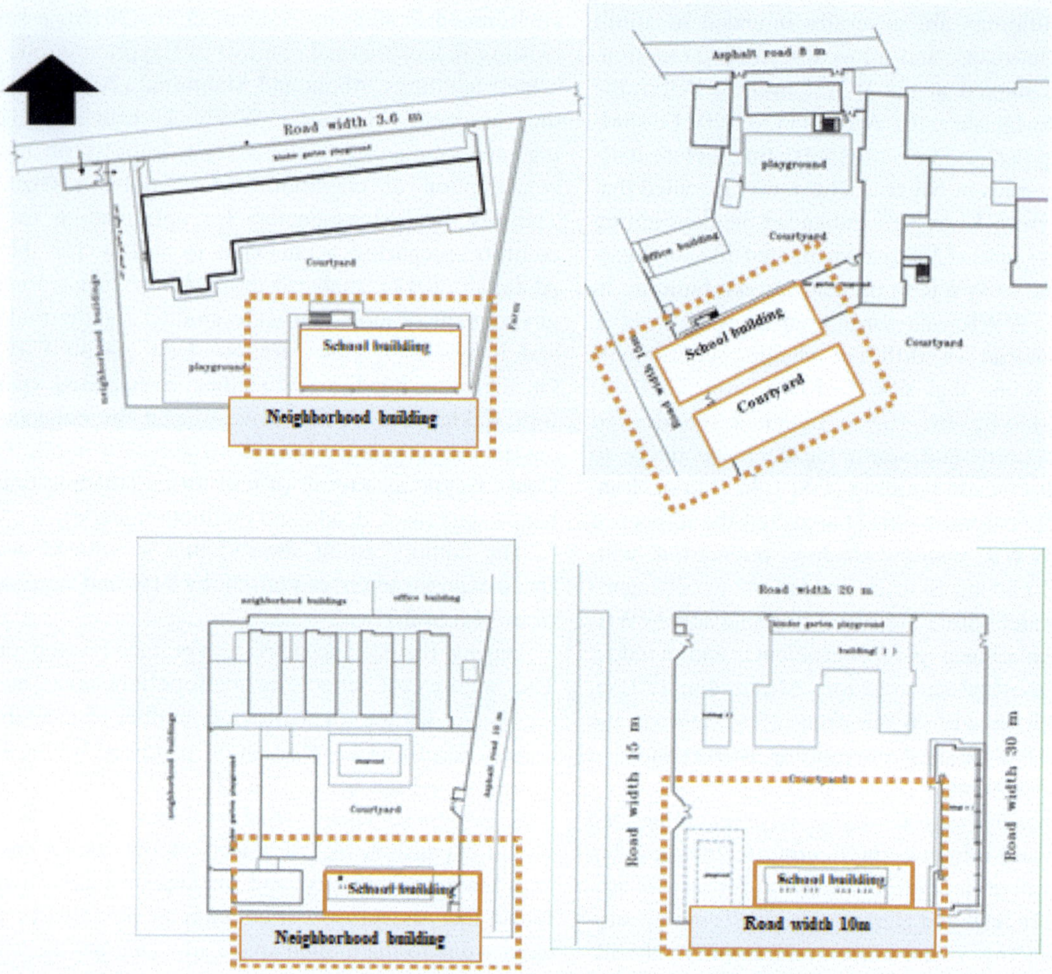

A) School building (with obstruction) B) School building with (no obstruction) C) School building (with obstruction) D) School building with (no obstruction)

Fig. 1 Four examples of a public school existing expansion: A, B, C, and D. *Source* (GAEB, 2021)

Objectives

- Highlighting the literature review concerning daylighting performance in school buildings.
- Defining the problems in public existing school expansion in Egypt. Considering the daylighting performance and selected case studies for examination and analysis.
- Providing daylight performance simulation to existing classroom expansion can be useful in further research to improve visual comfort.

3 Methodology

This study included three parts; the first is the literature review concerning daylighting performance in school buildings. The Second defines the problem and the selected case studies. The third part is the simulation process details concerning the tools introduced in this study.

3.1 Case Studies

The General Authority of Educational Buildings (GAEB, 2021) regulates public school buildings in Egypt and is responsible for designing and constructing school buildings. GAEB designed two prototypes (single and double-row linear forms) arranged along the corridor, as shown in Fig. 2. Furthermore, classrooms in the single model have typical windows on the corridor facing the school's inner courtyard; the prototypes are four stories. Orientation is one of the essential factors in building design to use daylight and conserve energy. Buildings with good orientation maximize daylight receiving through building facades and require less artificial lighting. In countries like Egypt, the sun's apparent position is mainly in the southern part of the sky hemisphere. Since the geographical locations are in the northern hemisphere, sunshine duration is relatively more extended at the south orientation. Hence, this study evaluated daylighting performance in a generic south-oriented classroom in two cities in Egypt; Cairo City has a hot desert climate (30.0444° N, 31.2357° E) (weather file EGY-Cairo. 623,660. IWEC), with more than 184 sunshine days/per year. And Aswan City (24.0889° N, 32.8998° E), with 210 sunshine days per year (the highest average of sun hours daily) (EGY-Aswan. 624,140. IWEC). Two prototypes of school buildings that GAEB commonly used are selected for investigation, single-loaded case (A) and double-loaded case (B) as shown in Fig. 2 (GAEB, 2021).

The dimensions and parameters in each type are the same, as shown in Fig. 3 and Table 1. In each prototype, the windows' external façade faces two situations; first, there is no external obstruction, while second, a neighborhood building is considered an external obstruction. External obstruction affects daylighting performance in two ways. First, the amount of sky that is blocked. Second, The external surface material (reflectivity), can be seen as the luminance reflected from the obstructed surfaces. The angle of obstruction could be used to show the impact of an external obstruction (Li et al., 2006) as shown in Fig. 4.

The obstructions in real cases may always be irregular in both the horizontal and vertical surfaces. This study extends to evaluate daylighting performance under irregularly obstructing building heights in urban regions of Egypt. By specifying various angles, it is possible to analyze the numerous potential combinations of building highest. Hence, the obstructing buildings' heights used for the simulation were (9, 12, and 21 m). Three angles of obstruction (Ø) are analyzed in this study; (36°, 61°, and 79°). GAEB prescribes 3m as a minimum distance between the extension school and obstruction buildings. Therefore, three alternatives were analyzed, as shown in Fig. 5.

Alternative 1: (Ø) = 36° from the windowsill.

Alternative 2: (Ø) = 61° from the windowsill.

Alternative 3: (Ø) = 79° from the windowsill as shown in Fig. 5.

3.2 Daylighting Evaluations

Annual simulations were run to analyze the daylighting performance inside classrooms; by calculating the following

Fig. 2 A typical plan of the two prototype models by GAEB pointed out the studied classroom and a section. *Source* (GAEB, 2021)

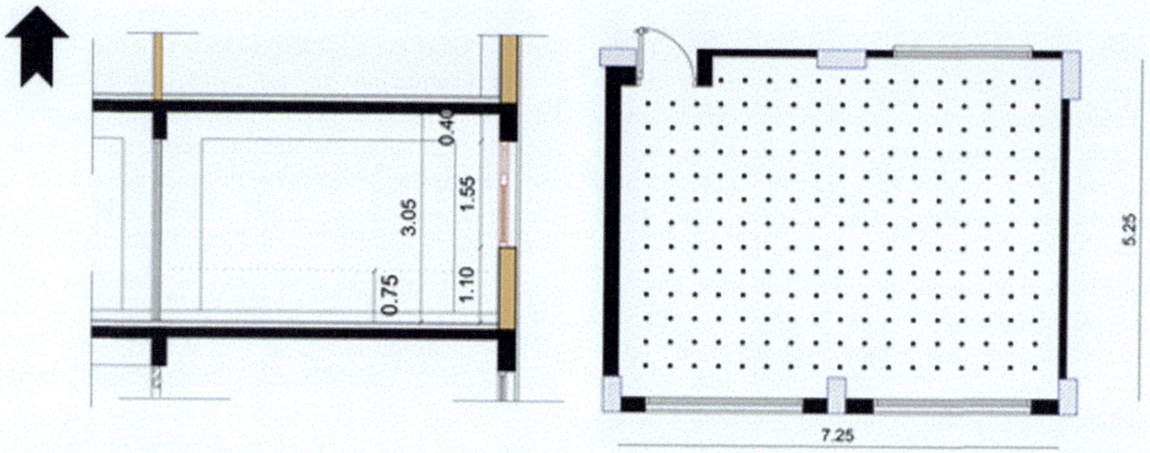

Fig. 3 A typical plan for the classroom for the two models, and section

Table 1 Classroom parameters, surface material, and its reflectance values. *Source* (GAEB, 2021)

Space dimension			
7.25*5.25*3.05 m			
Window dimensions			
Window sill	Window Height	Window width	Window dimensions
1.10 m	1.55	2.6*2	36%
Internal surfaces			
Surfaces		Materials	Reflectance percentage value
Wall		White plastic paintings	70
Ceiling		White plastic paintings on cement plater	70
Window frames		Aluminum sections	90
Window glazing		Single glazing 6 mm thick	88
Tiles of indoor floors		Cementitious terrazzo-mosaico-tiles	40
The chosen studied classroom is located on the second floor			

metrics (sDA), and (ASE) according to USGBC (2014) LEED v.4. The two metrics were chosen to be used in this study because they give information on both the availability of daylight and the possibility of glare from undesired direct sunlight. A validated software program, Rhinoceros v6 is applied for modeling the two prototype classrooms, it is a Non-Uniform b-Splines (NURBS) modeling program that uses Honeybee 1.6.0 as a plugin to link Radiance. All typical classrooms are supposed to be occupied all year from September to May, 5 days from 8:00 am to 2:00 pm. The daylighting evaluations were made for classrooms on the second floor, with grid sensor points placed on a "working plane" at the desk height 0.75 m above the floor. The analysis grid was set at 0.40*0.40 m.

The assessment through "Spatial Daylight Autonomy ($sDA_{300Lux,50\%}$) is defined as evaluating a space that receives adequate daylight. It specifically refers to the proportion of floor area that receives at least 300lx for at least 50% of the yearly occupied hours. While Annual Sunlight Exposure (ASE) outlines how much of the area is exposed to excessive direct sunlight" (Wagdy et al., 2017), which can cause visual discomfort and glare. Annual Sunlight Exposure ASE measures the percentage of floor area receiving at least 1000 lx for at least 250 occupied hours yearly. These metrics indicate daylighting sufficiency and visual conditions. Radiance parameters for (sDA), and (ASE) are shown in Table 2.

According to $LEED_{v4}$ requirements, two conditions for the acceptance criterion were adopted in this study. First, the percentage of Spatial Daylight Autonomy (sDA) should meet a minimum requirement of 75% of the whole classroom area. Second, the Annual Sunlight Exposure (ASE) should not exceed 10% to ensure efficient performance.

4 Simulation Results

4.1 Results of Case (a) Single-Loaded Prototype: In the First Situation

Without external obstruction, all classrooms under study have been simulated. Through assessment of the daylight, performance is given in Table 3 by means of dynamic daylight measurements. However, in all classrooms, the (sDA) values (88.4%, 89.32%) are higher than 75%, which means the classroom passes (sDA) requirements for $LEED_{v4.}$ It

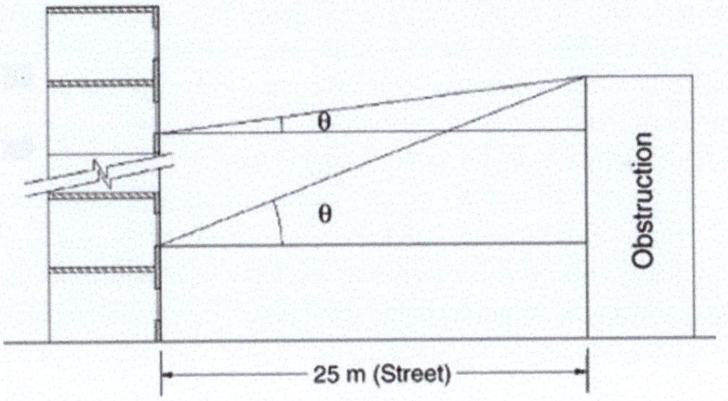

Fig. 4 The definition of the angle of external obstruction (Ø). *Source* (Li et al., 2006)

indicated that classrooms are fully daylighted throughout the year, with better daylight distribution. An image of daylighting simulation can be seen in Cairo City as shown in Fig. 6 and Aswan City in Fig. 7. In addition, as indicated by results as shown in Table 3, classrooms suffer from high (ASE) values of (58.85% and 66.14%) which exceed 10%, which means the classroom fails to achieve the (ASE) requirements for $LEED_{v4.}$ South-facing classrooms suffer from visual discomfort. This is due to variations in the solar position and the significant effect of south orientation. Figure 8 shows a comparison between (sDA & ASE) values in the selected two cities.

Table 2 Radiance parameters for sDA, and ASE. *Source* Author

Metrics	Ambient bounces	Ambient divisions	Ambient accuracy	Ambient resolution	Ambient sampling
sDA	6	1000	20	0.1	300
ASE	0	1000	20	0.1	200

Table 3 Results for case (A) without external obstruction

Metrics	Cairo (%)	Aswan (%)
sDA	88.40	89.32
ASE	58.85	66.14

4.2 Results of Case(a) Single-Loaded Prototype: In the Second Situation

Single-loaded prototype classrooms face a neighborhood building. There are three angles of obstructions; each angle has its results for the two cities shown in Table 4. Each classroom in the three alternatives has been simulated. It was observed that there is a relatively low daylight level for classrooms, as is evidenced by values of (sDA).

Figures 9 and 10 show daylighting simulation results for the three alternatives in two selected cities. The (sDA) values for classrooms are around (31.94, 28.04, and 25.52%) in Cairo and (33.61, 29.53, and 26.82%) in Aswan. However, (ASE) values in both cities meet the assessment

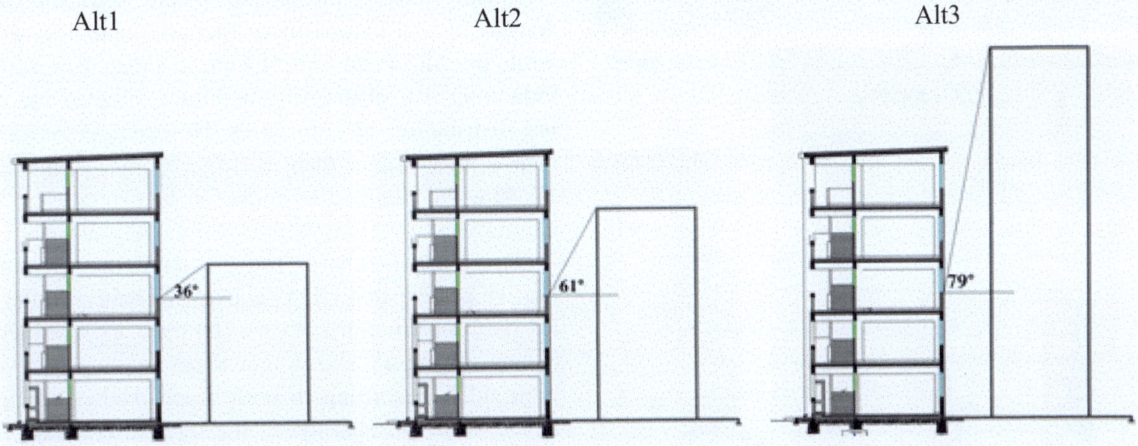

Fig. 5 The three alternatives for angles of obstruction. *Source* Author

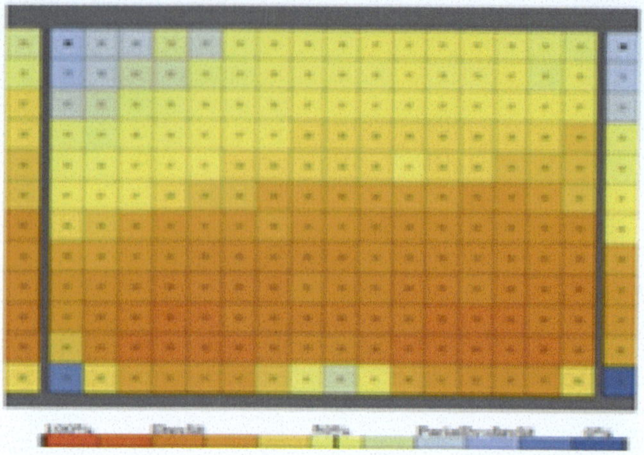

sDA= 89.32%

Fig. 6 Simulation results for case (A) in Cairo city (without obstruction). *Source* Author

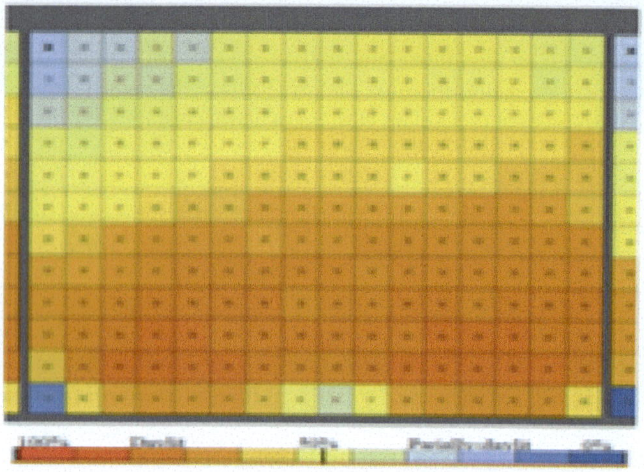

sDA= 88.4%

Fig. 7 Simulation results for case (A) in Aswan City (without obstruction). *Source* Author

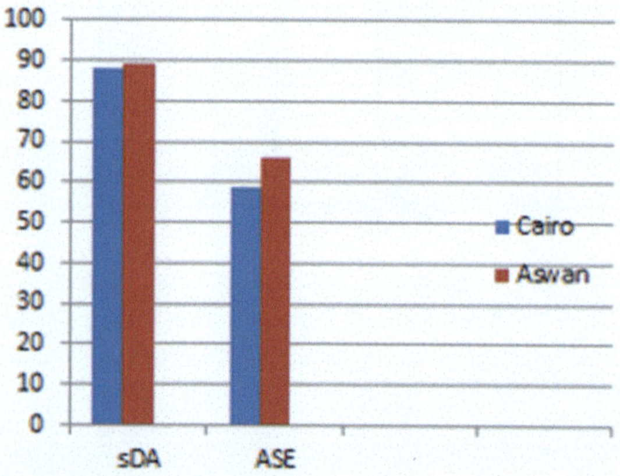

Fig. 8 Simulation results for the two cities for case (A) without obstruction. *Source* Author

criteria of not exceeding 10%. Figure 11 shows two charts of (sDA) and (ASE) values in the selected two cities which have similar performance. It also shows the presence of a neighbor has a positive effect in terms of improving the (ASE), while it declined the (sDA) values.

Figure 12 shows the graph chart that indicates slight differences on both sides, and the illuminance decreases dramatically. The three alternative daylight distribution levels don't meet the required range (300 lx).

4.3 Results of the Case (B) Double-Loaded Prototype: In the First Situation

Without obstructing buildings, (sDA) evaluation values of each classroom range from 88.79% in Cairo to 89.45% in Aswan, as represented in Table 5. This has a great implication for visual comfort. While eliminating glare seems challenging as (ASE) values of 55.61% in Cairo and 62.56% in Aswan. Figures 13 and 14 presents the simulation results. Figure 15 shows the sDA and ASE values in the two cities.

4.4 Results of Case (B) Double-Loaded Prototype: In the Second Situation

Table 6 shows the simulation results of the three alternatives. In this situation, the indoor illumination level was below the standards according to LEED_{v4}. With (sDA) values were (41.69, 35.84, and 32.5%) in Cairo and (48.56, 45.2, and 42.8%) in Aswan. Where ASE for Cairo and Aswan in (Atl.1) was more than 10% but less than 12%, while in (Alt.2) and (Alt.3) were less than 10% results also, indicating that obstructing buildings affected the daylighting distribution in both cities. However, in Aswan, (sDA) values were higher than in Cairo by 10%. Figures 16 and 17 show the simulation results for the two cities in the three alternatives.

The analysis charts for the daylight performance in Fig. 18 present that (sDA) and (ASE) values in the two cities (with obstruction) in case (B) have approximate performance. Figure 19 shows that there are small variances on both sides, according to both graph readings. The illuminance drops off sharply. In the three alternatives, daylight distribution levels don't meet the required range (300 lx).

5 Discussion

This study's results are divided into two groups; without obstruction (first situation) and the second with an obstacle to daylight (second situation). Results were tabulated, and an analysis was conducted to identify potential and

Table 4 Results for case (A) with external obstruction for the two cities. *Source* Author

Angle of obstruction	Cairo		Aswan	
	Sda (%)	ASE (%)	sDA (%)	ASE (%)
Alt.1 (Ø) = 36°	31.94	6.25	33.61	6.25
Alt.2 (Ø) = 61°	28.04	5.20	29.53	2.08
Alt.3 (Ø) = 79°	25.52	0	26.82	0

limitations. It was neglected if any differences reached less than or equal to 10% in the result.

In the First Situation

The analysis in the two cities indicated that the more daylight penetrates classrooms, the more sDA and ASE values increase. This result is more compatible with previous studies (Theodorson, 2008), and could be attributed to the effect of shading devices on southern classrooms. However, there is sufficient daylight in the two cases, accompanied by excessive brightness ratios and visual discomfort. That is related to the high value of ASE. When comparing the results of the two cases in the two cities, as shown in Fig. 20, it is observed that there's a slight difference between the results. It can be said that the two prototypes have similar daylighting performance.

In the Second Situation

(with obstruction), Cases (A) and (B) in the three alternatives revealed poor daylighting. The presence of obstructing buildings reduces the excess of daylight proximity to the window. The performance charts in Fig. 21 indicate that the neighbor has a positive effect in improving the ASE values; on the other hand, it has a negative impact in

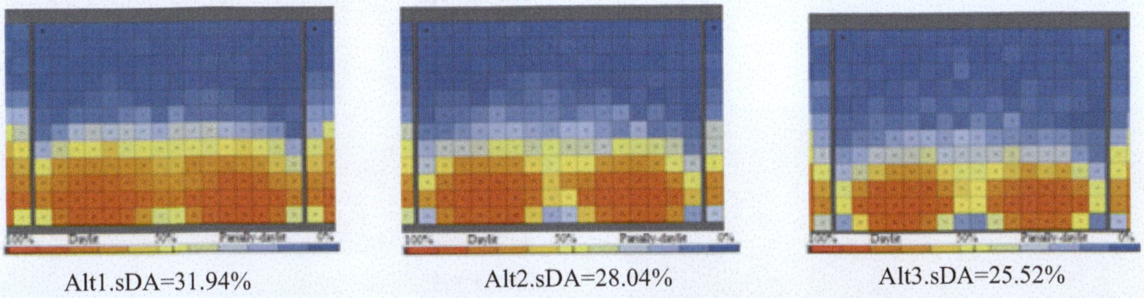

Fig. 9 The simulation results for the three alternatives in Cairo for case (A) (with obstruction). *Source* Author

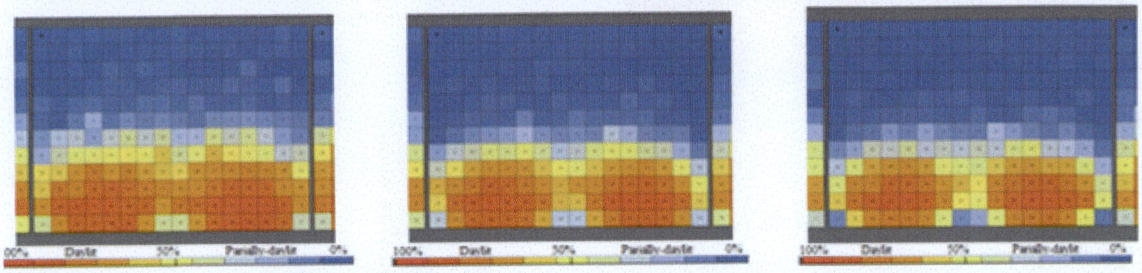

Fig. 10 The simulation results for the three alternatives in Aswan for case (A) with obstruction. *Source* Author

Fig. 11 Simulation results for the three alternatives of the two cities, case (A) (with obstruction). *Source* Author

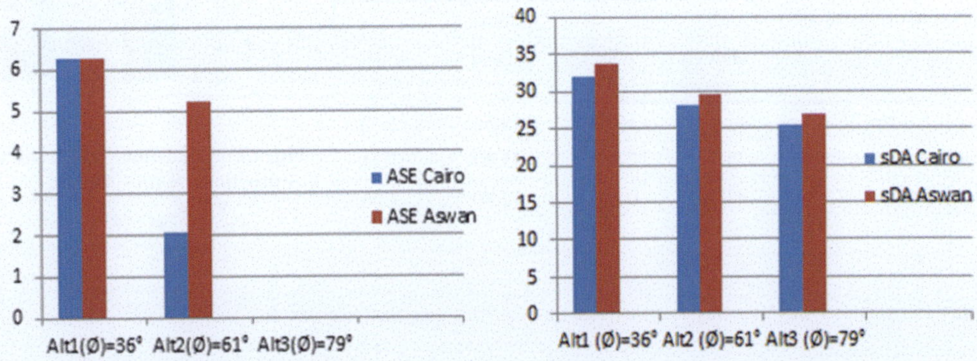

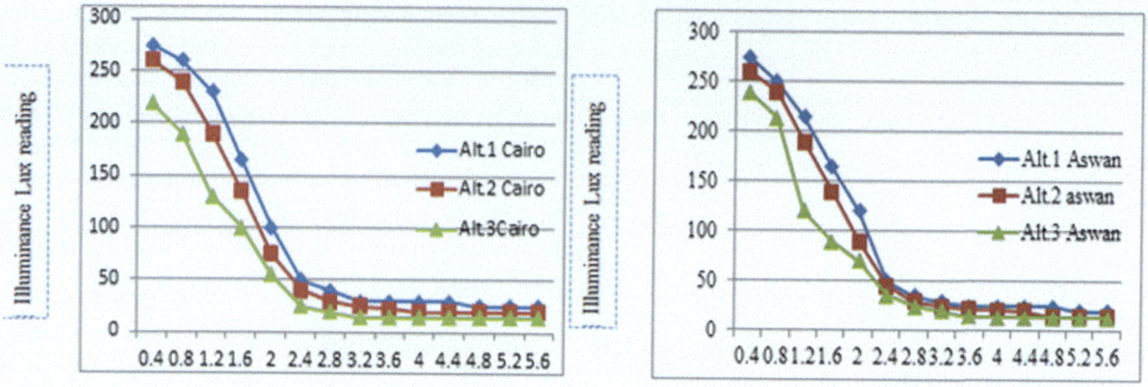

Fig. 12 Graph chart of the cumulative distribution of the hourly interior daylight illuminance with distance from window in the two cities for the whole year case (A) (with obstruction). *Source* Author

Table 5 Results of case (B) without obstruction. *Source* Author

Metrics	Cairo	Aswan (%)
sDA	88.79	89.45
ASE	55.61%	62.56

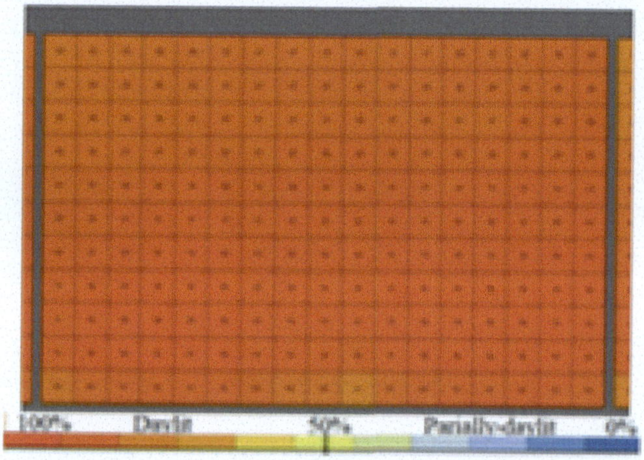

sDA=89.45%

Fig. 14 Simulation result for case (B) in Aswan city (without obstruction). *Source* Author

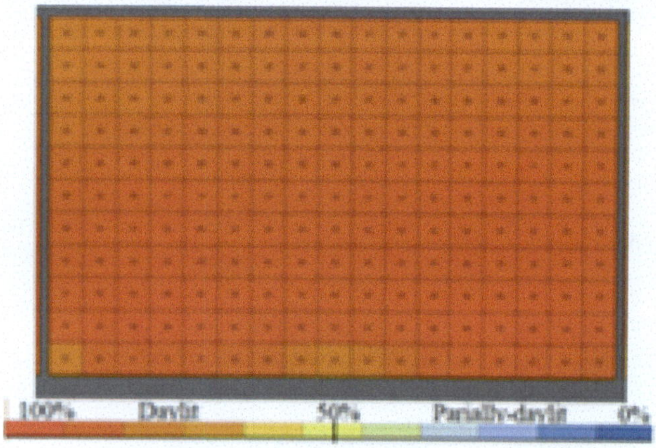

sDA=88379%

Fig. 13 Simulation result for case (B) in Cairo city (without obstruction). *Source* Author

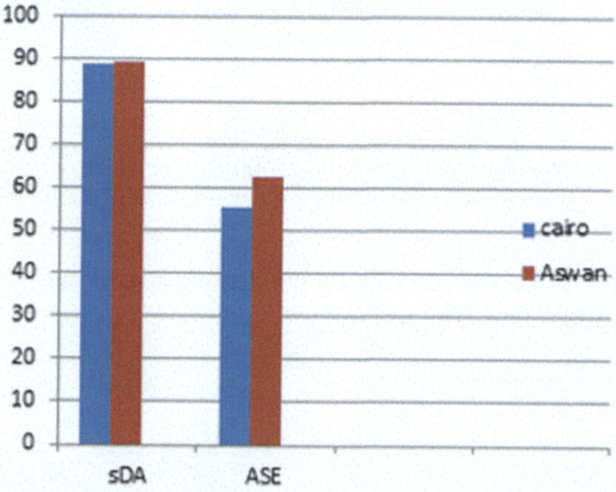

Fig. 15 Simulation results in the two cities in case (B) (without obstruction). *Source* Author

Table 6 Results of case (B) (with external obstruction). *Source* Author

Angle of obstruction	Cairo		Aswan	
	sDA (%)	ASE (%)	sDA (%)	ASE (%)
Alt.1 (Ø) = 36°	41.69	10.69	48.56	11.22
Alt.2 (Ø) = 61°	35.84	7.48	45.20	5.34
Alt.3 (Ø) = 79°	32.50	0	42.80	1.06

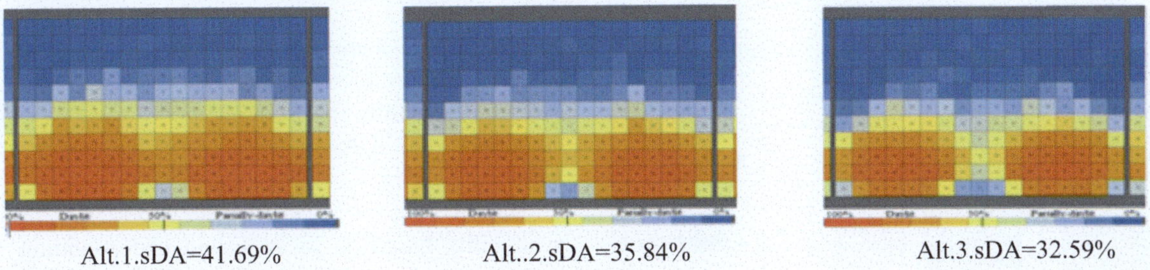

Fig. 16 The simulation results for the second situation for the three alternatives in Cairo City for case (B) (with obstruction). *Source* Author

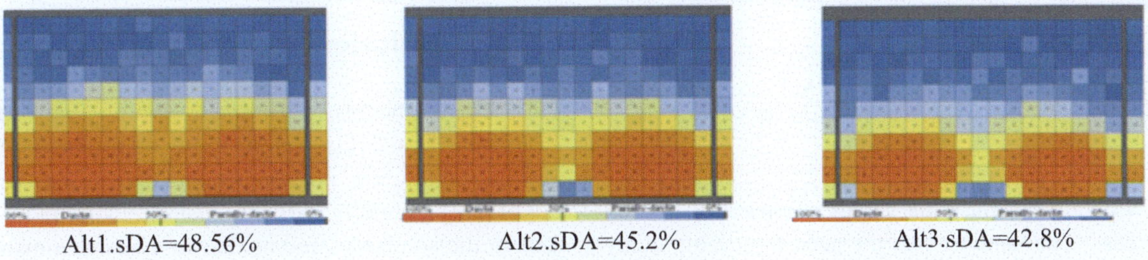

Fig. 17 The simulation results for the second situation for the three alternatives in Aswan City for case (B) (with obstruction). *Source* Author

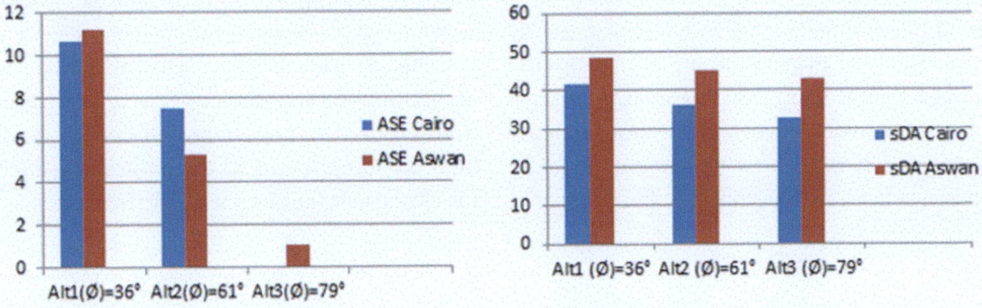

Fig. 18 Simulation results for the three alternatives of the two cities, case (B) (with obstruction). *Source* Author

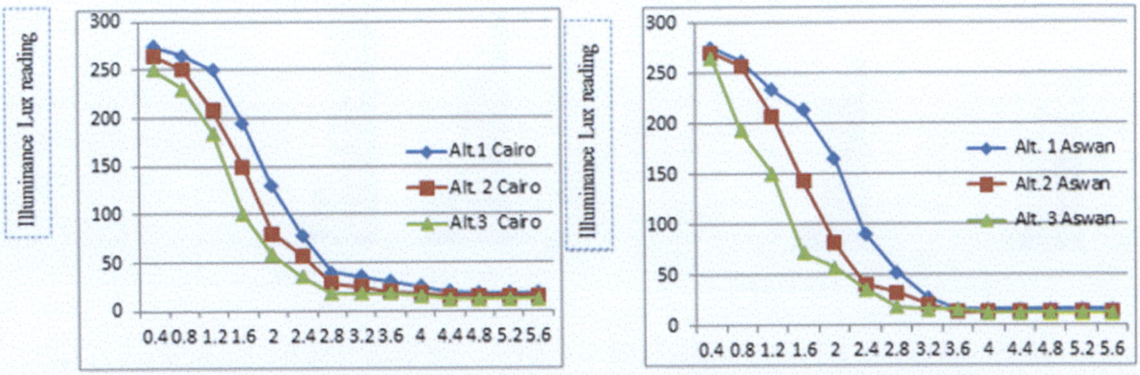

Fig. 19 Graph chart of the cumulative distribution of the hourly interior daylight illuminance with distance from window in the two cities for the whole year case (B) (with obstruction). *Source* Author

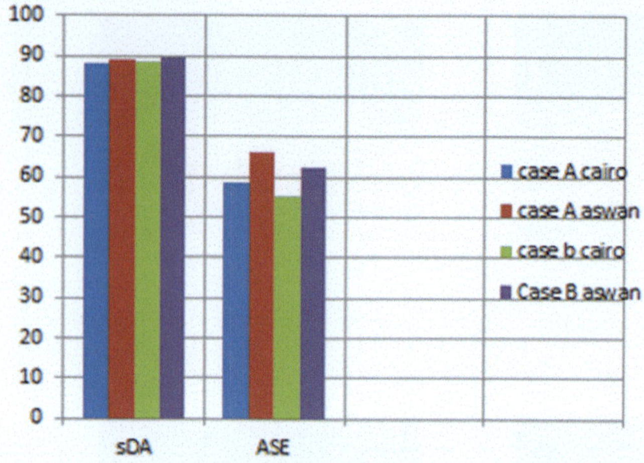

Fig. 20 Shows the difference of the Sda, and ASE values between the two cities in cases (A), and (B) (with no obstruction). *Source* Author

terms of a decline in the sDA values compared to the no-neighbor case. This underscores the need to develop a daylight system to enhance daylighting performance, such as a high reflecting façade surface material, varying WWR, etc. In general, the three alternatives in the two cases have an almost similar increasing rate of change; the daylight levels obtained from all classrooms appear to be inadequate. It is observed that the effects of orientation are not as significant. And it is noticed that at (Ø) of 36° or more, the daylighting performance was lower than $LEED_{v4}$ requirements considering sDA values. It relates to the severe distance between the extension school and neighboring buildings.

Location is a crucial consideration. It can be seen that Aswan's findings are higher than Cairo's; the discrepancies are greater since there are more sunny days in Aswan than in Cairo. Generally speaking, it can be seen that as latitudes decrement, the sDA values increase, but not by the same amount.

The findings demonstrate the importance of daylight availability based on location and understanding how the built environment affects interior conditions. To make decisions that maximize the use of daylight within illuminance ranges that are useful for performing visual tasks.

6 Conclusion

This research begins to document the tangible differences in south-oriented classrooms with and without obstruction that may occur in existing public school expansion in Egypt. A potentially difficult visual environment is created in the south classrooms due to dynamic sunshine patterns and significant proportional fluctuations in daylight distribution.

It is recommended that further research be done to improve daylighting performance in existing school expansion in Egypt, demonstrating the need for external shading systems to reduce glare. However, the shades should be orientation-appropriate, effective, and easily manipulated. Reinforcing the importance of correct control and application, blending modified daylight and reduced artificial light for classroom building with obstruction.

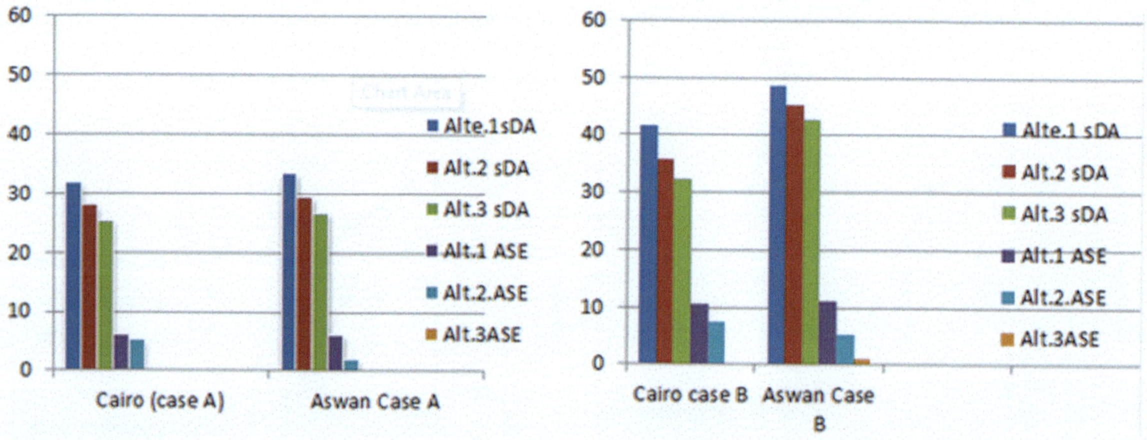

Fig. 21 Shows a comparison of the three alternatives in cases (A), and (B) in Cairo, and Aswan (with obstruction). *Source* Author

References

Abdallah, A. S. H. (2017). Thermal monitoring and evaluation of indoor CO2 concentration in classrooms of two primary governmental schools in New Assiut City, Egypt. *Procedia Engineering, 205*, 1093–1099.

Abdin, T., & Mahmoud, A. H. A. (2017). A checklist for the assessment of energy performance of public schools in Cairo, Egypt. In *International Conference for Sustainable Design of the Built Environment-SDBE London* (p. 269).

Akin, S., Ergun, O., Surer, E., & Gursel Dino, I. (2021). An immersive performative architectural design tool with daylighting simulations: A building information modeling-based approach. *Engineering, Construction and Architectural Management, 28*(4), 1319–1344.

Alvarez, S. P. (2020). Natural light influence on intellectual performance. A case study on university students. *Sustainability, 12*(10), 4167.

Ayoosu, M. I., Lim, Y. W., Leng, P. C., & Idowu, O. M. (2021). Daylighting evaluation and optimisation of window to wall ratio for lecture theatre in the tropical climate. *Journal of Daylighting, 8*(1), 20–35.

Bakmohammadi, P., & Noorzai, E. (2020). Optimization of the design of the primary school classrooms in terms of energy and daylight performance considering occupants' thermal and visual comfort. *Energy Reports, 6*, 1590–1607.

Boubekri, M., & Lee, J. (2017). A comparison of four daylighting metrics in assessing the daylighting performance of three shading systems. *Journal of Green Building, 12*(3), 39–53.

Boubekri, M., Lee, J., Bub, K., & Curry, K. (2020). Impact of daylight exposure on sleep time and quality of elementary school children. *European Journal of Teaching and Education, 2*(2), 10–17.

Costanzo, V., Evola, G., Marletta, L., & Pistone Nascone, F. (2018). Application of climate based daylight modelling to the refurbishment of a school building in Sicily. *Sustainability, 10*(8), 2653.

Dewi, O. C., Rahmasari, K., Hanjani, T. A., Ismoyo, A. D., & Dugar, A. M. (2022). Window-to-wall ratio as a mode of daylight optimization for an educational building with opaque double-skin façade. *Journal of Sustainable Architecture and Civil Engineering, 30*(1), 142–152.

GAEB. (2021). *Authority prototypes (Design standards, and construction specifications).* General Authority of Educational Buildings, Architectural Design Department Research and studies administration, Cairo

Hernández, F. F., López, J. M. C., Suárez, J. M. P., Muriano, M. C. G., & Rueda, S. C. (2017). Effects of louvers shading devices on visual comfort and energy demand of an office building. *A Case of Study. Energy Procedia, 140*, 207–216.

Li, D. H., Wong, S. L., Tsang, C. L., & Cheung, G. H. (2006). A study of the daylighting performance and energy use in heavily obstructed residential buildings via computer simulation techniques. *Energy and Buildings, 38*(11), 1343–1348.

De Luca, F., Kiil, M., Simson, R., Kurnitski, J., & Murula, R. (2019). Evaluating daylight factor standard through climate based daylight simulations and overheating regulations in Estonia. In *Proceedings of the 16th IBPSA Conference* (Vol. 2, p. 4).

Mangkuto, R. A., Koerniawan, M. D., & Yuliarto, B. (2022). On the interaction between the depth and elevation of external shading devices in tropical daylit classrooms with symmetrical bilateral openings. *Buildings, 12*(6), 818.

Maria, Y. S., & Prihatmanti, R. (2017). Daylight characterization of classrooms in heritage school buildings. *Planning Malaysia, 15*.

Michael, A., & Heracleous, C. (2017). Assessment of natural lighting performance and visual comfort of educational architecture in Southern Europe: The case of typical educational school premises in Cyprus. *Energy and Buildings, 140*, 443–457.

Mohelníková, J., Novotný, M., & Mocová, P. (2020). Evaluation of school building energy performance and classroom indoor environment. *Energies, 13*(10), 2489.

Mohelnikova, J., & Hirs, J. (2016). Effect of externally and internally reflective components on interior daylighting. *Journal of Building Engineering, 7*, 31–37.

Motazedian, F. (2019). Analysis of optimum window-to-wall ratio in horizontally expanded and vertically expanded windows in Tehran, Iran. *International Journal of Architectural Engineering & Urban Planning, 29*(1), 61–68.

Octarino, C. N. (2022). Evaluation of building envelope performance on the kanisius kalasan junior high school building Yogyakarta. In *International Webinar on Digital Architecture 2021 (IWEDA 2021)* (pp. 187–191). Atlantis Press.

Rahman, F., & Tuhin, M. M. H. (2019). Daylight impact on learning environment in classrooms of secondary high school at Ishwardi, Pabna, Bangladesh. *Science, 300*, 500.

Rakha, W. M., Nessim, A. A., & Ismail, M. R. (2023). Retrofitting shading units to optimize daylighting in the governmental school building in CAIRO. *Journal of Al-Azhar University Engineering Sector, 18*(66), 183–202.

Salary, S., Holliday, L., Keesee, M., & Wachter, H. P. (2018). Building features in schools that influence academic performance. *Journal of Civil Engineering and Architecture, 12*(3), 163–197.

Saleem, A. A., Abel-Rahman, A. K., Ali, A. H. H., & Ookawara, S. (2016). An analysis of thermal comfort and energy consumption within public primary schools in Egypt. *IAFOR J Sustain Energy Environ, 3*.

Samadi, S., Noorzai, E., Beltrán, L. O., & Abbasi, S. (2020). A computational approach for achieving optimum daylight inside buildings through automated kinetic shading systems. *Frontiers of Architectural Research, 9*(2), 335–349.

Santos, I. G. D., Auer, T., & Souza, R. V. G. D. (2017). Optimized indoor daylight for tropical dense urban environments. *Ambiente Construído, 17*, 87–102.

Syaheeza, R. N., Husini, E. M., Arabi, F., Ismail, W. N. W., & Kandar, M. Z. (2018). Secondary school classrooms daylighting evaluation in Negeri Sembilan, Malaysia. In *IOP Conference series: Materials science and engineering* (Vol. 401, No. 1, p. 012024). IOP Publishing.

Tharim, A. H. A., Abdullah, W. N. H. W., Ismail, A., & Ahmad, A. C. (2022). Determination of design solutions to overcome the daylighting design failure observed in existing educational building. *International Journal of Sustainable Construction Engineering and Technology, 13*(2), 153–167.

Theodorson, J. (2008). *North v. south: The impact of orientation in daylighting school classrooms*. SOLAR Conference. https://www.academia.edu/8568609/North_V_South_The_Impact_of_Orientation_in_Daylighting_School_Classrooms

USGBC. (2014). *LEED v4 Daylight*. http://www.usgbc.org/credits/healthcare/v4-draft/eqc-0

Wagdy, A., Sherif, A., Sabry, H., Arafa, R., & Mashaly, I. (2017). Daylighting simulation for the configuration of external sun-breakers on south-oriented windows of hospital patient rooms under a clear desert sky. *Solar Energy, 149*, 164–175.

Wen, S., Hu, X., Hua, G., Xue, P., & Lai, D. (2023). Comparing the performance of four shading strategies based on a multi-objective genetic algorithm: A case study in a university library. *Journal of Building Engineering, 63*, 105532.

Zhang, A., Bokel, R., van den Dobbelsteen, A., Sun, Y., Huang, Q., & Zhang, Q. (2017). Optimization of thermal and daylight performance of school buildings based on a multi-objective genetic algorithm in the cold climate of China. *Energy and Buildings, 139*, 371–384.

Ziaee, N., & Vakilinezhad, R. (2022). Multi-objective optimization of daylight performance and thermal comfort in classrooms with light-shelves: Case studies in Tehran and Sari, Iran. *Energy and Buildings, 254*, 111590.

Zomorodian, Z. S., & Tahsildoost, M. (2017). Assessment of window performance in classrooms by long-term spatial comfort metrics. *Energy and Buildings, 134*, 80–93.

Zomorodian, Z. S., & Tahsildoost, M. (2019). Assessing the effectiveness of dynamic metrics in predicting daylight availability and visual comfort in classrooms. *Renewable Energy, 134*, 669–680.

Topia—The City, the Livable Place to Be

Mustapha El Moussaoui and Kris Krois

Abstract

In any academic discourse where the term "utopia" is invoked, references to the philosophical and sociopolitical constructs within the likes of Thomas More's Utopia, Francis Bacon's New Atlantis, or Plato's Republic are almost inevitable. However, these historical understandings of utopia, often posited as ideal societies marked by perpetual bliss, may, when reimagined through a contemporary lens, sometimes distort into a specter of dystopia or a utopia hyper-focused on materialistic or technological satisfaction. This paper, instead, reframes the utopia discourse from an eco-social perspective, speculating a bottom-up dialogue where utopia seems less like a distant Sci-Fi fantasy and more like a feasible model for our cities and societies. Rather than envisioning futuristic cities dominated by data and technology or consumed by materialistic pursuits, it reimagines utopia through the lens of environmental stewardship, social equity, and sustainable development. Within this framework, we delve into the complexities of human-environmental-social interfaces and the transformative potential of care and collaboration in this dynamic. Central to this discourse are three utopian models: the "Consumer Utopia," "Collaborative Utopia," and "Eco-Social Utopia," each demarcating distinctive pathways for a sustainable and harmonious co-existence. The roles of government policies, community participation, and corporate engagement in facilitating this transition towards an eco-social utopia are scrutinized, revealing the indispensability of these actors in shaping a more sustainable future. Real-world initiatives have been explored that serve as tangible exemplars of this utopian vision in action. By reframing the utopian discourse in this manner, this paper aims to rekindle hope in the possibility of creating healthy and livable cities grounded in environmental care and respect.

Keywords

Utopia · Eco-social · Dystopia · Well-being · Future cities · Nature

1 Introduction—Utopia, Dystopia, and Eco-Social Topia

The word "utopia" refers to an imagined place in which all aspects of society are portrayed as being superior to those in real life. The word "utopia," which literally can be translated as "no place," has come to be used interchangeably with "idealism." The literary definition of ("no place") is derived from the combination of the Greek words ou (meaning "not") and topos (meaning "place"). The meaning is changed to "good place" when the Latin syllable eu, which means "good" or "well," is added to the word, followed by the suffix "-ia" (Sisk, 1997). Therefore, the meaning of utopia can be either "no place" or "exceptional place." In the Merriam-Webster Collegiate Dictionary, utopia is defined as "a hypothetical and endlessly faraway country; a place of ideal perfection notably in laws, government, and social conditions; and an impracticable scheme for societal reform" (Merriam, 2022) (Abdelbaky, 2016). However, to completely investigate the notion of utopia, one must also study the concept of dystopia. This is owing to the fact that the numerous ways in which these two concepts intersect conceptually define one another. Both are intricately connected to one another, with one having a positive outlook on the future and the other having a negative outlook.

M. El Moussaoui (✉) · K. Krois
Free University of Bolzano, Bolzano, Italy
e-mail: mustapha.elmoussaoui@unibz.it

K. Krois
e-mail: kris.krois@unibz.it

I. Pigliautile et al. (eds.), *Urban Resilience, Livability, and Climate Adaptation*,
Advances in Science, Technology & Innovation, https://doi.org/10.1007/978-3-031-54911-3_10

Realists and skeptics may question how utopia relates to daily life. One response is that literary interpretations of utopias and dystopias give us ideas to consider regarding a variety of social and ethical issues, such as those relating to care and the quality of life in the face of disease and mortality. They allow us to consider and discuss the positive and negative aspects of relationships between people, cities, and the environment. It might be difficult to think about the effects of regional and global actions that promote or obstruct the flourishing of cities towards utopia. The world is currently faced with some stark visions of dystopia and a highly ambitious program of utopian dimensions, which is another response to question the modern relevance of utopian philosophy. The potential for climate change brought on by human-caused global warming and the hazards of ecological collapse that go along with it constitute one dystopia. Another is the worry that the development of robotics and other forms of automation could lead to widespread unemployment. In contrast to these, the United Nations' program for achieving the Global Goals for Sustainable Development by 2030 presents a new "utopian" vision for the globe (WHO, 2015). As a result, both dystopian and utopian ideologies are becoming more prevalent in politics (Gallagher & Christie, 2017).

The existential dilemma, apocalyptic fiction, dystopias, serious reports, and publications on multiple crises, wars, global climate change, biodiversity loss, and other (perhaps) irreversible processes provide a perfect excuse for despair, creating an atmosphere of fear, hopelessness, and powerlessness. Big tech creates and extensively promotes "smart" futures of computer-driven comfort—disguising the fact that this (most likely) leads to convenience and control for the elites only. Imaginaries and narratives from nationalist and far-right forces are also becoming more popular. They provide overly simplistic, risky, but for many people alluring solutions to the numerous crises and (presumed) threats. Concrete utopias of inclusive, open, and sustainable societies can be co-created by writers, academics, filmmakers, designers, photographers, and other storytellers, which is both inspiring and encouraging. They can create settings in which individuals can envision, experience, and experiment with such futures. Positive future trajectories motivate and reinstall hope and the ability to "do something" now.

To ground our discussion in the current context, we propose the term "eco-social topia" as a more realistic and achievable vision that tackles the struggles we are facing in the present zeitgeist. The problems we raise are general struggles encountered in cities around the world and elsewhere. However, we mainly look at the problems from a social and ecological perspective, as we believe that city infrastructures can be addressed through empirical methods. As eco-social transformation has gained traction in the academic realm, and it is our research background, we present a topia—or the place that could be achieved derived from the linguistic background of Utopia—that is from a certain perspective, simulating our current zeitgeist, which is open to different perspectives, alterations, improvements, and criticism. We show tangible examples and real-world applications of eco-social concepts, drawing from applications from around the world that could be labeled as eco-social acupuncture projects towards "U"-topia.

1.1 Methodology

In order to conduct this review of eco-social utopias, we employed a systematic literature search methodology. We searched databases such as Web of Science, Scopus, and Google Scholar to identify relevant articles, books, and reports. The keywords used in our search included "utopia," "dystopia," and "eco-social utopia." We also searched for literature discussing specific topics related to our proposed eco-social topia, such as social justice, collaborative design, and well-being.

Once we identified an initial pool of sources, we applied inclusion and exclusion criteria to narrow down our selection. The inclusion criteria were: (1) articles published in peer-reviewed journals, (2) articles written in English, and (3) articles focusing on eco-social utopias or related concepts mentioned above. The exclusion criteria were: (1) articles not focused on the topic of eco-social utopias.

After applying these criteria, we conducted a thorough analysis of the selected sources, examining the main arguments, findings, and examples presented in each. We synthesized the information gathered from these sources to develop our discussion of eco-social topia, showing examples from real-world applications to ensure the relevance and practicality of our review.

1.2 Cities and Policy Makers that Make a Difference

Cities are the incubators and conduits of new ideas and information, just as they were since antiquity. Sassen (2018) argues that the desire to create a perfect society is fostered in the world's global cities, much like Thomas More did when he set his utopian society on a city-island. Cities are "Both the cancer and the foundation of our networked society, both virus and antibody" (Khanna, 2010). When it comes to issues like global warming, poverty, and inequality, urban areas are both the source of the problem and the potential remedy. Therefore, understanding the significance of urban dynamics is a must for every initiative to improve human cohabitation on Earth or to promote sustainable development and think of a Utopia. A sustainable and resilient society that is also thriving and dynamic, prosperous, and connected to the rest of the globe is today's goal. To rephrase, an eco-city is the built environment most akin to an ideal society (Albanese, 2020).

This ideal of greater harmony between humans and natural ecosystems seems at odds with the current paradigm of an economy driven by rising global consumerism, which is frequently fueled by fossil fuels and unsustainable energy resources (Baer, 2020). The burden on future generations and the environment will only rise if environmental services are not given the proper priority in global value chains, and if market economies struggle to account for the detrimental consequences industrial activity has on society as a whole (Dasgupta & Ehrlich, 2013).

If environmental management can be limited to externality management (Baumol & Oates, 1988), then there remains plenty that today's global cities can do to lessen their environmental impact and help usher in a completely new way of thinking about economic growth and urbanization. The concept of economies of scale can also be understood in light of Adam Smith, the "father of modern economic thought," who argued that the key to prosperity was not in unrestricted expansion but rather in the development of increasingly specialized labor that could carry out industrial tasks more effectively at larger scales (Foley, 2006). This exploitation of natural resources due to the current economic model led to climate change (Steffen et al., 2018).

Global climate change emerged as a clear and present risk in the decades that followed as a result of our way of life. Climate scientists produced and collected observations that demonstrated why Earth would be heading towards a permanent climate crisis if trends did not change through an unprecedented collective effort (IPCC, 2014).

These realizations inspired the thinking and comprehension that led to the SDGs (Sustainability Development Goals)[1] and the Paris Agreement[2] in 2015, the modern-day ideal. The SDGs' ultimate goal is to completely eradicate the environmental and climate challenges we have brought about, albeit in a variety of methods (United Nations, 2015). Though, it makes no mention of how we are to accomplish that enormous aim. Another significant problem humanity is dealing with is how to do it effectively.

The fact that governments from all over the world did not come to an agreement on a utopian vision that took nature into account until the relatively recent publication of the United Nations' 17 SDGs in 2015 is both surprising and startling.

The SDGs place nature in the spotlight, portraying it as the starting point from which everything else flows and as something that is inextricably linked to people. Modern civilization as a whole, not just particular cultures or countries, is in danger if the natural environment is harmed. This collective utopia is likely to become a reality quicker than previous ones since the motivation for change is no longer merely the desire for a better life, but rather for survival (Paasche & Österblom, 2022).

Sixty years ago, Carson (1962) published "Silent Spring," which highlighted how the western world's prosperity came with a terrible price, which was the destruction of the environment and all life therein. It predicted the arrival of a global dystopia and showed a planet devoid of the sounds that life creates as it develops and thrives.

Even in nations where environmental principles are largely considered, they frequently fall short when juxtaposed with other urgent requirements. Germany, for example, a country who has been leading the green energy methods, reverted to the use of coal mining (which is the least sustainable and environmentally friendly) after the Russian war on Ukraine (Dazem, 2022). Additionally, although marshes and mires are prized for their ability to store carbon, in the name of development, expansion, jobs, and innovation, it is frequently required to construct highways and wind farms prior to their being preserved (Paasche & Österblom, 2022).

Moreover, cities all over the world have been expanding at the expense of regional habitats for a considerable amount of time without accurate assessments of their environmental implications. Claims on formerly unspoiled (or rather, substantially less-impacted) natural areas are frequently made separately, which obscures the cumulatively harmful impact of the loss of this type of environment. The full impact of this process or the value of nature somehow never reached the highest levels of world politics until recently. In 2019, when the IPBES Global Assessment on Biodiversity and Ecosystem Services was released, more people became aware of the true and varied value of nature (IPBES, 2019).

Additionally, the EU Urban Agenda was created by the European Commission in 2016 as part of the Pact of Amsterdam to encourage collaboration and participation from a wide range of actors in the effort to innovate cities in a way that is both socially equitable and environmentally sustainable (EC, 2016).

---

[1] The United Nations (2015) introduced the Sustainable Development Goals (SDGs) as part of the 2030 Agenda for Sustainable Development. The SDGs consist of 17 goals aimed at ending poverty, protecting the planet, and ensuring prosperity for all.

[2] The Paris Agreement, adopted in 2015 by the United Nations Framework Convention on Climate Change (UNFCCC), is an international treaty aimed at limiting global temperature rise to well below 2 °C above pre-industrial levels and pursuing efforts to limit the temperature increase to 1.5 °C. The agreement seeks to strengthen global response to climate change and supports the transition to a low-carbon and climate-resilient future (UNFCCC, 2015).

Taking a global perspective on eco-city development, it is clear that governance will play a significant role in determining the future of eco-cities. Conversely, states generally view urban development as a tool to channel investment and attract highly educated communities, both of which are vital for a growing economy; thus, the rise of cities does not inevitably mean a loss of authority for central governments.

Furthermore, the fact that having a social license to operate will increasingly compel businesses to act responsibly is one reason why many businesses are responding to global sustainability concerns. They are also urged to comply because there are legal structures in place, such as the EU-taxonomy.

This may be the reason why there have been so many net-zero emission strategies developed during the past two to three years. Major players like the EU and the IEA have developed roadmaps, but funds and businesses also have their own strategies. The overall impact of the net-zero emission strategies is still being assessed, but in terms of raising awareness, they are beneficial because they demonstrate what constitutes socially acceptable behavior (Paasche & Österblom, 2022).

In addition to changes in individual "consumer" behavior, government initiatives aimed at protecting nature within national borders, and through international cooperation, including ambitious pledges by corporations, more fundamental changes are occurring in how people interact with each other and with nature.

Governments may be exploring new avenues for multilateralism to address common challenges, as evidenced by projects like the "High Level Coalition on Health and Energy" and the "High Ambition Coalition for Nature and People." Engagements occur at various societal scales, such as the expanding global network of city mayors combating climate change, but there are also countless other ways that people are collaborating to realize new possibilities (Bennet et al., 2016).

To open up discussions and provide solutions for the current situations in cities, the EU launched the New European Bauhaus initiative in 2021. This creative and interdisciplinary plan connects the European Green Deal to our living spaces and experiences. The NEB initiative calls on projects to envision and build a sustainable and inclusive future that is "Beautiful, Sustainable, and Together" (NEB, 2022).

Despite the ongoing tendency to consistently underestimate the value of nature in terms of its economic, biophysical, sociocultural, holistic, and health benefits, there is plenty of evidence that we are actively opting for more responsible solutions and behaving as if we care about the natural world around us.

Multilateral agreements to safeguard the environment are still necessary, but we also need to reconsider how we view nature and how we regard it as a source of resources. We need to accept that we remain reliant on crude oil and other unsustainable resources, for now.

2 Utopia of Care and Collaboration

As we discussed the significance of policies and their potential to bring about transformative changes in our relationship with nature, we now turn our attention to "care." We envision a world that emphasizes holistic well-being and collaboration, and by integrating these principles, we can benefit both species and the environment.

The "Utopia of Care" encompasses a vision where health care systems prioritize not only physical health but also mental and emotional well-being, considering the broader social and ecological determinants of health (Marmot & Bell, 2012). Such a utopia would promote equitable access to care, emphasize preventive measures, and encourage community-based approaches that foster social cohesion and environmental sustainability (Gupta & Vegelin, 2016). In this context, we discuss the ideas and perspectives of various authors and experts in the field, differentiating between their views and our interpretation of the "Utopia of Care."

Caring for the environment is one aspect, caring for one another is also crucial, as the two are intricately intertwined. If we care about each other, we will care about the environment that sustains us. Indeed, many of our existential problems and well-being issues are resolved once we live sustainably with ourselves and the natural environment around us (El Moussaoui, 2020).

The "Collaborative Utopia," on the other hand, envisions a society where cooperation and collective action are at the core of decision-making processes. This enables communities to tackle pressing challenges such as climate change and social inequality more effectively (Ostrom, 2010). This utopia is based on the principles of shared responsibility, participatory governance, and social innovation, which have been shown to yield positive outcomes in various real-life settings (Hawken, 2017; Westley et al., 2011).

This form of utopia underscores the importance of social capital and community engagement in promoting sustainable development (Putnam, 2000). As global challenges such as climate change and social inequality continue to intensify, a collaborative utopia suggests a shift from a competitive, individualistic mindset to one that fosters collective action and shared responsibility (Hawken, 2007).

Collaborative utopias recognize the power of networks, partnerships, and alliances among various actors, including governments, businesses, civil society organizations, and local communities (Ansell & Gash, 2008). By working together, these actors can develop innovative solutions

Fig. 1 An event were Farmfluencers joined to raise awareness on sustainable and organic farming occurring in South Tyrol Source: Farmfluencers website (2023)

to pressing social and environmental issues. For instance, community-based natural resource management initiatives globally have demonstrated the potential for collaboration among local communities, NGOs, and government agencies in conserving ecosystems and promoting sustainable livelihoods (Berkes, 2007).

The concept of a collaborative utopia also extends to urban planning and design, where participatory approaches are increasingly advocated to involve citizens in shaping their cities' future (Innes & Booher, 2004). By incorporating diverse perspectives, knowledge, and experiences, collaborative urban planning processes can yield more inclusive, equitable, and sustainable outcomes (Fung, 2006). FarmFluencers in South Tyrol—Italy is an example of a collaborative eco-social utopia due to its unique approach to promoting sustainable agriculture, fostering community engagement, and educating the public about the importance of environmental stewardship and sustainable living (Fig. 1). By connecting farmers, influencers, and consumers, FarmFluencers create a platform that encourages collaboration, knowledge sharing, and innovative solutions to pressing environmental and social challenges.[3]

In an eco-social utopia, sustainable practices are prioritized, and communities collaborate to establish a more balanced and harmonious relationship with the environment. FarmFluencers embody these principles by advocating organic farming, biodiversity, and responsible land management, all of which contribute to a healthier and more sustainable ecosystem. By supporting local farmers and encouraging the adoption of environmentally friendly practices, FarmFluencers help to create a more sustainable food system that benefits both people and the planet.

Specifically, one of the farmers on the platform is “Solos”[4] in Tramin, a shared greenhouse hosting multiple parties (Fig. 2). One section houses the start-up “Solos,” featuring floating lettuce and a small fish farm in an aquaponic system. This soil-less, first-of-its-kind system in South Tyrol enables local food production, particularly in urban areas. Co-founder Matthäus, along with other founding team members, sees the climate crisis as a pressing issue. They believe agriculture must abandon economically viable but ecologically damaging practices. Matthäus envisions aquaponics as a forward-looking solution in agriculture and urges action instead of mere discussions.

3 Eco-Social Topia

These new projects revitalize positive externalities in urban environments by enhancing connections and fostering innovation. Proximity facilitates network economies that address scarcity, transforming consumers into prosumers who co-create services, thereby increasing their value (Veronica, 2017). This counteracts the traditional issues of resource exhaustion and overconsumption, known as the “tragedy of the commons” (Hardin, 1968), with growing benefits resulting from shared and combined resources in cities.

[3] Farmfluencers. (n.d.). Retrieved April 21, 2023, from/.

[4] Solos. (n.d.). Retrieved April 21, 2023, from https://www.solos.farm/.

Fig. 2 Solos farmer controlling the products to keep them healthy. *Source* Solos website (2023)

Iaione (2017) describes this as the quintuple helix approach to the knowledge-based economy. Cities become spaces where public governance, corporations, and research institutions engage with the educated population, creating, commercializing, and sharing new technologies. While some research views eco-cities as mere technology showcases within the neoliberal global order (Cowley, 2016), cooperative governance is crucial for reimagining cities as physical and conceptual infrastructure (Araral, 2014). Eco-cities, or e-co-cities, combine technological and digital viability with a fundamentally different perspective on the government's role in resource allocation.

Three general strategies may be recognized concerning the interaction between developing eco-cities and national administrations. First, is the entrepreneurial state strategy (Mazzucato, 2011), where central governments typically act as the supervisors and pilots of eco-city initiatives, coordinating resources and investments from the private sector (Joss, et al., 2013). The second strategy is the enabling state method. As Iaione explained, many European towns are returning to their communal roots to lead cooperative governance initiatives that involve improved management of shared urban resources (Bianchi, 2018). NGOs and other third-sector organizations usually spearhead projects, with the government acting as an "enabler" by providing essential support and funding. The third is the dormant state strategy because, for the most part, governments lack the funds or expertise to take climate action.

In light of these strategies, let's examine a tangible example of an eco-social utopia in action. The project La Cañada Real, developed by Recetas Urbanas[5] (Fig. 3), can be described as an eco-social utopia for showcasing its potential to address various conflicts in public spaces and promote sustainable, equitable, and inclusive urban development. By understanding the root causes of social, cultural, environmental, political, and economic conflicts, this project contributes to the creation of a more harmonious urban environment that aligns with eco-social utopian principles (Fig. 4).

The methodology begins by engaging with environmental education centers, where most of the young residents of Cañada Real can be found. Furthermore, door-to-door visits are conducted to connect with families within the project's target area. Concurrently, efforts are made to identify and invite local social organizations and neighborhood associations to participate, along with any newly established entities. Recognizing that some neighbors, who are also potential future users, are currently incarcerated, collaboration is established with those at the Soto del Real prison. In this facility, as in other locations, various construction elements such as slabs, walls, trusses, roof panels, and window frames are prefabricated and later assembled at the official project site (Cirugeda, 2020).

Another example of an eco-social utopia is the work of "Cooperativa de Diseño." This interdisciplinary design collective contributes to the concept of an eco-social utopia through their innovative projects and concepts, which prioritize sustainability, social equity, and community

[5] Recetas Urbanas. (n.d.). Retrieved April 21, 2023, from https://recetasurbanas.net/.

Fig. 3 Collaborative work in action during the assembly of one of the structures. *Source* Recetas Urbanas (2023)

engagement. Their work demonstrates a commitment to creating a more balanced and harmonious relationship between humans, the environment, and social systems (Fig. 5).

The cooperative emphasizes sustainable design, social equity, and inclusivity, along with community engagement, education, and capacity building. They use a holistic approach to problem-solving. The interdisciplinary nature of Cooperativa de Diseño allows the collective to approach complex problems from multiple perspectives, ensuring that environmental, social, and economic factors are considered in the development of their projects.

4 Conclusion

The act of collectively envisaging desirable futures is a crucial step, but it does not end there. The real task lies in the endeavor to actualize these visions by constructing the necessary infrastructure, establishing supportive institutions, critically evaluating their efficacy, and disseminating their impact for wider adaptation and learning. These practices then fuel inspiration and contribute to narratives that spur further innovation and positive change.

Topias, or viable eco-social utopias, are being conceived and realized in diverse contexts around the world. To enable an ecological transition that is not only plausible but also socially equitable and desirable, it is essential to manage and harmonize potential conflicts within these utopias and their peripheries. These eco-social topias teach us a critical lesson: that progress is not solely reliant on technological advances or economic prosperity. Rather, it's about nurturing utopian dreams, fostering concerted efforts to enhance one's surroundings, and working towards improving society and the environment—a vision akin to Thomas More's island-city, albeit infused with a deep-seated respect for nature. We present the possibility of realizing an utopia that ensures the welfare of both current and future generations.

Admittedly, our prospects might seem daunting at times. Yet, amidst the grim scenarios, hope persists. History bears testament to the human ability to adapt to challenging circumstances and transformational change. Collaboration, empathy, collective learning, and cultural evolution form the bedrock of our resilience. These inherent human traits have allowed us to learn from and rectify our past mistakes, although gradually. The journey towards actualizing a "real" utopia commences when we start functioning as a global community, addressing shared global concerns.

Today's younger generation, our emerging leaders, demonstrate this ethos by displaying extraordinary resilience and leadership. We need more of this active involvement. We need more of these acupuncture-like small projects that function within a network, advocating multilateral agreements and pushing for regulations that facilitate cooperative initiatives to safeguard our cities and the environment.

Throughout our history, humanity has demonstrated a remarkable ability to rally together for the common good

Fig. 4 Plans of the community constructed complex. *Source* Recetas Urbanas (2023)

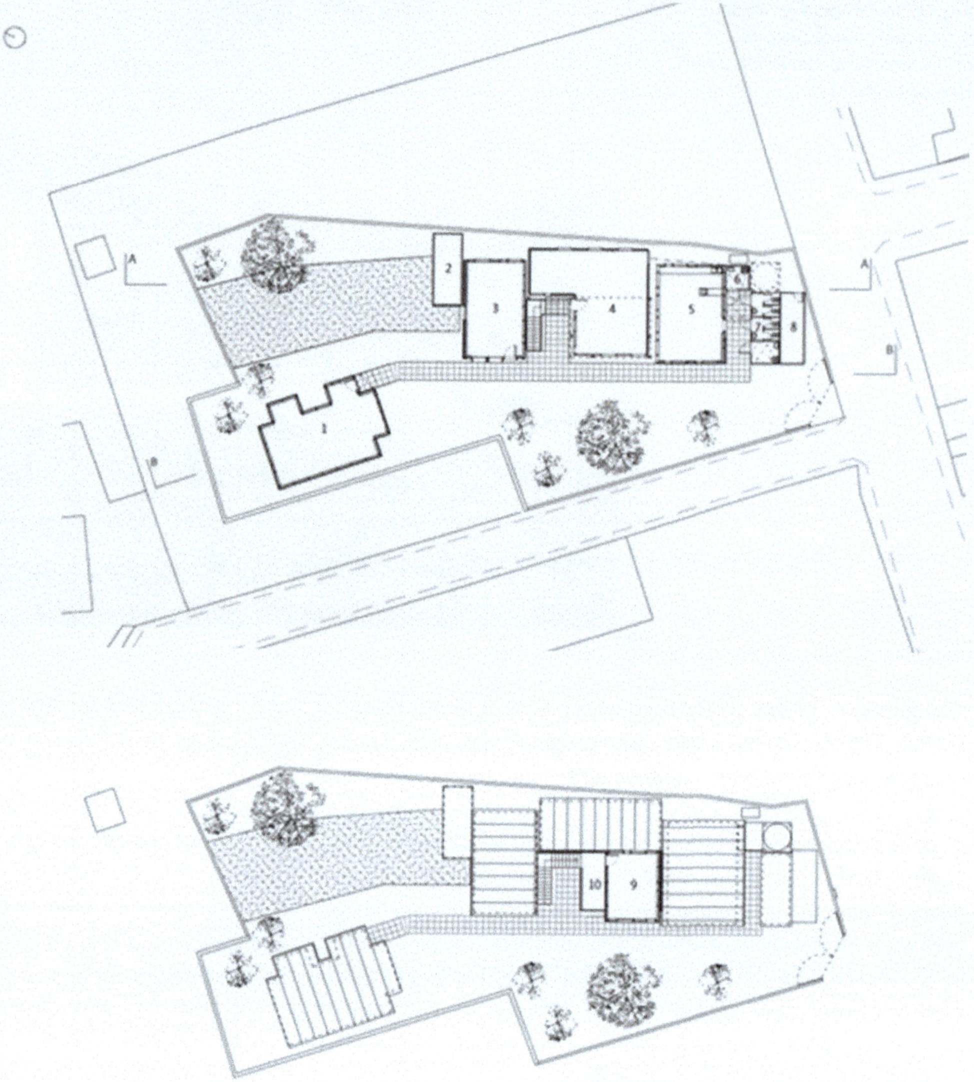

Legend

1. Multipurpose room
2. Storage room
3. Workshop
4. Office Space
5. Workshop and Kitchen
6. Accessible restrooms
7. Restrooms
8. Installations room
9. Workshops
10. Terrace

in the face of existential threats. This capacity has repeatedly propelled us forward, and it is now required more than ever as we face the mounting threats to our environment and biodiversity. With the same determination demonstrated in our past, we can confront these challenges and actualize our "Topias." In essence, the eco-social utopia is not just a dream, but a potential reality that embodies our collective aspirations for a sustainable and equitable future.

Fig. 5 An illustration of a collaborative work made to create artisanal products by community members. The goal of this initiative is to revalue artisanal production to achieve a fairer commercial exchange through comprehensive and participatory design framework. *Source* CD website (2023)

References

Abdelbaky, A. (2016). A perfect world or an oppressive world: A critical study of utopia and dystopia as subgenres of science, IJELLH, ISSN 231-7065.

Araral, E. (2014). Ostrom, Hardin and the commons: A critical appreciation and a revisionist view. *Environmental Science & Policy, 36*, 11–23.

Albanese, M. (2020). ISSN 2532-6570 IAI COMMENTARIES 20166-SEPTEMBER 20201 Global Eco-Cities: Pursuing the Utopia of Sustainable Urbanisation

Ansell, C., & Gash, A. (2008). Collaborative governance in theory and practice. *Journal of Public Administration Research and Theory, 18*(4), 543–571. https://doi.org/10.1093/jopart/mum032

Baumol, W., & Oates, E. (1988). *(1988): The theory of environmental policy* (2nd ed.). Cambridge University Press.

Baer, H. A. (2020). Democratic eco-socialism as a real utopia: Transitioning to an alternative world system (1st ed.). Berghahn Books. https://doi.org/10.2307/j.ctvw048m4

Bennett, E. M., Solan, M., Biggs, R., MacPhearson, T., Norström, A., Olsson, P., Pereira, L., Peterson, G. D., Raudsepp-Hearne, C., Beirmann, F., Carpenter, S. R., Ellis, E., Hichert, T., Galaz, V., Lahsen, M., Martin-Lopez, B., Nicolas, K. A., Preisser, R., Vince, G., … Xu, J. (2016). Bright spots: Seeds of a good Anthropocene. *Frontiers in Ecology and Environment, 14*(8), 441–448.

Berkes, F. (2007). Community-based conservation in a globalized world. *Proceedings of the National Academy of Sciences, 104*(39), 15188–15193. https://doi.org/10.1073/pnas.0702098104

Bianchi, I. (2018). The post-political meaning of the concept of commons: The regulation of the urban commons in Bologna. *Space and Polity, 22*(3), 287–306.

Carson, A. (1962). *Silent spring* (pp. 400). Mariner Books.

Cirugeda, S. (2020). Centro sociocomunitario Cañada Real Galiana, Madrid. *España. ARQ, 104*, 110–125. https://doi.org/10.4067/S0717-69962020000100110

Cowley, R. (2016). Eco-cities: Technological showcases or public spaces? Ph.D. thesis, University of Westminster. https://westminsterresearch.westminster.ac.uk/ item/9w780

Dazem, V. (2022). Germany bolsters coal-fired power to meet winter power demand. Retrieved December 14, 2022, from https://www.bloomberg.com/news/articles/2022-10-21/germany-bolsters-coal-fired-power-to-meet-winter-power-demand?utm_source=website&utm_medium=share&utm_campaign=copy

Dasgupta, P., & Ehrlich, P. (2013). Pervasive externalities at the population, consumption, and environment nexus. *Science, 340*(6130), 324328. https://doi.org/10.1126/science.1224664

El Moussaoui, M. (2020). The phenomenological significance of dwelling in architecture. The case of Eastern Beka'a Valley-Lebanon. Universitat Politècnica de València. https://doi.org/10.4995/Thesis/10251/152487

European Commission, Pact of Amsterdam establishing the Urban Agenda for the EU,agreed at the informal meeting of EU ministers responsible for urban matters on 30 May 2016 in Amsterdam, https://europa.eu/!hH69rT

Foley, D. (2006). *Adam's fallacy. A guide to economic theology* (p. 6). Belknap Press of Harvard University Press.

Fung, A. (2006). Varieties of participation in complex governance. *Public Administration Review, 66*(s1), 66–75. https://doi.org/10.1111/j.1540-6210.2006.00667.x

Gallagher, A., & Christie, I. (2017). Towards a Sustainable Care Utopia Ann Gallagher and Ian Christie University of Surrey, UK. sagepub.co.uk/journalsPermissions.nav https://doi.org/10.1177/0969733017712857

Gupta, J., & Vegelin, C. (2016). Sustainable development goals and inclusive development. *International Environmental Agreements: Politics, Law and Economics, 16*(3), 433–448.

Hawken, P. (2007). *Blessed unrest: How the largest social movement in history is restoring grace, justice, and beauty to the world*. Viking.

Hawken, P. (2017). *Drawdown: The most comprehensive plan ever proposed to reverse global warming*. Penguin Books.

Hardin, G. (1968). The tragedy of the commons. *Science* **162**(3859), 1243–1248. https://doi.org/10.1126/science.162.3859.1243

Iaione, C. (2017). The right to the co-city. *Italian Journal of Public Law* **9**(1), 80. http://www.ijpl.eu/archive/2017/issue-15/the-right-to-the-co-city

Innes, J. E., & Booher, D. E. (2004). Reframing public participation: Strategies for the 21st century. *Planning Theory & Practice, 5*(4), 419–436. https://doi.org/10.1080/1464935042000293170

IPCC. (2014). Climate change 2014: Synthesis report. Contribution of Working Groups I, II and III to the Fifth Assessment Report of the

Intergovernmental Panel on Climate Change [Core Writing Team, R.K. Pachauri and L.A. Meyer (eds.)]. IPCC, Geneva, Switzerland.

IPBES. (2019). *Global assessment report on biodiversity and ecosystem services of the Intergovernmental Science-Policy Platform on Biodiversity and Ecosystem Services*. (pp. 1148). IPBES Secretariat, Bonn, Germany. https://doi.org/10.5281/zenodo.3831673. (E. S. Brondizio, J. Settele, S. Díaz, & H. T. Ngo (Eds.)).

Joss, S., Cowley, R., & Tomozeiu, D. (2013). Towards the 'Ubiquitous Eco-City': An analysis of the internationalisation of EcoCity policy and practice. *Urban Research & Practice, 6*(1), 54–74.

Khanna, P. (2010). Beyond city limits. The age of nations is over. The New Urban Era Has Begun, in *Foreign Policy*, (No. 181, p. 128). https://www.jstor.org/stable/20753992.

Marmot, M., & Bell, R. (2012). Fair society, healthy lives. *Public Health, 126*, S4–S10.

Mazzucato, M. (2011). "The Entrepreneurial state", in soundings. *No., 4*, 131142.

Merriam-Webster. (n.d.). Utopian. In Merriam-Webster.com dictionary. Retrieved December 8, 2022, from https://www.merriam-webster.com/dictionary/utopian

New European Bahaus. Retrieved December 8, 2022, from https://new-european-bauhaus.europa.eu/index_en

Ostrom, E. (2010). Beyond markets and states: Polycentric governance of complex economic systems. *American Economic Review, 100*(3), 641–672.

Paasche, Ø., & Österblom, H. (2022). The utopia of a sustainable planet. *Elephant in the Lab*. https://doi.org/10.5281/zenodo.6556477

Putnam, R. D. (2000). Bowling alone: The collapse and revival of American community. Simon and Schuster

Sassen, S. (2018). *Cities in a world economy* (5th ed.). SAGE Publications.

Sisk, D. (1997). *Transformations of Language in Modern Dystopia*. Greenwood Press, 1997. (Print).

Steffen, W., Rockström, J., Richardson, K., Lenton, T. M., Folke, C., Liverman, D., Summerhayes, C. P., Barnosky, A. D., Cornell, S. E., Crucifix, M., & Donges, J. F. (2018). Trajectories of the earth system in the anthropocene. *Proceedings of the National Academy of Sciences, 115*(33), 8252–8259.

United Nations. (2015). Transforming our world: The 2030 Agenda for Sustainable Development. A/RES/70/1. Retrieved from https://sustainabledevelopment.un.org/content/documents/21252030%20Agenda%20for%20Sustainable%20Development%20web.pdf

United Nations Framework Convention on Climate Change. (2015). Paris Agreement. Retrieved from https://unfccc.int/sites/default/files/english_paris_agreement.pdf

Verónica, A. (2017). The age of prosumerism: some micro-economic analysis. *Journal of Theoretical and Applied Electronic Commerce Research* **12**(3), 1–12. (Sept2017). https://doi.org/10.4067/S071818762017000300002

Westley, F., Olsson, P., Folke, C., Homer-Dixon, T., Vredenburg, H., Loorbach, D., Thompson, J., Nilsson, M., Lambin, E., Sendzimir, J., Banerjee, B., Galaz, V., van der Leeuw, S. (2011). Tipping toward sustainability: Emerging pathways of transformation. Ambio. **40**(7), 762–780. https://doi.org/10.1007/s13280-011-0186-9. PMID: 22338714; PMCID: PMC3357751. (Nov)

Revisiting Scientific Theories, Towards Human Well-Being-Oriented Built Environments

Mohamed Sayed Hussein Ahmed Hussein,
Yasser Mohamed Mansour, and Shaimaa Mohamed Kamel

Abstract

The design paradigm in architecture and the built environment evolves with time. Each era was dominated by a distinct design paradigm. In the modern era, built-environment design paradigms began with an environmental behavior design approach. Furthermore, this method has shifted and designers have increasingly focused on the concept of environmental technology design. Subsequently, the concept of Green Architecture and sustainability became the main design paradigm, emphasizing limiting the impact of humans and built environments on nature and natural resources and creating green energy. A few years ago, architects, urban planners, and scientists realized and scientifically approved that the physical or built environment has a significant impact on human mental and physical health. As a result, they began to place greater emphasis on human psychological and physical health, as well as their interactions with their created environments, than in prior eras. Some design theories, design indexes, design patterns, and recently designed rating systems were created to direct the design and built markets towards the new design paradigm of human health and well-being-oriented built environments. Designers and scientists use different scientific terms to identify this design paradigm, such as human-centric design, healing built environments, restorative architecture/built environments, and positive built environments. The main objective of this new design paradigm is to enhance human physical, psychological, and physiological conditions to accomplish human well-being levels. Researchers work on six theories, indexes, and rating systems in ongoing research and these are attention restoration theory (ART), stress restoration theory (SRT), Biophilia theory, Biophilic design, neuroscience for architecture (neuro-architecture), and the well-building standards rating system. This study has two main objectives. The first was a deep analytical study to compare and integrate the three theories' approaches, objectives, methods, concepts, and findings to come up with a list of applied design principles and applied design considerations that will enable architects, urban planners, interior architects, and other designers to easily apply them in their design processes, a positive vital step towards transforming them to applied sciences. The second objective was to encourage and strengthen the integrated design process (IDP) between architects and other scientists in the design process and decisions. Researchers' findings emphasize the vital role that the science of human brain mechanisms, brain perception, mind–body connection, and human well-being needs to play in design drivers while designers design built environments. Researchers' comparative deep analysis and integration came up with a list of fifteen design considerations and principles, and they named it human well-being design principles (WBDP), which will enable designers to positively step forward towards creating positive built environments. Researchers have concluded that humans are part of nature. As a result, humans have to be reconnected with nature. Thus, built environments must be well-designed and integrated with nature to create livable, healthy, comfortable, and enjoyable places and spaces. The analysis shows that architects and other designers have no choice but to collaborate with other human minds and body scientists to create built environments that speak positively to individuals' brains and

M. S. H. A. Hussein (✉) · Y. M. Mansour · S. M. Kamel
Architecture Department, Faculty of Engineering,
Ain Shams University, Cairo, Egypt
e-mail: mharchitects.eg@gmail.com

Y. M. Mansour
e-mail: yasser_mansour@eng.asu.edu.eg

S. M. Kamel
e-mail: shaimaa_kamel@eng.asu.edu.eg

I. Pigliautile et al. (eds.), *Urban Resilience, Livability, and Climate Adaptation*,
Advances in Science, Technology & Innovation, https://doi.org/10.1007/978-3-031-54911-3_11

emotions. Architects must deeply study human brain mechanisms to understand how humans perceive their designs. Finally, a built environment cannot separate man from his native nature because nature has all the solutions.

Keywords

Human well-being · Built environment · Neuro-architecture · ART · SRT · Nature · Positive architecture · Restorative architecture · Design principles

1 Introduction

Modern design paradigms do not consider human mental or physical wellness. Human health is a primary concern; as chronic diseases are rising. Most mental problems and immune system ailments such as heart disease, high blood pressure, immune system diseases, mental concentration, and overall low performance are characterized by stress. Architects and designers are unaware of human brain mechanisms and emotions, which interact with what they design at all times. Stress causes death. Previous studies have found that physical environments mostly act as daily stressors for users. Nature or the natural environment in its different shapes, climatic characteristics, and ecosystems, all work in wonderful harmony. Man always has to have a very strong connection with their native nature, which has always been positively shaded in man's well-being. On the other hand, man has a negative impact on his native natural environment for many reasons, such as housing demand, energy demand, new materials demand, and more unplanned actions that destroy the natural environment. With the world's rapid population expansion and massive urban development, more than half of the world's population now lives in cities. As a result, the rapid increase of urban development is unplanned, and it breaks the deep bond between man and their native nature. Architects defined the man-nature relationship from a practical standpoint, resulting in a significant change in technological vision from man to machine. Furthermore, they exclusively judged user comfort in the built environment by building performance levels. Separating humans from their natural environment, on the other hand, has an impact on their physiological, psychological, behavioral, affective, and cognitive aspects (Giulia et al., 2017). Over the course of our evolution, people have established rules that regulate the connection between man and nature (Wilson, 1993). Several global organizations, such as the USA green building rating system known by LEED certificates, established well-known rating system certifications measuring building performance levels of energy consumption, water usage, materials used, and their impact on the natural environment. However, methods for human mental and physical health and well-being are critical in the growth of built-environment values (Betro & Barbiero, 2017). These created rating systems named the Green building rating systems are globally used to encourage the concept of reducing energy conception levels based on these certificates' regulations. However, they cannot measure the exact impact of the B.E. on the natural environment. (Gou & Xie, 2017). These rating systems mainly consider and focus on the building's performance as one of the sustainable design strategies, and human or user's health is not their main concern (Yeang & Spector, 2011). Essentially, there were no rating systems or official certifications known globally that measure our built environments and their impact on individual's mental and physical conditions a few years ago. Although the main objective of our built environments is supposed to be individuals' satisfaction and their well-being level, because individuals spend more than 90% of their time inside these built environments, rating systems were only energy-oriented. Recently, researchers have been working and making more efforts to create new architectural waves or concepts that mainly consider and focus on human health-oriented built environments. Researchers show a historical timeline of these actions taken towards the human well-being-oriented built-environment design paradigm, as shown in Fig. 1. The newly built-environment design paradigm focuses on accomplishing human well-being through designing built environments that have a positive impact on their health. In the 1980s, attention restoration theory and stress restoration theory both emphasized the role of a positive built environment, or nature play, to enable their users to restore their direct attention or put them in a fascinating mood in order to enable them to perform positively again (Ulrich, 1981) (Kaplan & Kaplan, 1989). Biophilia theory in the 1970s and biophilic design in the 1990s represent the restorative built environment based on the design concept of re-establishing the connection between humans and nature again (William et al., 2014). Moreover, in 2003, designers collaborated with neuroscientists to understand the human brain and the extremely complicated processes that occur inside it while perceiving the built environments and interacting with them. A New School of Architecture emerged in New York the school of neuroscience for architecture. Finally, in 2013, a well-building standards rating systems were created by a group of designers and scientists based on the human's twelve body systems. Various studies have shown that built environments affect the human brain and its psychological, physiological, and physical conditions. As a result, in this study, researchers brought together six

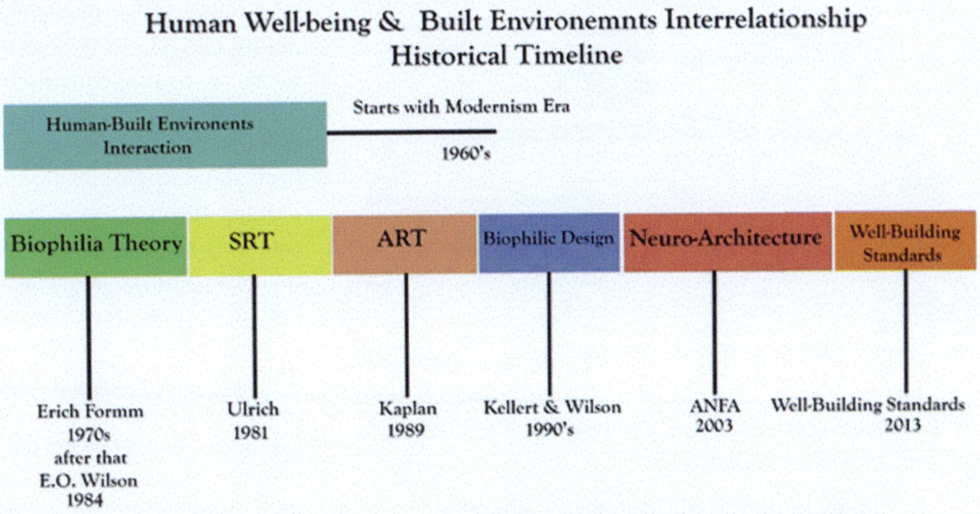

Fig. 1 The timeline shows all the actions that have been taken towards creating the new design paradigm by researchers

different design theories, indexes, design guidelines, and design standards to conduct a thorough deep comparative analytical study of their main approaches, objectives, methodologies, and findings to create a list of design principles and design considerations that architects and designers can use while they design built environments. Specifically, in this paper, researchers will apply the above objectives and methodologies to only three of them, ART, SRT, and neuro-architecture, and the other three will be in another paper because they are still under study.

2 Built Environment

The built environment is definitely universal, and its effects are more far-reaching than individuals realize, even if they live and interact with it every moment in their life on Earth. The United States Environmental Protection Agency stated that the built environment interacts with all aspects of human life, including the buildings we live, work, learn, and heal in; man-made systems that provide humans with water and electricity; and roads, bridges, and transportation systems. Tom J. Bartuska, in his book The Built Environment: Definition and Scope, defined the built environment on the basis of four interconnected qualities. First, it is enormous; it is omnipresent; it is everything generated or created by humans to satisfy their activities. The second characteristic is that all human minds create in order to meet their needs, desires, and comfort. Third, people were designed to protect themselves from their surroundings while also achieving comfort and well-being. The final characteristic is that context defines and shapes every element of the built environment; each and every individual element has a positive or negative impact on both the built and natural environments, as well as human–environment relationships (Bartuska, 2011). These impacts are on the microscale and extend to the macro and global scales. Figure 2 helps users to visualize these four characteristics below (see Fig. 2 in Bartuska, 2011).

*Built Environment = B.E.

The built environment includes places and spaces created or modified by humans. This includes buildings, parks, and transportation systems. Scientists in public health and well-being research have expanded the definition of built environment to include healthy food access, community activities, walkability, psychological states, and social interactions.

2.1 Components of the Built Environment

Tom J. Bartuska described the built environment as having seven interrelated components: Earth, region, cities, structure, landscape, interiors, and products, as shown in Fig. 3

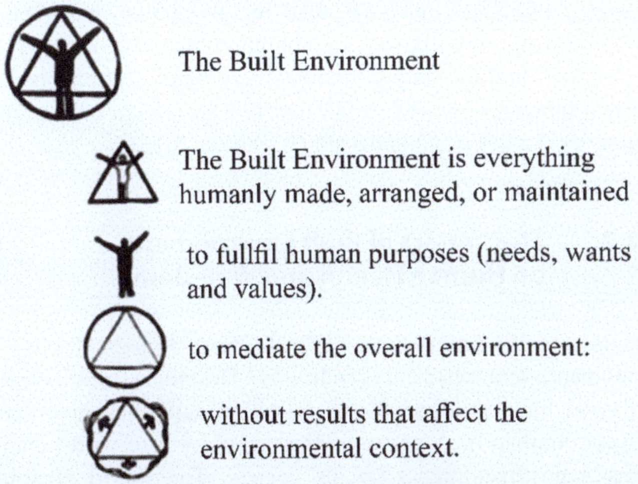

Fig. 2 The Definition of the built environment and its four related characteristics (Bartuska, 2011)

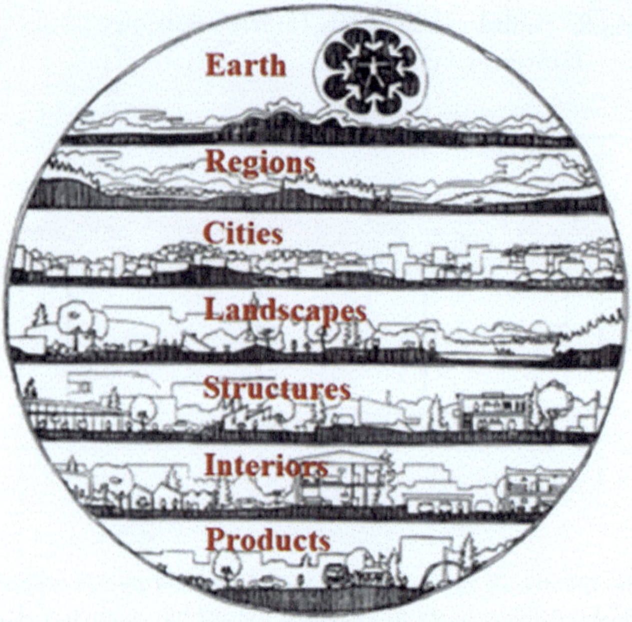

Fig. 3 The layers of the built environment, the integrative or composite symbol design after (S. Recken; the linear graphic formats by J. Singleton) and after (Bartuska, 2011)

(see Fig. 3 in Bartuska, 2011). The sum of the seven outlines the space of the built environment. The term built environment is presented as a broader framework formed by the interrelationships of independently designed components. It is essential to consider the interdependencies among various mechanisms, as well as between individuals and professions.

A strong cause-and-effect relationship exists between Man and Man-made. The built-environment design process is very problematic, and different subdivisions must be involved in creating a livable built environment. The English planners and architects, in his book on the built environment, emphasized the vital need for integration, participation, and interdisciplinary collaboration between designers, socialites, and the public to create better and healthier built environments now and in the future. The integrated design process is a key to good design (Bartuska, 2011).

2.2 The Impact of Built Environment on Human Health and Well-Being

Human well-being plays a vital role in psychological, physiological, and physical conditions. Designers must have a good understanding of its definition and rules in order to accomplish it. A person with a good level of well-being has a positive impact on his society. The World Health Organization (WHO) defined human health to mean the comprehensive condition of physical, mental, and social well-being, rather than simply the absence of disease or disability (WHO, 2023). On the other hand, there is no agreement on a single definition of well-being, but previous research has agreed that it includes a mixture of positive emotions, the absence of negative emotions, life satisfaction, and positive functioning. This can also be described as judging life positively. It is critical to understand the vital role that the man-made or the built environment plays in humans' positive or negative psychological conditions. All participants in the built-environment design process have to understand this interrelationship well to create healthier built environments that positively impact individuals' overall well-being. Environmental Psychology is a scientific term that refers to the science of the interrelationship between humans and the spaces and places in which they live, work, play, and learn. Human behavior and environmental behavior where humans exist have to be well analyzed and examined together in order to have proven knowledge about the real impact of these environments on humans' psychological, physical, and social interaction conditions. Environmental psychology provides science to enable all sectors to create B.E. considering the human behaviors inside it. Donald D. Patterson emphasized in his article, named "Contributions of Environmental Psychology to Visitor Studies," that some psychological studies were applied to understand the impact of the physical surroundings on employees' productivity levels inside their offices in the 1940s (Patterson, 1989). This was a crucial beginning that prompted architects to collaborate with psychologists to create better-designed environments that fulfill the requirements of individuals. "Environmental psychology" as a scientific term was created at a conference in the 1960s when scientists studied the impact of B.E. on the treatment development of psychiatric patients. In the mid-1970s, the first, thus, designers and researchers have begun to propagate new science all over the world (Patterson, 1989). This science escorted in a new era of design and design paradigms based on the interdependence of man's mental and body demands and the Man-made. This science enables designers to create B.E. that provides the psychological needs of their users not just their behavioral, or functional needs. Also, designers believe that this type of built environment can help to treat a variety of health problems because they reduce their users' stress levels. Moreover, B.E. plays a sophisticated role in individuals' emotional and physical health, including stress, depression, loneliness, and anxiety. There is also an urgent need to acknowledge that each individual has their own unique emotions and life experiences that result in a unique experience in a given space or place (Ulrich et al., 1991). Researchers and scientists study humans' interactions with different B.E. and natural environments to understand their behavioral and psychological outcomes with specific environments (Kopec,

2010). Physicians usually use operations and medications to heal physical disorders, but therapists use behavioral modifications to treat human mental and emotional issues. Environmental psychologists consider different design principles to create positive B.E. that have a positive impact on their users' mental conditions to reduce stress and undesirable behaviors; some of these principles are natural, sensory elements, and spatial alliance. B.E. is a vital and critical aspect of human life, which is why it be carefully designed. (De Botton, 2008).

3 Nature

Nature is the most valued and appreciated part of human and nonhuman lives. Nature can be defined as the living things and non-living things part of an ecosystem's general structure including magnificent objects, forests, seas, and oceans plus the man-made gardens, parks, plants, and various forms of vegetation. Humans refer to nature as green areas, but they are still natural when greens are replaced by other colors such as white, brown, red, and yellow. Environmental psychologist Joachim defines nature as a huge sphere of biological and inorganic matter unrelated to human activity or influence (Judith, 2012). Nature in many parts of the globe is without man's modifications, but in other parts has been modified, especially those parts inside or beside man's built environments named natural environments. Nature and natural environments play a vital role in the psychological and physical conditions of all living things on Earth. Landscapes, especially multifunctional ones in our built environments, are the breath, and they have been the subject of painting in all man-made environments. History has emphasized that the majority population has experienced an agricultural lifestyle, which built a strong inner relationship with their native nature and natural environments. A few centuries ago, man began to transform their native physical environment into urbanism; studies show by 2050, more than 60% of the world's population will be living in developed urban areas. Some researchers have stated that people who have an established relationship with nature are happier than those who do not. Max Fordham an environmental engineer stated that people have become separated from their individual responses to the physical atmospheres of the environment by relying on mechanical ventilation and artificial lighting. He believed that built environments must be designed considering natural ventilation and daylight, which are mainly based on local climate characteristics (Wernick, 2008). As an example, Max Fordham highlighted those individuals who experienced the skylight with its shades and different effects by their senses, which make them feel vibrant and alive and lead them to their comfort zone and happiness. He believed that when individuals feel a wide range of natural conditions, they can accomplish the meaning of happiness again. Integrating nature's benefits of lighting, heating, and cooling into man-made built environments has been vital since vernacular architecture began. Most design schools, however, have disregarded studying nature and how to integrate it into modern built environments. (Wernick, 2008). The Industrial Revolution led to a dramatic shift towards modern urbanization, technology, and manufacturing, which played an important role in separating man from his native nature. An American landscape architect, Frederick Law, proved the fact that landscape interacts with the human's mind without causing fatigue, but it provides the effect of refreshing and relaxation to the entire body systems based on mind and body connection (Olmsted, 1903). Human and nature's connection continues to be reduced and vanishes because man replaced nature with artificial shapes, which represent natural elements in his built environments. Separating man from nature or natural environments and its natural health value would have a negative impact on his health; the relation is not mutual nor equal. Wright's Broadacre City concept was to highlight the man-nature connection again and its positive impact on man opposing the B.E. impact on man. However, designers integrate nature elements with B.E. to increase the apparent value of the designed built environment without concentrating on the nature vital role it plays to accomplish humans' well-being (Browning et al., 2014).

4 Biophilia Theory, 1970s

The Biophilia hypothesis states that humans possess an innate tendency to seek connections with nature and other forms of life. Biophilia is a Greek word; "the passionate love of life and of all that is alive." German-born American psychoanalyst Erich Fromm created the term love nature or love living things in the anatomy of human destructiveness because he was trying to emphasize the great interrelationship between humans and their native nature or living things that surround them. Subsequently, American biologist Edward O. Wilson tried to create a new concept that focuses on re-establishing the connection between humans and their native nature. In addition Wilson, Kellert, Heerwagen, and Mador believe that the physical surroundings can play a noticeable part in enhancing a human's physical and psychological status (Kellert et al., 2008). Harvard et al. proposed the biophilia theory. Wilson in 1984 proposed that humans have an underlying biological bond with nature. He recognized the critical role that nature plays in the mental, bodily, and social well-being of humans (Wilson, 2009). Biophilia increases mental tension, which benefits the human body and helps the immune system battle infections. Nature has had a tremendous impact on man's natural healing process throughout history. Unfortunately, unrestrained urbanization caused by industrialization and technical progress has expanded the separation between man and his natural environment. Some people believe that the Biophilia theory is all about "green" elements. It has been scientifically proven that plants have a positive impact on the human healing process. However, B.E. themselves has to bring into line with healing principles. Biophilia theory involves the design of spaces and places with a bottom-up philosophy, and it must be involved in the early design process (Salingaros, 2015). Moreover, humans try to mimic nature to accomplish its benefits; most of the time biomorphic and biomimicry shapes and designs are the only solutions in the B.E. where nature does not exist. One of the solutions is to create visually intelligible ornamentation to apply to the form inside or outside the building (Salingaros, 2015). The human brain and emotions have their own natural mathematical equations in response to nature's visual patterns, order, symmetry, smile, and so on, and human survival mechanisms are automatically relaxed and reduce stress if they detect an impact from particular characteristics of the supportive surroundings. The human immune system responds positively, but more research is needed (Lakoff & Johnson, 1999).

5 Stress Recovery Theory, 1981

Built environment and its rapid growth have created great interest in studying different aspects of environmental stressors such as crowding, noise, and air pollution due to their negative impact and the stress they put on a large number of people globally (Cohen et al., 1983). Some researchers have focused on person-based variables or environmental characteristics. However, some other studies mainly focus on situations that cause stress. In addition, most studies have investigated extreme environmental characteristics such as too-hot zones, too-cold zones, and loud aircraft noise. In the modern period, stress is the leading cause of widespread chronic diseases. Stress has a negative impact on an individual's mental health and body core (immune system). Stress causes mental health disorders and cardiac conditions, which can lead to unexpected death. (Miller et al. 2011). The World Health Organization had emphasized that workspaces have to have an opportunity for their employees to relieve their stress; otherwise, they will suffer from different diseases and low performance (WHO, 2023). A critical question was emphasized by Ulrich and Simons; dose the non-extreme B.E. help to heal or recover the stress of their users? (Ulrich & Simons, 1986). The stress restoration theory by Ulrich proposes that natural environments promote recovery from stress, whereas urban environments tend to hinder the same process (Ulrich, 1981). Urban scenes: Understanding the concepts of stress and restoration was critical in order to get answers to the following two questions: Is it true that interaction with common situations has a favorable impact on individuals' restorative processes while others work against it? Baum described stress as the psychological, physiological, and often behavioral response of an individual to a circumstance that challenges or threatens well-being (Baum et al., 1985). The theory's second fundamental feature is stress restoration. These concepts can be interpreted as a broader idea that is not restricted to stress recovery scenarios or recovering from conditions characterized by intense psychological and physiological awakening, but can also be applied to healing, for example, extremely low encouragement (Ulrich, 1981). Urban scenes: Understanding the concepts of stress and restoration was critical in order to get answers to the following two questions: Is it true that interaction with common situations has a favorable impact on individuals' restorative processes while others work against it? Baum described stress as a psychological condition, or frequently human's behavioral response to a situation that impacts well-being in a negative way (Baum et al., 1985). ART can be a broader concept not only attached to stress recovery but can be applied to other psychological disorders like extremely low encouragement (Ulrich, 1981). The stress healing process corresponds to variables that extend to physiological energy recharge, which affects individuals' psychophysiological mobilization when exposed to a stressor. Recovery from stress involves various positive changes in psychological, physical, and physiological conditions that affect behavior, performance, and cognitive functioning. Furthermore, stress restoration has a direct connection to emotional states; decreasing negative emotions like fear or rage while raising positive emotions leads to a quicker recovery process (Ulrich, 1979).

6 Attention Restoration Theory, 1989

Human beings naturally belong to their native nature. Watching the sunset, gazing at the Ocean, spending time inside the greenery, and playing with the wind provide Man with the opportunity to refresh, relax, and restore their souls. Interacting with Nature enables individuals the opportunity to rest, reflect, and refresh their minds. Individuals have noticed this phenomenon. This phenomenon emphasized the vital role that nature plays in man's mental fatigue and physical conditions. It allows them to focus again effectively. In contrast, crowded urban environments have often been considered attention and energy burners. (Kaplan &

Kaplan, 1989) claims that urban environments' users suffer from an excess of bottom-up stimulation that serves to dramatically capture attention. In the last few decades, researchers focused on the concept that nature can re-establish humans' direct attention and keep their mental conditions healthy. As a result, Kaplan's Attention Restoration Theory is significant to be understood. Attention restoration theory or ART was developed by and popularized by Stephen and Rachel Kaplan between 1989 and 1995. At this period of time, the built environments were characterized by the great improvement in technology and ever-increasing indoor performance. Users were spending a vast of their time indoors and were concerned about the lack of time they spend interacting with nature. Attention Restoration Theory suggests that interacting with nature is not only enjoyable but can also assist humans to improve their focus levels and ability to concentrate and create (Roe & McCay, 2021). Nature, according to ART, has the ability to renew direct attention after exhausting mental energy, such as working hard on a project or an exam, or after restless nights. It is an example of a motivation theory that explains individuals' responses, particularly to physical features of their surroundings (Ohly et al., 2016). William in 1892 recognized two types of attention important to human mental activities, which the Kaplans based their work on: voluntary attention and involuntary attention (Kaplan, 1995). The direct attention part of the human brain is extremely important for daily actions and responses, because it is the part that solves the problems, understands actions and the surroundings, and feels everything. Moreover, it is a vital part of the brain to complete a task that needs concentration and focus. Accordingly, overusing it could lead to mental fatigue condition, which reduces cognitive functioning (Kaplan, 1993). This means that fatigue is the main enemy of voluntary attention; thus, finding a way to relax and restore direct attention is an essential task. Attentional fatigue is a serious health issue; it is associated with poorer decision-making and lower levels of self-discipline. It also has been linked to different diseases such as obesity through increasingly understood neural and behavioral pathways (Gu et al., 2013). Environment psychologists emphasized that urban lifestyles impose increasing stresses on our cognitive resources (Kaplan & Berman, 2010). Environments can contain any number of incentives to which any or all of our senses can react to them. Thus, understanding the critical role of built environments in human performance, mental health, and well-being is the crucial guide to a successful design strategy (Kaplan & Berman, 2010). The Kaplans provided the environmental characteristics needed in order to be a restorative fascination, being away, extent, and compatibility built environments (Kaplan, 1995).

7 Biophilic Design, 1993

Biophilic design was back in 1993 when Stephen Kellert and E.O. Wilson edited the biophilia hypothesis. Stephen Kellert and E.O. Wilson established a biophilic hypothesis based on the interrelationships between ecology, humans, and design. Stephen Kellert and E.O. Wilson in 1993 emphasized that "We are beings that have evolved with Nature, we are constrained to be connected to nature at multiple levels of our being." (Stephen & Wilson, 1993). Wilson and Kellert established a biophilic design concept in the praxis of ecology and design. Designers and researchers defined Biophilic design in the 1990s as a new design paradigm based on their understanding and work experiences. They defined it in the Biophilic Design book which had been published in 2011, highlighting some of their definitions; Kellert's definition of Biophilia is the human's inner thoughts that naturally want to be connected with natural systems, mainly the other living and non-living things. Moreover, Heerwagen and Gregory underline that biophilia, originally a Latin word that means "love of life," is essential for designing settings that evoke pleasant emotional experiences, which are predecessors to the human connection to and dedication to a place. Hilary Brown, but isn't biophilic design described as an artistic endeavor guided by or oriented to bio-logic that tries to safeguard and enhance our link with nature's forces and faces a logical and natural organizing element of all works of design (Stephen et al., 2011). The main design approach of biophilic design is to rebuild the destroyed connection between man and their native nature. Modern urbanism has separated man from their native nature, which leads to unhealthy and stressful built environments. B.E. seems out of place in natural environments, in particular that B.E. relies on advanced man's created technologies. Some scientists relate humans' separation from their native nature to the way they grow and their brains use to perceive their physical surroundings. Focusing on understanding the strong relationship between man and nature will encourage the designers to integrate nature with the B.E. Salingaros and Masden in Biophilic Design book, 2011 emphasized that humans are attached to nature due to their complexity, and they illustrated that Humans have a stronger physiological and psychological connection to physical surroundings that embody planned complexity than to bold or non-organized environments. Benyus as an author has worked meaningfully on aspects of biophilic design. He presents an example: "Biomimicry" is a design learning from nature, and it has an intimate relationship with biophilic design, in which a design is developed from natural characteristics. (Stephen et al., 2011). Biophilic design is a restorative B.E. designed to help humans restore their mental and physical conditions. Rose in the same book said "Restorative design has two intrinsic implications: to restore the human and natural environment and to restore the self." (Stephen et al. 2011). Biophilic design become a proven knowledge that professionals apply to reconnect man with nature, also, in research in different scales such as urban, architecture, and interior. It is not only related to the physical aspects of design but also users' psychological conditions (Giulia et al., 2017). Sustainable solutions and their technologies enlarge the gap between individuals and nature (William et al., 2014). This design concept illustrates the term restorative B.E. because of its unique approach, which aims both to reduce the impact of B.E. on natural environments and improve the positive impact of B.E. on humans' health and well-being in nowadays urbanism (Kellert et al., 2008). The positive impact of biophilic design on health increases the experimental facts for the human-nature connection and raises its priority level within both design research and design practice (William et al., 2014). The basic principle of biophilic design agrees with the idea and science emphasizing that humans naturally seek out natural characteristics that speak to their main emotions enhancing and encouraging their mental, emotional, physical, and even spiritual health (Kellert, 2005). Scientists, designers, and researchers have been working for decades to define the biophilic design concept and its positive impact on individual health. Thus, one of the latest outcomes provides design considerations to enable designers to apply them effectively. There are more than 500 publications on biophilic responses in order to provide these design considerations. In other words, as they name patterns those designers can use to design a better and healthier built environment. They created "The Fourteen Patterns of Biophilic Design" book that has applications in both macro-scale and microscale. They alienated Biophilic design into three main strategies. First, designers can approach to bring nature inside the designed B.E. in different strategies. Second, designers use the biomimicry concept, biomorphic shapes, and nature analogous. Third method, they can mirror nature in its characteristics such as scene, refuge, anonymous, and risk. In this book, they focus on man's mental health and well-being and their relationships with the built environments and nature. (William et al., 2014).

8 Neuroscience for Architecture, 2003

The aim is to examine how each environmental element influences certain brain processes, which in turn affect an individual measure (Edelstein, 2002). Neuro-architecture scientific aims to have proven knowledge that illustrates the impact of physical surroundings and designed B.E. on humans' brain mechanism, immune system, emotions, and neuroscience system. Researchers measure some vital parts and actions inside humans' brains such as the hippocampus which creates memories, the hypothalamus connected to other body systems, and the Cerebral Cortex which expresses emotions (Eberhard, 2009a). The development of architectural design methodologies has been enabled by the emerging school of neuroscience for architecture (neuro-architecture). Another objective of neuro-architecture is to explain why an architectural environment can evoke feelings of peace or wonder, and how humans can recall the smell of Sea sand, the feel of soft cloth, and the sound of breaking glass (ANFA, 2023). In other words, the study of the human brain mainly investigates its mechanism to store information, or obtain knowledge, and shape its emotions using its senses. This science enables designers to understand users' reactions and expected outcomes of a certain B.E., situation, or location. Built Environment that created considering the neuro-architecture sciences: it is known as the B.E. designed to encourage users' mental cognition and reduce stress levels (Eberhard, 2009b). Gage proved by the advanced science technologies that the human brain keeps producing new neurons in adults because changing environments influence the brain's neurogenesis (Badger, 2014). On the other hand, Architects such as Juhani Pallasmaa, Steven Holl, and Peter Zumthor have illustrated the process of perceiving the physical surroundings and converting these collective data inside the brain into images, memories, and experiences (Palasmaa, 2005). The American Institute of Architects (AIA) established the Academy of Neuroscience for Architecture (ANFA) in 2003 as a global center for multidisciplinary endeavors to build educated links between neuroscientific studies and B.E. designers and professionals.

9 Well-Building Standards 2013

In 2013, well-building standards emerged to design and build a global community. The well construction standard v1.0 was created after seven years of research and collaboration among top medical experts, designers, and building industry practitioners. The team's results turned into viable methods for building design and operations, in combination with notable architects, engineers, lighting specialists, sustainability consultants, and designers (Institute, 2023). A substantial collection of research has aided in the creation of the underlying ideas of well-built standards and the required standards. In early 2015, the consulted research was published in good bibliographies, providing material and references for the Well-Building Standard. A well-building standard is a performance-based methodology

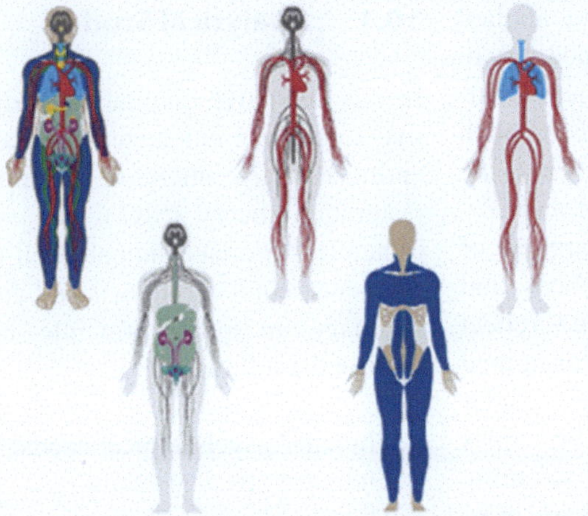

Fig. 4 The 12 human body systems on which Well-building standards are founded (Institute, 2023)

Fig. 5 The ten concepts of the Well-Building Standards Vol. 2. (Institute, 2023)

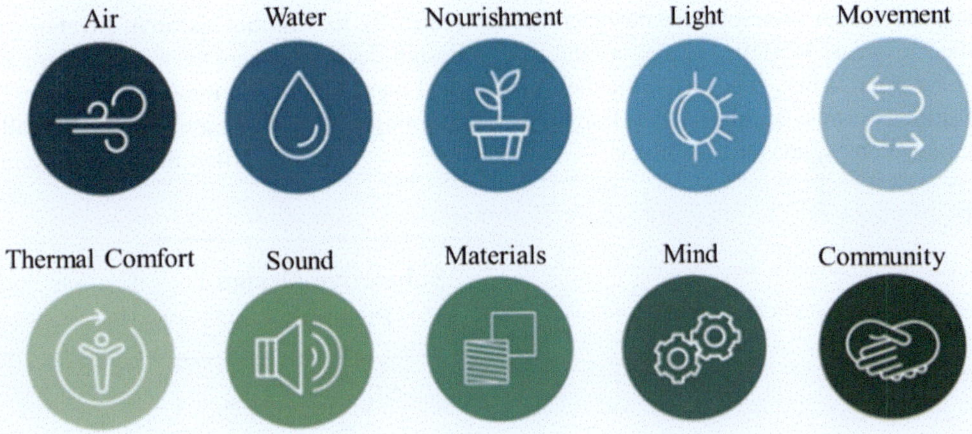

for measuring, certifying, and monitoring built-environmental characteristics that influence well-being via air, water, nutrition, light, fitness, comfort, and mind. The well-building standard is based on the foundation of medical research, which investigates the relationship between buildings in which individuals spend more than 90% of their time. Each well-built standard feature was assigned to one or more of the bodily systems listed below, as shown in Fig. 4 (see Fig. 4 in Institute, 2023). They created ten concepts air, water, mind, community, movement, innovation, light, sound, materials, nourishment, and thermal comfort, as shown in Fig. 5 (see Fig. 5 in Institute, 2023). In 2018, they developed well Vol.2 with more than 150 well concept advisors who contributed to the creation of Well v2. Concept advisers are experts in one or more of the ten well concepts and offer advice on solutions to health and well-being issues, methodologies for localization, case studies to address research gaps, and creative topic areas (Institute, 2023).

10 Methodology

This paper is part of the first phase of ongoing research. It is a deductive qualitative study in which researchers applied the thematic analytical technique and axial coding approach to compare and correlate different selected design theories, design concepts, design indexes, and design rating systems to create the WBDP, which will enable architects and designers to create better B.E. Its main focus is human health and well-being. In this study, ART, SRT, and neuro-architecture are scientifically proven theories and knowledge that link the sciences of the human brain's mechanism, emotions, and the way they perceive and interact with their B.E. with architects and designers who design the built environments. These selected scientific theories are part of continuing studies incorporating other theories and rating systems such as biophilia theory, biophilic design, and well-building standards rating systems. Researchers revisited these selected existing scientific theories and new design school approaches and applied thematic analysis techniques and axial coding methods to compare and correlate their main objectives, methodologies, design principles, and considerations to accomplish the two main objectives of this part of the study. First, the thematic analysis provides a detailed comparison and correlation analysis and results that enhance the awareness of architects and designers towards the human brain, body, and B.E./physical surroundings interrelationships, and reflect it in their design decisions. Second, researchers examined each component to identify differences and connect similar principles to develop Human Well-being Design Principles & Considerations (WBDP) based on the selected scientific theories in this study. These design factors, design principles, and design concepts, when merged, are very specific, clear, and new design drivers based on the concept of enhancing human health and reducing stress levels inside built environments. This is a positive step towards providing applied design principles for designers to apply directly to their projects. WBDP will play a vital role in design decisions, approaches, concepts, principles, factors, and considerations because architects and designers are designing for human beings, not for machines.

10.1 Analytical Studies

In this deductive qualitative study, an analytical comparative study was conducted between the selected three scientific theories: attention restoration theory (ART), stress restoration theory (SRT), and neuro-architecture. Design principles to create human well-being-oriented built environments require architects and designers to consider different main concepts or principles based on different aspects, such as the following:

- Humans' psychological needs.
- Humans' physical needs.
- Climate characteristics.
- Moral/cultural/ background needs.
- Financial conditions.
- Subjective needs.

In this research, psychological needs and physical needs in turn are the main concern as shown in Fig. 6.

Attention Restoration Theory (ART) primarily illustrates, based on scientifically proven brain sciences and theories, that the human brain has two parts: the frontal part,

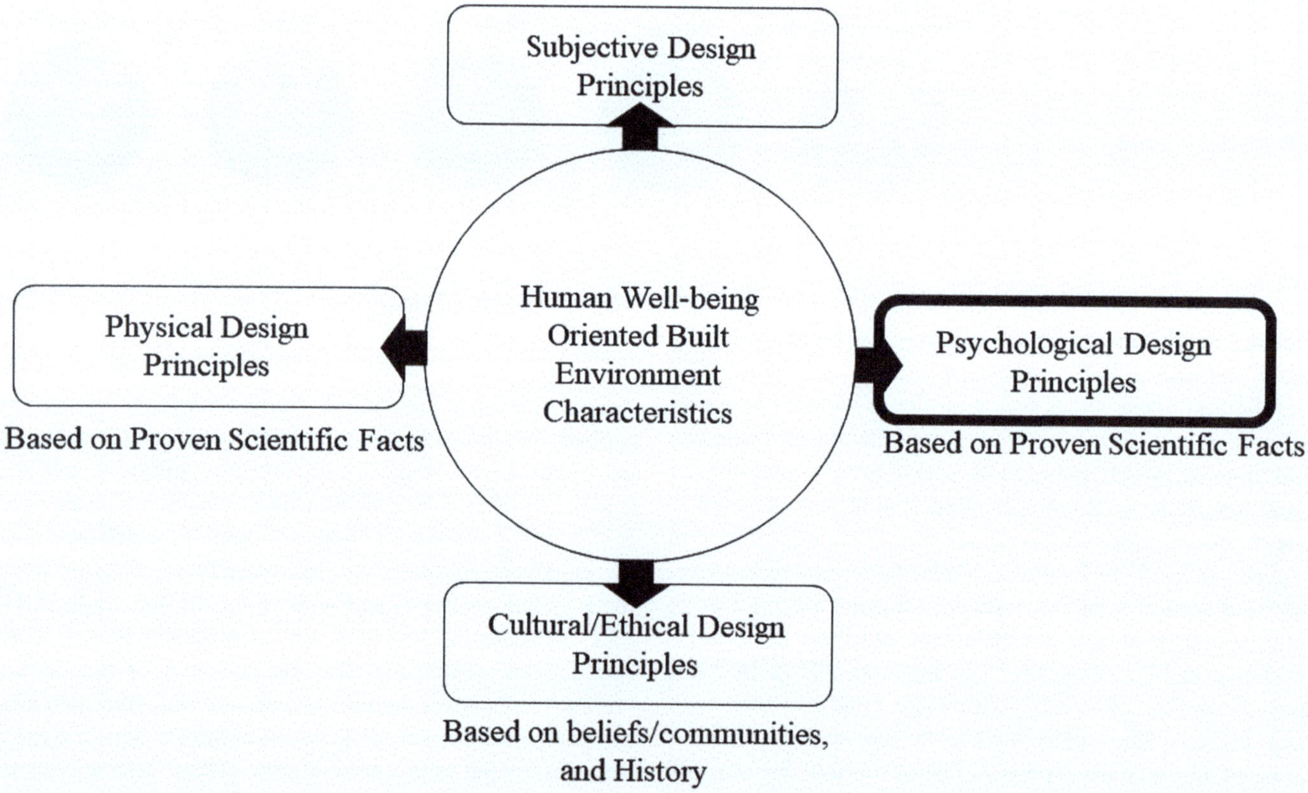

Fig. 6 Main design considerations towards enhancing Human/ user health and well-being level by researchers

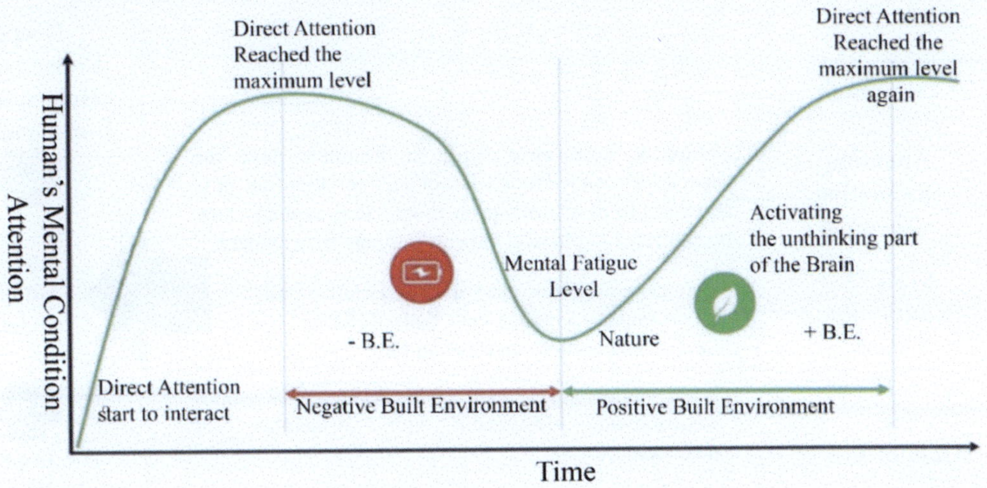

Fig. 7 Graph shows the main concept and the methodology of Attention restoration theory, and also, shows how is human brain's direct attention reacts to its surroundings by researchers

known as direct attention, which is responsible for all thinking activities and reactions all of the time, and the back part known as secondary attention, which is responsible for all secondary attention. The other part of the human brain is known as the unthinking part and is active when a person is calm or enjoying something. Thus, ART's primary concern is to encourage architects and designers to create comfortable, enjoyable, pleasant, and healthy built environments, which are referred to as restorative/positive built environments, because, as shown in Fig. 7, ART mainly concentrates on recharging human direct attention and encouraging the unthinking part to work, which automatically changes human psychological and mental conditions from negative to positive mood.

SRT, on the other hand, emphasizes that inadequately planned or poorly created settings constitute daily stressors for users. Furthermore, it was emphasized that being in nature reduces humans' stress levels and chronic diseases because nature is not only an attractive environment for humans, but also a part of it. Third, neuro-architecture directed architects and designers to consider very critical design principles that mainly consider the influence of the designed B.E. on humans' brains bodies, emotions, and well-being in turn. The graph below illustrates the key drivers of these three scientific theories as well as their added value and impact on the way built-environment designers think and make design decisions towards building good, restorative places, and settings Fig. 8.

Figure 9 depicts a comparison of the main objectives of ART, SRT, and positive and negative built environments, as well as their impact on the human brain and mental conditions. It illustrates the mental conditions in both B.E. different characteristics in both positive and negative ways. Furthermore, it demonstrates how the direct attention restoration process occurred as well as its impact on users' stress levels.

Neuroscience for architecture provides a good understanding of the human mind–body connection and is essential for designing a built environment that has a positive impact on people's mental health. The human brain perceives its surroundings through a complex process, as illustrated in Fig. 10. The image formation process and memory process are sophisticated mental processes that architects and designers have to learn about to be highly considered in their design principles and design decisions. Figure 11 shows the impact of light and lighting designs on human visual perception. Lighting design must be based on science rather than on art or function on its own.

11 Discussion and Results

Stress-rising levels and related chronic diseases are the main reasons why the human well-being-oriented built-environment design paradigm is becoming the dominant design paradigm nowadays. Moreover, most current urbanism increases the stress levels of users. Thus, Architects, urban planners, and interior architect designers must create a B.E. that reduces their users' stress to avoid causing mental disorders and chronic diseases. Built environments, as daily stressors, must become restorative places and spaces for their users.

11.1 Discussion

As shown in Fig. 9, a comparative analytical study demonstrated the cause and effect of stress on human performance, mental health, and physical health. Furthermore, fascination, extent, being away, and compatibility are restorative built environment characteristics, but researchers linked these concepts with SRT concepts and created design

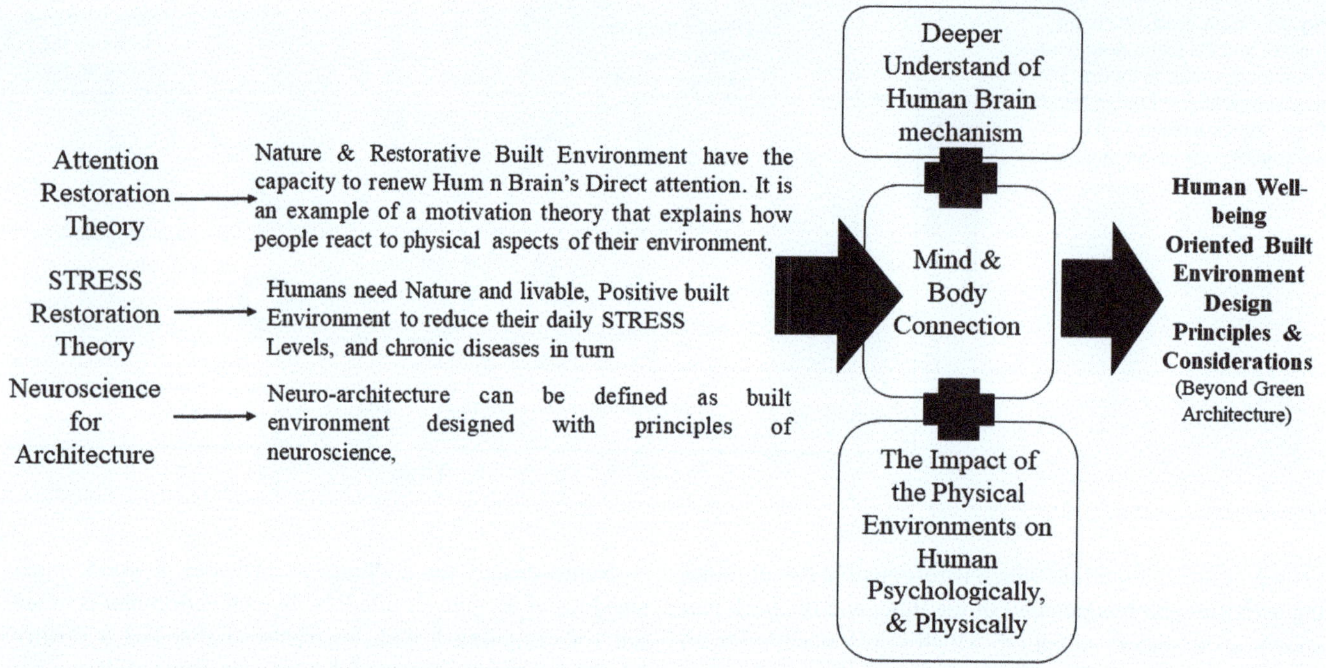

Fig. 8 The diagram shows the main objectives, drivers, and concerns of each of the selected theories, which enable designers to create livable, healthier, and enjoyable built environments by researchers

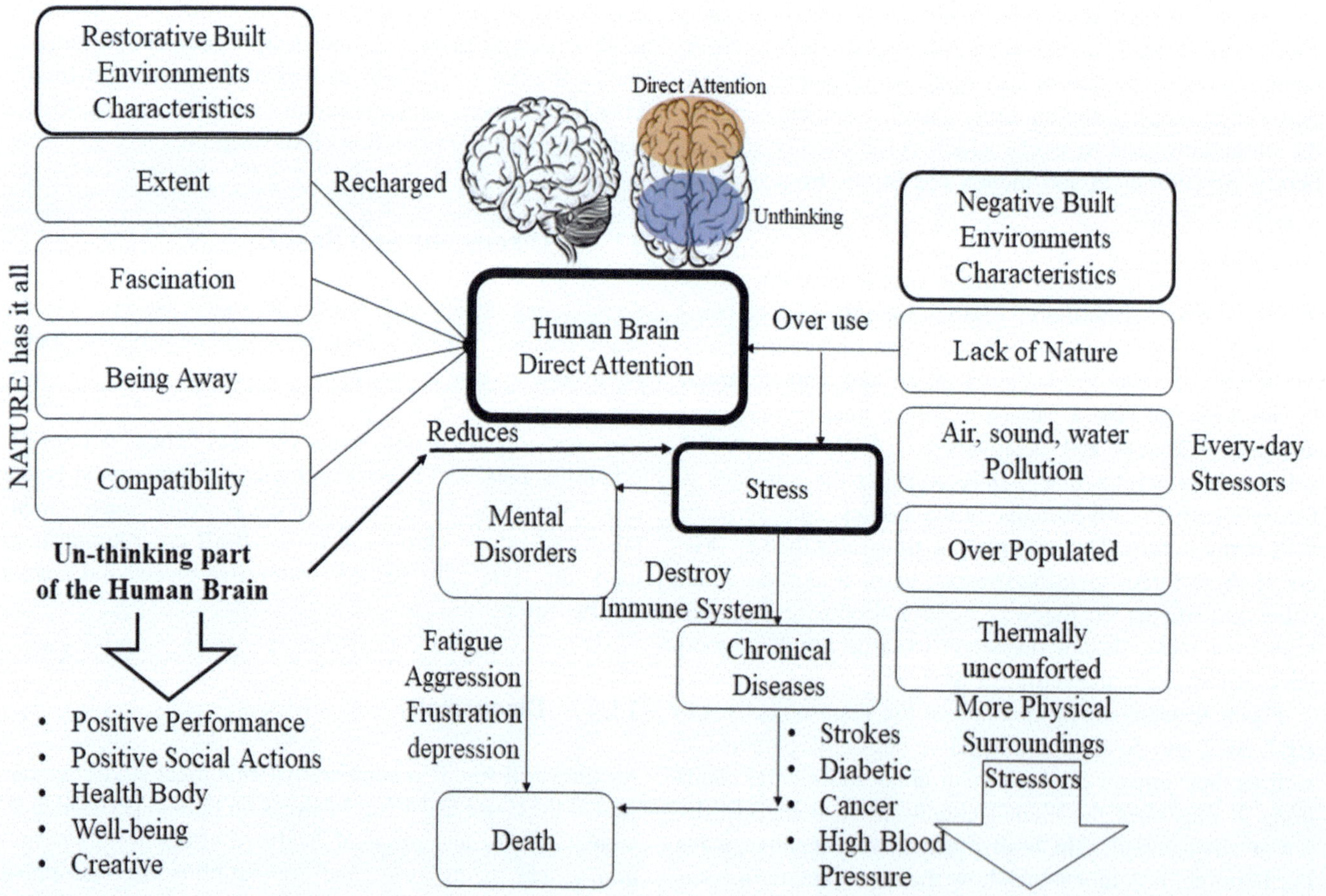

Fig. 9 Comparative analytical study, showing the different impacts of the restorative and negative built environments on the human brain and health by researchers

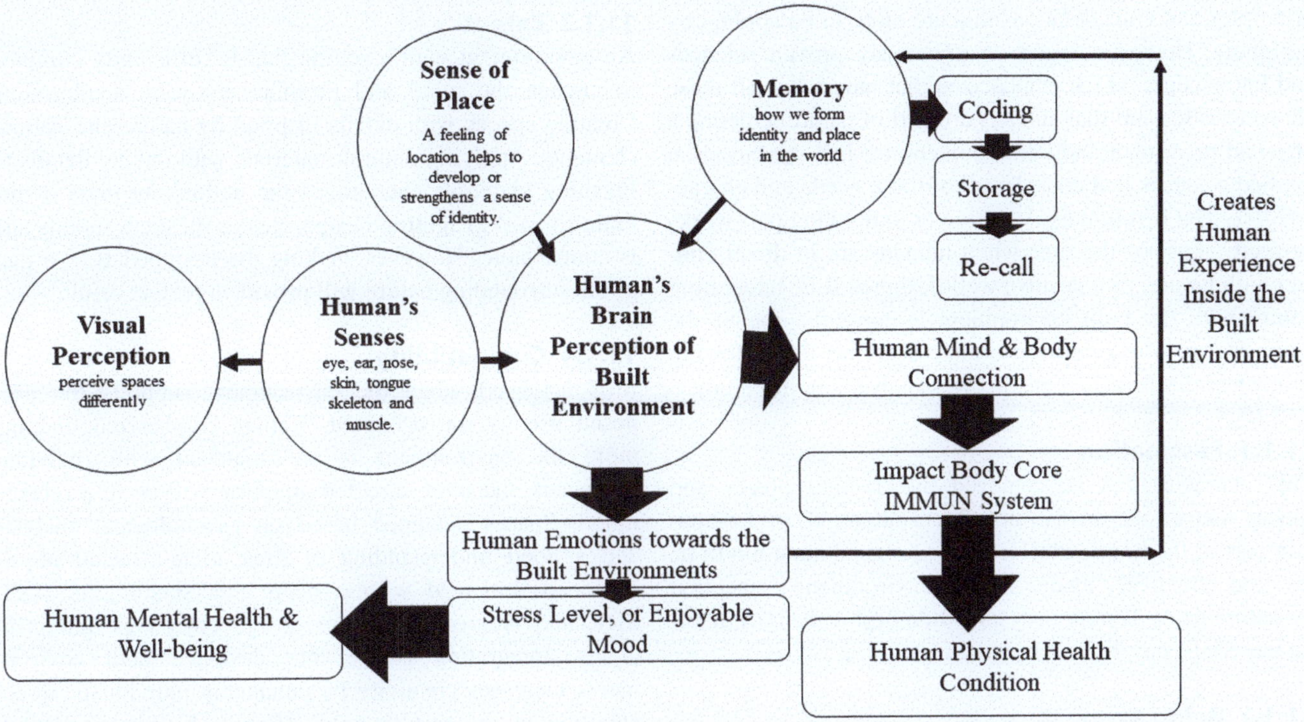

Fig. 10 The human brain Perception and visual Perception processes and their impact on the Human sense of place, memory, Stress level, Immune System, and Well-being level by researchers

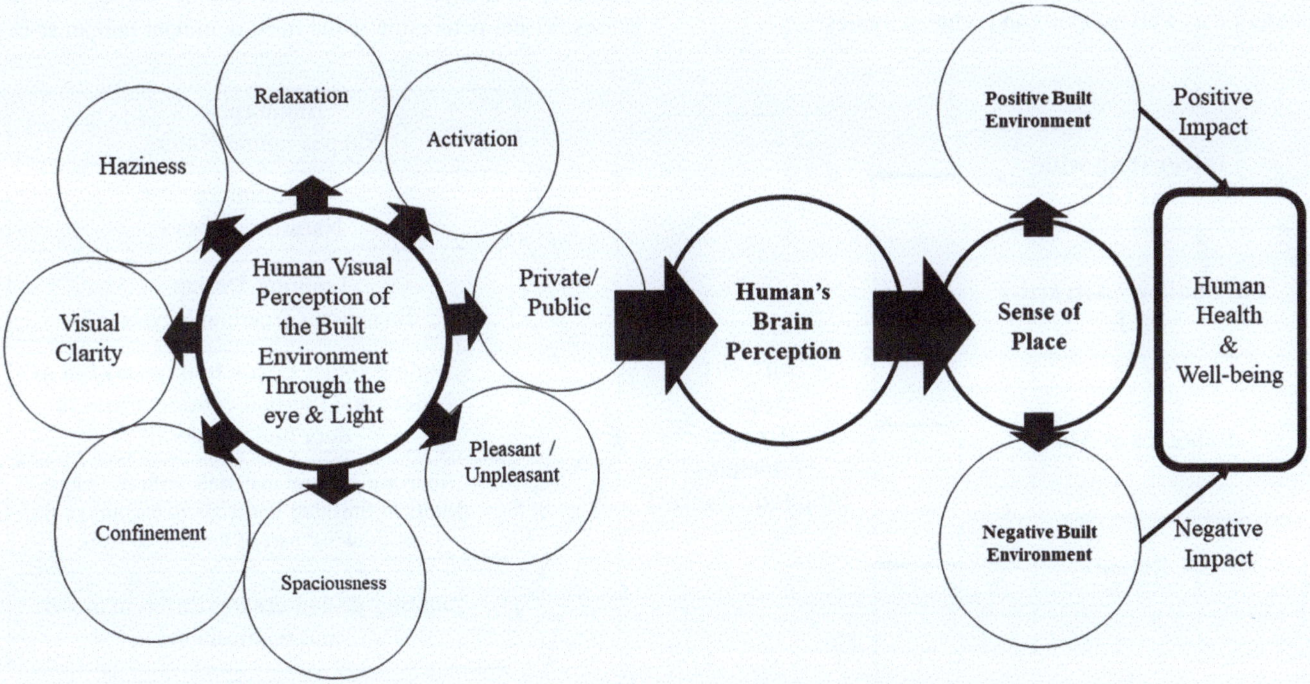

Fig. 11 The human visual perception turns into emotions that create the sense of the place and then create its own architectural and emotional memories by researchers

principles and considerations that are more to be applied by designers. However, based on previously proven sciences and knowledge, which emphasized that nature has all these characteristics and that humans are part of nature, creating a physical man-made built environment needs to be based on applied sciences that consider all of these needs and characteristics. The Involuntary Attention or unthinking part works perfectly most of the time when humans are in direct contact with nature, or a positive well-designed B.E. Integrating nature with the built environment is currently one of the main drivers of design. Restorative B.E. has the following characteristics based on the theories analyzed in this study.

11.1.1 Fascination

This character can be achieved by creating places and spaces that easily involve non-direct attention, or fascination part of the brain, which enables fatigued attention to be restored effectively again. A built environment is inspired by nature using Biomimicry and Biomorphic designs, natural materials, day-lit, and living things.

11.1.2 Being Away

Design based on mental restorative principles unique, enjoyable, and comfortable Built environment totally different than their everyday places. Creating social interaction spaces full of nature and thermal comfort zones indoors and outdoors, as well as sports and other activities.

11.1.3 Extent

An environment with a setting that is sufficiently complex to engage the mind and promote enjoyable exploration. Creating spaces with details inspired by nature and natural elements. Applying natural materials with unique details to finishing materials encourages the unthinking parts of the brain to work. Lighting design creates different moods and is controllable. However, making the designed B.E. a part of the surrounding nature will provide a perfect result.

11.1.4 Compatibility

It is achieved when the environment supports the proposed use by the occupant. Kaplan emphasized that the more the environment is well-matched with individuals' needs, the more directed attention will be in a relaxed mood. Spaces designed based on psychological analysis and a good understanding of their users' special physical needs engaged with a deep understanding of ART, SRT, and Neuro-architecture design objectives and principles. Integrating nature into designed built environments increases creativity by enhancing individuals' direct attention to creative thoughts (Runco & Pritzker, 1999). Figure 12 shows the applied ART and SRT design principles based on the above analysis and their main objectives.

Neuro-architecture, on the other hand, emphasizes that the human brain interprets physical surroundings through human senses. Visual perception is the most dominant human sense,

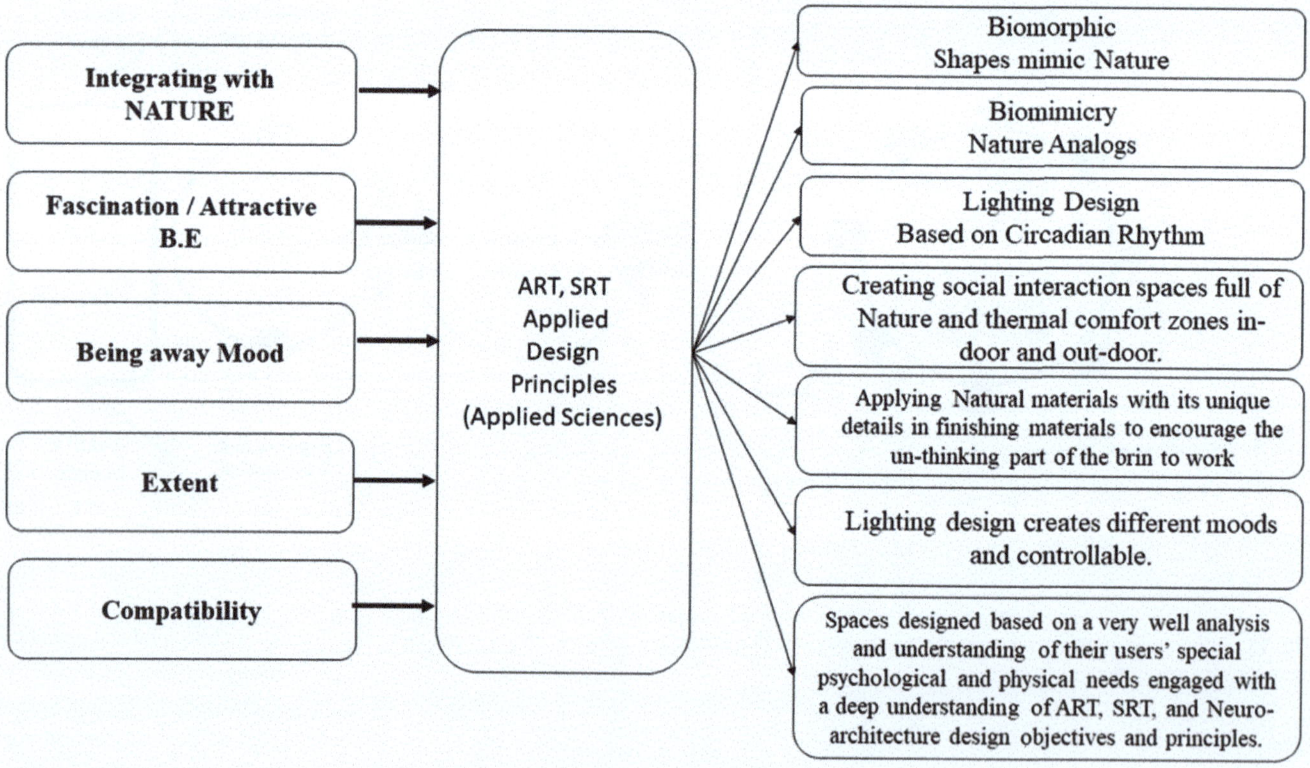

Fig. 12 Applied ART and SRT design principles based on the above analysis and their main objectives by researchers

transmitted directly to the brain via the eyes, while other senses are required in the perception process to totally mold the image and produce a sense of place. This sense communicates with human emotions and creates a favorable or negative reaction to a given location or region. This complicated process of perception and emotional transformation of physical surroundings, as well as interactions, culminates in a one-of-a-kind experience and impression of this place or space. Following that, the human brain builds a special memory for this location or space that can be retrieved as needed. This process, as well as the impact of stressful situations on individuals' mental and physical health, is depicted in Fig. 10. The neuroscience of architecture informs designers about the dynamic role that brain perception and visual perception play in generating a specific built environment's interior impression, the feeling of place, the memory of this location, and pleasures or non-pleasures. As seen in Fig. 8, lighting design and visual perception help the human brain and emotions create an impression and set their mental status to a specific mood, such as relaxation, activation, private/public, pleasant/unpleasant, spaciousness, confinement, visual clarity, or haziness. Thus, these brain states have major consequences for the human body, immune system, and neurobiology, culminating in a level of anxiety or Well-being.

11.2 Results

Research results show that a positive built environment, or restorative built environment, creates a built environment that speaks to the human brain and body positively. The built environment creates an enjoyable and energetic mood. As a result of the above comparative analysis, researchers found that a high level of integration between nature and the designed built environment will help designers, users, and decision-makers create more livable, healthy, and enjoyable built environments. Based on the selected theories and sciences, researchers have concluded that the human brain uses its five senses to understand its physical surroundings, especially its visual perception mechanism. Thus, for architects and designers to create human-well-being-oriented built environments, they need to have a very good understanding of the human mind, body, and B.E. interrelationships by collaborating with other scientists in their design process. Moreover, architects need to be knowledgeable about brain perception, visual perception, mind and body connections, and science beyond these vital design considerations. On the other hand, researchers show that nature elements speak to humans positively, but at the same time, integrating nature with built environments is a very sophisticated design process that needs to be based on sciences and other main considerations. Researchers have suggested the following design equations:

Art + Science + Ecology + design + Health + human

The following Fig. 13 shows that the three approved scientific theories and concepts agreed that Nature has it all and many design principles that speak to these senses positively, which Nature has it all.

Researchers have created human well-being design considerations and principles, WBDP Fig. 14. They emphasized the impact of B.E. on the human sympathetic and parasympathetic systems, which affect individuals' stress levels. Designers will need to study more about human brain mechanisms and reflect this in their designs, such as the human circadian rhythm to be considered in the lighting design process, especially in indoor environments, and the psychological impact of color has to be highly considered in design decisions, especially on the microscale. B.E. has to be pleasant, comfortable, healthy, and functional to its users. An enjoyable B.E. enables the human brain to produce new neurons. Moreover, it creates individuals' memories which are built by their emotions and impressions about places and spaces they interacted with. In addition, researchers have concluded that human emotions and moods are affected by brain activities, so encouraging positive social activities will put users' minds in a good mood. The vital part for designers to be well understood is human perception, which is mainly based on understanding the brain's process of creating images and impressions and interacting with its neuroscience system to create a specific emotion and visual perception. Physical surroundings assist users in decision-making. However, integrating with nature and creating natural analogs inside the B.E. will have an essential impact on individuals.

12 Conclusion

According to the researchers, it is critical to connect architects, urban planners, and designers with other sciences that primarily demonstrate humans' perception of their physical surroundings. They also emphasized the direct impact of the B.E. on humans' brains and emotions, which produce their impressions or senses of the locations and, as a result, their memories of the places and spaces. The researchers accentuated the key design principles upon which the chosen scientific ideas were founded. The developed or implemented design principles, on the other hand, would be a positive step for designers to consider in their design process and decisions to create and design built environments that are less stressful on their users. Because humans are a part of nature, nature, according to the selected theories, may enhance human psychological and physical circumstances. To develop the final WBDP, researchers are working on the other three design theories, ideas, and rating systems stated

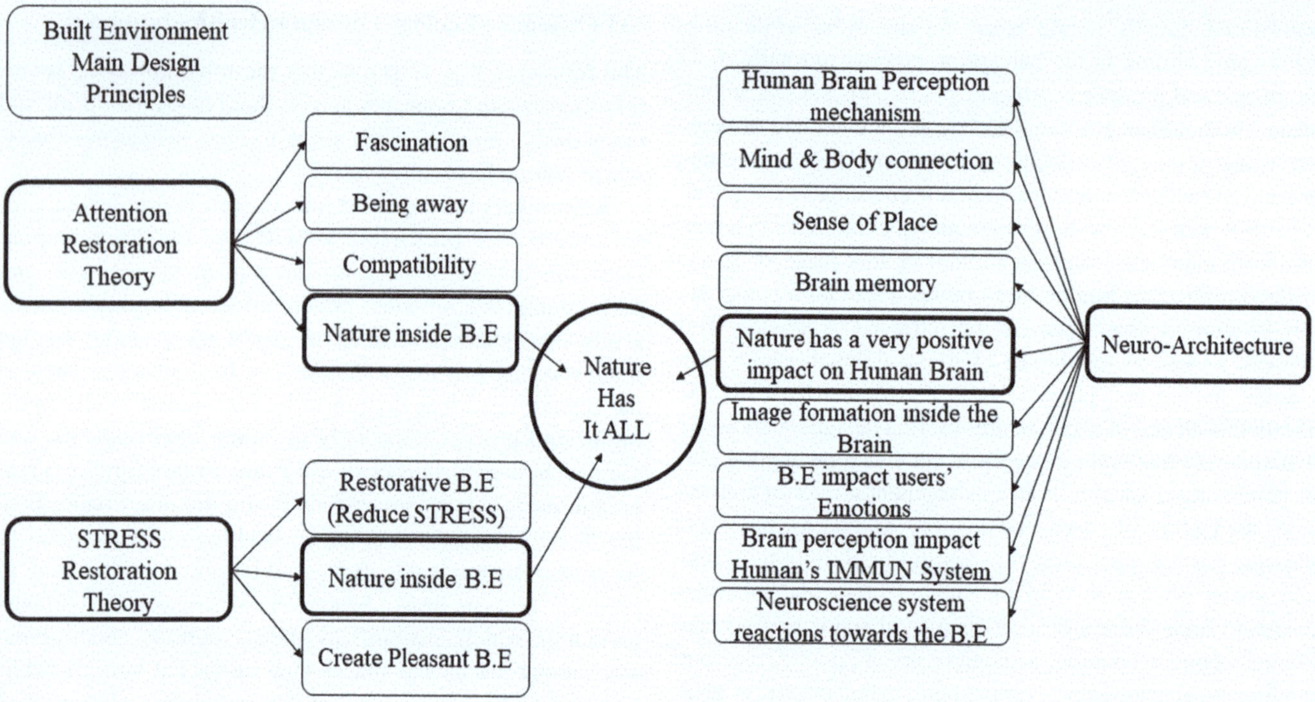

Fig. 13 Main concepts of ART, SRT, and neuroscience for architecture all emphasizing that nature has a great impact on human mental and physical health by researchers

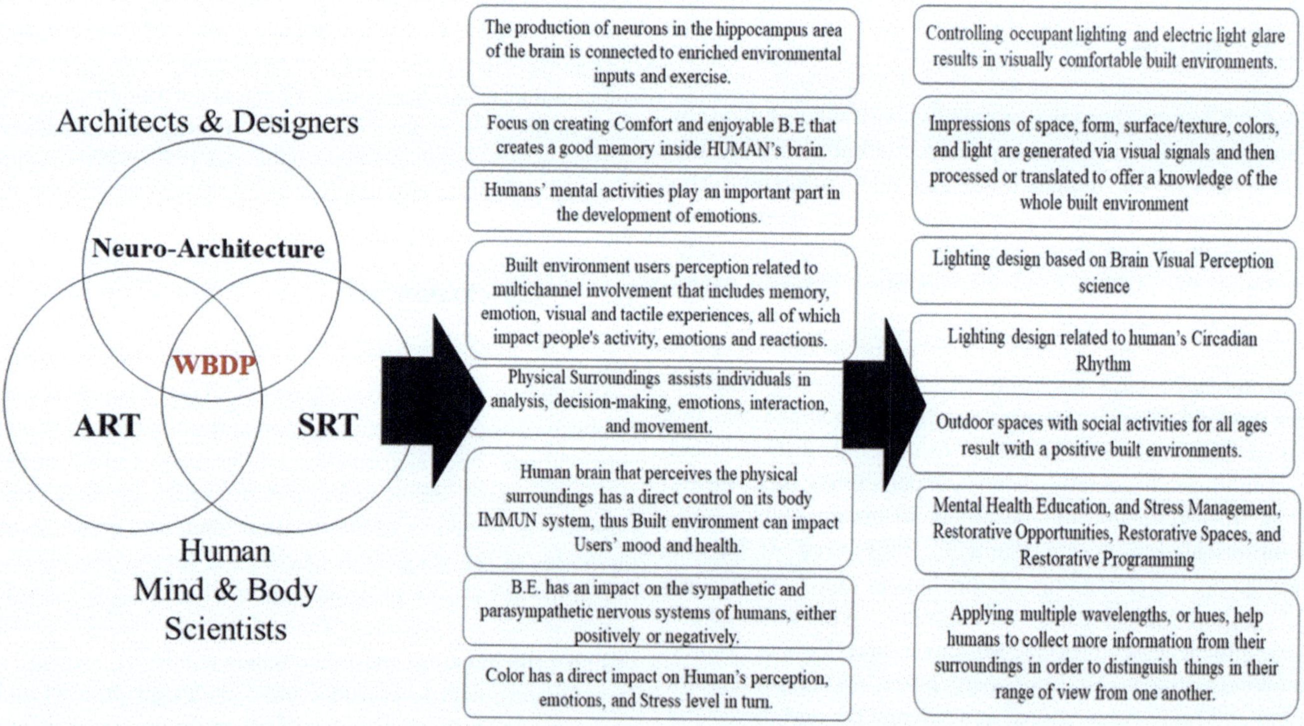

Fig. 14 The created design considerations and design principles out of the selected theories, WBD by researchers

earlier, such as biophilia theory, biophilic design, and well-building standards rating systems. Man-made environments are numerous, and disregarding nature and human well-being requirements will result in more bad constructed environments and chronic diseases. More partnerships between architects and scientists are needed, as is the incorporation of these fields into design institutions at both the undergraduate and postgraduate levels. More research in this critical area, according to researchers, is needed to enable designers to create habitable and healthy spaces.

References

ANFA. (2023). Academy of Neuroscience for Architecture. Retrieved June 06, 2023, https://anfarch.org/. (March 17).

Badger, E. (2014). Corridors of the Mind. Pacific Standard. (April 6).

Bartuska, T. J. (2011). The built environment: Definition and scope. In *Introduction: Definition, design, and development of the built environment*.

Baum, A., Fleming, R., & Singer, J. (1985). Understanding environmental stress: Strategies for conceptual and methodological integration. *Environmental Psychology*, 185–205.

Betro, R., & Barbiero, G. (2017). The biophilic quality index. A tool to improve a building from "Green" to restorative. *Visions for Sustinability*.

Browning, W., Ryan, C., & Clancy, J. (2014). *14 Patterns of biophilic design*. Terrapin Bright Green LLC.

Cohen, S., Kamarck, T., & Mermelstein, R. (1983). A global measure of perceived stress. *Journal of Health and Social Behavior*, 385–396.

De Botton, A. (2008). *The architecture of happiness*. Vintage Books.

Edelstein, E. A. (2002). Neuroscience and Health Care Facilities Workshop: Sponsored by the American Institute of Architects and the Vinyl Institute The National Academy of Sciences, Woods Hole MA, 13–15 Aug 2002. Detailed Report: chrome-extension://efaidnbmnnnibpcajpcglclefindmkaj. Retrieved from https://anfarch.org/workshop/PDFResources/NeuSciHCFclt_Workshop_Aug2002_DetailedReport.pdf

Eberhard, J. P. (2009a). Applying neuroscience to architecture. *Neuron, 62*, 753–756.

Eberhard, J. P. (2009b). *Brain landscape: The coexistence of neuroscience and architecture*. Oxford University Press.

Giulia, N., Roxana, M., Guido, C., Rita, B., & Giuseppe, B. (2017). When biophilic design meets restorative architecture: The strambinello project. *Visions for Sustainability*, 46–58.

Gou, Z., & Xie, X. (2017). Evolving green building: triple bottom line or regenerative design? *Journal of Cleaner Production*, 600–607.

Gu, X., Hof, P. R., Friston, K. J., & Fan, J. (2013). Anterior insular cortex and emotional awareness. *The Journal of Comparative Neurology*.

Institute, I. W. (2023). International Well Building Institute. Retrieved June 06, from https://standard.wellcertified.com/well

Judith, G. (2012). *Back to nature for good: using biophilic design and attention restoration theory to improve well-being and focus in the workplace*. University Digital Conservancy Home.

Kaplan, S. (1995). The restorative benefits of nature: Toward an integrative framework. *Journal of Environmental Psychology*, 169–182.

Kaplan, S., & Berman, M. G. (2010). Directed attention as a common resource for executive functioning and self-regulation. *Association of Psychological Science*.

Kaplan, R. (1993). The role of nature in the context of the workplace. *Landscape and Urban Planning*, 193–201.

Kaplan, R., & Kaplan, S. (1989). The experience of nature: A psychological perspective.

Kellert, S. R. (2005). Building for life: Designing and understanding the human-nature connection. Researchgate.

Kellert, S. R., Heerwagen, J., & Ma, M. (2008). *Biophilic design: The theory, science and practice of bringing buildings to life*. John Wiley & Sons.

Kopec, D. (2010). *Environmental psychology for design* (2nd ed.). Fairchild Books.

Lakoff, G., & Johnson, M. (1999). *Philosophy in the flesh: The embodied mind and its challenge to western thought*. Basic Books.

Miller, G. E., Chen, E., & Parker, K. J. (2011). Psychological stress in childhood and susceptibility to the chronic diseases of aging: Moving towards a model of behavioral and biological mechanisms. *PMC*, 959–997.

Ohly, H., White, M. P., Wheeler, B. W., Bethel, A., Ukoumunne, O. C., Nikolaou, V., & Garside, R. (2016). Attention restoration theory: A systematic review of the attention restoration potential of exposure to natural environments. *Journal of Toxicology and Environmental Health, Part B*.

Olmsted, F. L. (1903). Yosemite and the Mariposa Grove:. Library of Congress.

Palasmaa, J. (2005). *The Eyes of the Skin Architecture and the Senses*. Hoboken: John Wiley & Sons.

Patterson, D. (1989). Chapter 10: Contributions of environmental psychology to visitor studies. In D. Patterson (Ed.), Visitor Studies. Semantic scholar.

Roe, J., & McCay, L. (2021). *Restorative cities: Urban design for mental health and wellbeing*. Bloomsbury Publishing.

Runco, M. A., & Pritzker, S. R. (1999). *Encyclopedia of creativity*, (Vol. 2). Academic Press.

Salingaros, N. (2015). *Biophilia and healing environments: Healthy principles for designing the built world*. Terrapin Bright Green LLC.

Stephen, K. R., Judith, H., & Mador, M. (2011). *Biophilic Design*. N.Y: Wiley.

Stephen, R., & Wilson, E. O. (1993). *The biophilia hypothesis*. Island Press.

Ulrich, R. (1979). Visual landscapes and psychological well-being. *Landscape Research*, 17–23.

Ulrich, R. (1981). Natural versus urban scenes: some psychophysiological effects. *Environment and Behavior*, 523–556.

Ulrich, R., & Simons, R. (1986). Recovery from stress during exposure to everyday outdoor environment. In J. Wineman, R. Barnes & C. Zimring, (Eds.), *The Seventeenth Annual Conference of the Environmental Design Research Association* (pp. 115–122).

Ulrich , R. S., Simons, R. F., Losito, B. D., Fiorito, E., Miles, M. A., & Zelson, M. (1991). Stress recovery during exposure to natural and urban environments. *Journal of Environmental Psychology*, 201–230.

Wernick, J. (2008). Building happiness: Architecture to make you smile. Good Reads.

Wilson, H. E. (2009). *Biophilia*. Harverd University Press.

WHO.int. (2023). World Health Organization. Retrieved from. https://www.who.int/data/gho/data/major-themes/health-and-well-being#:~:text=The%20WHO%20constitution%20states%3A%20%22Health,of%20mental%20disorders%20or%20disabilities. (March 16)

William, B., Catherine, R., & Joseph, C. (2014). 14 Patterns of biophilic design. Terrapin Bright

Wilson, O. (1993). *The biophilia hypothesis*. Island Press.

Yeang, K., & Spector, A. (2011). *Green design: From theory to practice*. Black Dog.

Adaptation, Livelihood, and Social Dynamics

Assessing Community Adaptive Capacity for Climate Resilient Development: A Complex Adaptive Systems Case Study Analysis of Mathare Valley, Nairobi, Kenya

J. Metcalfe and A. Gianoli

Abstract

This research takes an urban resilience standpoint to explore the interconnections of resilience components within a community network. It identifies leverage points through an in-depth case study analysis, based on empirical evidence and structural equation modelling, to draw conclusions as to the mechanics of these relationships in the context of a post-colonial sub-Saharan African state, where informal settlements are in a state of deprivation and subject to significant risk.

Keywords

Climate resilient development · Urban resilience · Community adaptation · Adaptive capacity · Inequality · Resilience activities · Complex adaptive systems · Kenya · Mathare

1 Introduction

The compounding impacts of urbanization and climate change place increasing pressure on interconnected urban systems challenging communities on the frontline to develop the capacity to adapt (IPPC, 2021, 2022). These challenges are 'two sides of one coin' (Shi et al., 2021) as adaptation challenges are inextricably tied to development challenges (Global Centre on Adaptation, 2022). Throughout Sub-Saharan Africa contexts of informality present particularly complex developmental challenges, as the nature and organizational structure of these environments sit outside of the prevailing formalized governance models of contemporary cities (Leck & Roberts, 2015). Global structural inequality and negligent policies are barriers to securing sustainable urban environments in the future. Communities that face these increasing climate and developmental hazards are extremely vulnerable and must develop complex survival mechanisms to cope. Climate resilient development calls for robust urban systems capable of managing critical and complex interdependencies to maintain system function. Consequently, climate policy action must focus on building adaptation capacities at the city scale to increase urban resilience (Croese et al., 2020; Swilling & Annecke, 2012; Tyler & Moench, 2012). This research explores how to achieve climate resilient development in the informal settlement of Mathare in Nairobi across the community system.

This research investigates the range of existing capacities across three levels of the community in Mathare Valley, and the level of adaptive action undertaken to solve challenges in this urban complex adaptive system. It asks the question what is the influence of the current adaptive capacities and engagement in resilience activities on Community Resilience across three component levels of governance in Mathare for the transition towards climate resilient development? This enquiry offers an opportunity to investigate what is critical to support the governance of adaptation in complex environments.

The question has been examined through the development of an empirical case study which has measured, scored, and analysed multi-level variables to determine the level of adaptive capacity of each component of the community structure, as well as their interconnections to gain key insights into the overall resilience of the governance system. The aim of this research is to then integrate these findings through the framing of a Complex Adaptive System and identify the key mechanisms of community resilience that might enable transitions towards climate resilient development.

J. Metcalfe · A. Gianoli (✉)
Institute of Housing and Development Studies (IHS), Erasmus University, Rotterdam, Netherlands
e-mail: gianoli@eur.nl

I. Pigliautile et al. (eds.), *Urban Resilience, Livability, and Climate Adaptation*, Advances in Science, Technology & Innovation, https://doi.org/10.1007/978-3-031-54911-3_12

2 Climate Resilient Development in Urban Theory

Over the course of the twentieth century, fundamental shifts have occurred in approaches to urban development and planning to respond to growing environmental challenges resulting from associated systemic changes (Adhern 2010; Barrière et al., 2019; Berry, 1973; McGranahan & Satterthwaite, 2014; Rauws, 2017; Restemeyer et al., 2018). However, sustainability discourse in the field has been critiqued for not going far enough to address the underlying epistemological flaws on which the global development agenda is based (Bernstein, 2000; Dobson, 2003; Pieterse et al., 2010). It is now recognized that significant urban complexity results in 'wicked problems' which have no clear solution (Batty, 2008; Baynes, 2009; Bettencourt, 2013; Mele et al., 2010; Waddock, 2013) and which require a systemic view if we are to achieve tangible change and have a significant impact on our dual developmental and climate crises (Von Bertalanffy, 1972).

Urban resilience is conceptualized as a critical emergent feature of complex socio-ecological and socio-technical systems across temporal and spatial scales when they are exposed to stresses or shocks and which resist, absorb, or transform in response to external interferences to maintain their core structure and critical functioning (Adhern, 2011; Cao et al., 2021; Khatibi et al., 2021; Meerow et al., 2016; Shi et al., 2021; Tyler & Moench, 2012; Tzioutziou & Xenidis, 2021). Much of the urban resilience literature focuses on tracing and unpacking vulnerability (Dodman et al., 2018; Douglas et al., 2008; Ford et al., 2018; Kyprianou et al., 2022; Pieterse, 2022; Seebaß, 2017; UN-Habitat, 2020). Vulnerability can be understood as existing when a system does not have the capacity to 'engage' latent resilience mechanisms to withstand and rebound from the risks of incoming shocks (short-term), and stressors (long-term). Adaptation is the process of adjustment to moderate harm or the taking advantage of opportunities and "plays a key role in reducing exposure and vulnerability to climate change" (IPCC, 2022: 7). These processes can be anticipatory, reactive, incremental, and/or transformational and are subject to hard and soft limits (IPCC, 2022). Urban resilience, therefore, is the capacity of a system to cope with hazardous disturbances by reorganizing in a way that maintains essential functions, "while also maintaining the capacity for adaptation, learning, and transformation" (IPCC, 2022: 7).

Taking an urban resilience systems approach is a particularly useful characterization of urban functioning as it encourages a systematic understanding of disparity in urban experiences and recognizes that urgent action is needed at varying levels to increase a system's ability to transform. Significantly, it has been suggested that the indirect, complex, and unpredictable stresses of wicked problems and climate change may be unrecognized if climate adaptation plans continue to focus on addressing vulnerabilities and that a more strategic focus would be to build *adaptive capacities* across urban systems.

3 Adaptive Capacity

Adaptive capacity is defined as the latent ability of systems to adjust, take advantage of, or respond proactively or positively to the consequences of stressors or opportunities (Whitney et al., 2017). It is argued here to be the component of resilience that informs what adjustments are possible in a particular context based on the capacities and resources that exist across a particular system. In the social sphere, it has been defined as the ability of individuals or groups to self-organize, increase knowledge, and produce quality planning positing necessary characteristics of adaptive capacity to include ***diversity and flexibility***, ***access to assets***, ***learning and knowledge***, and ***governance and institutions*** (Adger, 2000; Whitney et al., 2017; Wulansari et al., 2022). ***Learning and knowledge***, a core aspect of resilience, is enhanced through the deployment of technologies that enable increased efficiencies in *self-organization* and innovation across a range of mechanisms (Tzioutziou & Xenidis, 2021) which contribute significantly to reconfiguring urban systems as dynamic, continuous, and inevitable processes (Egerer et al., 2021).

Societal adaptation is a highly complex and dynamic process where deep transformations across the cultures, structures, and practices within an urban system are required to successfully withstand and rebound from incoming shocks and stressors (Roo, 2016; Wolfram et al., 2019) and will largely be determined by the limitations and capacities of its various ***agents***, ***networks,*** and ***institutions*** to act collectively (Adger, 2003). Strategic action interacts across these multiple levels of governance to galvanize change, which Tyler and Moench, (2012) characterize as three critical components for resilience planning: ***systems***, ***agents***, and ***institutions.*** They state that, "Resilience is high where robust and flexible systems can be accessed by high-capacity agents and where that access is enabled by supportive institutions" (Tyler & Moench, 2012: 318).

4 Multilevel Network Governance and Complex Adaptive Systems

The transition towards climate resilient development will require integrated systemic change across multiple sectors at multiple levels. The scale of this task implies that urban systems will need to learn how to self-regulate in order to build the capacities required to respond to local conditions and support sustainable and resilient urbanization (Ribeiro & Goncalves, 2019), moving beyond short-term coping mechanisms to interrelated long-term adaptation strategies (Yang & Andriesse, 2021).

Complex Adaptive Systems (CAS) theory and resilience literature share their roots in Complex Systems Theory which conceptualizes capacity as relational and contingent on the context and systems in which the element of analysis is embedded (Befani, 2013). CAS are composed of networked non-linear interacting elements which can comprise sub-systems capable of adapting behaviours to their environments, while depending on mutual co-operation (Shi et al., 2021). Han et al. (2021) define six properties of CAS which are argued to be relevant to an analysis of urban systems: Emergence through 'self-organization'; open but distinct boundaries; complex components; non-linear dynamics; adaptability through dynamic interactions; and sensitivity to initial conditions. Each component of community resilience described in the section above has its own unique characteristics and functions which are all highly interdependent in supporting mutual development and therefore can be understood as a complex adaptive system.

'Management' of complex adaptive systems is therefore comprised of dynamic interactions of individual behaviours which are conditioned by institutions and enabled by networks. The ability to participate and manifest change therefore takes place within an ecosystem of constraining and enabling factors. To capture the potential for transition, or to measure adaptive capacity in one moment, requires framing how strategies/pathways are created from the interrelated actions of individuals performing functions at different levels.

Resilience literature outlines three types of response for system functions when exposed to disturbances including: (a) when critical functions persist and bounce-back; (b) when functions transition between states to absorb change; or (c) transformation into a new state of dynamic adaptation (McDaniels et al., 2008). The level of adaptive capacity within the system determines the system response and is mediated by decisions or developments regarding adaptation and mitigation efforts illustrated below.

5 Theoretical Framework

The resultant integrated framework (Fig. 1) assesses community resilience as an indirect dependent variable which is comprised of three critical components identified by Tyler and Moench, (2012): *systems*, *agents*, and *institutions*. Agents are understood as individuals; systems as networks; and institutions as organizations each with particular characteristics, strategies, and defining mechanisms. The level of agent capacity in the community system is critical and dependent on four variables, as well as the degree of interaction between each of the three resilience components, to facilitate increases in adaptive capacities. In other words, there is a feedback loop between the community resilience system components and variable indicators, as they are stimulated by the effectiveness of the components relationships (community resilience), while simultaneously affecting the effectiveness of that relationship. Therefore, the adaptive capacity of the Agent Component is directly

Fig. 1 Theoretical framework

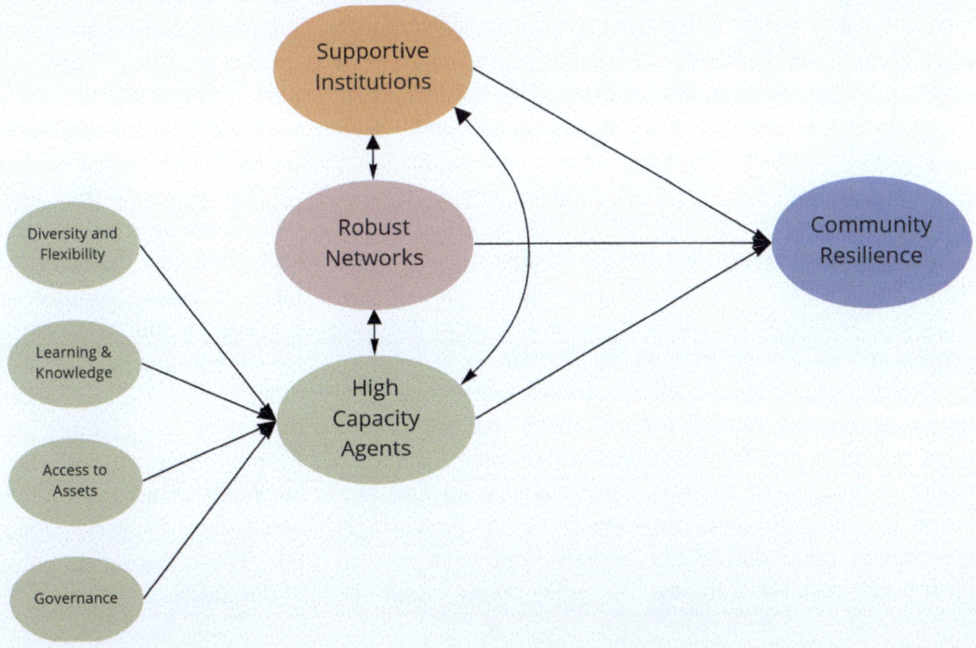

measurable through four associated variables drawn from literature: *Diversity & Flexibility*, *Learning and Knowledge*, *Access to Assets*, and *Governance*, which are further broken down into quantifiable sub-variables which are also drawn from the literature. The remaining two components are more complex constructs to measure and will require in-depth exploration to determine the level of functioning as well as the internal relationships which impact functioning.

6 Case Study Area

Urban development in Nairobi is fragmented and highly unregulated (Jimmy et al., 2020), which compounds numerous environmental management failures across the city (Mwaniki et al., 2019). Colonial legacies of spatial segregation have resulted in systems of exclusion and marginalization of large portions of the population who live in areas that have developed informally from the 1890's through forms of 'subsistence urbanization'. This has resulted in complex dynamics surrounding land ownership, discretionary power relationships, and a degraded urban environment which perpetuates the status quo as urban systems have been co-opted by cartels (powerful resource owning families who benefit from the system as it is) who dominate numerous aspects of life in informal settlements across the city. These systemic vulnerabilities have been exacerbated by climate change and continued unplanned urbanization and have resulted in several well-known risk factors that face residents daily which are explored through the case of Mathare (Douglas et al., 2008; Hendriks, 2010; Hirst & Lamba, 1994; Jimmy et al., 2020; Lamba, 1994; Myers, 2015; Wainaina, 2018).

Mathare is a large slum located 10 km from Nairobi's city centre (Hendriks, 2010; Hirst & Lamba, 1994) with a population of approximately 207,000. It is characterized by dense informal housing and business (UN-Habitat, 2020) in unsanitary conditions, with an inadequate and unsafe water supply; deficient drainage, sewage, and waste collection systems (Lamba, 1994); and frequent fires (Hendriks, 2010) and floods (Karisa, 2010; Owuor & Mwiturubani, 2021). The area is bisected by the Mathare River which flows through the settlement, carrying the contents of contaminated stormwater drainage from the city as well as raw sewage and solid waste from within Mathare.

The impacts of climate change further exacerbate these issues as variable rainfall patterns mean that the ground is increasingly dry for longer periods of time leading to health consequences as wastewater pits dry up and pollution becomes airborne. Riparian areas are increasingly encroached upon during dry periods, and unanticipated flash floods rush down through the valley taking everything in its path. These structural issues must be addressed for both the increased risks associated with predicted climate changes, as well as for unanticipated and unpredictable events that may push vulnerable communities over their tipping points. This is a major challenge in all contexts, but particularly so in contexts of minimal resources and high vulnerability.

There is an urgent need to build on the understanding of these contexts and situate this knowledge within the discourse of climate vulnerability to prepare for uncertain futures as multiple stakeholders in the urban realm are required to work together to find solutions and respond to these challenges and risks in a way that allows for transitions to more resilient forms of development. Individuals, households, organizations, and the institutions that support them interact through a set of complex, dynamic, contested, and interdependent relationships. These mechanisms can be framed within Complex Adaptive System (CAS) and Transition literature so that leverage points can be identified that might have the most positive impact on Climate Resilient Development.

There is growing awareness among urban planning practitioners of the complexities that coalesce around the impacts of climate change and urbanization dynamics and the increasing pressure being placed on urban communities at different scales around the world in the context of inequality. Empirical studies that synthesize theoretical developments and explore their relevance in real-world contexts have the potential to provide significant insights for the development of tangible responses. Understanding social change in this context requires research with specific aims and contextually embedded knowledge that can inform policy and action at multiple levels and assist in the development of strategies that enable individuals or groups to self-organize, increase knowledge, and produce quality planning. Therefore, the ability to learn from the past, examine the present, and predict future impacts from a theoretical and empirical point of view is needed. This research aims to contribute to these objectives.

The scope of this study is limited to one settlement with a unique history and a contemporary socio-political and economic character. All measures possible have been taken to ensure validity and reliability for generalization; however, a much larger study would be needed to identify fine-grained patterns of interdependency needed for a more nuanced understanding. This study is limited to the theoretical associations outlined below and would benefit from engagement with the vast literature that exists across the fields of urban development studies, urban geography, behavioural sciences, economics, and psychology to name a few. These theoretical considerations are beyond the scope of this study.

The first section of this research introduces the theoretical and contextual background motivation for this study and presents the research objectives and questions. This is followed by an outline of the trajectory of existing urban development literature regarding Climate Resilience, Adaptive Capacity, Community Governance, and Transitions as Complex Adaptive Systems. The Research Design and Methods present the strategies of data preparation and collection employed followed by an analysis and discussion of the quantitative and qualitative research and offer implications for climate resilient development policies and further research.

7 Research Design and Methods

The research makes use of empirical and deductive mixed methods comprised of primary data collection and analysis methods designed to explore, describe, and test the theoretical framework through an in-depth case study of the community in Mathare Valley. Operationalization of the key concepts outlined above formed the basis of quantitative and qualitative research and analysis as per Tables 1 and 2, respectively.

The research programme was supported through a partnership with the *Ghetto Foundation* and the *International Centre for Frugal Innovation* (ICFI) in Africa—two research organizations in Nairobi. The first is an organization of community researchers who are Mathare residents. ICFI is an academic research consultancy which supports development research internationally across multiple domains.

Household Survey Collection and Analysis Methods

A comprehensive survey conducted at the household level of 141 respondents used a closed ended questionnaire to capture the individual (agent) adaptive capacity and interaction with other components to evaluate the role of the *High-Capacity Agents* in the CAS. Statistical analysis will explore the relationships across variable and sub-variable levels. Component 1 is primarily explored through quantitative analysis of individual adaptive capacity, and engagement in resilience activities. The required sample size of 139 (Hsieh et al., 1998) was drawn from households living either along the river or along significant drainage paths within the settlement, as they would be most affected by flooding. This component provides a baseline measure of household adaptive capacity and participation in resilience activities employing both regression and multivariate modelling to account for different levels of detail. SEM is a multivariate analysis method that can be used in cross-sectional research to examine the relationships between latent variables (Streiner, 2006). The benefit of this method is that it captures multifaceted concepts, which are hard to measure, with less random error (NCRM, 2022), and allows the researcher to test the robustness of the postulated relationships as part of a hierarchical network (Fig. 3). However, in this study, sample size considerations require simplification of the SEM model, requiring the addition of multiple regression modelling to capture lost detail (see Fig. 2).

Community and Institutional Interviews and Analysis Methods

In-depth interviews with members of relevant organizations (qualitative) at community and institutional levels to capture the collective attitudes and roles of the *Robust Networks* and *Supportive Institutions* components in the CAS are conducted. Community organizations and institutional experts represent a body of respondents which together give insight into the level of functioning of Components 1 and 2 and are therefore analysed in parallel to determine the degree of functioning and internal mechanisms, as well as the interrelationships between each component. The sample method used was a mixture of purposive methods including maximum variation (Glaser & Strauss, 1967) to capture different perceptions across the community and convenience sampling (Merriam & Tisdell, 2016) for community organizations contacted through the Ghetto Foundation Network. The sample size was determined by a combination of availability restrictions and information saturation and redundancy (Merriam & Tisdell, 2016). Semi-structured interviews were conducted with representatives of community organizations and institutional experts from different domains working in Mathare. These components are measured through content analysis of the in-depth interview transcriptions. The results of both components are drawn from interview transcriptions, content coding, weighting, and relationship tracing. Content analysis of interviews and statistical outcomes to develop an integrated analysis of the relationships between each component as part of a CAS is performed.

Several challenges and limitations manifested in the given research context were unavoidable and, while mitigated as much as possible, there likely remains a degree of influence on the study. Limitations are primarily related coinciding with the runup to the 2022 National Elections, which hold significance due to the living memory of the infamous post-election violence experienced in 2007 (Myers, 2015).

Table 1 Detailed operationalization table for quantitative data

X1	Diversity and Flexibility		Y1	Actions to Increase Capacity	
x1.1	*Livelihood and Income Diversity*	Whitney et al. (2017)	y1.1	*Responsiveness*	Tyler and Moench (2012)
	Level of Employment			Level of problem solving	
	Level of Income	Whitney et al. (2017)		Level of participation in organizing	
	Level of Income Diversity	Whitney et al. (2017)		Level of participation in planning	
	Perception.–Level of resources adequate	Chaskin (2001)	y1.2	*Resourcefulness*	Tyler and Moench (2012)
	Perception.–Level of tradable skills			Level of Active search for alternatives	
	Perception.–Level of independence			Level of Active search for collaboration	
x1.2	*Economic Opportunities*	Whitney et al. (2017)	y1.3	*Capacity to Learn*	Tyler and Moench (2012)
	Perception–Level of economic opportunities			Level of participation in educational programme	
	Perception–Level of willingness distance travelled for income			Level of participation in training activities	
x1.3	*Level of Dependence on Natural Resources*	Whitney et al. (2017)		Level of experimentation with new solutions	
	Level of dependence on natural water source			Level of experimentation with new skills	
	Level of dependence on harvested firewood		y1.4	*Independence*	Ribiero and Goncalves (2019)
	Level of dependence on livelihood			Amount of income spent on others	
x1.4	*Occupational Mobility*	Whitney et al. (2017)		Amount of time spent on others	
	I can do my work anywhere?			Participation in household decisions	
x1.5	*Place Attachment*	Whitney et al. (2017)	Y2	*Interaction with Networks*	
	Perception–Sense of home		y2.1	*Flexibility and Diversity*	Tyler and Moench (2012)
	Perception–Sense of comfort/safety			Level of Participation in community activities	Chaskin (2001)
	Perception. Health of environment			Level of Participation in awareness activities	
x1.6	*Migration Patterns*	Whitney et al. (2017)		Level of Participation in political activities	
	Amount of time spent in Nairobi p/y			Level of Use of education apps	
x1.7	*Willingness to Change*	Whitney et al. (2017)		Level of Use of banking apps	
	Perception–benefit of personal change			Level of Use of social media	
	Perception–benefit of community change			Level of Use of news apps	
	Perception–benefit of gov support			Level of Use of weather apps	
	Perception–benefit of environmental change			Level of Use of flooding info apps	
	Level of willingness to increase education			Level of Use of health care apps	
	Level of willingness to attend training			Level of Use of water access apps	
	Level of Willingness to experiment			Level of Use of electricity access apps	

(continued)

Table 1 (continued)

X1	Diversity and Flexibility		Y1	Actions to Increase Capacity	
	Level of Willingness to try new skill			Level of Use of transport apps	
	Level of Willingness to participate in community			Level of Use of income generating apps	
	Level of Willingness to participate in political activities			I try to first solve my problems with technological solutions	
	Level of willingness to use technology			Level of use of e-citizen	
X2	*Learning and Knowledge*		y2.2	*Redundance and Modularity*	Tyler and Moench (2012)
x2.1	*Resource Monitoring and Feedback Mechanisms*	Whitney et al. (2017)		No. of sources of water used	
	Perception–Level of knowledge re. cell phone			No. of sources of health care used	
	Perception–Level of knowledge re. computer use	Chaskin (2001)		Participation in saving groups (merry-go-rounds)	
x2.2	*Knowledge of Disturbance*	Whitney et al. (2017)	y2.3	*Safe Failure*	Tyler and Moench (2012)
	Perception–Level of knowledge re.CC			Level of experimentation	
	Perception–Level of knowledge re. flooding			Level of Resource Monitoring	
	Perception–Level of knowledge re. drought			Level of Feedback Mechanisms	
	Perception–Level of knowledge re. water availability		y2.4	*Robustness*	Ribiero and Goncalves (2019)
x2.3	*Perceptions of Risk and ability to anticipate change*	Whitney et al. (2017)		Perception reliability–internet	
	Perception–Frequency of flooding			Perception reliability–water access	
	Perception–Frequency of drought			Perception reliability–community orgs	
	Perception–Level of safety (flooding) Perception–Level of future planning			Perception reliability–government & institutions	
				Access to water	
			y2.5	*Multiscalar Connectivity*	Adhern (2010), Ribiero and Goncalves (2019)
x2.4	*Diversity of Knowledge and information sources*	Whitney et al. (2017)		Distance travelled to earn income	
	Level of Access to Internet Level of Access to smart device Level of Knowledge of social support networks			Distance travelled to make everyday purchase	
			Y3	*Interaction with Institutions*	
			y3.1	*Rights and Entitlements*	Tyler and Moench (2012)
x2.5	*Recognition of Causality and Human Agency*	Whitney et al. (2017)		Applied for water card	
	Perception–Level of sense of commitment	Chaskin (2001)		Campaign for rights	
	Perception–Level of Agency			Campaign for services	
	Perception–Level of ability to solve problems		y3.2	*Decision-Making Processes*	Tyler and Moench (2012)
	Perception–Level of Value of education			Participation in Community Orgs	Whitney et al. (2017)

(continued)

Table 1 (continued)

X1	Diversity and Flexibility		Y1	Actions to Increase Capacity	
x2.6	*Intergenerational learning capacity*	Whitney et al. (2017)		Participation in voting	
	Perception–Level of learnt knowledge		y3.3	*Information Flows*	
	Perception–Knowledge to teach			Information from gov	
X3	*Access to Assets*			Information to gov	
x3.1	*Household Material Assets*	Whitney et al. (2017)		No. of visits from official	
	Level of construction material			No. of visits from researchers	
	Level of household assets		y3.4	*Application of New Knowledge*	Tyler and Moench (2012)
	Perception–No. of sources of water			Acquired new knowledge	
	Perception–No. of sources of health care			Used new knowledge	
x3.2	*Levels of Education and Training*	Whitney et al. (2017)	y3.5	*Inclusion*	Ribiero and Goncalves (2019)
	Level of education			Positive interaction with other genders	
	Further Training			Positive interaction with other ethnicity	
x3.3	*Financial Status and Access to sources of credit*	Whitney et al. (2017)		Positive interaction with other nationality	
	Access to banking		y3.6	*Adaptive Planning and Design*	Adhern (2010)
	Access to credit/loans			Participation in gov. schemes	
x3.4	*Access to Markets*	Whitney et al. (2017)			
	Access to international relative transfer				
	Access to electricity				
	Access to sanitation				
	Access to health care				
	Level of food security				
	Access to clean water				
x3.5	*Social Capital and Institutional Support*	Whitney et al. (2017)			
	Perception–Institutional Support				
	Proof of Address				
	Right to Vote				
	Level of bonding				
	Level of bridging				
	Level of linking				
x3.6	*Natural Capital*	Whitney et al. (2017)			
	Level of health				
	Perception. Health of Environment				

(continued)

Table 1 (continued)

X1	Diversity and Flexibility		Y1	Actions to Increase Capacity	
x3.7	*Cultural Memory, traditions, and assets*	Whitney et al. (2017)			
	Perception learnt from elders				
	Perception knowledge to teach				
X4	*Governance*				
x.1	*Levels of Trust, Networks*	Whitney et al. (2017)			
	Perception–Social Inclusion				
	Perception–Sense of community	Chaskin (2001)			
	Perception–Level of trust in community				
x4.2	*Gender and Race Relations*	Whitney et al. (2017)			
	Perception–Level of Gender Equality				
x4.3	*Levels of Participation and quality decision-making*				
	Perception–Quality of decision-making–COM				
	Perception–Quality of decision-making–GOV				
x4.4	*Planning Capacity*	Whitney et al. (2017)			
	Perception–Community ability to plan				
	Perception–Gov. ability to plan				
x4.5	*Presence of local Environmental Institutions and Social Norms*	Whitney et al. (2017)			
	Perception–Knowledge about env. Orgs				
	Perception–stewardship				
	Perception–conservation				
	Perception–waste management				
x4.6	*Quality of Governance and Leadership in environmental policies and agencies*	Whitney et al. (2017)			
	Perception–Leadership in flood management				
	Perception–Leadership in environmental issues				
	Perception–Leadership in infrastructure challenges				
x4.7	*Accountability of managers and governance bodies*	Whitney et al. (2017)			
	Perception–Government Responsibility–floods				
	Perception–Government Responsibility–water access				
x4.8	*Active Risk Management and Adaptive Governance Processes*	Whitney et al. (2017)			
	Perception–Government Action–floods				
	Perception–Government Action–water access				
	Perception–Community Action–floods				
	Perception–Community Action–water access				

Table 2 Detailed operationalization table for qualitative data

Independent Variables and Sub-variables For Community Adaptive Capacity (Components 2 & 3)		Dependent Variables and Sub-variables Robust Networks Component 2		Independent Variables and Sub-Variables Supportive Governance Component 3	
Diversity and Flexibility		*Actions to Increase Robustness*		*Actions to Increase Support*	
Livelihood and Income Diversity	Whitney et al. (2017)	Flexibility and Diversity	Tyler and Moench (2012)	Rights & Entitlements	Tyler and Moench (2012)
Economic Opportunities	Whitney et al. (2017)	Redundance and Modularity	Tyler and Moench (2012)	Decision-Making Processes	Tyler and Moench (2012)
Level of Dependence on Natural Resources	Whitney et al. (2017)	Safe Failure	Tyler and Moench (2012)	Information Flows	Tyler and Moench (2012)
Occupational Mobility	Whitney et al. (2017)	Robustness	Ribiero and Goncalves (2019)	Application of New Knowledge	Tyler and Moench (2012)
Place Attachment	Whitney et al. (2017)	Multiscalar Connectivity	Adhern (2010) & Ribiero and Goncalves (2019)	Inclusion	Ribiero and Goncalves (2019)
Migration Patterns	Whitney et al. (2017)				
Willingness to Change	Whitney et al. (2017)				
Learning and Knowledge		*Interaction with Agents*		*Interaction with Agents*	
Resource Monitoring and Feedback Mechanisms	Whitney et al. (2017)	Responsiveness	Tyler and Moench (2012)	Responsiveness	Tyler and Moench (2012)
Knowledge of Disturbance	Whitney et al. (2017)	Independence	Ribiero and Goncalves (2019)	Independence	Ribiero and Goncalves (2019)
Perceptions of Risk and ability to anticipate change	Whitney et al. (2017)	Capacity to Learn	Tyler and Moench (2012)	Capacity to Learn	Tyler and Moench (2012)
Diversity of Knowledge and information sources	Whitney et al. (2017)	Independence	Ribiero and Goncalves (2019)	Independence	Ribiero and Goncalves (2019)
Recognition of Causality and Human Agency	Whitney et al. (2017)				
Intergenerational learning capacity	Whitney et al. (2017)				
Access to Assets		*Interaction with Institutions*		*Interaction with Networks*	
Household Material Assets	Whitney et al. (2017)				
Levels of Education and Training	Whitney et al. (2017)	Rights & Entitlements	Tyler and Moench (2012)	Flexibility and Diversity	Tyler and Moench (2012)
Financial Status and Access to sources of credit	Whitney et al. (2017)	Decision-Making Processes	Tyler and Moench (2012)	Redundance and Modularity	Tyler and Moench (2012)
Access to Markets	Whitney et al. (2017)	Information Flows		Safe Failure	Tyler and Moench (2012)
Social Capital & Institutional Support	Whitney et al. (2017)	Application of New Knowledge	Tyler and Moench (2012)	Robustness	Ribiero and Goncalves (2019)

(continued)

Table 2 (continued)

Independent Variables and Sub-variables For Community Adaptive Capacity (Components 2 & 3)		Dependent Variables and Sub-variables Robust Networks Component 2		Independent Variables and Sub-Variables Supportive Governance Component 3	
Natural Capital	Whitney et al. (2017)	Inclusion	Ribiero and Goncalves (2019)	Multiscalar Connectivity	Adhern (2010) & Ribiero and Goncalves (2019)
Cultural Memory, traditions, and assets	Whitney et al. (2017)	Adaptive Planning & Design	Adhern (2010)	Multifunctionality	Adhern (2010)
Governance					
Levels of Trust, Networks	Whitney et al. (2017)				
Gender and Race Relations	Whitney et al. (2017)				
Planning Capacity	Whitney et al. (2017)				
Presence of local Environmental Institutions & Social Norms	Whitney et al. (2017)				
Quality of Governance and Leadership in environmental policies and agencies	Whitney et al. (2017)				
Accountability of managers and governance bodies	Whitney et al. (2017)				
Active Risk Management and Adaptive Governance Processes	Whitney et al. (2017)				

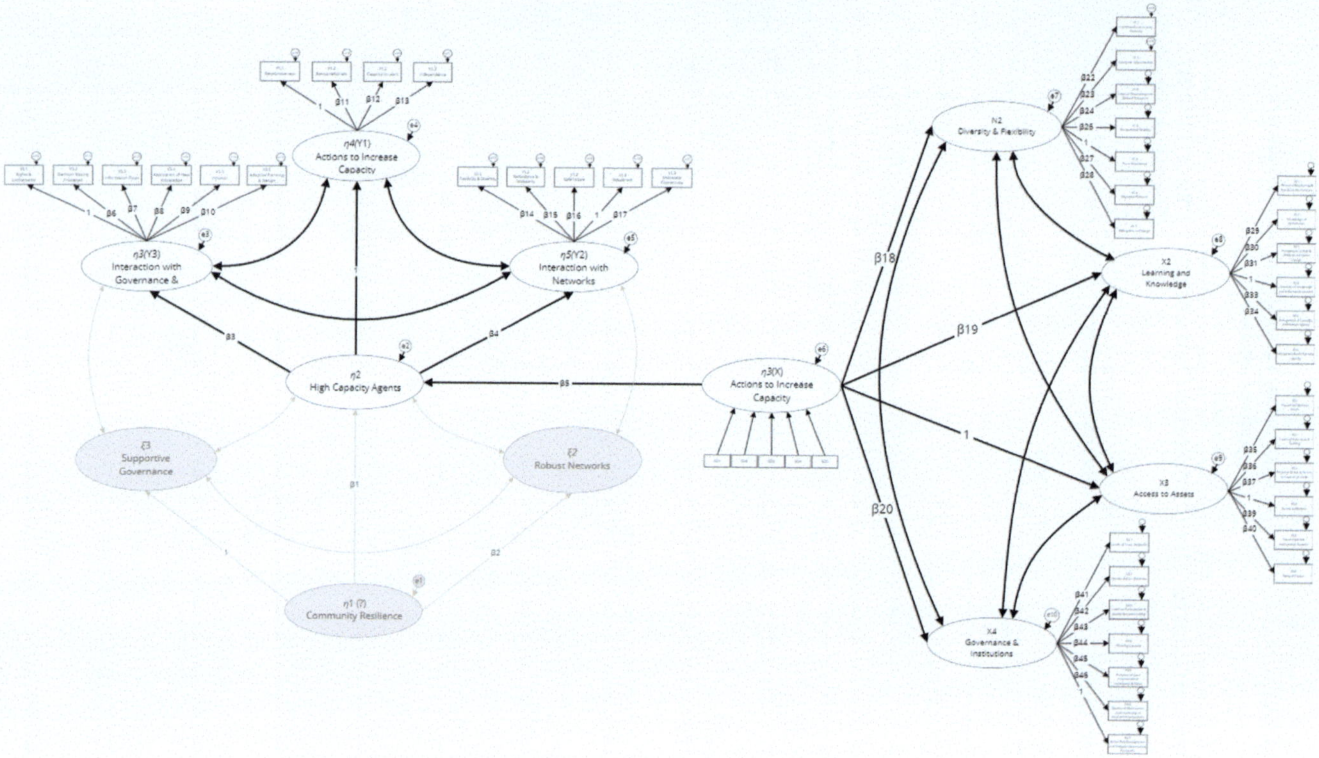

Fig. 2 General structural equation model with latent variables

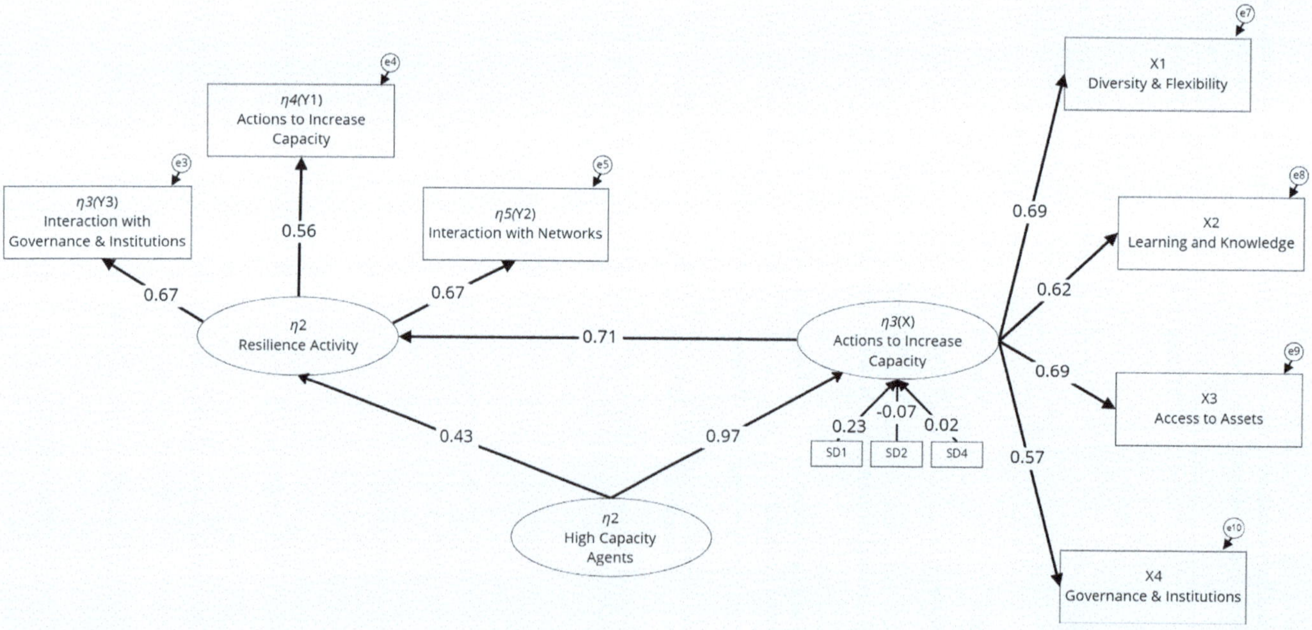

Fig. 3 General structural equation model with standardized coefficients

8 Data Analysis and Discussion

Summary of Results

Analysis reveals that within Component 1 perception of individual latent *Adaptive Capacities* and engagement with *Resilience Activities*. High correlation values reflect strong relationships between all independent and dependent variables, with the strongest relationships existing between *Learning & Knowledge* and *Access to Assets* with *Interaction with Networks*. Particularly strong relationships also exist between *Diversity & Flexibility* and *Governance* with *Interaction with Institutions*, followed by *Access to Assets* and *Interaction with Networks*. Socio-demographic control variables are largely negligible in relation to both independent and dependent variables.

Frequency analysis (Table 3) of all interviews reveals that, of the independent variables, *Access to Assets* was mentioned the most, followed by *Learning and Knowledge*, *Governance*, and *Diversity & Flexibility,* which was the lowest with a margin of almost four per cent—indicating the prominence of these variables in relation to perceptions of resilience for Components 2 and 3. Of the dependent variables, *High-Capacity Agent* was mentioned the most, followed by *Robust Networks* and *Supportive Institutions* the least. The two most frequent sub-variables for each variable are summarized as follows:

Structural Equation Modelling (SEM)

Due to the relatively small sample size, the model needed to be simplified to include only the main variable relationships which were subsequently compiled using item parcelling techniques. The model converged normally after 38 iterations, resulting in a proper solution, the outcome of which can be found in the annex.

Results of overall goodness of fit indices show that the model is a good fit to the data, and it can therefore be concluded that the SEM confirms the hypothesis that there is a relationship between the observed variables measured in the household survey and the latent constructs which they intend to measure (refer again to Fig. 3).

The parameter estimates (Table 4) present the quality of the solution (Gana & Broc, 2019). The non-standardized factor loadings indicate the nature of the predicative relationship and are all positive. Z-values, except for resilient activities, are all significantly higher than 1.96 at $p<0.05$ and are therefore significant.

All standardized factor loadings are positive and greater than 0.4 indicating a relevant link between all items and the factors on which they depend, therefore, estimating that the quality of the relationship between the latent variables and indicators is valid. The variance observed in *Adaptive Capacity* is estimated to account for 71% of the variance seen in *Resilience Activity*, which in turn accounts for 43%

Table 4 Standardized factor loadings

Independent variable	Dependent variable	Standardized factor loading β
Diversity and Flexibility	Adaptive Capacity	0.69***
Learning and Knowledge	Adaptive Capacity	0.61***
Access to Assets	Adaptive Capacity	0.69***
Governance	Adaptive Capacity	0.57***
Actions to Increase Capacity	Resilience Activity	0.67***
Interaction with Network	Resilience Activity	0.56***
Interaction with Governance	Resilience Activity	0.67***
Adaptive Capacity	Resilience Activity	0.71
Adaptive Capacity	High-Capacity Agent	0.97
Resilience Activity	High-Capacity Agent	0.42

Heteroskedasticity-robust standard errors are given in parenthesis under coefficients. The individual coefficient is statistically significant at the ***0% level, **0.1% level, *1% level, or '.'5% level

Table 3 Frequencies table

	Variable	First sub-variable	Second sub-variable
Adaptive capacities			
1	Access to assets	*Social Capital and Institutional Support (392)*	*Financial/Economic Status (216)*
2	Learning and Knowledge	*Perception of Risk and Change (374)*	*Recognition of Human Agency (284)*
3	Governance	*Planning Capacity (233)*	*Levels of Trust (203)*
4	Diversity and Flexibility	*Willingness to Change (163)*	*Livelihood and Income (141)*
Components and resilience activities			
1	High-Capacity Agent	*Responsiveness (407)*	*Independence (363)*
2	Robust Networks	*Robustness (437)*	*Safe Failure (141) and Redundancy and Modularity (141)*
3	Supportive Institutions	*Rights and Entitlements (205)*	*Information Flows (200)*

of the variance seen in *High-Capacity Agents*. The variance observed in Adaptive Capacity accounts for 97% of the variance seen in *High-Capacity Agent*. This model therefore indicates that the theoretical constructs and their relationships identified in the theoretical framework can be confirmed with a reasonable amount of certainty.

All the observed indicators have a statistically significant relationship (at 0.01%) with their respective latent variables and there is a large likelihood that the reported outcome is robust. In particular, it shows that Adaptive Capacity indicators identified in the literature are robust determinants for determining Resilience Activities as well as for High-Capacity Agents. This can be interpreted to evidence the argument that in order to have agents with high capacities operating within the resilience structure the adaptive capacity indicators are relevant measures. Furthermore, the lesser effect of *Resilience Activity* indicators on *High-Capacity Agents* is interesting as it confirms the strength of latent *Adaptive Capacity* over the role of interaction within the structure. As this study is primarily concerned with community adaptive capacity, identified in the literature as the latent ability within the community network to respond to change, this finding is significant. It provides a robust line of argument for the analysis to follow.

Adaptive Capacity and Resilience Levels

Resilience is understood as the total level of adaptive capacity in relation to risk (Tyler & Moench, 2012). Life in Mathare is characterized by individuals manifesting flexibility in the face of multiple risks and stresses associated with poverty; perseverance despite numerous socio-economic complexities; responsiveness through informal networking; and survival in the face of crippling daily challenges. The state of deprivation in Mathare increases residents' exposure to hazards, which take multiple forms and are often unavoidable, increasing the sensitivity of community members to their impacts. These hazards limit the capacity of community members to participate meaningfully in resilience activities.

In summary, the total measures of adaptive capacity below (Tables 5 and 6) show that the community of Mathare has a reasonably high level of adaptive capacity and a slightly lower level of engagement in resilience activities.

Table 5 Total score of all components

Independent|adaptive capacity			Dependent|resilience activities	
1	Agent	3	Agent	3
2	System	2	System	2
3	Institution	3	Institution	2
	Total	**3**	**Total**	**2**

*Scores same scale as Likert used in household survey questionnaire (1–4)

These results indicate that all components of the Mathare community governance system have both the capacity to act and are engaging in resilience activities to an adequate level as per the theoretical framework drawn from adaptation and resilience literature. According to this literature, we would expect the levels of deprivation in Mathare to be decreasing and for substantial gains to be made towards climate resilient development and sustainability goals (IPCC, 2022; UN, 2015, 2017). Although some progress has been made, it has been alarmingly slow and is not keeping pace with the rate of development that is needed (Hendriks, 2010; Hirst & Lamba, 1994; Siddiqi, 2022; UN-Habitat, 2020).

One way of framing this tension is to consider the bounds of soft limits to which adaptation is subject (IPCC, 2022). Poverty, inequity, and a lack of climate literacy lead to soft limits where financial, governance, institutional, and policy environments place constraints on adaptation activity and planning (IPCC, 2022). These constraints and their implications will be explored in the section to follow. However, it is important to note here that the results of this scoring suggest that despite significant constraints, the community has developed adequate mechanisms required to sustain adaptive capacities and resilience activity within the limits of what is possible.

8.1 Summary Analysis of Contextual Risk Factors and Relationships Between Variables

Corruption, Cartels, and Land Ownership|Institutions

Distinctive challenges surround the ownership of land in Mathare which coalesce around critical interrelationships

Table 6 Robust determinants by component

Component	Most robust sub-variable determinant	Second most robust sub-variable determinant
Community Resilience	High-Capacity Agent	
High-Capacity agent	Access to assets	Diversity and Flexibility
Robust Networks	Learning and Knowledge	Governance
Supportive Institutions	Diversity and Flexibility	Learning and Knowledge

between corruption, cartels, and institutional complexities. *Levels of Trust & Social Capital* is severely impeded by questions surrounding the legitimacy of land title and common perceptions of collusion between land administration officials and powerful individuals who have been granted land title and are benefitting from the status quo and actively maintaining the system as it is. Implicit in these dynamics is the significant role of access to assets and in particular social capital. *Social Capital* and *Financial Status* are key components of *Access to Assets,* which is a proven robust determinant of *High-Capacity Agents*. This elite who takes advantage of their position, to the detriment of the system as a whole, can be understood as a form of maladaptation of *Responsiveness*. Additionally, most of the Mathare community are informal renters who do not have security of tenure and remain exceptionally vulnerable living in a state of precariousness owing to numerous threats of eviction ranging from arson, which is used as an eviction method by landowners, to pervasive government evictions in the area have been pervasive (Hirst & Lamba, 1994). The implication of these findings is that the erosion of trust within the community has significant impacts on numerous *Adaptive Capacity* indicators. This lack of trust in Mathare is further expressed in multiple conflicts which surround land ownership; (Departmental Committee on Lands, 2020), and a consequent development stalemate has evolved stifling *Planning Capacity* due to the costs associated with resolution and the implicit considerations of rate-of-return considerations of making this investment.

Institutions are expected to play a supporting role in connecting agents to robust networks (Tyler & Moench, 2012); however, the dynamics set out above create a situation where the dominant institutions are not supportive but rather play a repressive role, limiting agents' interactions to fragile networks characterized by informality, illegality, and inadequacy. SEM reveals that despite supportive institutions being a critical determining factor, the overall interaction with this component is minimal, and institutions are perceived to be largely absent. In the context of a considerable lack of trust and complexities surrounding corruption, cartels, and land ownership, and where *Access to Assets* is the most significant determinant of *Resilience Activity*, it can be deduced that agents who perceive risk, and possess a strong will to change, might be more willing to engage with the most robust network that seems to exist in Mathare—that of the cartels. In the absence of significant institutional support, as we have noted, *High-Capacity Agents* develop slum ecosystem governance models with implications for community resilience. Together these findings indicate that individuals who exhibit a strong *Willingness to Change* might be more likely to interact with planning mechanisms within the community network, as opposed to interacting with institutions—as, in the context of scarcity, *Access to Assets* is particularly significant (Obaitor et al., 2021). Higher capacity agents have higher levels of access to assets and therefore higher adaptive capacities which enable them to be more responsive and independent—and able to interact with networks and institutions—albeit not necessarily positively. This is an important lesson that agency, assets, and adaptation can be deployed for negative outcomes.

Lack of Service Provision and Poverty as Violence | Networks

The development stalemate outlined above contributes to the severe shortage of service and infrastructure provision in Mathare which perpetuates the ongoing immiseration of the quality-of-life and of extreme poverty. Informality can be viewed as a coping mechanism and as a function of neglect (Azunre et al., 2021). In Mathare, informal and often illicit business is ubiquitous and exists across all systems including those around brewing, land, electricity, water, dumpsites, security, stealing, prostitution, drugs, and smuggling. Some of these forms of livelihood are known as 'Cash Crops' and are specific to each area. These activities constitute much of the economy, and multiple additional systems exist to buffer their legal fragility and ensure system functioning. As suggested above, the absence of supportive institutions contributes to the prominence of cartels and corruption.

This precarious state suggests that the introduction of alternate systems potentially threatens existing systems which provide a means of survival for many people, and that government intervention may not be wanted by the groups that are benefitting from its absence—and keeping the system 'functioning'. The majority of people are just trying to survive and are highly adaptable within their limits of power (as the community resilience scores suggest) but they do not have the capacity to change the broader context to which they are subject. Thus, gross injustice takes refuge in the complexities that exist at the nexus of poverty, neglect, and informality as they become normalized and coping mechanisms are developed at individual and systemic levels—not all of which are positive.

Interaction with Robust Networks is a key robust determinant for *Resilience Activities* and, building on the analysis of access to assets, strong correlations and numerous co-occurrences can be found between *Robust Networks* and *High-Capacity Agents* which confirms the significance of the relationship as identified in the literature (Tyler & Moench, 2012). Capacity building activities run by community organizations foster trust and build social capital—which analysis shows has a significant impact on building high-capacity agents who in turn are able to participate more fully in community resilience activities. A significant part of capacity building in Mathare is linked to digitization and the critical role that digital infrastructure plays in reducing 'lock-in', and multiple aspects of *Learning & Knowledge* in general (Boschma, 2005; Pahl-Wostle, 2009; Tzioutziou & Xenidis, 2021). *Learning & Knowledge* is the most robust determinant of Interaction with Network, as is estimated in the SEM to account for 67% of the variance seen in *Resilience Activity*. *Perception of Risk and Change (Learning & Knowledge)* is the most significant independent sub-variable due to its high co-occurrence with both *Robustness (Robust Networks) and Responsiveness (High-Capacity Agents)*. The ability to adjust and respond positively to stressors or opportunities (Whitney et al., 2017) relies on a component successfully recognizing them. This ability is an emergent property that manifests when a challenge is presented and is identified in the literature as a critical component of adaptive capacity (Chaskin, 2001; Whitney et al., 2017). Analysis indicates that the degree to which people are willing to frame challenges as risks depends on their ability to take action to avert them. If there is not enough capacity to address the risk, then it is reframed or ignored as a potential coping mechanism.

Environmental Risks and Normalization as Coping|Agents

Building on this understanding, it is imperative to recognize the second aspect of violence which is normalization. In Mathare poverty, neglect and survival are normalized to the extent that coping mechanisms developed in the context of deprivation are accepted as 'this is the way it is' (resp.1). This recognition implicitly acknowledges that many of the challenges are systemic and beyond the reach of individual action.

A consequence of this normalization is complacency, at all component levels, which mutually reinforces attitudes of neglect and perpetuates the status quo. The theoretical framework draws attention to the central role of agents within the resilience system as the primary actors and beneficiaries of resilience activities. It is consequential to consider the effects of normalization across the community resilience system and the impact that this numbing might have on key variables. This can be seen in multiple arenas of decision-making regarding risk in Mathare where decisions are made to meet immediate challenges. Where agents (individuals), lack sufficient access to assets to cope, independence is reduced and people are forced to rely on dubious sources of provision. Furthermore, normalization of these multiple issues affects how people think about their environment, the river, and the role of government which has implications for the development of larger resilience strategies for degraded urban spaces at institutional levels. Therefore, the consequences of neglect and subsequent vulnerability and its normalization in Mathare will be made worse by the effects of climate change as increasing pressure is placed on urban systems and resources become increasingly scarce and costly (Arsel, 2022; Azunre et al., 2021; Bankoff & Hilhorst, 2022; Berry, 1973; Chu et al., 2016; Dodman et al., 2018). Resignation to these challenges makes the issue of complacency a critical issue to address at all levels of governance in Mathare in order for needed systemic changes to be recognized as a city level priority.

8.2 Resilient Community Components: A Complex Adaptive System

The complexity of these interconnected and overlapping relationships is framed here by the six characteristics of urban Complex Adaptive Systems (CAS) and Transition literature to identify leverage points within the system that might have the most positive impact towards Climate Resilient Development.

One of the main aims of framing the dynamics above is to clarify how complex causes can produce simple effects or how simple rules can have unpredictable consequences (Anderson et al., 1999; Han et al., 2021). In CAS, three fundamental elements include agents, interaction, and the environment (Han et al., 2021) which is appropriate for an analysis of the community resilience system of Mathare described above as these three properties are prominent. The significant complexity within the system is framed as a CAS to provide insight into the mechanisms of cause and effect.

8.3 Mathare Community Resilience CAS

Emergence through Self-organization

In CAS theory, complexity is the result of a large number of heterogeneous agents who interact according to individually defined features and objectives (Han et al., 2021). Across Mathare, the community is comprised of diverse individuals and includes not only the residents, members of community level networks, engaged institutions, and governmental representatives, but also individuals within the neighbouring edges and daily visitors. In the context of informality and lack of governance capacity, the emergent system of organization has been described as a 'slum ecosystem' as a form of quasi-government with its own distinctive dynamics and way of control.

Open but Distinct Boundaries

CAS are open and fluid systems engaged in constant exchange with their environment (Han et al., 2021). Mathare is an open system comprised of complex arrangements of heterogeneous agents mentioned above who are engaged at differing levels. As a marginalized community in the centre of the city, the Mathare boundary, while permeable, is defined by a definite state of 'lock-in' (Boschma, 2005). This has produced detrimental normalization of coping mechanisms that perpetuate the stigmatization and further marginalization of the community system.

Complex Components

In CAS, a key consideration is the degree of multiplicity of agents as well as of their interactions and relationships (Han et al., 2021). The population of Mathare residents is large, and as has been discussed, the community includes the broader community of integrated stakeholders. In the system defined components, each is comprised of internal interactions taking place between levels *Adaptive Capacity* and *Resilience Activities*, but which are highly influenced by the interdependent relationships that exist across all components.

Non-linear dynamics

A significant effect of *component complexity* is the degree to which non-linearity—that is the degree of disproportionality between inputs and outcomes—exists across a system (Han et al., 2021). One way of assessing these dynamics is to account for positive and negative feedback loops, which can multiply the effect of particular inputs (Han et al., 2021).

A feedback loop exists between *Robust Networks* and *High-Capacity Agents* through which both components directly benefit from each other's development. *Diversity & Flexibility* is a key adaptive capacity in both components and shares significant relationships with aspects of *Learning & Knowledge,* making inputs at the intersection of these two sub-variables a key leverage point for outcomes that will affect both components. Their associated primary sub-variables, *Perception of Risks & Change* and *Willingness to Change,* have significant implications for the ability of individuals to raise their level of capacity in *Diversity & Flexibility* to interact meaningfully with *Robust Networks* and increase overall *Community Resilience*. Additionally, the most significant determinant of *High-Capacity Agents* is *Access to Assets*, which shares a strong correlation with *Interaction with Networks. Social Capital*, the primary sub-variable of *Access to Assets*, is highly positively correlated with *Willingness to Change*. However, *Social Capital* also changes *Willingness to Change* in the same direction. Consequently, a feedback loop exists at the intersection of *High-Capacity Agents* and *Interactions with Networks* where inputs at the level of the sub-variables *Social Capital* or *Willingness to Change* will multiply effects throughout the system, with potentially significant outcomes on *Resilience Activities* of both components. In this case, the feedback can be positive or negative, as the components move in the same direction.

Unfortunately, more complex feedback loops also exist across the system which can produce unproductive cycles of lock-in, and which require more significant system changes to restructure the effects. Using the same example of *Robust Networks* and *High-Capacity Agents* and focussing on the aspect of *Access to Assets* and the relationship between its primary sub-variable, *Social Capital* with *Perception of Risk & Change*, a very different picture is produced. *Social Capital*, the primary sub-variable of *Access to Assets*, has a significant negative correlation with *Perception of Risk and Change* where, as *Social Capital* changes, *Perception of Risk and Change* moves in the opposite direction. Consequently, an unproductive feedback loop exists surrounding *High-Capacity Agents* where, as *Access to Assets* might increase, *Learning and Knowledge* and *Diversity and Flexibility* would decrease—reducing the interaction of *High-Capacity Agents* with *Robust Networks* and *Resilience Activities*. This can be understood in the context of cartel influence and the specific context characterized by misunderstanding, lack of trust, and normalization of coping mechanisms to larger socio-economic dynamics which contribute to the complexities surrounding intervention.

Adaptability through dynamic interactions

CAS are, by definition, adaptive. This means that these systems are capable of adjustment necessary to internal struggles and external threats while maintaining their core functions through continual transformation and dynamic adaptation (Han et al., 2021; McDaniels et al., 2008). In Mathare, the level of adaptability emerges through the engagement of adaptive capacities in resilience activities within each component, as well as by the relationships and interactions between each component, the structure of which has been confirmed. The mechanisms of these actions are determined by this structure of the system, as well as in response to external disturbances, and reflect the non-linear dynamics identified as feedback loops. This can be the case for intentional external inputs as well as unexpected disturbances.

Sensitivity to initial conditions

A combination of feedback loops and non-linear relationships creates conditions of uncertainty, which are described as sensitivity to initial conditions. Through this lens, this characteristic describes the current state of a CAS via a combination of historical system structure and new pathway developments. The unintended consequences of feedback loops, dynamic interactions, and adaptability affected the ability of the institutions to meaningfully support networks at the community level and connect agents who are subject to these forces of slum ecosystem dynamics. In this sense, cartels are significant posing a challenge to the theory as they are an incredibly adaptive and resilient subsystem within the community and are a disrupting and destabilizing factor. This subsystem highlights the significance of ensuring an assessment of the entire system is conducted—not just as components in isolation—in order to measure the key measurables and assess how the system works together.

8.4 Resilient Community Components: A Complex Adaptive System

In order to enable Climate resilient development in Mathare, transformational adaptation is required. The level of adaptive capacity within a system determines the system response (resilience activities), when exposed to disturbances (McDaniels et al., 2008). 'Management' of this system requires an acknowledgement of the complexity of this environment and the autonomy of the actors within the system. The dynamic interactions, non-linear relationships, and numerous feedback loops have resulted in fractured 'institutional norms' and complex development pathways. Marginalization and the lack of supportive institutions have the effect of producing various emergent properties as coping mechanisms at the systemic level of the community interactions. This peripheralization and neglect embeds a 'lock-in' effect which normalizes survival mechanisms that strip individuals of their dignity. By framing this research on community resilience as a CAS, key conclusions regarding the level of resilience can be drawn.

This analysis shows that the adaptive capacity of the current system has met its threshold for what is feasible to be addressed at the community scale, indicating that soft limits have been reached. If real transformation to a state of increased resilience is to be made, then a range of constraints (which primarily consist of financial, governance, institutional, and policy constraints) will need to be systematically addressed. Institutions need to play a far more supportive role by removing the key barriers identified and supporting community level networks and agents to build the required adaptive capacities. Increased systemic 'robustness' and 'adaptation' to the realities of urbanization impacts and climate risks that go beyond the immediate challenges of everyday, call for policy shifts that reframe system capacities for transition towards patterns of Climate Resilient Development in Mathare.

This will require bold action. The IPCC states that, "transitioning from incremental to transformational adaptation can help overcome soft adaptation limits" (IPCC, 2022: 29) and that the window to enable climate resilient development pathways is rapidly narrowing (IPCC, 2022: 32). To overcome the challenges in Mathare, transformational approach is needed to leverage key points in this system which revolve around the issues of embedded cartels, poverty, and normalization, which translates into an acute need to address institutional neglect, to recognize coping mechanisms of poverty, and to develop strategies to decrease isolation.

9 Conclusions and Recommendations

This research has demonstrated the applicability of the key current theoretical constructs to this locality. The study was able to establish factors associated with, and mechanisms which influence, community resilience in Mathare. It explored the interconnections of resilience components within a community network from an urban resilience standpoint.

Key findings include the following:

Institutions

There is an acute need to address soft limits on institutional reform around land management and service delivery practice in Mathare. The impact has been shown to significantly impact key community resilience indicators.

Networks

Multiple coping mechanisms have been developed in response to an acute need for robust networks across all domains, many of which compound the existing risks and introduce new ones. System functioning is maintained through precarious systems that are fragile and symptomatic of larger structural issues. These networks need to be recognized for the role that they play and be supported to build core capacities to support community resilience. Learning and knowledge is critical and must be supported as it is connected to multiple positive feedback loops between Robust Networks and High-Capacity Agents.

Agents

Neglected individuals have subsequently normalized negative coping mechanisms that often perpetuate violence of different forms. Support initiatives contribute to integrating Mathare into the city to break normalization and lock-in, as it is shown to have detrimental effects across multiple adaptive capacity indicators. Communication infrastructure is critical to support mechanisms that have the potential to break the effects of cognitive, organizational, and social lock-in present in Mathare.

Resilient Community Components: A Complex Adaptive System

The Community Resilience System in Mathare is a complex adaptive system which is characteristically unpredictable. However, key leverage points do exist and have been identified. Transformative resilience is needed. Soft limits need to be removed, and institutions need to play a far more supportive role to connect agents to systems in Mathare so that the system can benefit from the feedback loops identified. Climate resilient development involves questions of equity and systems transitions (IPCC, 2022) which go beyond city or even national scale, and the findings of this research can be seen to have implications at a broader scale. Global inequities and the climate crisis require transformational transitions in our approach to the development of cities around the world. Institutions at a global scale will need to play a far more supportive role.

Further research could include testing these results in a similar context, or at a different scale, to compare findings and identify points for clarification or refinement. Quantitative analysis of the institutional and community network capacities and resilience activities would add value for cross-comparison. This study could be developed into a more comprehensive and detailed assessment of the adaptive capacities and resilience strategies on each component for policy development.

These findings may assist in the development of strategies that will enable individuals or groups to self-organize, increase knowledge, and contribute to the appropriate planning in this urban context. Specific recommendations in this regard follow.

The most critical issue to be addressed is land ownership and security of tenure. Bold action and significant investment are needed to transform the limiting factors that shape life in Mathare. Climate change will compound the challenges being faced and is expected to increase with every degree of warming (IPCC, 2022). The best time to act is now. Planning policy must be developed to address these challenges and break the status quo.

The significant role of community networks in the provision of critical services must be recognized and incorporated into the transformational development strategy for Mathare. These networks not only provide necessary services, but also engage agents in networks which provide opportunities for learning and diversification of capacities. Specific policies that support their development and integration would benefit the community as a whole and would include supporting mechanisms of ICT infrastructure, the dissemination of powers, capacity building, and increased funding. Learning and Knowledge is a key leverage point and must be capitalized on to break lock-in and improve integration.

A3.3.2 Detailed Dependent Operationalization Table for Quantitative Data

Y1	**Actions to Increase Capacity**	
y1.1	*Responsiveness*	Tyler and Moench (2012)
	Level of problem solving	
	Level of participation in organizing	
	Level of participation in planning	
y1.2	*Resourcefulness*	Tyler and Moench (2012)
	Level of Active search for alternatives	
	Level of Active search for collaboration	
y1.3	*Capacity to Learn*	Tyler and Moench (2012)
	Level of participation in educational programme	
	Level of participation in training activities	
	Level of experimentation with new solutions	
	Level of experimentation with new skills	
y1.4	*Independence*	Ribiero and Goncalves (2019)
	Amount of income spent on others	
	Amount of time spent on others	
	Participation in household decisions	

Y2	**Interaction with Networks**	
y2.1	*Flexibility and Diversity*	Tyler and Moench (2012)
	Level of Participation in community activities	Chaskin (2001)
	Level of Participation in awareness activities	
	Level of Participation in political activities	
	Level of Use of education apps	
	Level of Use of banking apps	
	Level of Use of social media	
	Level of Use of news apps	
	Level of Use of weather apps	
	Level of Use of flooding info apps	
	Level of Use of health care apps	
	Level of Use of water access apps	
	Level of Use of electricity access apps	
	Level of Use of transport apps	
	Level of Use of income generating apps	
	I try to first solve my problems with technological solutions	
	Level of use of e-citizen	
y2.2	*Redundance and Modularity*	Tyler and Moench (2012)
	No. of sources of water used	
	No. of sources of health care used	
	Participation in saving groups (merry-go-rounds)	
y2.3	*Safe Failure*	Tyler and Moench (2012)
	Level of experimentation	
	Level of Resource Monitoring	
	Level of Feedback Mechanisms	
y2.4	*Robustness*	Ribiero and Goncalves (2019)
	Perception. reliability–internet	
	Perception. reliability–water access	
	Perception. reliability–community orgs	
	Perception. reliability–government & institutions	
	Access to water	
y2.5	*Multiscalar Connectivity*	Adhern (2010) & Ribiero and Goncalves (2019)
	Distance travelled to earn income	
	Distance travelled to make everyday purchase	
Y3	**Interaction with Institutions**	
y3.1	*Rights and Entitlements*	Tyler and Moench (2012)
	Applied for water card	
	Campaign for rights	
	Campaign for services	
y3.2	*Decision-Making Processes*	Tyler and Moench (2012)
	Participation in Community Orgs	Whitney et al. (2017)
	Participation in voting	
y3.3	*Information Flows*	
	Information from gov	
	Information to gov	
	No. of visits from official	
	No. of visits from researchers	
y3.4	*Application of New Knowledge*	Tyler and Moench (2012)
	Acquired new knowledge	
	Used new knowledge	
y3.5	*Inclusion*	Ribiero and Goncalves (2019)
	Positive interaction with other genders	
	Positive interaction with other ethnicity	
	Positive interaction with other nationality	
y3.6	*Adaptive Planning and Design*	Adhern (2010)
	Participation in gov. schemes	

A3.3.3 Detailed Operationalization Table for Qualitative Data

Independent Variables and Sub-variables For Community Adaptive Capacity (Components 2 & 3)		Dependent Variables and Sub-variables Robust Networks Component 2		Independent Variables and Sub-Variables Supportive Governance Component 3	
Diversity and Flexibility		*Actions to Increase Robustness*		Actions to Increase Support	
Livelihood and Income Diversity	Whitney et al. (2017)	Flexibility and Diversity	Tyler and Moench (2012)	Rights & Entitlements	Tyler and Moench (2012)
Economic Opportunities	Whitney et al. (2017)	Redundance and Modularity	Tyler and Moench (2012)	Decision-Making Processes	Tyler and Moench (2012)
Level of Dependence on Natural Resources	Whitney et al. (2017)	Safe Failure	Tyler and Moench (2012)	Information Flows	Tyler and Moench (2012)
Occupational Mobility	Whitney et al. (2017)	Robustness	Ribiero and Goncalves (2019)	Application of New Knowledge	Tyler and Moench (2012)
Place Attachment	Whitney et al. (2017)	Multiscalar Connectivity	Adhern (2010), Ribiero and Goncalves (2019)	Inclusion	Ribiero and Goncalves (2019)
Migration Patterns	Whitney et al. (2017)				
Willingness to Change	Whitney et al. (2017)				
Learning and Knowledge		*Interaction with Agents*		*Interaction with Agents*	
Resource Monitoring and Feedback Mechanisms	Whitney et al. (2017)	Responsiveness	Tyler and Moench (2012)	Responsiveness	Tyler and Moench (2012)
Knowledge of Disturbance	Whitney et al. (2017)	Independence	Ribiero and Goncalves (2019)	Independence	Ribiero and Goncalves (2019)
Perceptions of Risk & ability to anticipate change	Whitney et al. (2017)	Capacity to Learn	Tyler and Moench (2012)	Capacity to Learn	Tyler and Moench (2012)
Diversity of Knowledge and information sources	Whitney et al. (2017)	Independence	Ribiero and Goncalves (2019)	Independence	Ribiero and Goncalves (2019)
Recognition of Causality and Human Agency	Whitney et al. (2017)				
Intergenerational learning capacity	Whitney et al. (2017)				
Access to Assets		*Interaction with Institutions*		*Interaction with Networks*	
Household Material Assets	Whitney et al. (2017)				
Levels of Education & Training	Whitney et al. (2017)	Rights & Entitlements	Tyler and Moench (2012)	Flexibility & Diversity	Tyler and Moench (2012)
Financial Status & Access to sources of credit	Whitney et al. (2017)	Decision-Making Processes	Tyler and Moench (2012)	Redundance & Modularity	Tyler and Moench (2012)

Independent Variables and Sub-variables For Community Adaptive Capacity (Components 2 & 3)		Dependent Variables and Sub-variables Robust Networks Component 2		Independent Variables and Sub-Variables Supportive Governance Component 3	
Access to Markets	Whitney et al. (2017)	Information Flows		Safe Failure	Tyler and Moench (2012)
Social Capital & Institutional Support	Whitney et al. (2017)	Application of New Knowledge	Tyler and Moench (2012)	Robustness	Ribiero and Goncalves (2019)
Natural Capital	Whitney et al. (2017)	Inclusion	Ribiero and Goncalves (2019)	Multiscalar Connectivity	Adhern (2010), Ribiero and Goncalves (2019)
Cultural Memory, traditions, and assets	Whitney et al. (2017)	Adaptive Planning & Design	Adhern (2010)	Multifunctionality	Adhern (2010)
Governance					
Levels of Trust, Networks	Whitney et al. (2017)				
Gender and Race Relations	Whitney et al. (2017)				
Planning Capacity	Whitney et al. (2017)				
Presence of local Environmental Institutions and Social Norms	Whitney et al. (2017)				
Quality of Governance and Leadership in environmental policies and agencies	Whitney et al. (2017)				
Accountability of managers and governance bodies	Whitney et al. (2017)				
Active Risk Management and Adaptive Governance Processes	Whitney et al. (2017)				

A.3.4.10 Correlation Table with Variables

See (Table 7).

A.3.4.11 Detailed Correlation Table with Variables and Sub-variables

See (Table 8).

Table 7 Descriptive statistics and correlations

Variable	Code	mean	V	SD	CV	SD.1	SD.2	SD.3	SD.4	SD.5	X1	X2	X3	X4	Y1	Y2	Y3
Control Variables																	
Gender	SD.1	2.72	2.22	1.49	0.55												
Age	SD.2	2.09	0.41	0.64	0.31	0.05											
Distance from River	SD.3	2.66	1.14	1.07	0.4	-0.1	0.04										
Religion	SD.4	2.82	0.22	0.47	0.17	-0.14	0	-0.15									
Area	SD.5	4.96	6.12	2.47	0.5	-0.11	0.03	0.08	0.04								
Independent Variables																	
Diversity & Flexibility	X1	2.83	0.13	0.36	0.13	0.25**	-0.05	-0.01	-0.02	0.05							
Knowledge & Learning	X2	2.86	0.19	0.43	0.15	0.15	-0.05	0.01	0.01	0.03	0.7****						
Access to Assets	X3	2.61	0.3	0.55	0.21	0.18*	0.03	0.03	-0.02	-0.22**	0.59****	0.66****					
Governance	X4	2.28	0.63	0.79	0.35	0.27**	-0.06	0.04	-0.01	-0.19*	0.48****	0.46****	0.66****				
Dependent Variables																	
Action to Increase AC	Y1	3.1	0.22	0.46	0.15	0.12	0.08	-0.03	0.09	-0.12	0.52****	0.5****	0.61****	0.34****			
Interaction with Networks	Y2	2.32	0.41	0.64	0.28	0.16	-0.01	0.02	-0.02	0	0.65****	0.75****	0.67****	0.61****	0.48****		
Interaction with Governance & Institutions	Y3	2.65	0.39	0.63	0.24	0.03	-0.05	0.05	0.02	0.19*	0.64****	0.57****	0.51****	0.45****	0.5****	0.62****	

****. Correlation is significant at the 0.0001 level (2-tailed).
***. Correlation is significant at the 0.001 level (2-tailed).
**. Correlation is significant at the 0.01 level (2-tailed).
*. Correlation is significant at the 0.05 level (2-tailed).
N = 141; V= Variance; SD = Standard Deviation; CV= Coefficient Variance

Table 8 Descriptive statistics and correlations

Variable	Code	mean	V	SD	CV	SD.1	SD.2	SD.3	SD.4	SD.5	X1	X2	X3	X4	X.v1.1	X.v1.2	X.v1.3	X.v1.4	X.v1.5	X.v1.6
Control Variables																				
Gender	SD.1	2.72	2.22	1.49	0.55															
Age	SD.2	2.09	0.41	0.64	0.31	0.05														
Distance from River	SD.3	2.66	1.14	1.07	0.4	-0.1	0.04													
Religion	SD.4	2.82	0.22	0.47	0.17	-0.14	0	-0.15												
Area	SD.5	4.96	6.12	2.47	0.5	-0.11	0.03	0.08	0.04											
Independent Variables																				
Diversity & Flexibility	X1	2.83	0.13	0.36	0.13	0.25**	-0.05	-0.01	-0.02	0.05										
Knowledge & Learning	X2	2.86	0.19	0.43	0.15	0.15	-0.05	0.01	0.01	0.03	0.7****									
Access to Assets	X3	2.61	0.3	0.55	0.21	0.18*	0.03	0.03	-0.02	-0.22**	0.59****	0.66****								
Governance	X4	2.28	0.63	0.79	0.35	0.27**	-0.06	0.04	-0.01	-0.19*	0.48****	0.46****	0.66****							
Independent Sub - Variables																				
Diversity & Flexibility																				
Livelihood and Income Diversity	X.v1.1	2.34	0.37	0.61	0.26	0.11	0.09	0.03	-0.02	0.01	0.67****	0.54****	0.55****	0.22**						
Economic Opportunities	X.v1.2	2.87	0.62	0.79	0.27	0.2*	0.11	-0.03	0.06	-0.22**	0.64****	0.46****	0.6****	0.44****	0.52****					
Level of Dependence on Natural Resources	X.v1.3	1.54	0.71	0.84	0.54	0.18*	-0.03	-0.03	-0.17*	0.21*	0.41****	0.17*	-0.06	0.33****	0.03	0.08				
Occupational Mobility	X.v1.4	3.59	0.84	0.92	0.26	0.02	-0.1	-0.01	0.08	-0.17*	0.46****	0.4****	0.4****	0.34****	0.2*	0.48****	-0.06			
Place Attachment	X.v1.5	2.71	1.49	1.22	0.45	0.07	-0.11	0.04	0.03	0.28***	0.7****	0.45****	0.3***	0.27**	0.42****	0.24**	0.29***	0.26**		
Migration Patterns	X.v1.6	1.43	0.28	0.53	0.37	-0.06	0.06	-0.02	-0.09	-0.22**	-0.31***	-0.26**	-0.12	-0.07	-0.14	-0.18*	-0.12	-0.18*	-0.43****	
Willingness to Change	X.v1.7	3.61	0.19	0.44	0.12	0.14	-0.11	0	0.02	0.02	0.63****	0.49****	0.41****	0.18*	0.2*	0.25**	-0.01	0.33****	0.33****	-0.28***
Knowledge & Learning																				
Resource Monitoring & Feedback Mechanisms	X.v2.1	2.79	1.41	1.19	0.42	0.14	-0.07	0.06	-0.07	-0.12	0.39****	0.7****	0.41****	0.26**	0.26**	0.3***	0.12	0.28***	0.11	-0.02
Knowledge of Disturbance	X.v2.2	2.27	1.22	1.1	0.49	0.14	-0.04	0.07	-0.04	0.23**	0.49****	0.63****	0.33****	0.54****	0.28***	0.15	0.41****	0.19*	0.4****	-0.2*
Perceptions of Risk & ability to anticipate change	X.v2.3	2.24	0.53	0.73	0.33	-0.16	-0.03	-0.06	0.14	0.37****	0.04	0.01	-0.25**	-0.41****	0.22**	-0.05	-0.04	-0.21*	0.22**	-0.21*
Diversity of Knowledge and information sources	X.v2.4	2.54	0.83	0.91	0.36	0.16	-0.01	0.1	0.02	-0.11	0.34****	0.7****	0.48****	0.28***	0.29***	0.26**	0.02	0.23**	0.05	-0.02
Recognition of Causality and Human Agency	X.v2.5	3.64	0.21	0.46	0.13	0.08	0.04	-0.07	0.03	-0.07	0.55****	0.63****	0.55****	0.33****	0.34****	0.37****	-0.03	0.47****	0.42****	-0.25**
Intergenerational learning capacity	X.v2.6	3.23	1	1	0.31	0.14	-0.02	-0.06	-0.04	-0.26**	0.4****	0.5****	0.62****	0.5****	0.33****	0.45****	0	0.34****	0.21*	-0.14
Access to Assets																				
Household Material Assets	X.v3.1	2.78	0.46	0.67	0.24	0.03	0.11	0.03	0.01	-0.37****	0.25**	0.33****	0.73****	0.37****	0.27**	0.44****	-0.31***	0.31***	0.04	0.04
Levels of Education & Traning	X.v3.2	2.62	1.08	1.04	0.4	0.26**	0.07	-0.09	-0.01	-0.11	0.59****	0.46****	0.58****	0.27**	0.64****	0.54****	0.09	0.26**	0.27**	-0.08
Financial Status & Access to sources of credit	X.v3.3	2.23	1.55	1.25	0.56	-0.02	0.07	0.04	-0.07	0.29***	0.38****	0.47****	0.32****	0.02	0.44****	0.11	0.16	-0.01	0.3***	-0.18*
Access to Markets	X.v3.4	2.34	0.69	0.83	0.35	0.17*	0.03	0.06	-0.05	-0.3***	0.35****	0.49****	0.87****	0.52****	0.41****	0.52****	-0.26**	0.32***	0.09	-0.01
Social Capital & Institutional Support	X.v3.5	2.94	0.61	0.78	0.27	0.15	-0.1	-0.01	0	-0.11	0.6****	0.52****	0.78****	0.69****	0.45****	0.49****	0.12	0.42****	0.45****	-0.27**
Natural Capital	X.v3.6	2.21	0.62	0.79	0.36	0.11	0.02	0.05	0.06	0.04	0.27**	0.29***	0.31***	0.48****	0.1	0.18*	0.25**	0.17*	0.12	0.01
Governance																				
Levels of Trust, Networks	X.v4.1	2.51	1.35	1.16	0.46	0.24**	-0.05	0.03	-0.06	-0.32***	0.27***	0.33****	0.57****	0.8****	0.1	0.37****	0.2*	0.28***	0.02	0.04
Gender & Race Relations	X.v4.2	1.54	1.05	1.02	0.67	0.2*	-0.13	0.18*	-0.07	0.1	0.24**	0.17*	0.2*	0.49****	0.02	0.13	0.26**	0.12	0.28***	-0.05
Levels of Participation & quality decision making	X.v4.3	2.2	1.16	1.08	0.49	0.19*	-0.03	-0.03	-0.06	-0.32***	0.21*	0.22**	0.48****	0.83****	0.07	0.29***	0.19*	0.27**	0	0.06
Planning Capacity	X.v4.4	2.45	1.28	1.13	0.46	0.22**	-0.08	-0.06	0.02	-0.29***	0.3***	0.29***	0.52****	0.83****	0.1	0.3***	0.18*	0.31***	0.12	0.02
Presence of local Environemental Institutions & Social Norms	X.v4.5	2.45	0.53	0.73	0.3	0.2*	-0.09	0.07	0.08	-0.03	0.64****	0.66****	0.64****	0.76****	0.45****	0.45****	0.29***	0.34****	0.46****	-0.21*
Quality of Governance and Leadership in	X.v4.6	2.18	1.09	1.04	0.48	0.26**	0.02	0.11	-0.06	0.04	0.55****	0.43****	0.55****	0.79****	0.32****	0.43****	0.35****	0.23**	0.45****	-0.19*
Active Risk Management and Adaptive Govenance Processes	X.v4.7	2.2	0.99	0.99	0.45	0.19*	-0.03	-0.02	0.04	-0.2*	0.3***	0.32****	0.51****	0.9****	0.05	0.33****	0.32***	0.28***	0.13	-0.01
Dependent Variables																				
Action to Increase AC	Y1	3.1	0.22	0.46	0.15	0.12	0.08	-0.03	0.09	-0.12	0.52****	0.5****	0.61****	0.34****	0.49****	0.39****	-0.09	0.27**	0.31***	-0.11
Interaction with Networks	Y2	2.32	0.41	0.64	0.28	0.16	-0.01	0.02	-0.02	0	0.65****	0.75****	0.67****	0.61****	0.46****	0.38****	0.31***	0.28***	0.37****	-0.18*
Interaction with Governance & Institutions	Y3	2.65	0.39	0.63	0.24	0.03	-0.05	0.05	0.02	0.19*	0.64****	0.57****	0.51****	0.45****	0.43****	0.26**	0.2*	0.26**	0.61****	-0.38****

****. Correlation is significant at the 0.0001 level (2-tailed).
***. Correlation is significant at the 0.001 level (2-tailed).
**. Correlation is significant at the 0.01 level (2-tailed).
*. Correlation is significant at the 0.05 level (2-tailed).
N = 141; V= Variance; SD = Standard Deviation; CV= Coefficient Variance

Table 8 (continued)

Code	X.v1.7	X.v2.1	X.v2.2	X.v2.3	X.v2.4	X.v2.5	X.v2.6	X.v3.1	X.v3.2	X.v3.3	X.v3.4	X.v3.5	X.v3.6	X.v4.1	X.v4.2	X.v4.3	X.v4.4	X.v4.5	X.v4.6	X.v4.7	Y1	Y2	Y3
X.v2.1	0.3***																						
X.v2.2	0.26**	0.25**																					
X.v2.3	-0.03	-0.22**	-0.1																				
X.v2.4	0.29***	0.8****	0.3***	-0.28***																			
X.v2.5	0.53****	0.24**	0.29***	-0.14	0.25**																		
X.v2.6	0.23**	0.12	0.21*	-0.29***	0.17*	0.45****																	
X.v3.1	0.26**	0.22**	0.03	-0.2*	0.24**	0.35****	0.47****																
X.v3.2	0.28***	0.31***	0.22**	-0.12	0.36****	0.29***	0.43****	0.3***															
X.v3.3	0.26**	0.37****	0.29***	0.28***	0.36****	0.19*	-0.05	-0.04	0.28***														
X.v3.4	0.24**	0.32***	0.15	-0.3***	0.45****	0.4****	0.59****	0.69****	0.44****	0.11													
X.v3.5	0.38****	0.18*	0.35****	-0.26**	0.21*	0.58****	0.67****	0.44****	0.43****	0.07	0.55****												
X.v3.6	0.14	0.21*	0.41****	-0.03	0.22**	0.13	-0.04	0.13	-0.05	0.17*	0.13	0.13											
X.v4.1	0.08	0.23**	0.36****	-0.57****	0.25**	0.34****	0.51****	0.39****	0.27**	-0.16	0.54****	0.56****	0.32****										
X.v4.2	0.07	0.07	0.2*	-0.09	0.01	0.18*	0.15	-0.02	0.05	0.07	0.16	0.24**	0.3***	0.34****									
X.v4.3	0.02	0.17*	0.32***	-0.51****	0.19*	0.19*	0.37****	0.36****	0.13	-0.17*	0.4****	0.54****	0.4****	0.74****	0.33****								
X.v4.4	0.13	0.19*	0.33****	-0.46****	0.24**	0.3***	0.38****	0.41****	0.21*	-0.18*	0.4****	0.59****	0.33****	0.69****	0.29***	0.83****							
X.v4.5	0.33****	0.33****	0.6****	-0.17*	0.4****	0.45****	0.51****	0.27**	0.42****	0.23**	0.46****	0.7****	0.3***	0.52****	0.31***	0.43****	0.49****						
X.v4.6	0.24**	0.2*	0.52****	-0.1	0.18*	0.19*	0.35****	0.25**	0.3***	0.28***	0.35****	0.56****	0.39****	0.44****	0.42****	0.48****	0.54****	0.67****					
X.v4.7	0.05	0.19*	0.47****	-0.4****	0.2*	0.2*	0.38****	0.29***	0.06	-0.05	0.45****	0.5****	0.52****	0.67****	0.4****	0.77****	0.7****	0.57****	0.63****				
Y1	0.44****	0.21*	0.23**	-0.03	0.28***	0.57****	0.39****	0.47****	0.43****	0.3***	0.45****	0.47****	0.19*	0.29***	0.17*	0.21*	0.3***	0.44****	0.29***	0.16			
Y2	0.44****	0.63****	0.57****	-0.16	0.65****	0.43****	0.28***	0.28***	0.39****	0.5****	0.46****	0.5****	0.45****	0.41****	0.29***	0.4****	0.44****	0.66****	0.53****	0.51****	0.48****		
Y3	0.51****	0.22**	0.47****	0.04	0.24**	0.57****	0.28***	0.22**	0.25**	0.34****	0.24**	0.62****	0.31***	0.23**	0.23**	0.25**	0.38****	0.58****	0.45****	0.28***	0.5****	0.62****	

****. Correlation is significant at the 0.0001 level (2-tailed).
***. Correlation is significant at the 0.001 level (2-tailed).
**. Correlation is significant at the 0.01 level (2-tailed).
*. Correlation is significant at the 0.05 level (2-tailed).
N = 141; V= Variance; SD = Standard Deviation; CV= Coefficient Variance

References

Adger, W. N. (2000). Social and ecological resilience: Are they related? *Progress in Human Geography, 24*(3), 347–364. https://doi.org/10.1191/030913200701540465

Adger, W. N. (2003). Social capital, collective action, and adaptation to climate change. *Economic Geography, 79*(4), 387–404. https://doi.org/10.1111/j.1944-8287.2003.tb00220.x

Adhern, J. (2010). Planning and Design for sustainable and resilient cities - theories, strategies and best practices for green infrastructure. *Water Centric Sustainable Communities, 100.*

Adhern, J. (2011). From fail-safe to safe-to-fail. Sustainability and resilience in the new urban world. *Landscape and Urban Planning, 100*(4), 341–343. https://doi.org/10.1016/j.landurbplan.2011.02.021

Anderson, A., Meyer, K., Carley, A., Eisenhardt, P., & Pettigrew, K. (1999). Introduction to the special issue: Applications of complexity theory to organization science. *Organization Science*, 233–236.

Arsel, M. (2022). Climate change and class conflict in the anthropocene: Sink or swim together? *The Journal of Peasant Studies, 0*(0), 1–29. https://doi.org/10.1080/03066150.2022.2113390

Azunre, G. A., Amponsah, O., Takyi, S. A., & Mensah, H. (2021). Informality-sustainable city nexus: The place of informality in advancing sustainable Ghanaian cities. *Sustainable Cities and Society, 67*, 102707.

Bankoff, G., & Hilhorst, D. (2022). *Why vulnerability still matters: The politics of disaster risk creation* Routledge.

Barrière, O., Morand, S., Seyler, V. R. F., Prost, L. P. C., Loireau, T. L. M., Fargette, V. D. M., & David, M. B. G. (Eds.). (2019). *Coviability of social and ecological systems: Reconnecting mankind to the biosphere in an era of global change*. Springer. https://doi.org/10.1007/978-3-319-78497-7

Batty, M. (2008). Generating cities from the bottom-up: Using complexity theory for effective design. *Cluster Editions*, 1–14. Retrieved from http://www.casa.ucl.ac.uk/ASU/LectureB.pdf

Baynes, T. M. (2009). Complexity in urban development and management: Historical overview and opportunities. *Journal of Industrial Ecology, 13*(2), 214–227. https://doi.org/10.1111/j.1530-9290.2009.00123.x

Befani, B. (2013). Between complexity and generalization: Addressing evaluation challenges with QCA. *Evaluation, 19*(3), 269–283. https://doi.org/10.1177/1474022213493839

Bernstein, S. (2000). Ideas, social structure and the compromise of liberal environmentalism. *European Journal of International Relations, 6*(4), 464–512. https://doi.org/10.1177/1354066100006004002

Berry, B. J. L. (1973). *The human consequences of urbanization: Divergent paths in the urban experience of the twentieth century.* St. Martin's Press.

Bettencourt, L. M. A. (2013). *The kind of problem a city is (SFI working paper: 2013-03-008)*. Santa Fe: Santa Fe Institute. Retrieved from https://sfi-edu.s3.amazonaws.com/sfi-edu/production/uploads/sfi-com/dev/uploads/filer/fa/f6/faf61418-fc4f-42d5-8c28-df1197a39018/13-03-008.pdf

Boschma, R. (2005). Proximity and innovation: A critical assessment. *Regional Studies, 39*(1), 61–74.

Cao, Y., Lovell, E., Colenbrander, S., Wilkinson, E., & Pettinotti, L. (2021). *A decade of urban resilience: An analytical review*. UNDP.

Chaskin, R. J. (2001). Building community capacity: A definitional framework and case studies from a comprehensive community initiative. *Urban Affairs Review, 36*(3), 291–323. https://doi.org/10.1177/10780870122184876

Chu, E., Anguelovski, I., & Carmin, J. (2016). Inclusive approaches to urban climate adaptation planning and implementation in the global south. *Climate Policy, 16*(3), 372–392. https://doi.org/10.1080/14693062.2015.1019822

Croese, S., Green, C., & Morgan, G. (2020). Localizing the sustainable development goals through the lens of urban resilience: Lessons and learnings from 100 resilient cities and cape town. *Sustainability, 12*(2), 550. https://doi.org/10.3390/su12020550

Departmental Committee on Lands. (2020). *Report on the consideration of a petition by residents of hospital ward, mathare constituency regarding immenent eviction of residents of hospital ward by the Kenya police service*. Retrieved from http://www.parliament.go.ke/sites/default/files/2022-06/Report%20on%20the%20consideration%20of%20a%20petition%20by%20residents%20of%20Hospital%20Ward%20%2C%20%20Mathare%20Constituency%20regarding%20imment%20eviction%20by%20the%20Kenya%20Police%20Service.pdf

Dobson, A. (2003). Social justice and environmental sustainability: Ne'er the twain shall meet. *Just Sustainabilities: Development in an Unequal World*, 83–95.

Dodman, D., Archer, D., & Mayr, M. (2018). *Pro-poor climate action in informal settlements*. United Nations Human Settlements Program (UN-Habitat).

Douglas, I., Alam, K., Maghenda, M., Mcdonnell, Y., Mclean, L., & Campbell, J. (2008). Unjust waters: Climate change, flooding and the urban poor in Africa. *Environment and Urbanization, 20*(1), 187–205. https://doi.org/10.1177/0956247808089156

Egerer, M., Haase, D., McPhearson, T., Frantzeskaki, N., Andersson, E., Nagendra, H., & Ossola, A. (2021). Urban change as an untapped opportunity for climate adaptation. *Npj Urban Sustainability, 1*(1), 1–9. https://doi.org/10.1038/s42949-021-00024-y

Ford, J. D., Pearce, T., McDowell, G., Berrang-Ford, L., Sayles, J. S., & Belfer, E. (2018). Vulnerability and its discontents: The past, present, and future of climate change vulnerability research. *Climatic Change, 151*(2), 189–203.

Gana, K., & Broc, G. (2019). *Structural equation modeling with lavaan*. Wiley.

Glaser, B. G., & Strauss, A. (1967). *The discovery of grounded theory: Strategies for qualitative research*. Aldine.

Global Centre on Adaptation. (2022). *State and trends in adaptation report 2022: Executive summaries and synthesis*. Global Centre on Adaptation. Retrieved from https://gca.org/wp-content/uploads/2022/11/GCA_State-and-Trends-in-Adaptation-2022_Fullreport.pdf

Han, J., Ruan, Y., Wang, Y., & Zhou, H. (2021). Toward a complex adaptive system: The case of the zhongguancun entrepreneurship ecosystem. *Journal of Business Research, 128*, 537–550. https://doi.org/10.1016/j.jbusres.2019.11.077

Hendriks, B. (2010). *Urban livelihoods, institutions and inclusive governance in Nairobi: 'Spaces' and their impacts on quality of life, influence and political rights* Amsterdam University Press.

Hirst, T., & Lamba, D. (1994). *The struggle for Nairobi*. Mazingira Institute.

Hsieh, F. Y., Bloch, D. A., & Larsen, M. D. (1998). A simple method of sample size calculation for linear and logistic regression. *Statistics in Medicine, 17*(14), 1623–1634.

IPCC. (2022). *Summary for policy makers in climate change 2022: Impacts, adaptation, and vulnerability. Contribution of working group II to the sixth assessment report of the intergovernmental panel on climate change.*Cambridge University Press.

IPPC. (2021). Summary for policymakers. In V. Masson-Delmotte (Ed.), *Climate change 2021: The physical science basis. Contribution of working group I to the sixth assessment report of the intergovernmental panel on climate change* (pp. 1–42). In Press: Cambridge University Press. Retrieved from https://www.ipcc.ch/report/ar6/wg1/downloads/report/IPCC_AR6_WGI_Full_Report_smaller.pdf

Jimmy, E. N., Martinez, J., & Verplanke, J. (2020). Spatial patterns of residential fragmentation and quality of life in nairobi city, kenya. *Applied Research in Quality of Life, 15*(5), 1493–1517.

Karisa, C. (2010). (2010). A negotiated framework for rehabilitation of riparian zones in nairobi city: The case of mathare river valley (Kenya). Paper presented at the *46th ISOCARP Congress*, Retrieved from http://erepository.uonbi.ac.ke/handle/11295/45770

Khatibi, H., Wilkinson, S., Baghersad, M., Dianat, H., Ramli, H., Suhatril, M., Javanmardi, A., & Ghaedi, K. (2021). The resilient–smart city development: A literature review and novel frameworks exploration. *Built Environment Project and Asses Management, 11*(4).

Kyprianou, I., Carlucci, S., & Serghides, D. (2022). Urban vulnerability in the EMME region and sustainable development goals: A new conceptual framework. *Sustainable Cities and Society*, 103763.

Lamba, D. (1994). The forgotten half; environmental health in Nairobi's poverty areas. *Environment and Urbanization, 6*(1), 164–173.

Leck, H., & Roberts, D. (2015). What lies beneath: Understanding the invisible aspects of municipal climate change governance. *Current Opinion in Environmental Sustainability, 13*, 61–67. https://doi.org/10.1016/j.cosust.2015.02.004

McDaniels, T., Chang, S., Cole, D., Mikawoz, J., & Longstaff, H. (2008). Fostering resilience to extreme events within infrastructure systems: Characterizing decision contexts for mitigation and adaptation. *Global Environmental Change, 18*(2), 310–318.

McGranahan, G., & Satterthwaite, D. (2014). *Urbanisation concepts and trends (working paper: Urban: June 2014)*. International Institute for Environment and Development (IIED). Retrieved from http://pubs.iied.org/pdfs/10709IIED.pdf

Meerow, S., Newell, J. P., & Stults, M. (2016). Defining urban resilience: A review. *Landscape and Urban Planning, 147*, 38–49. https://doi.org/10.1016/j.landurbplan.2015.11.011

Mele, C., Pels, J., & Polese, F. (2010). A brief review of systems theories and their managerial applications. *Service Science (Hanover, Md.), 2*(1–2), 126–135. https://doi.org/10.1287/serv.2.1_2.126

Merriam, S. B., & Tisdell, E. J. (2016). Qualitative research: A guide to design and implementation (4th edn.). Jossey-Bass.

Mwaniki, B., Gakuya, D., Mwaura, A., & Muthama, N. (2019). Divergent perceptions in open spaces provision in nairobi: Towards new beginning in collaborative approaches. *International Academic Journal of Information Sciences and Project Management, 3*(3), 59–72.

Myers, G. (2015). A world-class city-region? Envisioning the Nairobi of 2030. *American Behavioral Scientist, 59*(3), 328–346. https://doi.org/10.1177/0002764214550308

NCRM. (2022). NCRM online learning resources.

Obaitor, O. S., Lawanson, T. O., Stellmes, M., & Lakes, T. (2021). Social capital: Higher resilience in slums in the Lagos metropolis. *Sustainability, 13*(7), 3879.

Owuor, M. O., & Mwiturubani, D. A. (2021). Nexus between flooding impacts and coping strategies in Nairobi's settlements. *International Journal of Disaster Risk Reduction, 64*, 102480. https://doi.org/10.1016/j.ijdrr.2021.102480

Pahl-Wostle, C. (2009). A conceptual framework for analysing adaptive capacity and multi-level learning processes in resource governance regimes. *Global Environmental Change, 19*(3), 354–365. https://doi.org/10.1016/j.gloenvcha.2009.06.001

Pieterse, E. (2022). Inhabiting the edge. In J. Engle, J. Agyeman & T. Chun-Tiam-Fook (Eds.), *Sacred civics building seven generation cities*. Routledge.

Pieterse, E., & Parnell, S. (2010). The 'Right to the city': Institutional imperatives of a developmental state. Retrieved from https://onlinelibrary-wiley-com.eur.idm.oclc.org/doi/epdf/https://doi.org/10.1111/j.1468-2427.2010.00954.x

Rauws, W. (2017). Embracing uncertainty without abandoning planning: Exploring an adaptive planning approach for guiding urban transformations. *disP-The Planning Review, 53*(1), 32–45. https://doi.org/10.1080/02513625.2017.1316539

Restemeyer, B., Vd Brink, M., & Woltjer, J. (2018). Resilience unpacked-framing of 'uncertainty' and 'adaptability' in long-term flood risk management strategies for London and Rotterdam. *European Planning Studies, 26*(8), 1559–1579. https://doi.org/10.1080/09654313.2018.1490393

Ribeiro, P., & Goncalves, L. (2019). Urban resilience: A conceptual framework. *Sustainable Cities and Society, 50*, 1–11.

Roo, G. D. (2016). Spatial planning, complexity and a world 'out of equilibrium': Outline of a non linear approach to planning. In G. D. Roo, J. Hillier, & J. V. Wezemael (Eds.), *Complexity and planning: Systems, assemblages and simulations* (pp. 129–165). Ashgate. Retrieved from https://www.math.vu.nl/~eliens/serious/local/report/planning.pdf

Seebaß, K. (2017). Who is feeling the heat?: Vulnerabilities and exposures to heat stress-individual, social, and housing explanations. *Nature + Culture, 12*(2), 137–161. https://www.jstor.org/stable/26430639

Shi, Y., Zhai, G., Xu, L., Zhou, S., Lu, Y., Liu, H., & Huang, W. (2021). Assessment methods of urban system resilience: From the perspective of complex adaptive system theory. *Cities, 112*, 103141. https://doi.org/10.1016/j.cities.2021.103141

Siddiqi, A. (2022). Disaster studies and its discontents: The postcolonial state in hazard and risk creation. In G. Bankoff, & D. Hilhorst (Eds.), *Why vulnerability still matters: The politics of disaster risk creation*. Routledge.

Streiner, D. L. (2006). Building a better model: An introduction to structural equation modelling. *The Canadian Journal of Psychiatry, 51*(5), 317–324. https://doi.org/10.1177/070674370605100507

Swilling, M., & Annecke, E. (2012). *Just transitions: Explorations of sustainability in an unfair world*. UCT Press.

Tyler, S., & Moench, M. (2012). A framework for urban climate resilience. *Climate and Development, 4*(4), 311–326. https://doi.org/10.1080/17565529.2012.745389

Tzioutziou, A., & Xenidis, Y. (2021). A study on the integration of resilience and smart city concepts in urban systems. *Infrastructures, 6*(24).

UN. (2015). *Transforming our world: The 2030 agenda for sustainable development*. United Nations.

UN. (2017). *The new urban agenda*. United Nations.

UN-Habitat. (2020). *Informal settlements vulnerability mapping in Kenya: The case of Mathare*. United Nations.

Von Bertalanffy, L. (1972). The history and status of general systems theory. *Academy of Management Journal, 15*(4), 407–426. https://doi.org/10.5465/255139

Waddock, S. (2013). The wicked problems of global sustainability need wicked (good) leaders and wicked (good) collaborative solutions. *The Journal of Management for Global Sustainability, 1*(1), 91–111. https://doi.org/10.13185/JM2013.01106

Wainaina, G. K. (2018). The case for non-revenue water: Kenya.

Whitney, C. K., Bennett, N. J., Ban, N. C., Allison, E. H., Armitage, D., Blythe, J. L., & Yumagulova, L. (2017). Adaptive capacity: From assessment to action in coastal social-ecological systems. *Ecology and Society, 22*(2). Retrieved from https://www.jstor.org/stable/26270135

Wolfram, M., VdJeijden, J., Juhola, S., & Patterson, J. (2019). Learning in urban climate governance: Concepts, key issues and challenges. *Journal of Environmental Policy and Planning, 21*(1), 1–15.

Wulansari, I., Abdoellah, O., Gunawan, B., & Parikesit. (2022). Identification of adaptive capacity assessments to improve collective adaptation of farmers to climate change. *Asian Journal of Agriculture and Rural Development, 12*(1), 1–9.

Yang, J., & Andriesse, E. (2021). Cumulative adaptation and linkages among adaptation, coping and vulnerability: A case of riverside slum households in Phnom Penh, Cambodia. *Environment and Urbanization, 33*(2), 539–559.

Assessing Street Vendors Liveability—A Comparative Study in Hosur and Vellore

V. Veenaa Sri and Jagadisan Sharmila

Abstract

Markets are social institutions which are quintessential public spaces, and they are considered to be an integral part of our neighbourhood which reflects the urban vibrancy and diversity and evoke senses and memories. Markets are engines of economic activities and play a key role in developing city's images where they encourage social cohesion. In markets Street vendors contribute in multiple ways towards the development of the society as they are a major source of affordable goods which lures many to them. The street vendors create a robust market space where they provide job opportunities to millions of youth and create array cost-effective goods of various kinds which cannot be accommodated in a sprawling hypermarket. The presence of street vendors in the markets of Hosur and Vellore brought changes in the character of urban fabric as these places acts as a catalyst for the emergence of internal linkage activity which is influenced by the emergence of many people who come and visit these places. Research on the liveability aspect of street vendors has not been addressed in greater detail. The Government of India has passed "The Street Vendors Protection of Livelihood and Regulation of Street Vending Act" in 2014 to protect the livelihood, rights and social security of urban street vendors in the country. As per the SDG (Sustainable development goals) enhancing the liveability and quality of marginalized workers like street vendors is crucial for creation of more inclusive, safe, resilient, and sustainable cities. The objective of the research is to do a comparative analysis of street vendors who resides in the market of Hosur and Vellore through comprehensive criteria (objectives ad subjective attributes in terms of physical, social, economic, health, and environmental) of liveability indices which would foster a congenial environment that fit their characteristics with respect to different context.

Keywords

Contextual · Liveability indicators · Street markets · Sustainable · Street vendors

1 Introduction

In the 21st century, there has been a mass migration of population from rural to urban, and it has increased globally in recent decades, with India holding the second position in this global demographic transition (Singh, 2016). The lack of employment in rural villages, especially in the context of shrinking agriculture activities, forces youths to prefer jobs in urban areas in search of livelihood opportunities. These migrants, having low levels of educational attainment and skills, enter the less profitable informal sector for their survival (Bhowmik, 2012). In India, the informal sector contributes to half of the GDP of our country and constitutes 93% of the workforce (Indira, 2014). About 60% of the share in net domestic product (NDP) is held by the informal sector, which is crucial for the urban economy's growth (Saha, 2010). Informal sectors include domestic workers, cobblers, artisans, labourers, street vendors, hawkers, peddlers, etc. During COVID-19, the travel restrictions along with successive lockdowns were a major hit and devastated the livelihood of street vendors, pushing the economy into the deepest contraction (Nitya Maniktala, 2021).

The Street Merchandisers (Protection of Livelihood and Regulation of Street Vending) Act, 2014, provides the description of road vendor as "a person engaged in the

V. Veenaa Sri · J. Sharmila (✉)
School of Architecture, Vellore Institute of Technology (VIT), Vellore, India
e-mail: sharmila.jagadisan@vit.ac.in

V. Veenaa Sri
e-mail: veenaasri.v2018@vitstudent.ac.in

I. Pigliautile et al. (eds.), *Urban Resilience, Livability, and Climate Adaptation*,
Advances in Science, Technology & Innovation, https://doi.org/10.1007/978-3-031-54911-3_13

vending of papers, goods, food particulars, or merchandise of everyday use or extending services to the general public, in a road, streetlane, sidewalk, path, pavement, or any other public or private spaces, from a temporary constructed structure or by moving from place to place" (The Gazette of India, 2014). It also mentions that hawkers, peddlers, and squatters are also terms often used interchangeably in terms of local and regional context. The Indian Government also provides various schemes such as DAY-NULM (Deendayal Antyodaya Yojana-National Urban Livelihoods Mission), NASVI (*National Association of Street Vendors of India)*, SEWA (Self-Employed Women's Association), PM SVANidhi (Pradhan Mantri Street Vendors Atma Nirbhar Nidhi), etc. to enhance the liveability of street vendors. Since they are one of the most disadvantageous sections of society, whether these street vendors are aware of these government schemes is a question to be asked. As per the SDGs (sustainable development goals), enhancing the liveability and quality of marginalised workers like street vendors is crucial for the creation of more inclusive, safe, resilient, and sustainable cities. The objective of the research is to do a comparative analysis of street vendors who reside in the markets of Hosur and Vellore through comprehensive criteria (objectives and subjective attributes in terms of physical, social, economic, health, and environmental) of liveability indices that would foster a congenial environment that fits their characteristics with respect to different contexts.

1.1 Objectives

i. To understand the concept of liveability and its parameters in the lives of street vendors.
ii. To understand the existing government policies of street vendors in Indian context (in relation to liveability aspect).
iii. To identify the issues/problems faced by the street vendors in terms of liveability quotient in the selected places i.e., Vellore and Hosur, Tamil Nadu.
iv. To compare, analyze and evaluate two case studies with respect to the most important liveability indicators (frequently applied in different literature) such as physical, environmental friendliness and sustainability, socio-cultural aspects, economic vibrancy, and competitiveness.
v. To propose design strategies in terms of SDG goals to enhance the quality of life, through improving and increasing livability in streets.

1.2 Significance

i. Street vending which is essential part of urban development often left unconsidered in urban planning.
ii. Street vending recognized as one of the potential contributor to Indian economy.
iii. In India the street vending sector provides goods and services at affordable prices for the poor strata of urban society.
iv. Street vendors should be aware of micro credit and other social welfare schemes provided by the government

1.3 Scopes and Limitations

The paper limits to propose designing strategies to enhance the liveability of street vendors and help them improve the overall street character comprising of street vendors and pedestrians, thus strengthening their spatial and social correlation. The study area of the paper is limited to only two places in Tamil Nadu, Hosur, and Vellore. Figure 1 shows the first phase of research methodology.

1.4 Methodology

In our research, a mixed-methods study approach has been adopted in two markets in Tamil Nadu, Vellore, and Hosur, respectively. The research focuses on studying the impact of the *l*iveabi*l*ity indicators on the lives of street vendors. Therefore, the research demands a combination of qualitative and quantitative studies. Qualitative methods were necessary to understand the nature of spaces, while quantitative methods were useful in terms of user agreements and needs in today's context.

The methods used to carry out the study were physical documentation, natural observation, and a structured survey. Surveys were conducted in two markets and were deliberately carried out in the presence of the users. The surveys and observations helped in understanding the scale of the space and the way the parameters (socio-cultural, economic, and environmental) impact the spatial liveability of street vendors in both markets. A structured survey is conducted as part of the mixed method. Figure 4 covers the interview questions that identify the issues or problems faced by the street vendors in terms of the liveability quotient in the selected places, i.e., Vellore and Hosur, Tamil Nadu.

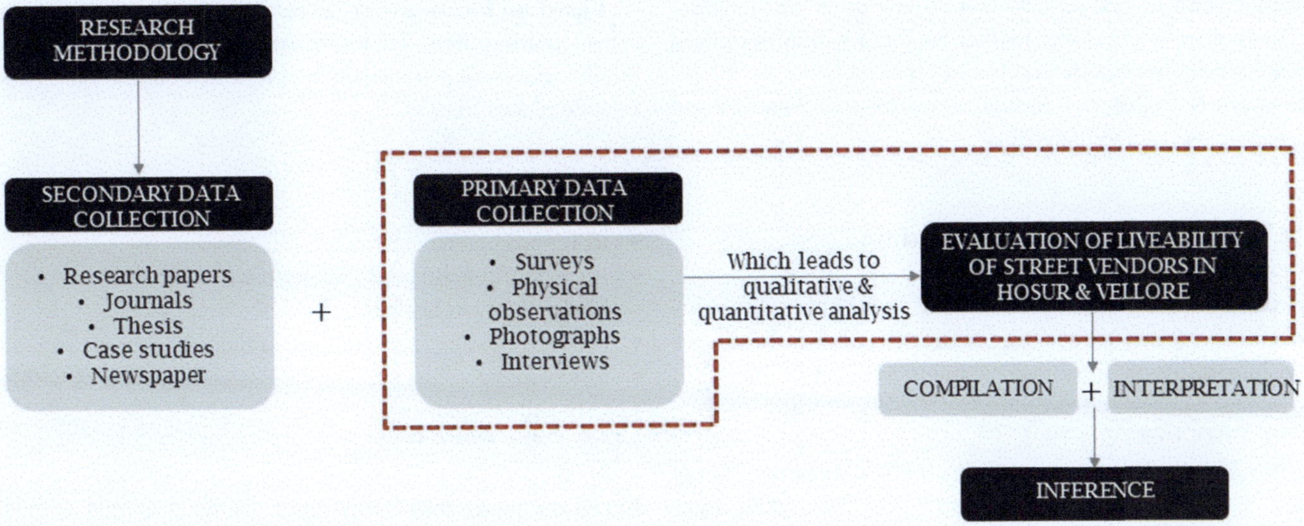

Fig. 1 Research methodology phase 1. *Source* Authors

2 Literature Study

2.1 What is the Meaning of Liveability?

The term "liveability" was first known in the year 1850. Liveability is an individual's notion of his or her role in life with respect to culture. It can be defined as "the ability of a megacity to provide favourable conditions to its inhabitants and others based on the sustainability parameters such as social, natural environment, and economic" (IFC, 2022). Wellness, environment and culture, stability, literacy, and infrastructure facilities were the listed parameters by EIU as the Global Liveability Index in 2022. Liveability is where one can live in conditions that have health outcomes and a long life expectancy (AUO, 2020). "Liveability" is a term used to express the living condition of a place. The term is also synonymous with 'well-being', 'quality of life', 'happiness," and 'life satisfaction.' But because the term is broad and varies with context, there is no agreed definition of this term globally. In urban form, it is a built landscape that ensures healthy and sustainable living and also enhances quality of life. Brook Lyndhurst states that liveability varies with different economies, societies, cultures, and local influences (Brook Lyndhurst, 2004). In the Indian context, "liveability" can be stated as a sense of identity, a clean, safe community, and well-maintained infrastructure and amenities (Pandey et al., 2013).

2.2 Understanding of the Concept Street Market and Vendors

Traditional street markets or bazaars which are the hubs for retail, bartering and trade is well known in India since ancient times. These bazars majorly contributes to India's urban economy development, and it has transformed into shopping stretches, streets, roads, and junctions (Kauser, 2018).

2.3 Types of Market/Vendors

"The street vendors (protection of livelihood and Regulation of street vending) act", 2014 has classified the markets under the following categories:

Natural market—A place where buyers and sellers have traditionally gathered to exchange goods and services and which has been designated as such by the local government on the advice of the Town Vending Committee.

Street vendor—A person who sells in a temporary built-up structure or travelling from one location to another to sell vital goods or providing services to the general public in a street, lane, sidewalk, footpath, pavement, public park, or any other public or private area.

Stationary vendor—A person who conducts his/her business regularly at a particular location.

Mobile vendor—A person who moves from one location to another while doing his/her business within permitted areas only to provide goods and services.

Hawkers & Peddlers—Similar terms to street vendors and can be used interchangeably.

2.4 Evolution of Street Vending

As mentioned earlier, Street vending is an important component of any city's economy. Nasreen Kauser in her paper "Sustainability of Traditional shopping streets of India" in 2018 has categorized evolution of streets vending as Pre-1950, 1950–1990, and 1991.

Pre-1950's: The majority of pre-1950's shopping streets may be traced back to the colonial era, when retail trade grew in influence and took the form of commercial establishments along the arterial roads within the city. There were very few market places in this category that were developed specifically to serve as retail shopping destinations in those early years. Connaught Place in New Delhi and Hogg Market in Kolkata are two well-known examples (Kauser, 2018).

1950–1990: During this period several networks of shopping streets adopted organic development to meet the needs of the consumer base and trends in demographic change. This was initially motivated by the concept to establish socialistic pattern along with economic growth after independence. For most of this period, local or domestic brands and thriving unorganized commercial sectors were flourishing at an exponential rate (Kauser, Sustainability of Traditional shopping streets of India, 2018). National brands began to emerge in the later stages as this dynamic business began to thrive with more market economies in play and as consumers sought to refresh their shopping experience through new age malls and shopping complexes which are air conditioned that had started dotting the retail landscape by that time.

Post 1991-The emergence of shopping streets in the suburbs and the physical extension of existing shopping streets or cities into the suburbs are reflected in the development of suburban markets. Street markets continue to be important despite changes in shopping experiences caused by urbanization (Kauser, 2018).

2.5 Types of Street Vending

As per the TISS and YAVA Tata Institute of Social Sciences (TISS) and Youth for Voluntary Action (YUVA)) survey the author has categorized the street vendors (locally referred as hawkers) based on locations, different goods sold, different installations/display units and storage units.

Based on locations—Markets are usually formed in the crowd pulling area. They are usually formed along major public spaces like transit hub, religious centres, etc. They include places like,

- Railway stations
- Bus terminus
- Religious centres
- Important junctions of the city/town
- Under fly over, bridges
- Outside existing market/commercial/retail shops
- Side walks
- City squares/plazas.

Based on goods sold—Broadly it can be classified as agricultural and non-agricultural goods. The list of categorize are as follows,

- Agricultural products
- Consumables (Cooked at site/cooked and brought to site/treated)
- Fish and poultry
- Non- agricultural products.

Based on installations/display units

- Open pitch (Baskets, carpet, tables)
- Stalls (umbrellas, bamboo kiosk, corrugated sheets, tarpaulin sheets, iron kiosks)
- Stationary cart (4 wheeled carts, cycles, autos).

Based on storage—Based on their spatial allotment and activities vendors store their goods to be sold in the following spaces,

- Residences of the vendor
- In shop itself
- Nearby shops.

2.6 Liveability and Street Vendors

Ambati, in his paper titled "Working Conditions and Quality of Life of Street Vendors in the City of Ahmedabad," evaluated the quality of street vendors lives through socio-economic and working conditions (Ambati, 2015). The researchers analyse the financial aspects of the street vendors, their working hours, facilities, legal sides, and the amount of bribes they have to pay in order to sustain themselves in the market, etc. The findings of the study

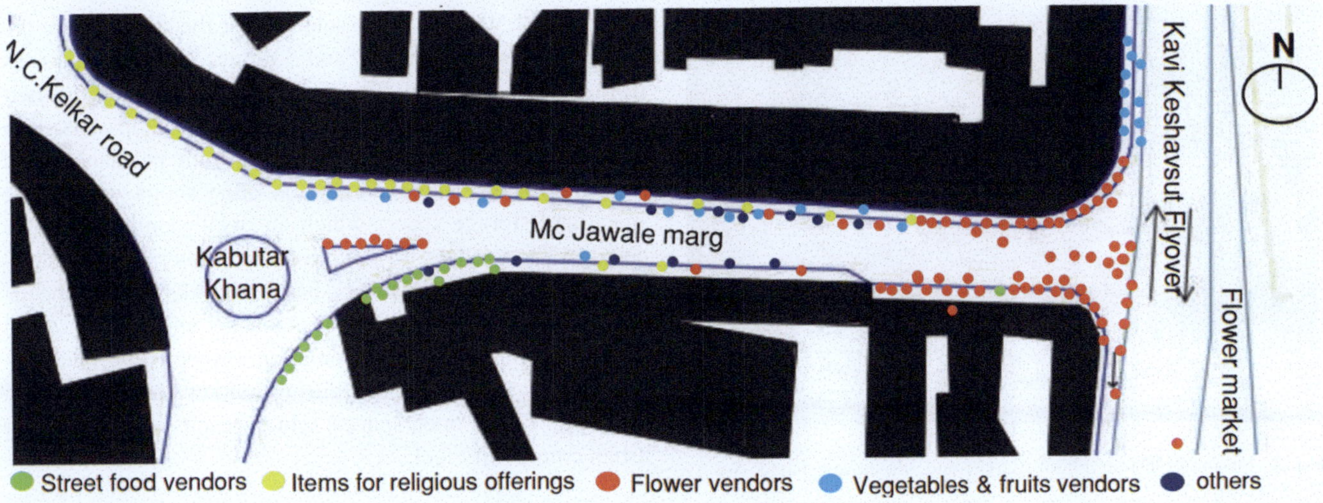

Fig. 2 Mapping of Mc Jawale road street vendors. *Source* Shah (2019)

revealed precarious conditions of work and life for street vendors, involving an extensive workload, very low sales and profits, and no guarantee for their safety and security. But the paper didn't focus on their housing, their mode of transport, or their ease of accessibility to vending places.

In Chap. 5 of the thesis report on "Livability of informal street vendors in Aizawl City," Thangluah (2022) has calculated the liveability index of street vendors in five different markets within the city. According to Thangluah (2022) the size and location of the study sites were the primary selection factors. With the help of relevant sustainability indicators like social, economic and physical environment of the street vendors in their workplace, the liveability aspect was evaluated. This study helps to develop 3 important paradigms such as workplace environment, safety and security and economic elements. The authors have not taken the economic aspects into consideration for evaluation purposes due to time constraints.

Sharit K. Bhowmik, in his report for NASVI in 2001, studied street vendors and hawkers in seven cities within India by dividing them into six sections (Bhowmik, 1998). The focus of the study was on municipal laws in each city, the inclusion of street vendors in urban planning, socio-economic status, street vendor perceptions, and issues faced by them in the seven listed cities. This paper didn't include infrastructure and facility-related issues.

The authors R. Widjajanti and H. Wahyono, in their paper titled "Space Livability of Street Vendors in Simpang Lima Public Space, Semarang," analysed spatial liveability by categorising road seller spaces based on their activities (type of products, trading places' size, trade place assessment, spatial dimension, trading time, duration, and period) and space conditions (accessbility, natural elements, safety and security, and availability of parking lots) (Widjajanti & Wahyono, 2018). The government's policies focus more on street vendors accessibility and safety aspects and the threat of eviction, but unfortunately, the infrastructure and facilities aspects were not taken into consideration.

3 Case Studies in Indian Context

3.1 Dadar Phool Market, Mumbai (Figs. 3 and 4)

Dadar is a densely populated residential area in the heart of Mumbai city. It has excellent connectivity to other parts of Mumbai. Dadar railway station is the most crowded area and is also known for its strong whole-sale market, retail shops, and commercial shops. About 3–4 markets in Dadar serve a population density of 2,00,000/km^2. Dadar is also known for its oldest retail market, Veer Sawakar Market. In spite of these markets, plenty of street vendors are found outside the market and railway station. The three roads Mc Jawale (Fig. 2), D Silva Road, and Ranade Road opposite the Dadar West Station are filled with street vendors and shops. The case study chosen is located at Mc Jawale Road Street Market, which branches from Tulsi Pipe Road under Kavi Keshavsut Flyover, where street vendors occupy the pavement. The 11 m-wide road with a 3.5 m walkway stretches up to 160 m until Kabuthar Khanna (Sharma, 2007). The flower market's activities start from 5 a.m. to 9 a.m. on a daily basis. After 9 a.m., other shops like vegetable, fruit, and food shops start. The vendors here use locally available materials for their shop structure. Most of the structures are made of bamboo sticks, jute bags, umbrellas, and plastic sheets. Around 17% of the hawkers stay in their vending place itself, 15% use pavements to sleep,

Fig. 3 Photos of Dadar Phool market and Mc Jawale Road. *Source* No-Mad (2022)

Fig. 4 Photos of Dadar Phool market street. *Source* No-Mad (2022)

and others have their homes in close proximity to their vending areas. These hawkers usually eat in nearby shops, use public toilets, and sleep on walkways. The footfall on the road is high, so the vendors, in order to increase their sales, not only utilize the sides of the pavement but also the road. This, in turn, causes a lot of vehicular and pedestrian congestion.

3.2 Mallick Ghat Market, Kolkata (Figs. 5, 6 and 7)

Mallick Ghat Market, popularly known as Howrah Phool Market, is located in Kolkata, the city of Joy in India. This market is located on the banks of the Hoogly River and Howrah Bridge on the west. This market has heritage ghats in close proximity, which attracts a lot of consumers as tourists. The market entrance is from Strand Road, and vending zones are along Strand Bank Road; thus, the market can be split into three zones. Zones 1 and 3 are street vending zones, and zone 2 is a market near the ghats. Zones 2 and 3 of 2,500 sq. m. are selected for the study. There are about 80 shops in zones 2 and 3 alone. The market works from 5 a.m. to 3 p.m. The vendors start assembling their products by 3 a.m. The vendors built their shops based on their sales, economy, etc. There are both permanent and temporary shops, single and double-story shops with resting space and storage space, which are well planned and executed by the vendors for their comfort. Materials like corrugated sheets, tarpaulin sheets, bamboo sticks, plastic sheets, and clothes the streets located in zones 2 and 3 are 2 m and 10 m wide for pedestrian circulation, respectively. The market place is very congested during market hours. The market doesn't have open spaces, which results in a lack of an urban community feel.

3.3 Comparison Study of Dadar Phool Market and Mallick Ghat Market

The authors have compared Dadar phool market with Mullick Ghat in terms of spatial organization, user's perspective and urban design principles (Figs. 8, 9, and 10).

On comparing two India's famous market based on spatial, market facilities and urban level, the inferences made could be used in the proposal of design strategies. The organized grid layout used in Mullick ghat market enables to allocate more number of shops and better circulation. On the other hand both the markets needs separate lanes for consumers and vendors. The street vendors in both markets were using similar materials to arrange their own economic friendly installations. The infrastructure facilities need to be enhanced in both the market. Mullick ghat market is comparatively legible than Dadar market. Both the markets lack walkability.

4 Case Studies in International Context

4.1 Farmers Market in Xiafu Village, China (Fig. 11)

In the City of Zhuji, China, the farmer's market in Xiafu village was an important communal space. The market was the place to trade crops, cattle, poultry, and others from 5 a.m. to 8 a.m. Due to the economic changes, the vendors avoided using the market space provided and started vending outside the market. The market also lacked facilities for vendors. In 2018, Bengo Studio took on the project of redeveloping the market. Their first step towards design was to analyze villager behavior, vendor behavior, and market needs. The 800 m^2. area followed a new layout designed to segregate vendors and buyers. This gives the vendors a

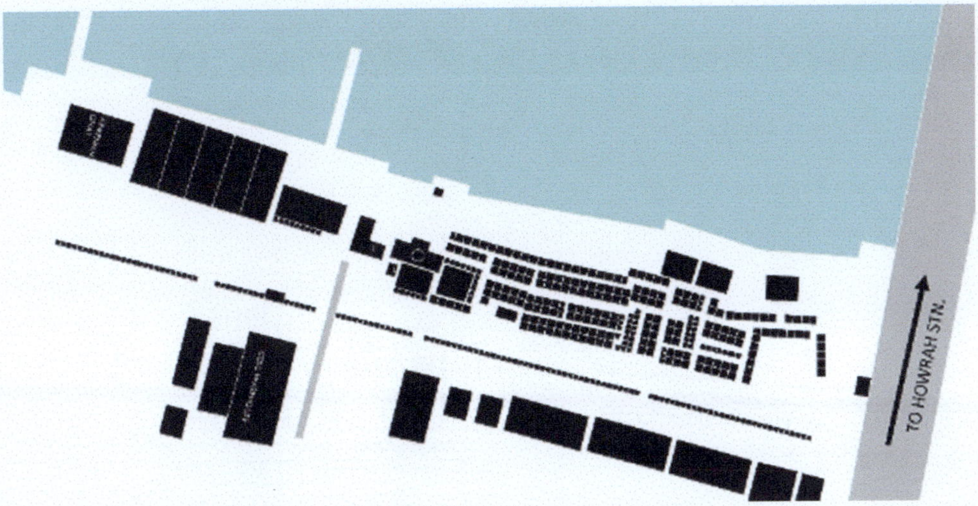

Fig. 5 Plan of Mallick ghatmarket. *Source* Author

Fig. 6 Photos of Mallick ghatmarket. *Source* Klook (2022)

Fig. 7 Photos of Mallick ghatmarket vendors selling garlands. *Source* Klook (2022)

sense of ownership to keep their allotted space clean and enables easy maintenance. The lighting was the issue with the previously used sloped roof. Thus, the new folded roof with timbers, steel girders, and clay tile is designed to lighten the space. The new design enabled the vendors on the street to return to the redesigned market.

4.2 Manifesto Market Andel

The 2600 m^2. market hub is exclusively designed for foods and drinks in 2021 in the Czech Republic (Fig. 12). The key idea is to adopt a place-making strategy to utilize the unused urban gap. The site is located in the core area of the Andel district, which is known for its shopping. The site is also surrounded by densely populated residential areas, thus enabling the creation of a vibrant, robust public space for them. The plan followed the well-known "courtyard strategy" to create a gathering space to promote social activities and communal participation. The materials, like aluminum sheets and furniture, used already on that site are reused for a sustainable solution. The reusable containers are used as temporary, reversible structures. The steel blue scaffolding and aluminum cladding on the containers on the façade attracted a lot of consumers and became an image for the neighborhood.

DESCRPTION	DADAR PHOOL MARKET	MULLICK GHAT	INFERENCE
Location	Mc Jawale road, Near Kavi Keshvsut fly over, Dadar west, Mumbai	Strand bank road, near Howra bridge, Kolkata	
Study/Site area	1500 sq.m (Street length – 160m)	2500 sq.m (Street length – 175 m)	
Climate	Tropical, wet and dry	Tropical, wet and dry	
SPATIAL			
Layout	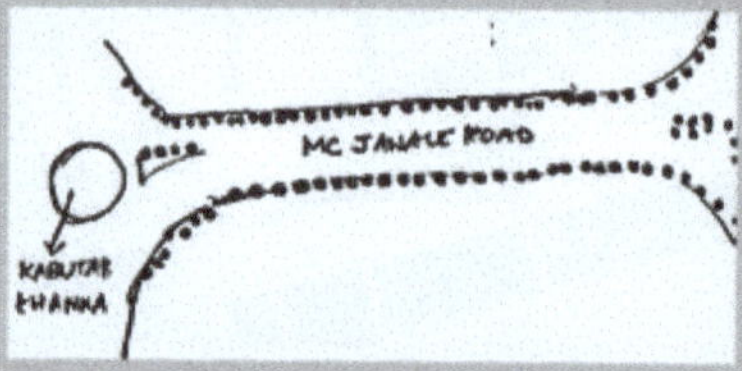 Shops along the street, lacks connectivity, provides chance to explore	 Shops in grid layout. Random grid layout could make the shops in periphery inactive and loose customers	Planned grid layout with lesser and clear circulation can enhance user experience.
Entrance & exits	Enter from Tulsi pipe road or from Kabuthar Khanna	Enter from Strand road (18m)	
Road access	11m Mc Jawale road branching from primary road (road under flyover)	10m Strand bank road (two way traffic road) branching from Strand road. 3m of hawkers on both sides.	Connectivity with primary roads are important to load and unload.
Pedestrian walkway	3.5m (approximately)	No separate pedestrian.	Separate walkways and vendor lanes could reduce congestion.

Fig. 8 Comparison of Dadar Phool market and Mallick ghat in terms of spatial planning. *Source* Author

4.3 Comparative Study of Xiafu Market and Manifesto Market

The comparative study of Xiafu market and Manifesto market (refer Figs. 13 and 14) and the inferences made can be used in Indian markets with respect to their context. The spatial layout with distinct vendors and consumer zone in Xiafu market and introverted layout with courtyard as a nucleus is a god strategy to create breakout spots. The reusable materials can be adopted in the Indian context with modifications based on climatic conditions. The planning in both markets has enabled good walkability.

4.4 Methodology Used to Arrive at Indicators and Sub-Indicators of Liveability

Having compared 4 markets at the national and international level, the authors could arrive at lists of Indicators and sub indicators with respect to liveability. This methodology arrived through literature has been used to evaluate the live case studies of Hosur and Vellore (Figs. 15 and 16).

5 Live Case Studies Site Study

5.1 Hosur

Hosur, "the industrial city of Tamil Nadu," is famous for its floriculture because of its suitable weather. The flower crops grown here are marketed to different parts of the country. Flowers and flower markets are one of the significant "images of the city". The existence of a flower bazaar opposite the bus terminus enables the street vendors to encroach on the road and start their vending (Figs. 17, 18, and 19). The street vendor encroachment was not only on the roadside but also under the flyover. There are about 60 street shops under the flyover and on the 1.5 m hawker's lane on the service road side. This causes the main Bagalur junction and the service road to be very congested. The market starts at 5 a.m. in the morning and continues until 9 p.m. in the night. The vendors here sell flowers and fruits exclusively. The vendors in these zones lack preparation space for them to sell garlands, so they use space under the flyover for that. Some vendors also mentioned that they are homeless and stay in the underpass of the flyover. This underpass,

DESCRPTION	DADAR PHOOL MARKET	MULLICK GHAT	INFERENCE
STREET MARKET			
Types of products available	70% are flower shops, others are vegetables, fruits, religious offerings and few street food	Exclusively flowers like roses, lilies, orchids, tuberoses, gladiolas etc.	
Number of shops	50 (Approx.)	80 (Approx. in Chosen zone 2 & 3)	
Types of structure for shops	 The bamboo sticks tied using ropes and umbrellas are the structures used here. Other use jute bags as carpet on floor.	 Corrugated sheets, tarpaulin sheets on bamboo sticks. Has G+ 1 structure also where the top part is for resting space.	Reusable materials like containers could be used for covered structure.
Materials	Jute bags, wooden board, bamboo stick, corrugated sheets, umbrellas, crates	Tarpaulin sheets, plastic sheets, bamboo stick, corrugated sheets, clothes	
Area per shop	5 sq.m (Approx.)	9 sq.m in Zone 2, 5 sq.m in zone 3	
Facilities available	Railway station, lot of religious and educational buildings, night shelter	Ware houses, police station, dump ground, tram service. Lack of drinking water and infra structure need to be developed	Public toilets, drinking water and cheap food for the vendors could enhance their quality of living.
Foot fall/ Traffic	Approx 5 lakhs per day / High pedestrian and vehicular traffic	- / very congested due to huge rush and small area	
Timings	Flower market from 5am – 9am. After 9am other shops like vegetable, fruits, food starts.	5am – 3pm	

Fig. 9 Comparison of Dadar Phool market and Mallick ghat. *Source* Author

which is currently the non-vending zone, also has a few food shops in carts; the vendors are depending on the food in nearby shops, and Amma Unavagam is next to the bus stand, where they use the toilets. The vendors here use umbrellas, tarpaulin sheets on bamboo sticks, carton boxes, and plastic baskets for their installations. The flower litter and waste are dumped in the underpass of the flyover.

5.2 Vellore Market

The fort city with its variety of architecture has the authentic Clock tower of the British era. The King George V and Queen Mary coronation Clock tower was built in memory of 277 villagers who were part of the 1914–1919 Great War. This exposed red brick structure with a segment arch and pediment acts as an entrance to the Nethaji flower bazar. The long bazaar road behind Vellore's old bus stand has this architectural marvel. But the structure, due to a lack of maintenance, is in poor condition. The 9 m-wide one-way Long bazar (refer to Figs. 20, 21, and 22) has street vendors occupying the shops frontage of 2 m on one side and two-wheeler parking on the other side. This makes the one-way road more congested. The street vendors encroach on the roadside because of a lack of space inside the Nethaji market. These vendors sell a variety of goods, but the majority of shops sell fruits and flowers. The vendors inside the market have structures made of corrugated sheets and steel columns. The flower street vendors outside the clock tower use economical structures like bamboo stick installations with tarpaulin sheets as their roof to vend. These street vendors also attract a lot of consumers due to their location. The rest of the street vendors mostly use wheel carts. Every vendor uses carton boxes or wooden or plastic boxes as display

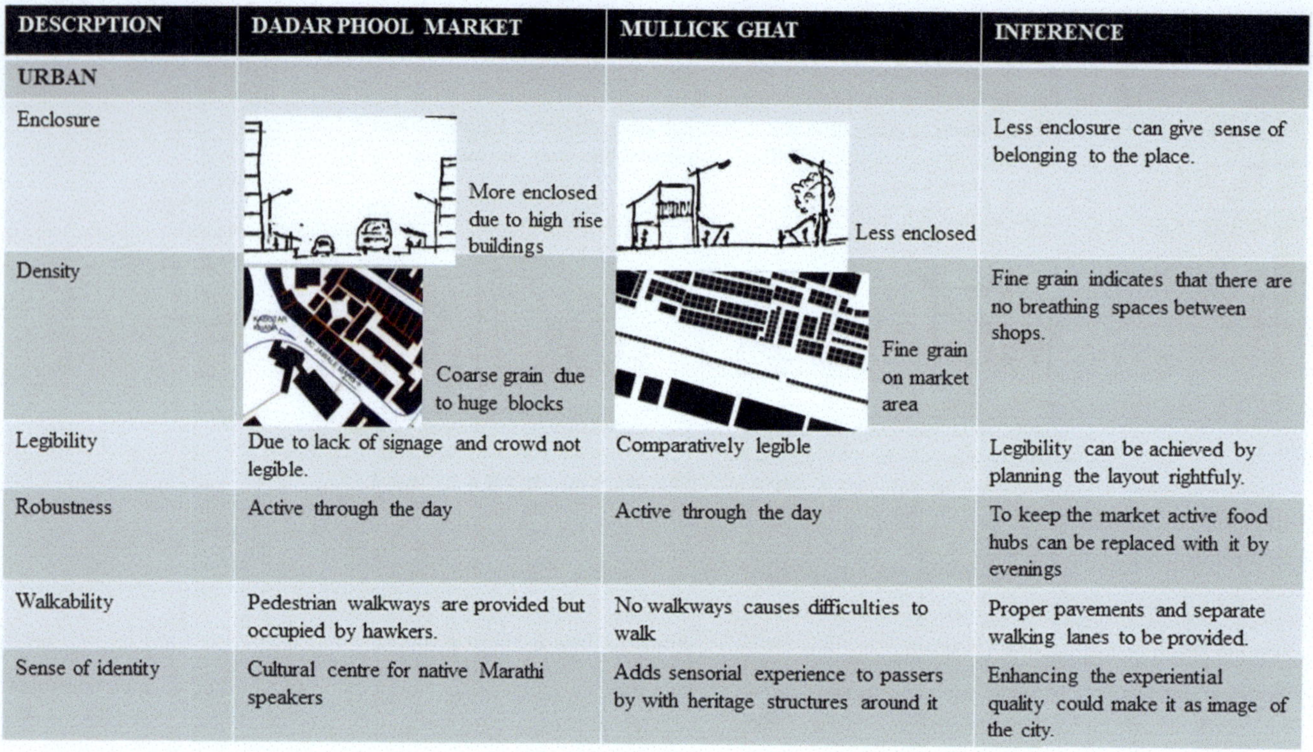

DESCRPTION	DADAR PHOOL MARKET	MULLICK GHAT	INFERENCE
URBAN			
Enclosure	More enclosed due to high rise buildings	Less enclosed	Less enclosure can give sense of belonging to the place.
Density	Coarse grain due to huge blocks	Fine grain on market area	Fine grain indicates that there are no breathing spaces between shops.
Legibility	Due to lack of signage and crowd not legible.	Comparatively legible	Legibility can be achieved by planning the layout rightfuly.
Robustness	Active through the day	Active through the day	To keep the market active food hubs can be replaced with it by evenings
Walkability	Pedestrian walkways are provided but occupied by hawkers.	No walkways causes difficulties to walk	Proper pavements and separate walking lanes to be provided.
Sense of identity	Cultural centre for native Marathi speakers	Adds sensorial experience to passers by with heritage structures around it	Enhancing the experiential quality could make it as image of the city.

Fig. 10 Shows the comparison between Dadar Phool Market and Mallick Ghat Market in terms of urban design principles. *Source* Author

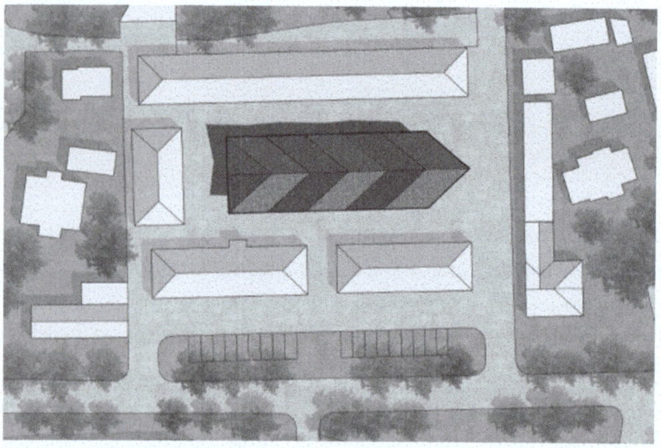

Fig. 11 Master plan of farmers market in Xiafu market, China. *Source* BengoStudio (2018)

units; some even display the goods in plastic sheets on the road itself. The street vendors here do not have any basic facilities for them, like public toilets, drinking water, etc. To use the public toilet at the bus stand, they have to walk for 300 m.

5.3 Comparitive Study of Hosur and Vellore Market Based on Observation

Using the above methodology, the authors have evaluated the live case studies—Hosur and Vellore (Figs. 23, 24, and 25).

5.4 Comparitive Study of Hosur and Vellore Market Based on the Survey Results

Figures 26, 27, 28, and 29 show the survey results collected on site.

Socio-cultural
See Fig. 26.

Economic
See Fig. 27.

Environment
See Fig. 28.

Physical
See Fig. 29.

Fig. 12 Photos of Manifesto market Andel Courtyard (Kristof, 2021)

Inference

From the inferences in Fig. 30, the layout of the street markets in both Hosur and Vellore is linear. This linear layout on one-way roads reduces the chances of exploring and is also not interesting. The layout can be redesigned, and pedestrianizing a particular stretch can be done to improve the experiential quality for vendors and consumers. The climatic conditions of both places make the roof or shade mandatory, so a semi-shaded or complete shaded structure made of reusable, sustainable materials must be considered. Comparatively, Hosur has more rainy days, so the chances of flooding are higher, and drains need to be added.

The chosen stretch in Hosur is filled with flower and fruit shops, so the opportunity to bring different goods under the flyover and make that vending area can be taken. Compared to Hosur, Vellore has harsh summers. But the existing structures are a few tarpaulin sheet structures and umbrellas; the rest are open carts. As per internet case studies, the reusable containers can be used for shops. Police booths, sufficient street lights, and CCTV cameras must be installed for the safety of vendors. To improve the liveability of vendors in both places, public toilets, drinking water fountains, and resting shelters can be added. The similar strategy used in the G + 1 structures of Mullick Ghat Market can be used to ensure store and resting space for vendors. To avoid bribes and harassment, vendors should be made aware of existing government policies such as the Street Vending Act, SEWA, and others. This could enhance the liveability of street vendors.

DESCRPTION	XIAFU MARKET	MANIFESTO MARKET	INFERENCE
Location	Xiafu village, City of Zhuji, China	Andel, a district of Prague, Czech republic	
Study/Site area	800 sq.m	2600 sq.m	
Climate	Temperate climate	Temperate climate	
SPATIAL			
Layout	 Layout resembling one of the Chinese character. This differentiates seller and buyers and provides sense of owning to the sellers.	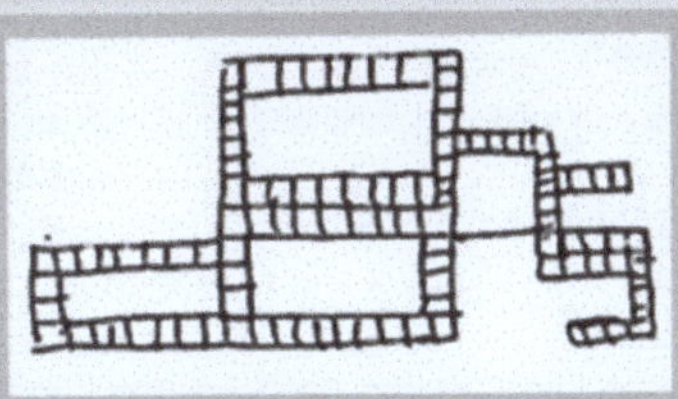 Introverted layout focusing courtyards. 4 different courtyards has different functions.	The central courtyard as communal space and shops around it with proper differentiation of buyer and seller zones.

Fig. 13 Shows the comparison between Xiafu Market and Manfesto Market. *Source* Author

DESCRPTION	XIAFU MARKET	MANIFESTO MARKET	INFERENCE
STREET MARKET			
Types of products available	Trade crops, livestock, and other goods	Food	
Number of shops	31	65 (Approx.)	
Types of structure for shops	Folded roof with long-span steel structure and clay tile 	Old or used containers with steel scaffolding 	Reusable materials like containers could be used for covered structure.
Materials	Timber roof frame, Steel girders, clay tile	Corrugated aluminium sheets to clad units, steel	
Area per shop	12sq.m/ shop	10sq.m / shop	
Timings	5am – 8am	9am – 12am	
URBAN			
Enclosure	 Not enclosed gives the feel of openness	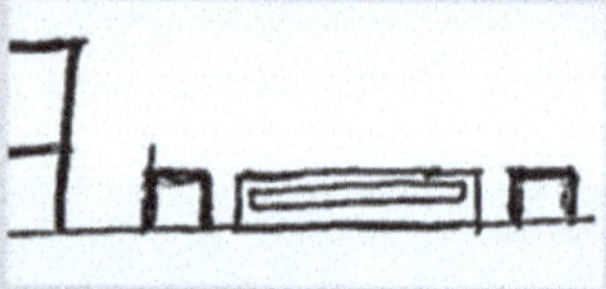 Less enclosed	Less enclosure can give sense of belonging to the place.
Legibility	The shops are very few thus legible	Legible	Planning could make the space legible
Robustness	Active through the day	Active through the day	
Walkability	The differentiation of buyers and sellers make its easily walkable.	Walkable	Interesting and separate walkways
Sense of identity	Important public space of the Village, starts the day of villagers live there.	The front façade with scaffolding acts as an imageability to the neighbourhood.	Enhancing the experiential quality could make it as image of the city

Fig. 14 Shows the comparison between Xiafu Market and Manfesto Market in terms of structure and urban design principles. *Source* Author

Both study areas (Fig. 29) are less closed, giving a sense of openness. The size of street vending structures should be handled well to make the retail shops behind them legible. Separate walking, hawking, and parking lanes are to be provided. This will encourage consumers to shop freely and increase the revenues of vendors. Universal design strategies like ramps and tactility to be adopted to ensure better vending and shopping. The proper maintenance and improved experiential quality can make these markets "images of the city," which will increase tourist customers. Thus, it could improve the liveability of street vendors in the chosen cities, Hosur and Vellore.

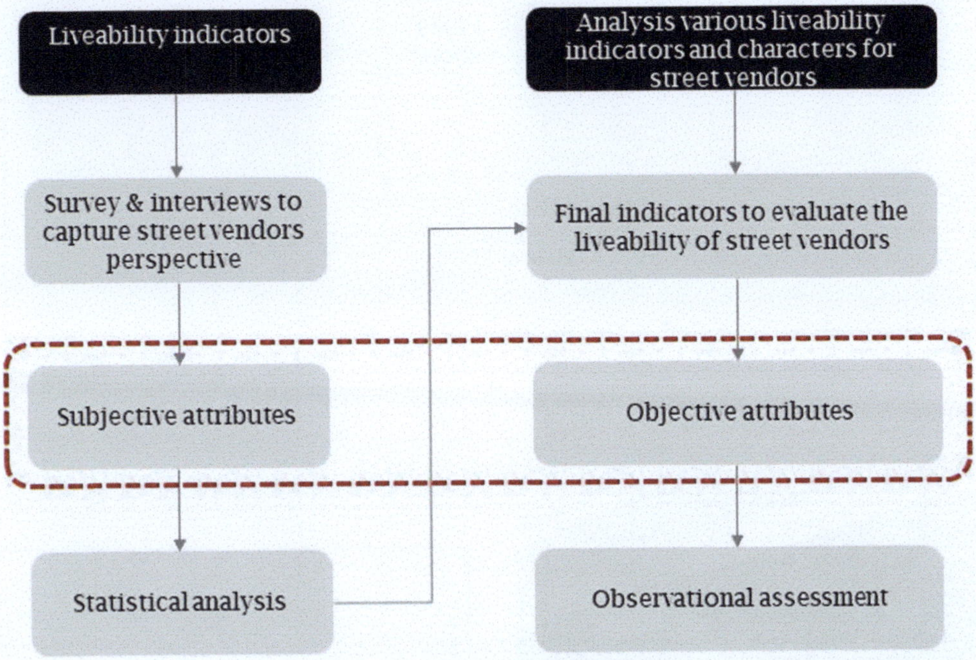

Fig. 15 Methodology used to arrive indicators and sub-indicators based on literature case studies. *Source* Author

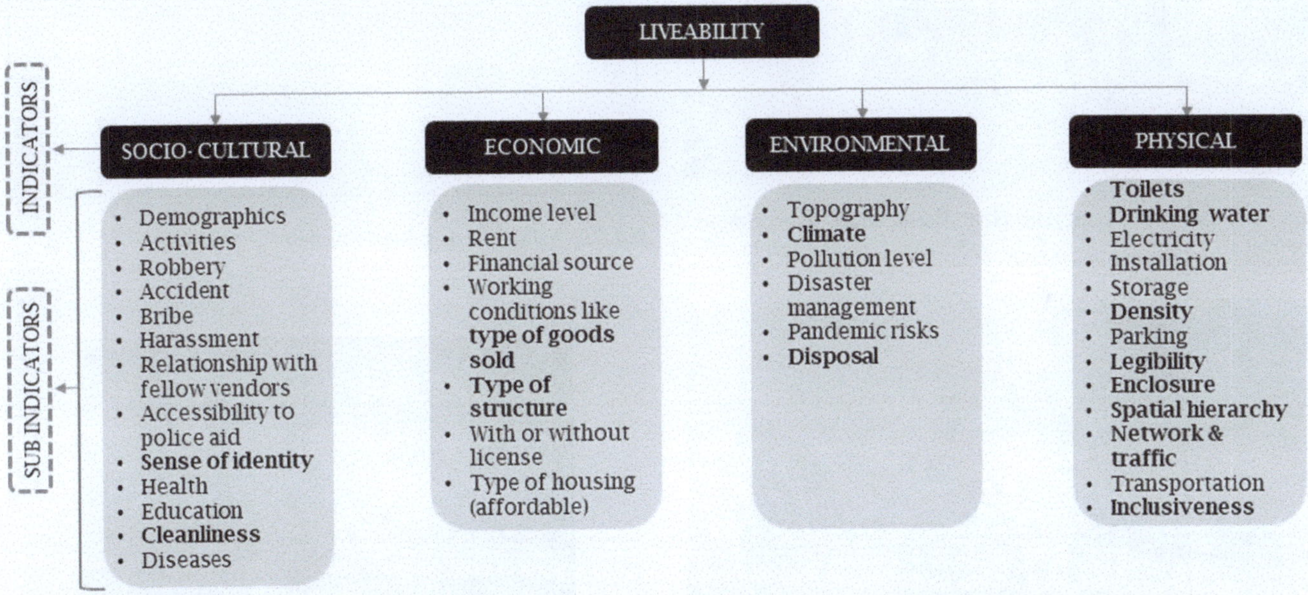

Fig. 16 Methodology used to arrive indicators and sub-indicators based on literature case studies. *Source* Author

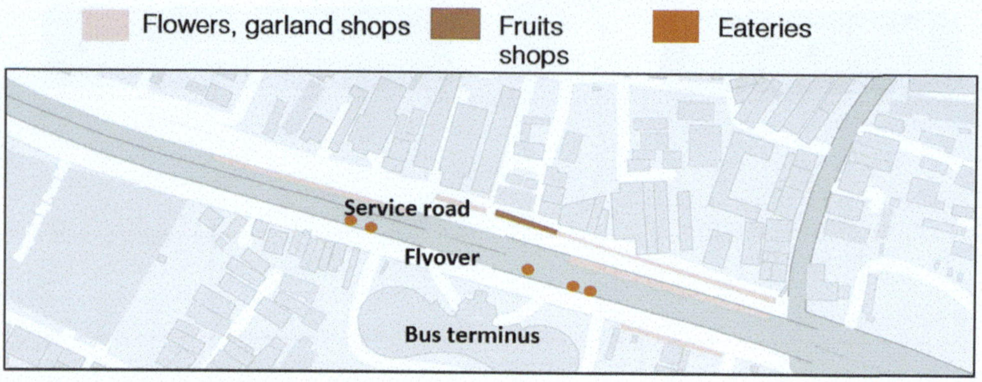

Fig. 17 Mapping of Hosur road street vendors

Fig. 18 Photos of Street vendors in Hosur under flyover. *Source* Author

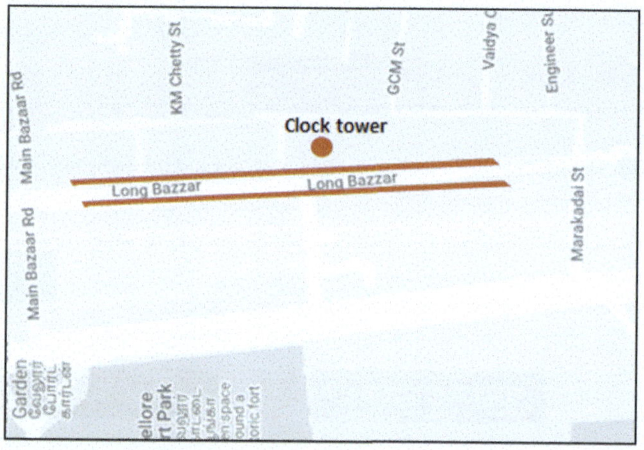

Fig. 20 Vellore market. *Source* Author

Fig. 19 Photos of Street vendors selling flowers in Hosur under flyover during rainy season. *Source* Author

Fig. 21 Photos of street vendors in Vellore market. *Source* Author

Fig. 22 Photos of shops in Vellore market. *Source* Author

6 Conclusion

From the analysis, the informal sectors of street vendors in Hosur and Vellore prove that they provide affordable, accessible products to their people within their context, acting as suppliers in the country. From the case studies, we are aware that bribes collected from these informal sectors can generate a lot of money, and these figures converted to taxes will turn into a revenue-generating sector. The only method for ensuring a good standard of living without using illegal means is the formalization of markets. The government and vendor organization must adopt the procedures which are clearly specified in the national policy for street sellers

DESCRPTION	HOSUR	VELLORE	INFERENCE
Location	Service road, Opposite to the Bus terminus, Hosur.	Long bazar, Outside clock tower, Vellore	
Study/Site area	250 m stretch of the road	150 m stretch of the road	
Climate	Savanna	Tropical savanna	
SPATIAL			
Layout	 Stalls are arranged in linear layout.	 Stalls are arranged in linear layout.	This linear layout on one way roads reduces the chances to explore and also not interesting.
Entrance & exits	6 meter one way Service road	7 meter one Long bazar road	Exclusive entry and exits for vendors to be given.
Road access	The 6m wide service road with 2m hawker's lane	7 m wide road with 2 m hawker's lane on one side and 2 m parking on one side.	The separate lines for vending, walking and vehicles to be provided.
Pedestrian walkway	No exclusive pedestrian walkway	No separate pedestrian.	

Fig. 23 Comparitive study of Hosur and Vellore market. *Source* Author

DESCRPTION	HOSUR	VELLORE	INFERENCE
STREET MARKET			
Types of products available	75% are flower shops, 25% fruit shops	Flowers, fruits, vegetables, fashion accessories, usable junks and others	
Number of shops	60 shops	40 shops	
Types of structure for shops	 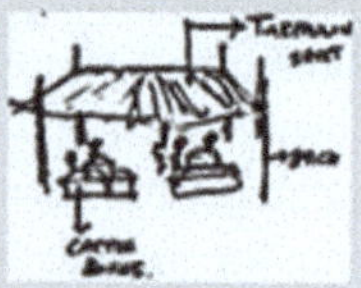 Umbrellas, tarpaulin sheets on service road side. Bamboo sticks tied with rope under flyover	  Tarpaulin sheets on bamboo sticks, wheel carts, carpet on floor	Sustainable and reusable materials can be used for the stalls.
Materials	Umbrellas, bamboo sticks, carton boxes, plastic baskets, sacks	Umbrellas, tarpaulin sheet, bamboo sticks, carton boxes, plastic baskets,	Reusable materials can be assembled as display unit.
Area per shop	4 sq.m (Approx.)	3-4 sq.m (Approx.)	The storage, preparation area to be included.
Facilities available	Public toilet in bus stand, foods on near by shops. No such facilities were given exclusively for street vendors.	No such facilities were given exclusively for street vendors.	Basic facilities like toilets, drinking water, food to be made available in closer proximity.
Foot fall/ Traffic	1200 foot fall/ day and heavy traffic congestion due to the hawker's lane in busy service road	2000 foot fall/ day due to other retails in this area	
Timings	5am – 9pm	5am – 3pm	
Lights	Sufficient street lights and commercial façade lightens the street	Lesser street lights	Street lights can be added to make place more vibrant and secure.
Security / surveillance	Police booth near, CCTV surveillance near bagalur circle signal.	Police station in 500 m distance. No surveillance	Police booth in closer proximity.
Robbery	Pick pockets were recorded	More robbery chances	Police booth in closer proximity
Bribe	Rents were collected by councillor in non- vending zones also.	-	Awareness on government acts and policies.

Fig. 24 Comparitive study of Hosur and Vellore market in terms of materials and structure. *Source* Author

DESCRPTION	HOSUR	VELLORE	INFERENCE
URBAN			
Enclosure	 Less enclosed	 Less enclosed	Gives sense of belonging
Density	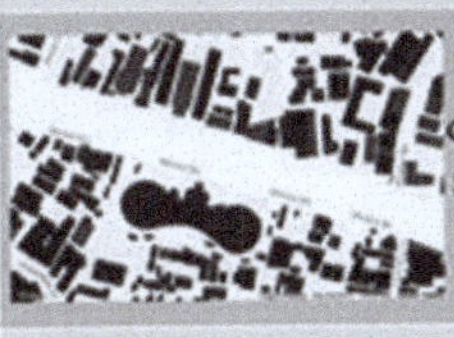 Huge block on commercial zone makes the grain coarse	 Mixed grain	
Legibility	Legible.	Street shops are legible but the market and the commercial shops are not legible as the street shops blocks the view.	The policy level strategies to be made to make the commercial shops except street shops be legible.
Robustness	Active through the day	Active through the day	The market should be adaptive to changes trough out the day.
Walkability	The divider divides vehicular and pedestrian. No separate walkway.	Exclusive pedestrian walkways are not provided.	Separate lanes and legible shops could make the street walkable.
Sense of identity	One of the significant image of Hosur	Poorly maintained Clock tower and market fails to give sense of identity	Improving experiential quality could make this image of the city

Fig. 25 Comparitive study of Hosur and Vellore market in terms of urban design principles. *Source* Author

		HOSUR					VELLORE				
SUB-INDICATORS		Very poor	Poor	Average	Good	Very good	Very poor	Poor	Average	Good	Very good
Safety & security	Safety from robbery/ harassment	10	25	50	15	0	25	60	15	0	0
	Safety in accident prone areas	25	40	20	10	5	10	10	50	20	10
Accessibility for help during emergencies	Accessibility to police aids	5	15	40	35	5	0	5	35	45	15
Fellow street vendors connectivity	Connection with the neighboring vendors in the street	5	10	35	30	20	5	10	30	50	5
Proximity	Proximity from living place to vending place	0	25	35	35	5	45	25	20	5	5
Sense of Belonging	Sense of belongong to the market	0	10	35	40	15	0	15	50	30	5
Health											
Cleanliness	Cleanliness & hygiene of vending area	15	40	25	15	5	5	25	35	30	5

Fig. 26 Survey results based on socio-cultural. *Source* Author

		HOSUR					VELLORE				
SUB-INDICATORS		Very poor	Poor	Average	Good	Very good	Very poor	Poor	Average	Good	Very good
Economic condition	Income level	5	20	55	20	0	10	55	30	5	0
Location	Benefits of shop location in the vending area	15	5	30	45	5	10	25	60	5	0
Type of structure for shop	Self built structure and its durablity	15	10	35	35	5	15	55	20	10	0
	Resilence of the economic friendly structures from sun/ rain	25	35	20	20	0	30	55	15	0	0
	Size of the shop alotted	10	35	40	15	0	15	50	25	10	0
Regulations	Benefits of holding license	15	0	45	35	5	20	15	40	25	0
	Contribution of demarcating vending, non- vending and restricted vending zones	45	25	30	0	0	40	45	15	0	0

Fig. 27 Survey result based on economics. *Source* Author

		HOSUR					VELLORE				
SUB-INDICATORS		Very poor	Poor	Average	Good	Very good	Very poor	Poor	Average	Good	Very good
Climate & topography	Contribution climate in vending	0	15	45	25	15	70	25	5	0	0
Surrounding context	Exposure of pollution in the context	15	45	30	10	0	15	20	50	10	5
Disaster management	Disaster management in vending area during heavy rains/ flood	30	45	20	5	0	25	70	5	0	0
Pandemic	Resilence of vending area during pandemics	60	35	5	0	0	75	25	0	0	0
User experience	Infrastructure experiencing by vendor in vending place during nights	5	30	45	20	0	5	20	50	25	0
	User experience of physical infra structure	25	35	20	20	0	15	15	60	5	5

Fig. 28 Survey results based on environment. *Source* Author

		HOSUR					VELLORE				
SUB-INDICATORS		Very poor	Poor	Average	Good	Very good	Very poor	Poor	Average	Good	Very good
Layout	Contribution of market layout in liveability of vendors	5	40	20	35	0	10	20	65	5	0
Facilities available	Parking facilities	35	40	15	10	0	25	10	45	20	0
	Facilities like public toilets, driniking water	55	25	25	5	0	50	45	5	0	0
	Preparation & storage space for each shop	20	35	30	15	0	20	60	20	0	0
Legibility	Legibility of vending area	5	40	20	35	0	10	65	20	0	5
Walkability	Walkability of vending area for user & vendor	5	40	35	20	0	25	50	20	5	0
	Width of pedestrian for user	30	40	20	10	0	30	70	0	0	0
	Maintainence of pedestrian	25	45	25	10	0	55	40	5	0	0
Robustness	Robustness of vending area	5	20	30	35	5	10	15	30	45	0
Inclusivity	Inclusivity in vending area for both vendors & consumers	20	20	45	10	5	60	20	10	5	0

Fig. 29 Survey results based on UD principles. *Source* Author

in order to formally recognize the vendor community as a sector. The limiting presumption that areas are restricted to their intended use only by the government bodies is one of the main obstacles to resolving this dilemma. For instance a flyover underpass can only be used as cross over and cannot be used for anything else.

If we remove these strict regulations and take the initiative to redesign urban space, it may be used to its fullest extent if we are receptive to the idea of flexible uses for urban spaces. When attempting to plan unofficial activities, it can be helpful to suspend rigid programming rules. By creating "free zones" (Koolhaas, 1989), where street vendors use public space to inform new and innovative design approaches, the tension between programme and use of space will be significantly reduced to a larger extent. This will allow for the equitable and inclusive integration of informal markets into the newly emerging modern paradigms of urban design.

INDICATORS	SUB-INDICATORS	HOSUR	VELLORE
Socio cultural	Safety & security		
	Accessibility for help during emergencies	✓	
	Fellow street vendors connectivity	✓	✓
	Proximity	✓	
	Sense of Belonging	✓	
	Health		
	Cleanliness		
Economy	Economic condition	✓	
	Location	✓	
	Type of structure for shop	✓	✓
	Regulations		
Environmental	Climate & topography	✓	
	Surrounding context prone to pollution	✓	✓
	Disaster management		
	Pandemic		
	User experience		
Physical	Layout		✓
	Facilities available		
	Legibility	✓	
	Walkability		✓
	Robustness	✓	✓
	Inclusivity		

Fig. 30 Inferences. *Source* Authors

References

Ambati, N. R. (2015). Working conditions and quality of life of street vendors in city of Ahmedabad. *International Journal of Social Science & Interdisciplinary Research,* 20–31.

AUO. (2020). *Australian urban observatory—Sustainable development goals—Guidance notes*. RMIT University.

BengoStudio. (2018, November 22). *Xiafu farmers' market*. Retrieved from Arch Daily. https://www.archdaily.com/912100/xiafu-farmers-market-bengo-studio

Bhowmik, S. K. (1998). *Street vendors in Mumbai*. Urban design Rearch Institute.

Bhowmik, S. K. (2012). Street vending in ten cities in India. *National Association of Street Vendors of India,* 1–174.

Brook Lyndhurst. (2004). *Liveability & sustainable development: Liveability & sustainable development*. Brook Lyndhurst Ltd.

IFC. (2022). *Liveability report*. Retrieved from Institute for Competitiveness. https://competitiveness.in/liveability-report/

Indira, D. (2014). A study on the organizing of street hawking business. *International Journal of Management and Commerce Innovations, 2*(1), 280–288.

Kauser, N. (2018). Sustainability of traditional shopping streets of India. *Journal of Emerging Technologies and Innovative Research (JETIR), 5*(11), 670–681.

Klook. (2022, October 30). *Mallick Ghat flower market morning walk tour*. Retrieved from Klook. https://www.klook.com/en-IN/activity/11956-mallick-ghat-flower-market-morning-walk-tour-kolkata/

Koolhaas, R. (1989). *Toward the comntemporary city.*

Kristof, C. (2021, November). *Manifesto Market Andel*. Retrieved from Arch Daily. https://www.archdaily.com/985267/manifesto-marke-andel-chybik-plus-kristof

Nitya Maniktala, T. J. (2021). State of street vendors in India: Pre and post COVID-19 analysis. *International Journal of Policy Sciences and Law, 1*(2).

No-Mad. (2022, September 22). *Dadar flower market*. Retrieved from No-Mad. https://www.no-mad.in/dadar-flower-market/?v=c86ee0d9d7ed

Pandey, R. U., Garg, Y. K., & Bharat, A. (2013). Understanding qualitative conceptions of livability: An Indian perspective. *International Journal of Research in Engineering and Technology,* 374–380.

Saha, D. (2010). Decent work for the street vendors in Mumbai, India—A distant vision! *Journal of Workplace Rights, 14*(2), 229–250.

Shah, A. (2019). *Organizing Chaos—Finding the 'Invisible Infrastructure' for Urban Street Vendors in India.* Thesis Report, University of Mumbai.

Sharma, S. (2007). *A matter of understanding: Urban design strategies to integrate street vendors in Mumbai.* Massachusetts Institute of Technology.

Singh, H. (2016). Increasing rural to urban migration in India: A challenge or an opportunity. *International Journal of Applied Research,* 447–450.

Thangluah, L. (2022). *Livability of informal street vendors in Aizawl City*. Mizoram University.

The Gazette of India. (2014). *The street vendors (Protection of livelihood and regulation of street vending) act, 2014.* Ministry of Law and Justice.

Widjajanti, R., & Wahyono, H. (2018). Space livability of street vendors in simpang lima public space, Semarang. In *IOP Conference Series: Earth and Environmental Science* (pp. 1–9). IOP.

Restorative Streets for Healthy Cities: A Critical Review Dissenting the Conventional Narrative of Restorative Environments

Amany Ramadan Arisha

Abstract

Cities are becoming the primary ecosystem for humankind's life. At the same time, the contemporary urban lifestyle can undermine individuals' mental health and psychological well-being. Consequently, establishing a healthy urban environment promoting restorative experiences is increasingly appreciated in urbanized cities. However, research on restorative environments has mainly focused on natural environments, which are now severely challenged by urban developments associated with the forces of urbanization trends. A limited and biased body of scholarly research has specifically considered alternative urban options that can promote restorative processes in everyday urban life. Unfortunately, the public health measures related to the Covid-19 pandemic have also challenged and restricted access to these restorative opportunities. This critical review explores the need to apply the concept of restorative urbanism to the utmost utilized urban spaces, the streets, particularly in future public health crises. A key finding and recommendation from this research are acknowledging street restorativeness as an environmental quality that must be fulfilled to achieve sustainable urban and health developments.

Keywords

Restorative urbanism · Restorative environments · Urban streets · Healthy cities · Psychological well-being · Everyday spaces

1 Introduction

Within the foreseeable future, demographic and economic trends on a global scale are propelling cities to be the principal habitat of a vast share of the worldwide population (UN DESA, 2019a). In 2018, cities accounted for an estimated 55.3% of the world's population; this figure is slated to rise to approximately two-thirds of the global population by 2030 (UN DESA, 2022). Consequently, the number and scale of urbanized cities will keep increasing, particularly in the global South (UN DESA, 2019b).

Urban life has a considerable appeal for offering variations and alternatives that no other social arrangement permits (Milgram, 1970). Cities provide individuals with various career and entertainment prospects, accessible and affordable educational and healthcare facilities, diverse housing options, and opportunities for interacting with different kinds of people in vibrant public places (APA, 2005). Unfortunately, preliminary evidence has suggested that the positive excitement and attraction of urban life can become urban toxicity simultaneously (Gruebner et al., 2017; Peen et al., 2010). Socioeconomic disparities, inconvenient housing, poor basic infrastructure, inadequate transport facilities, traffic congestion, and pollution, as well as the increment of environmental stressors, can riddle cities with undesirable outcomes that negatively influence the quality of urban life (Krefis et al., 2018; Neiva et al., 2020). As an outcome, the socio-spatial changes accompanying urbanization can impinge on individuals' health and behavior (Wirth, 1938).

The World Health Organization's Constitution, which came into force on April 7, 1948, defines health as "*an integrated condition of complete physical and mental prosperity, as well as social well-being, as opposed to merely the absence of disease or disability*" (WHO, 2006, p. 01). In this context, health is the absence of diseases and impairments or the capacity to meet the demands of daily life while maintaining a mental equilibrium within one's self

A. R. Arisha (✉)
Department of Architectural Engineering and Environmental Design, Arab Academy for Science, Technology and Maritime Transport, College of Engineering and Technology, Smart Village Campus, Giza, Egypt
e-mail: amanyramadan_ar@yahoo.com

I. Pigliautile et al. (eds.), *Urban Resilience, Livability, and Climate Adaptation*,
Advances in Science, Technology & Innovation, https://doi.org/10.1007/978-3-031-54911-3_14

and the external socio-physical environment (Sartorius, 2006). It is not exclusively a matter of the individual's health status but rather an interaction between the individual, the social structure, and the conditions of contemporary urban life (Lindström & Eriksson, 2005). Urbanization is currently regarded as one of the most influential global trends of the twenty-first century that significantly influences public health (Bai et al., 2012; Kuddus et al., 2020). Given the pace and scale of mass urbanization, it is not surprising that the unprecedented transformations, along with the degradation in the quality of urban life and the increment in the complexity of social structures, would give rise to global health risks and disparities within and across cities (Ramirez-Rubio et al., 2019).

The majority of urbanized cities are now confronting significant health hazards, including climate change, traffic injuries, and the emergence of infectious diseases (Dye, 2008; McMichael, 2000), as well as non-communicable diseases that are disproportionately prevalent in urban areas (WHO & UNDP, 2016). Non-communicable diseases (NCDs), which include obesity, physical inactivity, stress, and poor mental health, constitute a public health threat and have substantial economic repercussions for urbanized cities (WHO & UN-Habitat, 2016). It is worth mentioning that most well-known international and national initiatives relating health outcomes to urban life focus on promoting physical health by limiting prolonged sedentary behaviors (e.g., WHO, 2010b, 2016b, 2017, 2018a). Promoting physical well-being through urban design and planning policies is fundamental and laudable (Iravani & Rao, 2020); however, it is insufficient because mental and physical well-being are intrinsically interlinked (Kolappa et al., 2013). The World Health Organization considers the mental aspect of the health definition crucial; that is to say, "*there can be no true physical health without mental health*" (WHO, 1954, p. 03).

The concept of mental health (MH) is holistic and multidimensional and is linked to individuals' optimal experience and functioning; it is conceptualized as "*a dynamic state of well-being where individuals realize their potentials, adapt to regular life pressures, work effectively and fruitfully, and contribute to the community*" (WHO, 2004, p. 12). Flourishing mental health contributes to life satisfaction, the achievement of meaningful goals, and social wellbeing (Seligman, 2011). It does not necessitate the absence of mental disorders (such as anxiety, dementia, depression, and schizophrenia) but incorporates psychological wellbeing (Keyes, 2002). In another way, any individual with a mental illness can benefit from experiencing elevated levels of psychological well-being (WHO & Calouste Gulbenkian Foundation, 2014). In this context, psychological wellbeing (PW) is defined as "*the combination of feeling well and functioning productively*" (Huppert, 2009, p. 01), encompassing multiple affective, cognitive, and functional components (Ryan & Deci, 2001). As a consequence, it can enhance individuals' pleasure, coping abilities, interactions, educational and occupational accomplishments, as well as their housing and economic potential (Galderisi et al., 2015). It can also improve the overall physical health and social well-being of individuals, develop social capital and cohesion, and reduce the costs of healthcare and social care (McCay et al., 2017). Conceived in this way, thriving psychological well-being can be viewed as an individual and collective resource that can enhance the quality of life (WHO, 2004).

Establishing a healthy urban environment conducive to mental health and psychological well-being has recently become increasingly prominent in urbanized cities (Bornioli & Subiza-Pérez, 2022; Weber & Trojan, 2018). Nevertheless, numerous research studies have focused on the aspects of the urban environment that can be considered risk factors for developing mental health issues, providing ample evidence of the negative associations between mental health and the degradation in the quality of urban life in urbanized cities (Ancora et al., 2022; Barros et al., 2019; Núñez-González et al., 2020; Rose, 2020). Subsequently, there is a widespread conviction that the contemporary urban lifestyle can also undermine individuals' mental health and psychological well-being (Collado et al., 2017). From this perspective, everyday urban environments are usually associated with an elevated risk of poor mental health, including mental fatigue and acute or chronic stress, which can lead to mental disorders such as anxiety and depression (Chauvet-Gélinier et al., 2013; Lederbogen et al., 2011; Penkalla & Kohler, 2014). Respectively, sustaining a healthy life in everyday urban environments (i.e., balancing all aspects of life that are social, physical, spiritual, economic, and mental) is regarded as a psychological conundrum (Ellard, 2017).

Indeed, the healthy city of the future requires recognizing the role of urban design and planning policies in promoting mental health and psychosocial well-being (Bhugra et al., 2019; Glaeser, 2011; Montgomery, 2013). The everyday urban environment, consequently, can be of the utmost significance for reframing the mental health agenda, which is relevant to resilient and sustainable cities regardless of socioeconomic status (Patel et al., 2018; Pelgrims et al., 2021). It is, therefore, crucial to focus on the factors of the everyday urban environment that can enhance the quality of life for urban individuals by emphasizing individual and collective resources (Chen et al., 2021; Mouratidis, 2021; Pelgrims et al., 2021; Ventriglio et al., 2021). Accordingly, restorative environments are considered valuable assets and one of the public health resources that can promote and maintain mental health and psychological well-being in urban areas (APA, 2005). In some urbanized cities, establishing an everyday urban environment that promotes restorative experiences is increasingly appreciated (Bornioli & Subiza-Pérez, 2022; Weber & Trojan, 2018).

The above overview elucidates the emerging need for more robust research on the positive associations between mental health, psychological well-being, and the quality of urban life in urbanized cities. It is, therefore, crucial to focus on promising directions in urban design and planning policies that can positively associate mental health outcomes with urban life by promoting psychological well-being. The author of the current work argues that implementing and integrating the principles of restorative urbanism in everyday urban environments can contribute to breaking new ground in the intricate relationship between mental health, psychological well-being, and contemporary urban lifestyles. The following chapter has a threefold aim. Firstly, to present an overview of the literature in the field of restorative environments; secondly, to highlight the rationale behind the emerging need for restorative experiences in urban streets by casting light on the challenges that might face conventional restorative environments in urbanized cities. Thirdly, to pinpoint the principles of restorative urbanism that can be implemented and integrated at the scale of urban streets to foster mental health and psychological well-being.

2 Towards an Urban Environment Good for Mental Health (MH) and Psychological Well-Being (PW)

Ironically, the challenging living conditions that may predispose an individual to develop mental health issues are also the advantages that attract individuals to the contemporary urban lifestyle in pursuit of better prospects (D'Acci, 2021). Antecedent risk factors (e.g., poverty, unemployment, homelessness, previous trauma, personal crises, addiction, and immigration); social factors (e.g., feelings of injustice, experiences of discrimination, and crime victimization); and environmental factors (i.e., overload of stimuli or erosion of protective factors) can be affiliated with mental health issues (UD/MH, 2017). Improving socioeconomic conditions and promoting healthier lifestyles have sparked a growing interest in preventing mental health concerns (McEwen, 1998; Sluiter, 2000). Given the potential for competing interests (e.g., working more for higher income versus leisure activities, mental health expenditure versus taxation), it remains ambiguous how feasible such prevention efforts will be during the current turbulence in some urbanized cities. In the global South's low- and middle-income countries, for instance, the current environmental, social, political, and economic turbulence, as well as the inherited cultural, religious, and social norms, are associated with an increase in the prevalence of mental illness and the burden of poor mental health (Okkels et al., 2017). Specifically, there is a substantial treatment deficiency in these nations, meaning that the majority of individuals with mental illness lack access to proper medication and frequently encounter discrimination, segregation, as well as infringements of human rights (Kohn et al., 2004).

Unfortunately, failure to address the burden of mental health issues collectively can hinder the growth of individuals and the development of urban societies (Diener, 2000). The increasing burden of mental health issues among affluent and poorer countries is, directly and indirectly, responsible for enormous expenditures on healthcare systems and the global economy (Martin, 2012; WHO, 2001; World Mental Health Survey Consortium, 2004). By 2030, it is anticipated that the global economic cost of mental health issues will reach \$16 trillion (Patel et al., 2018). Given that one out of every four urban individuals will encounter a mental health issue at some point in their lives (WHO, 2001), this cost may be underestimated (Vigo et al., 2016). Given the scale of the global population, it would be challenging to provide access to adequate mental health facilities for all urban individuals (Patel et al., 2018), particularly in low- and middle-income nations (Vardoulakis & Kinney, 2019). Access to and the adequacy of mental health facilities differ according to various barriers and facilitators, such as the socioeconomic status of the population and the urban lifestyle, which may also result in inequalities within and across cities (Butsch et al., 2014).

Ultimately, the health of urban individuals is the most valuable asset for any city (WHO, 2019), which is fundamental for fostering decent living standards, constructing efficient workforces, boosting social interactions, creating resilient and vibrant societies, and giving a voice to those with vulnerabilities (WHO, 2016a). Correspondingly, urbanized cities should be acknowledged as the 'melting vessels' of evolving public health ideas and practices (WHO, 2010a). The principal dimensions of the New Urban Agenda (NUA) integrate health into a vision for a more favorable and sustainable future (UN-Habitat, 2017). On that account, guaranteeing the health of individuals should serve as an indicator for monitoring the progress of policies regarding the quality of life, in addition to being a primary driver of cities' sustainable development (UN-Habitat & WHO, 2020). The UN's 'SDG 11', which seeks to "*make cities and human settlements inclusive, safe, resilient, and sustainable*" (UN, 2016, p. 24), is presented as a prerequisite and an outcome of public health. Moreover, health and well-being are explicitly addressed in the UN's 'SDG 3', which seeks to "*ensure healthy lives and promote well-being for all people of all ages*" (UN, 2016, p. 18). Nevertheless, the 2030 Sustainable Development Goals (SDGs) related to health and cities are not explicitly linked; focusing on the synergies between health development and urban development will likely yield urban advantages (WHO & UN-Habitat, 2016), which will require an innovative and comprehensive approach to health governance.

Referring to the mental health agenda, the burden of mental illnesses is acknowledged as an impediment to sustainable development by the United Nations (UN), which also declared mental well-being a global development priority (UN, 2015). In this regard, the UN's 'SDG 3' refers to mental health three times. First, explicitly in target (3.4), to diminish premature mortality due to non-communicable diseases by one-third through prevention and treatment while simultaneously promoting mental health and well-being (UN, 2015, p. 18). Second, in target (3.5), in order to strengthen substance abuse prevention and remediation, especially for narcotic addiction and hazardous consumption of alcohol (UN, 2015, p. 18). Third, mental health is implicitly included in the target (3.8) to achieve universal health coverage for all people through financial risk protection, access to high-quality healthcare services, as well as efficient medicines and cost-effective vaccines (UN, 2015, p. 18). Having a healthy society with thriving mental health, however, has implications for various development factors, such as poverty, economic growth, equity, justice, and safe and resilient cities (Votruba et al., 2016). In that case, focusing solely on resolving mental health issues by guaranteeing universal access to healthcare facilities would be a misguided strategy that would not sustain prosperous societies (Keyes, 2007; Rutter et al., 2017). In this regard, the UN's 'SDG 11.3' emphasizes the significance of enhancing inclusive and sustainable urbanization, as well as the capacity for participatory, integrated, and sustainable planning and management of human settlements, in order to establish healthy cities (UN, 2015, p. 21).

In framing any potential sustainable development for healthy cities, a holistic understanding of the quality of the everyday urban environment and its influence on individuals' mental health and psychological well-being is essential (Panagopoulos et al., 2015). As a response, some high-income countries in the global North, such as Australia, England, New Zealand, and Scotland, are shifting away from measuring the progress of their policies in terms of wealth and economic growth (GDP) to figuring the Gross National Well-being/Wellness (GNW) or Gross National Happiness (GNH) (Anderson & Mossialos, 2019). Conversely, mental health issues are still hardly recognized as a concern in the global South's low- and middle-income nations (WHO & Calouste Gulbenkian Foundation, 2014). To overcome this underestimation and neglect, mental well-being should be acknowledged as a cross-sectorial issue that is critically influenced by multiple aspects of urban design and planning policies. Subsequently, there is an emerging global necessity to prioritize incorporating new directions in urban design and planning policies that can ameliorate the adverse aspects of contemporary urban life, enhance individuals' mental health and psychological well-being, and tackle mental health inequalities within and across urbanized cities (APA, 2005).

This chapter presents two distinct directions within the broader field of environmental psychology. The first direction is the salutogenic field, a new way of considering everyday urban environments regarding mental health and psychological well-being (Karaca, 2018). It is an asset-oriented approach that emphasizes establishing environments that promote positive health outcomes rather than preventing or managing disease (Antonovsky, 1979). It includes strategies that can enhance the quality of life for urban individuals by emphasizing individual and community capabilities and resources rather than risk factors (Lindström & Eriksson, 2005). The second direction is the field of restorative environments, which emphasizes creating environments that promote relaxation and recovery from stress, mental fatigue, and other negative experiences. It is concerned with resources whose availability can vary significantly over time and according to the different circumstances across everyday situations (von Lindern et al., 2022). Both fields share some fundamental concepts and objectives and can complement one another theoretically and practically.

Nonetheless, research on restorative environments can augment that on salutogenic attributes by highlighting the dynamics of resource depletion and renewal required for the maintenance and promotion of mental health and psychological well-being and by demonstrating the role of the socio-physical and cultural environments on individuals' ongoing endeavors to manage diverse resources (von Lindern et al., 2022). The socio-physical and cultural contexts that can enhance psychosocial conditions, build mental health resilience, and support healthy behaviors are essential to drawing attention to the health-improving attributes of the everyday urban environment (Chen et al., 2021; Clark et al., 2007; Groenewegen et al., 2006; Sullivan & Chang, 2011). Further, restorative environments are the foundation of restorative urbanism, a novel concept that expands on a global urban agenda that integrates mental health and equity within the broader context of urban design and planning policies (Roe & McCay, 2021). Restorative urbanism draws on the research of restorative environments and robust scientific evidence from various fields, such as the salutogenic field, to demonstrate how everyday urban environments can support human flourishing in a rapidly expanding world. Therefore, it can present possible interventions to improve individuals' mental well-being and the quality of urban life.

3 Restorative Urbanism (RU)

Restorative urbanism is a human-centered approach to a future healthy city that places mental health and psychosocial well-being, as well as quality of life, at the forefront of urban design and planning policies (Roe & McCay, 2021). This approach prioritizes the creation of urban environments that promote physical and mental health by taking into account the socioeconomic and environmental factors that affect urban health and well-being. Therefore, it can function within a broader system that overlaps with other significant urban frameworks, such as regenerative and resilient cities. On that account, the restorative city framework integrates the concept of an inclusive city with six additional typologies: the green city, the blue city, the sensory city, the neighborly city, the active city, and the playable city (Roe & McCay, 2021), as illustrated in Fig. 1. These typologies can be interrelated and applied to various types of development projects in urbanized cities for a healthier urban future by focusing on particular settings (e.g., open spaces, urban parks, squares, and urban streets) or target populations (e.g., disabled, elderly, women, and children).

In the contemporary urban environment filled with cognitive, emotional, and practical demands that can deplete people's psychological and social resources, leveraging the everyday urban environment to enhance individuals' mental health and psychological well-being resilience is crucial. The author of the current work argues that implementing and integrating the principles of restorative urbanism in everyday urban environments can contribute to breaking new ground in the intricate relationship between mental health, psychological well-being, and contemporary urban lifestyles. However, there are various ways to approach restorative urbanism. The framework presented here comprises seven key pillars that should be incorporated into any everyday urban environment's urban design and planning policy. In this regard, general design principles for each pillar are outlined, as illustrated in Fig. 2.

4 Methods and Materials

This chapter is intended to be a critical illustrative review, not a comprehensive one, in the field of restorative environments. This type of literature review synthesizes and evaluates existing research from multiple sources on a specific topic, identifies research gaps in the literature, and highlights areas for future research and practice. It can also bridge the divide between research and practice and inform decision-making in the real world.

Initially, an illustrative literature review was conducted by searching Google Scholar (GS), a scientific database frequently employed across different disciplines with extensive

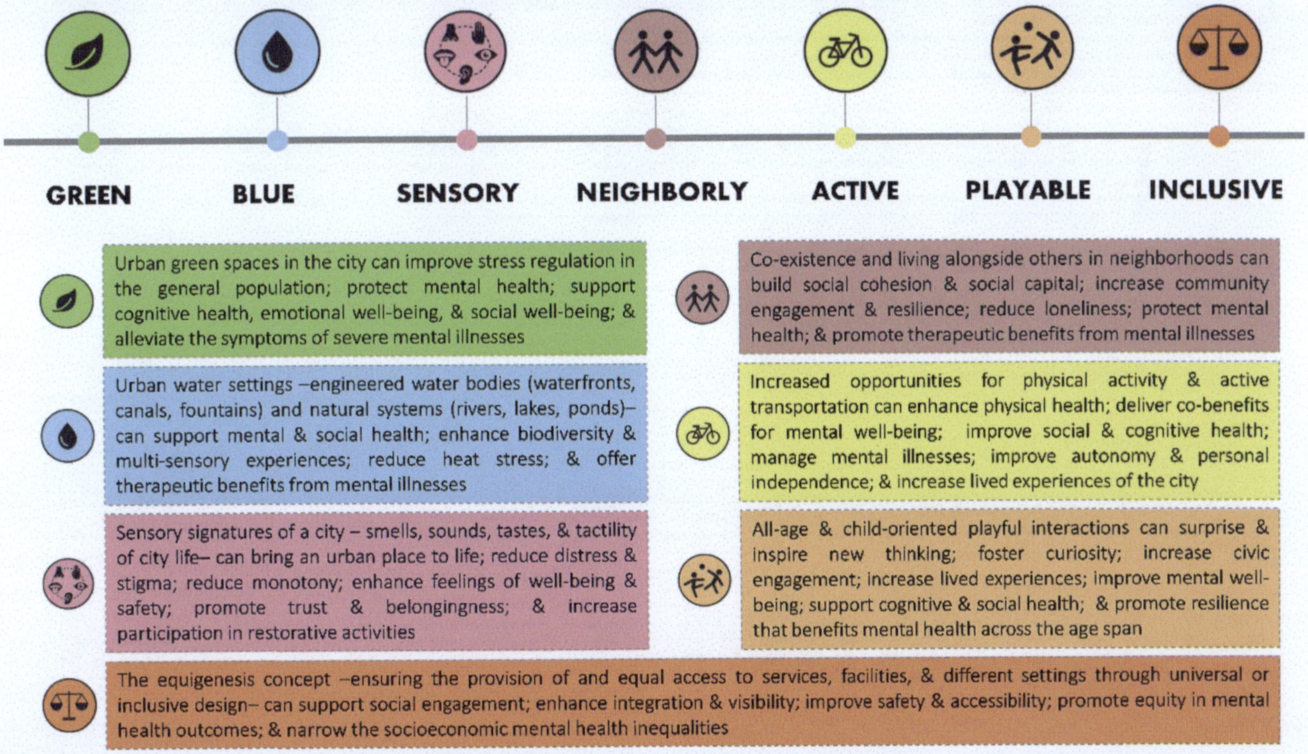

Fig. 1 The restorative city framework was adapted from Roe and McCay (2021). *Source* The Author

General Design Principles of Restorative Urbanism

GREEN

- Consult with local communities on the type of, sufficient amount of, perceptions of the quality of nearby green space
- Maximize year-round usage of nearby green space across cities, including both direct & indirect exposures
- Maximize the amount of time individuals spend in green space
- Maximize green space in close proximity to all homes
- Maximize physical & perceived & visual accessibility to green space
- Maximize investment for children, youth, & older people
- Maximize investment in biodiversity and ecological diversity

BLUE

- Maintain both residential visibility & public access to waterfronts & other aquatic settings, particularly within easy reach of home
- Provide children with opportunities to experience safe water play within easy reach of home
- Allow opportunities to walk/cycle/socialize alongside waterways & engage in water sports activities
- Consider hybrid parks/water features supporting sports/social activities in dry weather and capture water to prevent flooding
- Control rents in areas undergoing regenerative blue transformation
- Social & health impact analysis, combined with economic analysis, to ameliorate concerns about equitable access to blue space & identify health gains & losses under such scenarios

SENSORY

Seize opportunities to include salutogenic sensory design with evidence of promoting mental health and psychosocial well-being

- Use nature-based 'tranquil' sounds to create 'sonic refuges'; and use gentle sounds around residential contexts to instill belongingness
- Position public spaces to take advantage of smells appropriate to the location that are clean, natural, fresh, and familiar, particularly pleasant food smells
- Use symmetry, scale, proportion, order, unity, balance, rhythm, contrast, harmony, and color to increase aesthetic pleasantness, way-finding, local character
- Use fine-grain shop-fronts, natural features, buildings with a visually clear purpose, and other structures of visual interest, including public art, to increase diversity, reduce monotony and encourage fascination in the landscape; avoid featureless expanses
- Use LED lighting at night to increase individuals' use of urban spaces and to improve safety
- Use textural variation, including nature-based textures, underfoot and in structures' composition to encourage healthy use of urban spaces
- Ensure that all urban spaces have convenient access to nutritious foods with a diversity of stores, outdoor markets, and food trucks catering to different tastes, as well as access to urban farming and allotments
- Design markets to incorporate bumping places and opportunities for social interaction
- Provide urban spaces for food festivals and cultural celebrations that involve food
- Apply the principles of sensory garden design more widely to increase multisensory engagement and immersion in urban parks and other public spaces

Reduce exposure to unpleasant sensations by decreasing production and transmission

- Reduce individual's exposure to unpleasant sensory output, particularly near residential contexts, public spaces, and pedestrian/cycling routes
- Reduce traffic density by investing in pedestrian, cycling infrastructure, and green public transport to reduce unpleasant sights, sounds, smells, and atmosphere associated with traffic
- Reduce traffic noise by reducing traffic locations, speed, density, and noise transmission (e.g., speed bumps/platforms, gateway infrastructure for low-speed zones, and natural traffic-calming features such as humpback bridges and winding roads) and by using quiet-pavement road-surfacing technology
- Mask traffic noise with barrier walls, berms, trees, sound-masking facades, and water
- Focus on reducing rather than masking unpleasant smells through industrial odor-management processes, zoning, and separation-distance buffer approaches

NEIGHBORLY

- Public participation should be part of urban design and planning policies to ensure that the design is human-scaled and human-centered
- Ensure good stewardship, maintenance, and accessibility to attract all individuals to linger and use urban spaces, especially individuals with different abilities and needs
- Provide adequate, available housing alternatives that are easily accessible to wider social infrastructure and diverse amenities, such as recreational, educational, and economic
- Use design to demarcate public and private space for privacy, safety, and positive social interaction and engagement
- Co-locate public social spaces such as schools and religious buildings to give purpose to the space, harness regular use and familiarity, and enhance social vitality
- Invest in third spaces to create bumping places that are welcoming to all, require no money to be spent, and facilitate lingering, meeting, and encountering one another in attractive, safe, and informal settings; allow for flexibility in the use of urban spaces, such as temporary pedestrianization
- Provide fine-grain, permeable facades such as storefronts that are lively and engaging; and pet and animal amenities and infrastructure that promote social interaction
- Incorporate participatory physical features suitable for all ages, including flexible settings for cooperative community activities such as benches for chat and seated views, facilities for exercise and games, and places that can host celebratory events and various types of volunteering
- Integrate social infrastructure and diverse amenities, including playful and recreational opportunities that inspire communal and multi-generational interactions
- Avoid hostile architecture and rules that exclude specific individuals; and prioritize inclusive facilities (e.g., safe shelters that meet homeless people's safety and hygiene needs)

ACTIVE

- Provide multiple modes of transportation and be creative in the design of details to produce a distinctive sense of place; improve place keeping to minimize crime/vandalism
- Provide information about active transit routes (e.g., green bus and walking maps); encourage active transportation incentives
- Subsidize public transit for vulnerable groups that is easily accessible, regular, and integrated; provide public transportation that is convenient, affordable, frequent, safe, comfortable, and equitably distributed
- Provide pedestrianized zones around key 'attractors' and daily destinations; a diverse mix of facilities within a twenty-minute walk of where people live; mixed-income and multi-unit housing around activity centers with shops, services, and transport hubs
- Design walkable and bikeable cities that include pedestrianized areas, well-maintained pavements, traffic-calming measures, and shared transport services such as good lighting, signposting, and street furniture
- Ensure that frontages are open to the street, accessible for all, and welcoming, particularly for public facilities
- Design connectors (e.g., bridges and public stairwells) as 'events' to encourage use; reclaim underutilized infrastructure (e.g., elevated highways, disused lines, and railways)
- Make navigation intuitive (e.g., by distinctive landmarks and nodes); limit garage walls facing streets; reorganize parking lots to provide parklets and prioritize bike parking and provide protected bike storage; reclaim underutilized infrastructure as elevated highways, disused railways lines, disused industrial buildings; use vacant lots for pop-ups, parklets, community gardens, urban farming
- Foster pedestrian connectivity between facilities via short block lengths with safe and frequent intersections; and provide cut-throughs in cul-de-sacs and direct routes
- Use tactical urbanism to test out various design scenarios on health behaviors prior to making longer-term structural changes

PLAYABLE

- The playable principle should be approached through an all-age lens and facilitate both play and playfulness
- Provide 'on-the-doorstep' designated play facilities adjacent to individuals' homes; provide open play and recreational facilities to the wider population
- Install designed outdoor play facilities for all ages; develop intergenerational play facilities in key city locations
- Maximize informal nature play opportunities by leveraging pockets of green space, meadows, and vacant plots
- Design contemplative spaces for 'daydreaming', particularly near water and in other natural settings; design labyrinths, mazes, and murals to engage curiosity
- Implement 'play streets' (temporary closure of a street to cars), offering play right outside a child's home
- Provide pop-up spaces for self-expression by artists/entrepreneurs; open play and recreational facilities as school grounds to the wider population and expand operation hours
- Maximize flexible use of structures that support activities like parkour, skateboarding, and free running
- Explore urban digital gaming opportunities to link play to engagement with street artefacts/historic city landmarks/points of interest connected to a digital game (e.g., the artefacts are 'tagged' for points)

INCLUSIVE

- Engage with the full target community, involving diverse current and future users of the place, throughout all stages of inclusive planning, design, implementation, and maintenance
- Mandate inclusive design and development by involving diverse groups at all stages and by recognizing, engaging, and empowering resident leadership
- Ensure connectivity for all neighborhoods to high-quality housing, education, employment, local businesses, transport, culture, healthcare, and restorative features
- Plan mixed-use neighborhoods that enable walkability between homes, shops, and local facilities; provide facilities and commercial offers attractive to a diverse population
- Provide accessible pedestrian infrastructure and full access to public transportation for people with different physical abilities and mobility aids
- Design potential subdivisions of urban public spaces to encourage use by different groups and to avoid one group's taking over the whole space, creating opportunities for people of diverse and shared characteristics to get together
- Enhance way-finding safety by reducing the risk of confusion; increase biking safety infrastructure; provide walking trails that connect different neighborhoods
- Reduce crime vulnerability by ensuring natural surveillance with bustling streets, good lighting, and good maintenance; avoid opaque boundaries; ensure that public spaces have visual connections with the outside area
- Avoid stigmatizing different groups through the use of 'poor door' and other methods of socioeconomic segregation; encourage a welcoming attitude toward all
- Celebrate and reflect diversity throughout the city, including prominent and positive representations of the diverse population (e.g., through public art, places names, dedicated facilities, places for festivals and parades, and facilities to meet the needs of diverse groups)

Fig. 2 General design principles of restorative urbanism outlined from Roe and McCay (2021). *Source* The Author

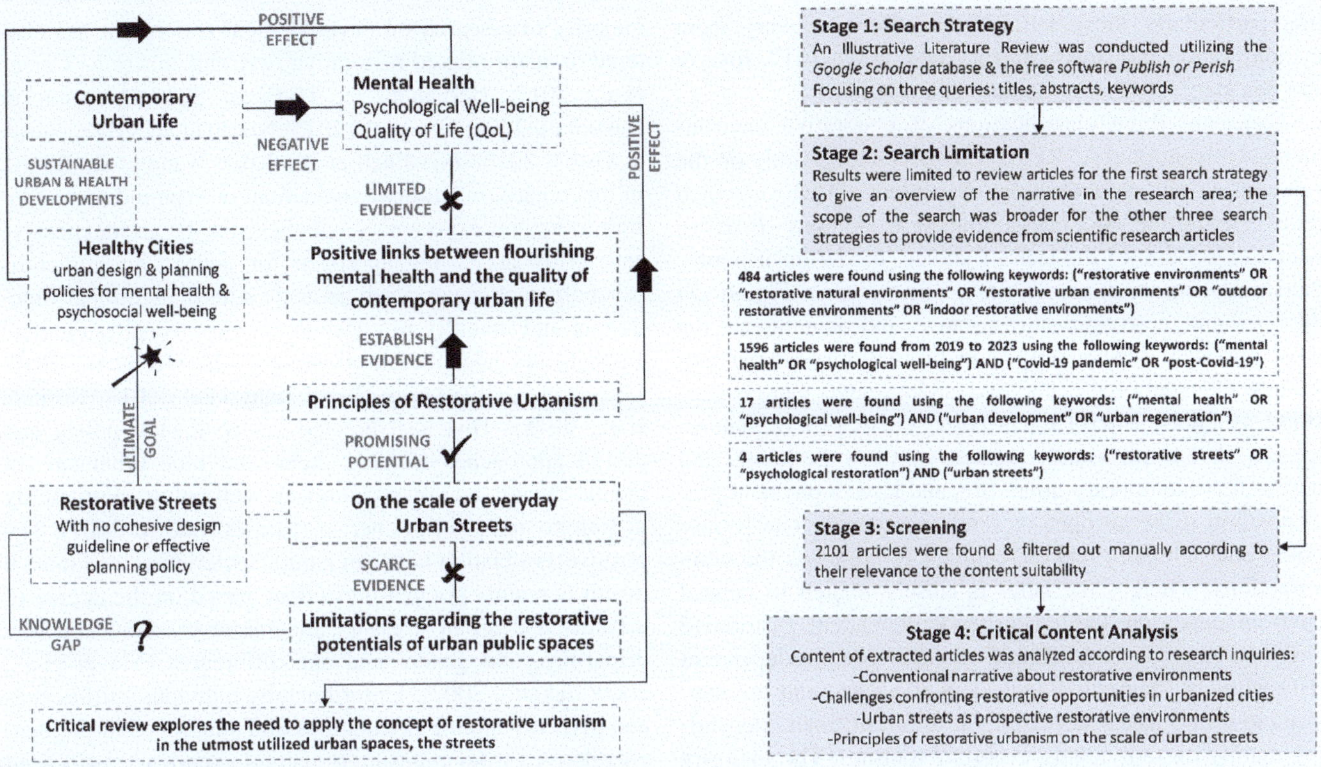

Fig. 3 The rationale behind the research argument and the methodology applied. *Source* The Author

source coverage and citation-tracking capability (Martín-Martín et al., 2018, 2021), utilizing the open-source software *Publish or Perish* (Harzing, 2007). It was a four-step search strategy focusing on three queries: titles, keywords, and abstracts, as illustrated in Fig. 3. Results were limited to review articles for the first search strategy to give an overview of the narrative in the research area; however, the search scope was broader for the other three search strategies to provide evidence from scientific research articles. The selection criteria included the most relevant articles based on content suitability through screening documents' titles and abstracts. The inclusion and exclusion of articles did not follow any specific protocol. Afterwards, a qualitative critical analysis method was performed to analyze the content of the extracted articles according to the research inquiries.

It is essential to highlight that this critical illustrative review emphasizes the conceptual contribution of each included article rather than any formal quality assessment. The strength of this review is that it aggregates the available literature in the field of restorative environments; however, the interpretative process is entirely subjective, and the resulting information is merely a starting point for further evaluation and investigation. Unfortunately, this review may be subjected to bias because the author has an interest in a particular perspective, namely restorative urban streets. It may also be limited by the quality and scope of existing research in this field.

5 Findings

The content of the 106 extracted articles was analyzed according to the four research questions that guided this chapter: What is the conventional narrative about restorative environments? What are the current challenges that confront restorative opportunities in urbanized cities? Why shift the narrative towards urban streets as prospective restorative environments? What are the principles of restorative urbanism that can be implemented and integrated on the level of urban streets?

5.1 Conventional Narrative About Restorative Environments

Since the 1980s, research on restorative environments has had formative roots in environmental psychology; however, the significance of this research has been exponentially appreciated outside of this field recently (Hartig, 2011; Staats, 2012). The restoration process is defined as "*renewing, recovering, or re-establishing physiological, psychological, and social resources or capacities that have become diminished or depleted in meeting the ongoing demands of everyday life*" (Hartig, 2004, p. 273). Accordingly, a restorative environment (RE) is any setting that can permit restoration from the demands of everyday life, particularly those settings that allow recovery from mental fatigue and stress (Hartig, 2004, 2007, 2017; Joye & van den Berg, 2019; von Lindern et al., 2017).

There are three classifications of restorative environments: conventional (C-REn), which depends mainly on the value of visual attractiveness; dependent (D-REn), which relies on routine and interaction; and enclosed (E-REn), which exists due to the lack of choice, rendering it unpredictable (Sibarani et al., 2023), as illustrated in Fig. 4. On the one hand, conventional restorative environments are the most prevalent type identified in the majority of existing studies. Conventional restorative environments can be classified as either entertainment-based (EB-CEn) or nature-based (NB-E), are usually located far from the hustle and bustle of contemporary urban life, and have a physical predisposition to be an open or semi-open space. The former offers entertainment and recreational activities as the main attractions, whereas the latter is closely related to natural environments. On the other hand, dependent restorative environments can be classified as either routine-dependent (RD-REn) or interaction-dependent (ID-REn) and are usually close and easily accessible at different times, depending mainly on individuals' everyday routines. The majority of routine-dependent restorative environments are indoor urban environments, such as workplaces, educational facilities, shopping malls, houses of worship, and traditional markets. Interaction-dependent restorative environments, such as cafes and restaurants, can be identified as indoor or outdoor settings for social interactions or reflection. The enclosed restorative environment is the last type, which was predominantly observed during the Covid-19 pandemic and represented individuals' safe zones for recreational or social activities.

Natural environments and flourishing mental health have been intrinsically associated for centuries; thus, the majority of research on psychological restoration has centered on nature-based restorative environments (e.g., Carrus et al., 2017; Hernández & Hidalgo, 2005; Ivarsson & Hagerhall, 2008; Lee et al., 2015; Nordh et al., 2009; Nordh & Østby, 2013; Tyrväinen et al., 2014; Wang et al., 2016). In this regard, intentional, incidental, or even indirect exposure to natural experiences is believed to be an inexpensive and easily accessible approach for addressing numerous psychological issues, such as acute stress, attentional functioning and inability to concentrate, elevated physiological arousal, and negative emotions (e.g., Bowler et al., 2010; Cole et al., 2017; Gillis, 2020; Jiang et al., 2014; Korpela et al., 2015). This has directed environmental policy and design approaches towards green and blue urbanism for mental health and psychological well-being, particularly in Europe and North America (Roe & McCay, 2021). The positive association between natural experiences and mental health is mainly context-dependent, based on the degree of urbanity and urban stressors, income level, socioeconomic disparities, and gender and age differences (Mayer et al., 2009; Takano, 2002). Unfortunately, individual differences and diversity among populations and subgroups in the benefits derived from exposure to natural experiences still need to be explored (Hartig et al., 2014). Nevertheless, the conventional narrative has been criticized for oversimplifying the relationship between the surrounding environment and individuals' psychological well-being and overlooking the restorative potential of non-natural environments, such as urban public spaces.

The latest research refutes the misconception that urban contexts are portrayed as less restorative than natural contexts (Joye & Dewitte, 2018; Stigsdotter et al., 2017; von Lindern et al., 2017, 2022). Recent studies on psychological restoration have simultaneously emphasized entertainment-based and dependency-based environments. Supporting

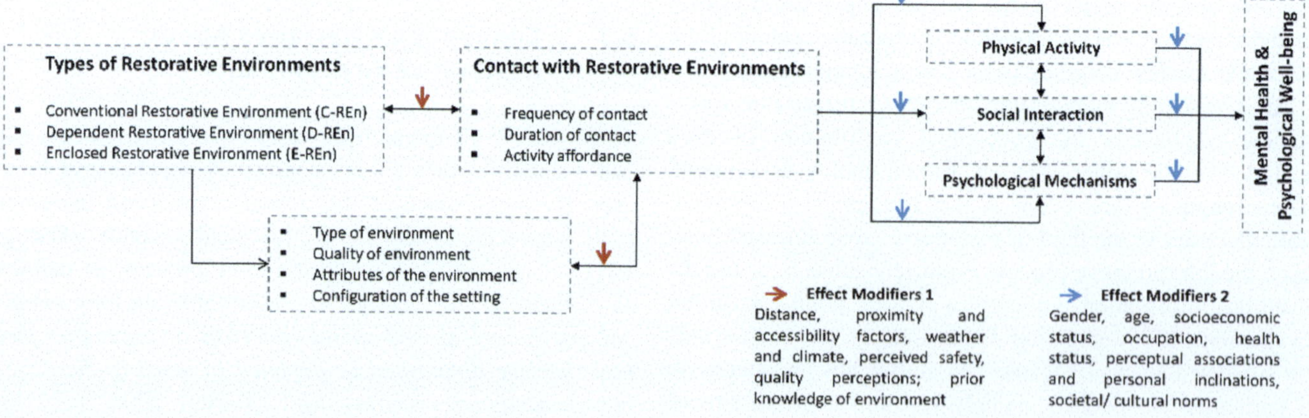

Fig. 4 Types of restorative environments. *Source* The Author

this, numerous studies have examined the perceived restorative potential of diverse urban contexts, concentrating primarily on indoor settings (e.g., Galindo & Hidalgo, 2005; Hartig et al., 1996, 1997; Hidalgo et al., 2006; Hipp et al., 2016; Luís et al., 2020; Nordh et al., 2017; Ouellette et al., 2005; Packer & Bond, 2010; Staats et al., 2010). These studies revealed that some urban environments, such as residential contexts, work and educational settings, care settings, and other leisure contexts, might have a restorativeness value comparable to or greater than that of natural environments (e.g., Herzog et al., 2003, 2009; Kang & Kim, 2019; Nasar & Terzano, 2010; Staats, 2012; Staats et al., 2016). Unfortunately, a limited and biased body of scholarly research has explicitly examined urban options for outdoor psychological restoration (Karmanov & Hamel, 2008; Polajnar Horvat & Ribeiro, 2023; San Juan et al., 2017; Stigsdotter et al., 2017).

This misconception can be related to the fact that research on restorative environments exhibits various conceptual knowledge gaps and methodological limitations regarding urban-based restorative environments, which inhibit the exploration of the restorative potential of the full spectrum of urban public spaces (Bornioli & Subiza-Pérez, 2022; Hartig, 2021; Weber & Trojan, 2018). There is still a need for additional research on the complete spectrum of urban public spaces and their optimal design. There is also concern that the vast majority of evidence regarding psychological restoration is derived from Western societies, namely European and US-American citizens (Weber & Trojan, 2018). In contrast, little attention has been paid to populations in other areas of the world, such as Asia and Africa (Sibarani et al., 2023). There is a need for more research on the impact of restorative environments on specific populations, such as low-income communities and marginalized groups. Furthermore, most of the research on restorative environments has focused on discrete restoration at the individual level (Hartig et al., 2013), whereas leveraging everyday outdoor urban environments to be effective restorative environments requires a focus on cumulative restoration at the collective level. There is also a need for more research on the long-term impact of restorative experiences on mental health and psychological well-being. Subsequently, psychological restoration in urban outdoor environments is a promising discipline in which numerous intriguing phenomena remain to be uncovered.

5.2 Challenges Confronting Restorative Opportunities in Urbanized Cities

In urbanized cities, urbanization, resource exploitation, and contemporary urban lifestyles have undermined individuals' mental well-being (Hartig et al., 2014). Concerns exist that urban developments associated with urbanization trends may not address aspects of the urban experience necessary to accomplish the desired mental health development (Thwaites et al., 2011; Walzer, 1986). Unfortunately, the ongoing urbanization is trading off the presence of natural world experiences for housing, commercial, transportation, and other infrastructure services, reducing the availability of natural contexts (Hartig & Kahn, 2016; Soga & Gaston, 2016). Moreover, the majority of urbanized cities have limited financial resources for the creation and maintenance of natural environments (Subiza-Pérez et al., 2019). In other words, implementing green infrastructure (GI) and nature-based solutions (NBS) in urbanized cities cannot be the only way forward for healthy cities due to the decreasing quantity of available free space (Cox et al., 2018) or other practical considerations. Thus, the first challenge confronting restorative opportunities in urbanized cities is that urban development processes may reduce the availability and accessibility of restorative natural contexts (Hartig, 2017; van den Berg et al., 2007).

The second challenge is the existing inequalities in individuals' exposure to and access to natural experiences that influence mental well-being (Jimenez et al., 2021). Numerous research studies have demonstrated that marginalized populations have less access to urban nature and greenspaces (Astell-Burt et al., 2014; Dahmann et al., 2010). Nonetheless, existing evidence suggests that exposure to nature avails marginalized populations disproportionately, a phenomenon known as the 'equigenic' effect of green urban space, which contradicts the predictable correlation between lower socioeconomic status and an increased hazard of adverse health outcomes (Mitchell et al., 2015). In other words, those with a lower socioeconomic status experienced more significant positive effects than those with a higher socioeconomic status (Rigolon et al., 2021). Conversely, the provision of natural environments in marginalized urban environments may contribute to 'green gentrification' (Gould & Lewis, 2017), a process characterized by changes in socioeconomic demographics due to the influx of new wealthier inhabitants, as well as inflation in real estate prices and consumption patterns (Anguelovski, 2016; Cole et al., 2017; Pearsall, 2010). This process paradoxically increases chronic stress by increasing the local expense of living, altering the social environment, transforming the local culture, and forcibly displacing long-term inhabitants, as well as diminishing the positive psychological effects of living in proximity to green space, especially for vulnerable populations (Cole et al., 2017; Rigolon et al., 2018). Unfortunately, in this scenario, restorative natural experiences become only accessible to urban inhabitants when their economic and social mobility permits (Thwaites et al., 2011). If providing natural environments for urban individuals seeking psychological recovery is no longer feasible or limited by practical considerations, it is crucial to consider other urban possibilities for restorative experiences at this time.

The third challenge is the implications of accelerated urban development on individuals' health and behavior. Some urban developments are associated with a professional culture focused on 'quick fix' development solutions (Jacobs & Appleyard, 1987), which can be susceptible to undermining the psychosocial potentials of the urban experience (Thwaites et al., 2011). There are numerous facets to these concerns, including a tendency to cater to a relatively narrow socioeconomic demographic, which consequently restricts possibilities for community development; international architectural styles that undermine local identity; as well as a preference for short-term commercial benefits and aesthetic appeal over long-term sustainable urban development (Thwaites et al., 2007, 2011). In this context, "*the vast majority of people live in cities where events take place without their knowledge or participation*" (Jacobs & Appleyard, 1987, p. 115). Rapid urban development accompanied by investments in automotive infrastructure, for instance, can create a culture of vehicular dependency, which may contribute to the planning and designing of urban streets with a fundamental function of accommodating excessive traffic volume (Kaw et al., 2020). In that way, the urban development solution exposed pedestrians to an increased chance of traffic accidents; marginalized them into narrow sidewalks; and promoted an inactive urban lifestyle that contributes to the risk of cardiovascular disease (Newman & Kenworthy, 1999). Therefore, the extent to which urbanized cities will continue to be sustainable as they grow will depend on how these cities manage the ramifications of the trade-offs of their development on the urban lifestyle and the health of individuals (WHO, 2018b).

The fourth challenge is environmental degradation and the quality of urban life. In this regard, restorative experiences can be impacted by environmental degradation, such as pollution, deforestation, and climate change (Gurevich, 2020). Moreover, the quality of the environment can reduce its restorative potential (Gillis, 2020). Unfortunately, most urbanized cities, especially in the global South's low- and middle-income countries, may not prioritize access to high-quality restorative environments in everyday urban life due to a lack of awareness of their importance (Arisha & El-Moneim, 2019; Mandeli, 2019). This ignorance reduces the demand for restorative environments and the resources allocated to their creation and maintenance. Consequently, the need for substantial funding may be the fifth obstacle confronting urbanized cities with restorative environments.

The sixth and final challenge is public health crises, such as the emergence of the global pandemic caused by the Covid-19 outbreak, which has adversely influenced all aspects of urban societies (Singhal, 2020; Wu et al., 2020). In addition to physical illness, the context of the Covid-19 pandemic has had a prevalent psychological impact on the majority of the world's population (Sibarani et al., 2023). The public health catastrophe induced by the pandemic is regarded as a worldwide-shattering crisis that ruptured contemporary urban lifestyles in a singular and unprecedented scenario (Bil et al., 2021). In this regard, urban individuals were confronted with a radical mass transformation of the space-time of their daily lives, in which public and private spaces and social and occupational roles converged. Simultaneously, the essential objective of the majority of public health policies implemented was to protect individuals and communities from the virus's rapid spread by altering the public realm to enable the mandated protective measures (social distancing, quarantine, and self-isolation) (McCay, 2020). This alone is detrimental to mental health, psychological well-being, social relations with communities, and interaction with the surrounding environment (Campion et al., 2020; Moreno et al., 2020).

Furthermore, the pandemic was characterized by varying degrees of uncertainty and fear of losing loved ones, livelihoods, security, and life plans, combined with the social exclusion of the prolonged quarantine and the restricted interaction with the outside world, which can be conceptualized as a psychological burden. Unfortunately, the number of individuals affected by psychological disorders was significantly greater than the number of individuals infected with the virus (Benke et al., 2020; Luo et al., 2020). It is now safe to assume that the primary source of the high prevalence of psychological disorders was not only the pandemic itself but also the indirect consequences of government policies and containment measures (e.g., Benke et al., 2020; Brooks et al., 2020; Facciolà et al., 2021; Lahav, 2020; Xiong et al., 2020; Ye et al., 2020). Providing access to restorative environments is one of the measures that can be taken to lessen this psychological burden. With increasing work-from-home practices and concerns about the compatibility of social gatherings, however, some cultures, ethnicities, religions, and other groups were impacted more than others by public health measures and policy decisions. This alone has revealed the inequities in individuals' access to the various restorative environments. In the liminal space that arose during the Covid-19 restrictions, the demand for equitable access to high-quality public spaces within the urban realm was elevated (Craig et al., 2022; Gillis, 2020).

Paradoxically, all the mentioned challenges that can be utilized to dissenting the conventional narrative of restorative environments are also the catalysts that can be utilized to shift the narrative towards urban streets as prospective restorative environments. In this regard, closing or restricting access to commercial, employment, entertainment, cultural, sporting, and academic facilities has curtailed

Fig. 5 A restorative urban street offers access to the seafront & supports mental health in Alexandria, Egypt. *Source* The Author

opportunities to engage in activities that enhance individuals' psychological well-being (Craig et al., 2022; Portegijs et al., 2021). Although the majority of Western and Asian cultures were more comfortable with their enclosed restorative environments within their homes or around the residential contexts, or within semi-private spaces, such as balconies and front porches (Sibarani et al., 2023), other cultures in the African continent have acquired new restorative habits that gave life back to urban streets (Ghanem et al., 2021). For instance, the pandemic outbreak has altered the street life of Cairo and increased the livability of urban public spaces, where the street was the only gateway for Egyptians' restorative experiences, regardless of the lockdown or the poor physical characteristics of the streets (Ghanem et al., 2021). It is, therefore, crucial to consider diverse restoration possibilities for different cultures in everyday outdoor urban environments, particularly in future public health crises (Polajnar Horvat & Ribeiro, 2023). Overall, addressing these six challenges will require a multi-faceted approach that addresses systematic inequalities, promotes environmental sustainability, raises awareness about the restorative potential of environments, and secures the necessary funding to create and maintain restorative experiences in everyday urban environments.

5.3 Urban Streets as Prospective Restorative Environments

Urban streets are the fundamental portion of urban spaces that take on multiple roles in urban life (Jacobs, 1961), with two primary simultaneous functions: place and mobility (von Schönfeld & Bertolini, 2017). Unlike any other urban public space, streets determine the spatial and visual characteristics of cities (Carmona et al., 2003). Urban streets are multi-dimensional and dynamic urban spaces that can influence individuals' experiences of everyday urban environments based on their function and significance in cities (Arisha & El-Moneim, 2019; Garvin, 2016). In tandem with increasing urbanization and decreasing quantities of natural environments, urban streets can be a practical solution for enhancing the population's mental health and psychological well-being, as shown in Fig. 5. Following this, urban streets should be acknowledged as potentially valuable resources for psychological restoration (Zhao et al., 2020).

The restorative potential of urban streets, as the most utilized public spaces in urban life, has recently attracted academic attention (Yin et al., 2022a, 2022b). The majority of these studies, however, were primarily concerned with street vegetation (e.g., Lin et al., 2014; Lindal & Hartig, 2015). In this regard, the presence of natural elements, their arrangement, and the extent of street vegetation can influence the restorative potential of streets (e.g., Hernández & Hidalgo, 2005; Todorova et al., 2004; Weber et al., 2008). A limited number of research studies have focused on architectural elements, such as building height and its variation on both sides of the street (Lindal & Hartig, 2013); traffic and historical elements (Bornioli et al., 2018); city busyness and street aesthetics (Bornioli et al., 2019). Even though the concept of restorative urbanism can be an excellent candidate for evaluating the psychological effects of urban streets, unfortunately, the restorative potential of urban streets has not been adequately investigated within the field of restorative environments. There is an absence of sufficient evidence regarding the associations between flourishing mental well-being and urban streets, which hinders the development of a cohesive design guideline or an effective planning policy for restorative streets.

Streets are intricate urban contexts with diverse functions; hence, the capacity of urban streets to deliver restorative experiences as one of their common functional attributes may be limited. Moreover, the restorative potential of urban streets may fluctuate according to their function and categorization, as well as their tangible and intangible characteristics. Therefore, additional research is required on the restorative values of different typologies of urban streets, such as commercial streets, residential streets, and transportation corridors. There is also a requirement for more research on the restorative impact of different modes of transportation, such as walking, biking, and public transportation. Because high levels of noise pollution frequently characterize urban streets, there is a need for more research on the role of noise pollution in the restorative potential of urban streets. Moreover, research on restorative environments has mainly focused on physical and natural features; there is a need for more attention to social factors that may contribute to the restorative potential of urban streets, such as social support and community engagement. Finally, additional research on the economic advantages of psychological restoration in the context of urban streets is also required.

5.4 Principles of Restorative Urbanism on the Level of Urban Street

Based on the general design principles of restorative urbanism outlined previously, this chapter presents several strategies and recommendations that can be interrelated and applied on the urban street scale in urbanized cities for a healthier urban future, as illustrated in Fig. 6. These strategies can improve individuals' mental health, psychological well-being, and the quality of urban life. From this viewpoint, the strategies presented here can be considered powerful tools or recommendations for future urban developments, particularly in the event of future public health crises. For instance, incorporating greenery to reduce air pollution, provide shade, and promote relaxation and stress reduction; reducing noise pollution through installing noise barriers, using low-noise pavements, and reducing traffic volume for a more restful environment; and providing opportunities for physical activity, including bike lanes, pedestrian crossings, and outdoor exercise equipment to promote physical health. Moreover, incorporating public art and cultural elements to promote social interaction and community engagement and provide a sense of identity; promoting safety through strategies such as improved lighting, traffic-calming measures, and crime prevention strategies to reduce stress and promote a sense of security; and incorporating water features such as fountains or reflecting pools to promote relaxation and stress reduction. Finally, and most simply, providing seating areas such as benches or outdoor cafes to provide opportunities for social interaction and relaxation.

Overall, designing urban streets to be more restorative requires a holistic approach that considers the physical, social, and cultural aspects of the urban environment. In addition, the seven main pillars of restorative urbanism should not be independent but interconnected, with evident crossovers and synergies. Therefore, further robust scientific research is still required to evaluate the impacts of these strategies on individuals' mental health and psychological well-being in the real world and to investigate their implementation in different urban streets. Respectively, future studies need to explore the restorative potential of the different typologies of urban streets in urbanized cities.

6 Discussion

Undoubtedly, urbanization is a sophisticated irreversible phenomenon that transforms the built environment and the spatial distribution of populations and alters societies' demographic and social structure by altering the dominant occupations, lifestyles, cultures, and behaviors (Montgomery et al., 2003). Thus, it presents a tremendous opportunity and poses various economic, environmental, social, and spatial challenges for sustainable urban development in cities (UN-Habitat, 2020). On the one hand, urbanized cities should prioritize "*development that meets current needs without compromising future generations' ability to meet their own demands*" (WCED, 1987). On the other hand, unplanned and unmanaged forms of extensive urbanization, as well as unsustainable production and consumption patterns, add immense pressure on strained land and natural resources (Kaw et al., 2020). Considering that the manner in which cities are planned, designed, developed, and managed influences urban lifestyles and individuals' health (Rydin et al., 2012). The rapid transformations associated with urbanization trends can influence and structure many of the critical determinants of the health of urban individuals, in addition to influencing policy development (Barton & Grant, 2013; Macintyre et al., 2002; Rice, 2020).

Throughout history, public health concerns have been the crucial impetus for urban design and planning policies, with the ultimate goal of reducing the harmful effects of rapid industrialization, notably the urban outbreak of infectious diseases (Duhl & Sanchez, 1999; Melosi, 2000; Porter, 1999). Following this, the WHO Healthy Cities Program has promoted a comprehensive policy and a systematic

Fig. 6 Strategies that can be applied on the level of urban streets based on restorative urbanism principles. *Source* The Author

planning approach for public health (WHO, 1996). In addition, the twenty-first-century Health for All (HFA) policy acknowledged health as an objective of urban and economic development (WHO, 1999). This approach, Healthy Urban Planning (HUP), has achieved significant strides to mitigate the negative health ramifications of urbanization through actions based on community participation, empowerment, and institution building (WHO, 1999), which can contribute to enhancing the quality of urban life. Yet, this approach should have been more focused on mental health outcomes (WHO, 2004).

Tailoring the discussion to the mental health benefits of restorative experiences, restorative urbanism and Sustainable Development Goals (SDGs) are closely related. Restorative urbanism can assist in achieving several SDGs related to urban and health developments in contemporary urban life. By addressing the UN's 'SDG 3', restorative urbanism can promote physical and mental health by fostering environments that encourage physical activity, reduce stress and mental fatigue, and promote social interaction and community engagement. By addressing the UN's 'SDG 11', restorative urbanism can support the development of sustainable and livable cities by fostering green infrastructure, reducing the effect of urban heat islands, and improving public transportation access. By addressing the UN's 'SDG 13', restorative urbanism can also contribute to climate action by promoting green infrastructure, reducing greenhouse gas emission levels, and enhancing energy efficiency. Lastly, by addressing the UN's 'SDG 15', restorative urbanism can support the conservation of biodiversity and natural ecosystems by encouraging the establishment of green experiences and corridors that connect urban contexts to natural contexts.

Considering that inequities are vividly reflected in measures of health determinants (Montgomery et al., 2003), policy and decision-making processes accord them a lower priority (Rice, 2020). Inequity in health can impede the capacity of society to flourish, with affluent individuals at the top obstructing the potential of poorer individuals. This can be exacerbated on a global scale by racial segregation, poor quality of housing and education, inadequate and inefficient basic infrastructure, and poor welfare. In this context, tackling health inequities and their socioeconomic determinants is crucial to achieving the synergy between health development and urban development in urbanized cities (WHO & UN-Habitat, 2016). On that account, the Urban Health Index (UHI) acknowledged public health and health equity as central components of city planning and governance beyond the provision of healthcare services (UN-Habitat & WHO, 2020; WHO, 2016a). Fortunately, health equity is a critical aspect of restorative urbanism, as it recognizes that not all urban societies have equal access to the advantages of contemporary urban life. Therefore, it acknowledges the significance of addressing social and environmental justice from the perspective of urban design and planning policies. Creating more sustainable, livable, and equitable cities that promote mental health and psychological well-being for all individuals, regardless of socioeconomic class or background, is feasible by prioritizing restorative urbanism in urban design and planning policies.

With a new global conversation on public health and the role of public spaces within the urban realm, upon which the Covid-19 pandemic has shed even more light, there is an opportunity to capitalize on more comprehensive urban global strategies for mental health and psychological well-being. The Covid-19 experience has led urbanized cities to realize that their resilience, including economic resilience, depends on their adaptability to use their assets best, particularly their urban streets and outdoor public spaces (Roe & McCay, 2021). Consequently, awareness of the role of everyday environments in promoting and enhancing individuals' mental health and psychological well-being has increased, particularly during the pandemic. Considering the challenges confronting conventional and dependent restorative environments in urbanized cities, street restorativeness as an environmental quality can be a step towards achieving sustainable urban and health developments.

Restorative streets prioritize the needs of urban individuals to create more livable and equitable communities. In this regard, implementing and integrating the principles of restorative urbanism in everyday urban streets can contribute to breaking new ground in the intricate relationship between mental health, psychological well-being, and contemporary urban lifestyles. It can also be used to address mental health inequities in urbanized cities. There are several ways in which restorative streets can contribute to achieving the SDGs related to urban and health developments in contemporary urban life. By addressing the UN's 'SDG 3', restorative streets can provide safe and accessible spaces for physical activity and active transportation, thereby improving the overall health and well-being of individuals. By addressing the UN's 'SDG 10', restorative streets can promote more significant equity by creating more inclusive and accessible public spaces for the benefit of all community members, regardless of income, age, or ability. By addressing the UN's 'SDG 11', restorative streets can contribute to the creation of livable and sustainable communities by reducing vehicle emissions, encouraging active transportation, and enhancing public transit. By addressing the UN's 'SDG 12', restorative streets can contribute to responsible consumption and production by promoting environmentally friendly transportation modes. By addressing the UN's 'SDG 16', restorative streets can promote peace and justice by establishing more inclusive and accessible public spaces that encourage social cohesion and community engagement. In summary, restorative streets can play a critical role in achieving the SDGs by promoting mental health and psychological well-being, reducing inequalities, and establishing more sustainable urbanized cities.

7 Conclusion

This chapter elucidates the emerging need for more robust research on the positive associations between mental health, psychological well-being, and the quality of urban life in urbanized cities, especially in the global South's low- and middle-income nations. It also illuminates the need for more promising directions in urban design and planning policies that can ameliorate the adverse aspects of contemporary urban life, enhance the mental health and psychological well-being of individuals, and tackle mental health inequalities within and across urbanized cities. Restorative urbanism as a human-centered approach is an excellent candidate for future healthy cities that prioritize mental health, psychosocial well-being, and quality of life in urban design and planning policies. Further, the everyday urban environment can be of the utmost importance in redefining the mental health agenda.

On that account, restorative urbanism relies on the research of restorative environments for scientific evidence demonstrating how everyday urban environments can support human flourishing in a rapidly expanding world. Unfortunately, there is still a dearth of research on restorative urban environments, and as a result, it has been of little practical use. Considering the challenges confronting restorative experiences in urbanized cities, urban streets can be a practical solution for enhancing the population's mental and psychological well-being. However, in the realm of restorative environments, the restorative potential of urban streets has not been adequately investigated. This critical review explored the emerging need to apply the concept of restorative urbanism to the utmost utilized urban spaces, the streets, particularly in future public health crises.

In an increasingly polarized and unequal world, restorative streets can help reconcile the divide and promote equity in mental health outcomes. Even though implementing and integrating the principles of restorative urbanism on the scale of urban streets can contribute to breaking new ground in the intricate relationship between mental health, psychological well-being, and contemporary urban lifestyles, it may prove difficult, considering the overall paucity of evidence. Unfortunately, the absence of sufficiently established evidence pertaining to the correlations between flourishing mental well-being and urban streets has hindered the development of a cohesive design guideline or an effective planning policy for restorative streets. Thus, further robust scientific research is still required to explore the full range of restorative urban streets across the age spectrum, diverse socioeconomic groups, and a wide range of nationalities. Yet, a key finding and recommendation of this research are acknowledging street restorativeness as an environmental quality that must be fulfilled to achieve sustainable urban and health developments.

References

Ancora, L. A., Blanco-Mora, D. A., Alves, I., Bonifácio, A., Morgado, P., & Miranda, B. (2022). Cities and neuroscience research: A systematic literature review. *Frontiers Psychiatry, 13*(983352). https://doi.org/10.3389/fpsyt.2022.983352

Anderson, M., & Mossialos, E. (2019). Beyond gross domestic product for New Zealand's wellbeing budget. *Lancet Public Health, 4*(7), e320–e321. https://doi.org/10.1016/S2468-2667(19)30109-4

Anguelovski, I. (2016). From toxic sites to parks as (Green) LULUs? New challenges of inequity, privilege, gentrification, and exclusion for urban environmental justice. *Journal of Planning Literature, 31*(1), 23–36. https://doi.org/10.1177/0885412215610491

Antonovsky, A. (1979). *Health, stress, and coping*. Jossey-Bass.

APA. (2005). *Toward an urban psychology: Research, action, and policy. Report of the task force on urban psychology*. American Psychological Association. Retrieved January 14, 2023, from https://www.apa.org/pi/ses/resources/publications/urban-taskforce.pdf

Arisha, A. R., & El-Moneim, N. M. A. (2019). Space syntax beyond Cairo street markets: An application of a theory on socio-spatial contemporary practices. In: H. Abusaada, C. Vellguth, & A. Elshater (Eds.), *Handbook of research on digital research methods and architectural tools in urban planning and design* (pp. 250–281). IGI Global. https://doi.org/10.4018/978-1-5225-9238-9.CH012

Astell-Burt, T., Feng, X., Mavoa, S., Badland, H. M., & Giles-Corti, B. (2014). Do low-income neighbourhoods have the least green space? A cross-sectional study of Australia's most populous cities. *BMC Public Health, 14*(292). https://doi.org/10.1186/1471-2458-14-292

Bai, X., Nath, I., Capon, A., Hasan, N., & Jaron, D. (2012). Health and wellbeing in the changing urban environment: complex challenges, scientific responses, and the way forward. *Current Opinion in Environmental Sustainability, 4*(4), 465–472. https://doi.org/10.1016/j.cosust.2012.09.009

Barros, P., Ng Fat, L., Garcia, L. M. T., Slovic, A. D., Thomopoulos, N., de Sá, T. H., Moraisg, P., & Mindell, J. S. (2019). Social consequences and mental health outcomes of living in high-rise residential buildings and the influence of planning, urban design and architectural decisions: A systematic review. *Cities, 93*, 263–272. https://doi.org/10.1016/J.CITIES.2019.05.015

Barton, H., & Grant, M. (2013). Urban planning for healthy cities: A review of the progress of the European healthy cities programme. *Journal of Urban Health, 90*(S1), 129–141. https://doi.org/10.1007/s11524-011-9649-3

Benke, C., Autenrieth, L. K., Asselmann, E., & Pané-Farré, C. A. (2020). Lockdown, quarantine measures, and social distancing: Associations with depression, anxiety and distress at the beginning of the COVID-19 pandemic among adults from Germany. *Psychiatry Research, 113462*. https://doi.org/10.1016/j.psychres.2020.113462

Bhugra, D., Ventriglio, A., Castaldelli-Maia, J., & McCay, L. (Eds.). (2019). *Urban mental health. Part of the series Oxford cultural psychiatry*. Oxford University Press.

Bil, J. S., Buława, B., & Świerzawski, J. (2021). Mental health and the city in the post-COVID-19 era. *Sustainability, 13*(14), 7533. https://doi.org/10.3390/su13147533

Bornioli, A., & Subiza-Pérez, M. (2022). Restorative urban environments for healthy cities: A theoretical model for the study of restorative experiences in urban built settings. *Landscape Research*, 1–12. https://doi.org/10.1080/01426397.2022.2124962

Bornioli, A., Parkhurst, G., & Morgan, P. L. (2018). Psychological wellbeing benefits of simulated exposure to five urban settings: An experimental study from the pedestrian's perspective. *Journal of Transport & Health, 9*, 105–116. https://doi.org/10.1016/j.jth.2018.02.003

Bornioli, A., Parkhurst, G., & Morgan, P. L. (2019). Affective experiences of built environments and the promotion of urban walking. *Transportation Research Part A: Policy and Practice, 123*, 200–215. https://doi.org/10.1016/j.tra.2018.12.006

Bowler, D. E., Buyung-Ali, L. M., Knight, T. M., & Pullin, A. S. (2010). A systematic review of evidence for the added benefits to health of exposure to natural environments. *BMC Public Health, 10*(1), 456. https://doi.org/10.1186/1471-2458-10-456

Brooks, S. K., Webster, R. K., Smith, L. E., Woodland, L., Wessely, S., Greenberg, N., & Rubin, G. J. (2020). The psychological impact of quarantine and how to reduce it: Rapid review of the evidence. *The Lancet*. https://doi.org/10.1016/S0140-6736(20)30460-8

Butsch, C., Kroll, M., & Bork-Hueffer, T. (2014). Megaurban health in developing and newly industrializing countries. In: A. C. Michalos (Ed.), *Encyclopedia of quality of life and well-being research* (pp. 3981–3985). Springer, Dordrecht. https://doi.org/10.1007/978-94-007-0753-5_4060

Campion, J., Javed, A., Sartorius, N., & Marmot, M. (2020). Addressing the public mental health challenge of COVID-19. *The Lancet Psychiatry, 7*(8), 657–659. https://doi.org/10.1016/S2215-0366(20)30240-6

Carmona, M., Heath, T., Oc, T., & Tiesdell, S. (2003). *Public places, urban spaces*. Architectural Press.

Carrus, G., Scopelliti, M., Panno, A., Lafortezza, R., Colangelo, G., Pirchio, S., Ferrini, F., Salbitano, F., Agrimi, M., Portoghesi, L., Semenzato, P., & Sanesi, G. (2017). A different way to stay in touch with "Urban Nature": The perceived restorative qualities of botanical gardens. *Frontiers in Psychology, 8*. https://doi.org/10.3389/fpsyg.2017.00914

Chauvet-Gélinier, J. C., Trojak, B., Vergès-Patois, B., Cottin, Y., & Bonin, B. (2013). Review on depression and coronary heart disease. *Archives of Cardiovascular Diseases, 106*(2), 103–110. https://doi.org/10.1016/j.acvd.2012.12.004

Chen, K., Zhang, T., Liu, F., Zhang, Y., & Song, Y. (2021). How does urban green space impact residents' mental health: A literature review of mediators. *International Journal of Environmental Research and Public Health, 18*(22), 11746. https://doi.org/10.3390/ijerph182211746

Clark, C., Myron, R., Stansfeld, S., & Candy, B. (2007). A systematic review of the evidence on the effect of the built and physical environment on mental health. *Journal of Public Mental Health, 6*(2), 14–27. https://doi.org/10.1108/17465729200700011

Cole, H. V. S., Garcia Lamarca, M., Connolly, J. J. T., & Anguelovski, I. (2017). Are green cities healthy and equitable? Unpacking the relationship between health, green space and gentrification. *Journal of Epidemiology and Community Health, 71*, 1118–1121. https://doi.org/10.1136/jech-2017-209201

Collado, S., Staats, H., Corraliza, J. A., & Hartig, T. (2017). Restorative environments and health. In: G. Fleury-Bahi, E. Pol, & O. Navarro (Eds.), *Handbook of environmental psychology and quality of life research* (pp. 127–148). Part of the book series *International handbooks of quality-of-life* (IHQL). Springer International Publishing. https://doi.org/10.1007/978-3-319-31416-7_7

Cox, D. T. C., Shanahan, D. F., Hudson, H. L., Fuller, R. A., & Gaston, K. J. (2018). The impact of urbanization on nature dose and the implications for human health. *Landscape and Urban Planning, 179*, 72–80. https://doi.org/10.1016/j.landurbplan.2018.07.013

Craig, C. M., Neilson, B. N., Altman, G. C., Travis, A. T., & Vance, J. A. (2022). Applying restorative environments in the home office while sheltering-in-place. *Human Factors: The Journal of the Human Factors and Ergonomics Society, 64*(8), 1351–1362. https://doi.org/10.1177/0018720820984286

D'Acci, L. S. (2021). Preferring or needing cities? (Evolutionary) psychology, utility and life satisfaction of urban living. *City, Culture and Society, 24*, 100375. https://doi.org/10.1016/j.ccs.2021.100375

Dahmann, N., Wolch, J., Joassart-Marcelli, P., Reynolds, K., & Jerrett, M. (2010). The active city? Disparities in provision of urban public recreation resources. *Health & Place, 16*, 431–445. https://doi.org/10.1016/j.healthplace.2009.11.005

Diener, E. (2000). Subjective well-being: The science of happiness and a proposal for a national index. *American Psychologist, 55*(1), 34–43. https://doi.org/10.1037/0003-066X.55.1.34

Duhl, L. J., & Sanchez, A. K. (1999). *Healthy cities and the city planning process: A background document on links between health and urban planning*. WHO Regional Office for Europe. Retrieved January 28, 2023, from https://apps.who.int/iris/handle/10665/108252

Dye, C. (2008). Health and urban living. *Science, 319*(5864), 766–769. https://doi.org/10.1126/science.1150198

Ellard, C. (2017). A new agenda for urban psychology: Out of the laboratory and onto the streets. *Journal of Urban Design and Mental Health*, 2(3). Retrieved January 28, 2023, from https://www.urbandesignmentalhealth.com/journal2-ellard.html

Facciolà, A., Laganà, P., & Caruso, G. (2021). The COVID-19 Pandemic and its implications on the environment. *Environmental Research, 201*, 111648. https://doi.org/10.1016/j.envres.2021.111648

Galderisi, S., Heinz, A., Kastrup, M., Beezhold, J., & Sartorius, N. (2015). Toward a new definition of mental health. *World Psychiatry, 14*(2), 231–233. https://doi.org/10.1002/wps.20231

Galindo, M. P., & Hidalgo, M. C. (2005). Aesthetic preferences and the attribution of meaning: Environmental categorization processes in the evaluation of urban scenes. *International Journal of Psychology, 40*(1), 19–27. https://doi.org/10.1080/00207590444000104

Garvin, A. (2016). *What Makes a Great City?* Island Press.

Ghanem, S., Ahmad, A. F., & Aboualy, S. (2021). COVID-19 bringing Cairenes back to their streets. *Journal of Urban Management*. https://doi.org/10.1016/j.jum.2021.06.001

Gillis, K. (2020). Nature-based restorative environments are needed now more than ever. *Cities & Health*, 1–4.10. https://doi.org/10.1080/23748834.2020.1796401

Glaeser, E. (2011). *Triumph of the city: How our greatest invention makes us richer, smarter, greener, healthier, and happier*. Penguin Group.

Gould, K. A., & Lewis, T. L. (2017). *Green gentrification*. Routledge.

Groenewegen, P. P., van den Berg, A. E., de Vries, S., & Verheij, R. A. (2006). Vitamin G: Effects of green space on health, well-being, and social safety. *BMC Public Health, 6*(1), 149. https://doi.org/10.1186/1471-2458-6-149

Gruebner, O., Rapp, M. A., Adli, M., Kluge, U., Galea, S., & Heinz, A. (2017). Cities and mental health. *Deutsches Aerzteblatt International Online., 114*(8), 121–127. https://doi.org/10.3238/arztebl.2017.0121

Gurevich, R. (2020). Restorative commons as an expanded ethical framework for public health and environmental sustainability. *The New Bioethics: A Multidisciplinary Journal of Biotechnology and the Body*, 1–16. https://doi.org/10.1080/20502877.2020.1767917

Hartig, T. (2004). Restorative environments. In: C. D. Spielberger (Ed.), *Encyclopedia of applied psychology* (Vol. 3, pp. 273–279). Academic Press. https://doi.org/10.1016/B0-12-657410-3/00821-7

Hartig, T. (2007). Three steps to understanding restorative environments as health resources. In C. W. Thompson & P. Travlou (Eds.), *Open space: People space* (1st ed., pp. 163–179). Taylor & Francis.

Hartig, T. (2011). Issues in restorative environments research: matters of measurement. In *Proceedings of the 11th Conference on Environmental Psychology 2011: Between Urban Studies and the Analysis of Sustainability* (pp. 41–66). University of Almería & the Spanish Association of Environmental Psychology.

Hartig, T. (2017). Restorative environments. In *Reference module in neuroscience and bio-behavioral psychology*. Elsevier. [An update of the 2004 Encyclopedia entry]. https://doi.org/10.1016/B978-0-12-809324-5.05699-6

Hartig, T. (2021). Restoration in nature: Beyond the conventional narrative. In A. R. Schutte, J. C. Torquati, & J. R. Stevens (Eds.), *Nature and psychology: Biological, cognitive, developmental, and social pathways to well-being* (pp. 89–151). Part of the book series *Nebraska Symposium on Motivation* (NSM, Vol. 67). Springer. https://doi.org/10.1007/978-3-030-69020-5_5

Hartig, T., & Kahn, P. H. (2016). Living in cities, naturally. *Science, 352*(6288), 938–940. https://doi.org/10.1126/science.aaf3759

Hartig, T., Böök, A., Garvill, J., Olsson, T., & Gärling, T. (1996). Environmental influences on psychological restoration. *Scandinavian Journal of Psychology, 37*(4), 378–393. https://doi.org/10.1111/j.1467-9450.1996.tb00670.x

Hartig, T., Korpela, K., Evans, G. W., & Gärling, T. (1997). A measure of restorative quality in environments. *Scandinavian Housing and Planning Research, 14*(4), 175–194. https://doi.org/10.1080/02815739708730435

Hartig, T., Catalano, R., Ong, M., & Syme, S. L. (2013). Vacation, collective restoration, and mental health in a population. *Society and Mental Health, 3*(3), 221–236. https://doi.org/10.1177/2156869313497718

Hartig, T., Mitchell, R., de Vries, S., & Frumkin, H. (2014). Nature and health. *Annual Review of Public Health, 35*(1), 207–228. https://doi.org/10.1146/annurev-publhealth-032013-182443

Harzing, A. W. (2007). *Publish or perish*. Retrieved February 14, 2023, from https://harzing.com/resources/publish-or-perish

Hernández, B., & Hidalgo, M. C. (2005). Effect of urban vegetation on psychological restorativeness. *Psychological Reports, 96*(3_suppl), 1025–1028. https://doi.org/10.2466/pr0.96.3c.1025-1028

Herzog, T. R., Colleen, Maguire, P., & Nebel, M. B. (2003). Assessing the restorative components of environments. *Journal of Environmental Psychology, 23*(2), 159–170. https://doi.org/10.1016/S0272-4944(02)00113-5

Herzog, T. R., Ouellette, P., Rolens, J. R., & Koenigs, A. M. (2009). Houses of worship as restorative environments. *Environment and Behavior, 42*(4), 395–419. https://doi.org/10.1177/0013916508328610

Hidalgo, M. C., Berto, R., Galindo, M. P., & Getrevi, A. (2006). Identifying attractive and unattractive urban places: Categories, restorativeness and aesthetic attributes. *Medio Ambiente y Comportamiento Humano, 7*(2), 115–133.

Hipp, J. A., Gulwadi, G. B., Alves, S., & Sequeira, S. (2016). The Relationship between perceived greenness and perceived restorativeness of university campuses and student-reported quality of life. *Environment and Behavior, 48*(10), 1292–1308. https://doi.org/10.1177/0013916515598200

Huppert, F. A. (2009). Psychological well-being: Evidence regarding its causes and consequences. *Applied Psychology: Health and Well-Being, 1*(2), 137–164. https://doi.org/10.1111/j.1758-0854.2009.01008.x

Iravani, H., & Rao, V. (2020). The effects of new urbanism on public health. *Journal of Urban Design, 25*(2), 218–235. https://doi.org/10.1080/13574809.2018.1554997

Ivarsson, C. T., & Hagerhall, C. M. (2008). The perceived restorativeness of gardens: Assessing the restorativeness of a mixed built and natural scene type. *Urban Forestry & Urban Greening, 7*(2), 107–118. https://doi.org/10.1016/j.ufug.2008.01.001

Jacobs, J. (1961). *The death and life of great American cities*. Vintage Books.

Jacobs, A., & Appleyard, D. (1987). Toward an Urban design manifesto. *Journal of the American Planning Association, 53*(1), 112–120. https://doi.org/10.1080/01944368708976642

Jiang, B., Li, D., Larsen, L., & Sullivan, W. C. (2014). A dose-response curve describing the relationship between urban tree cover density and self-reported stress recovery. *Environment and Behavior, 48*(4), 607–629. https://doi.org/10.1177/0013916514552321

Jimenez, M. P., DeVille, N. V., Elliott, E. G., Schiff, J. E., Wilt, G. E., Hart, J. E., & James, P. (2021). Associations between nature exposure and health: A review of the evidence. *International Journal of Environmental Research and Public Health, 18*(9), 4790. https://doi.org/10.3390/ijerph18094790

Joye, Y., & Dewitte, S. (2018). Nature's broken path to restoration. A critical look at attention restoration theory. *Journal of Environmental Psychology, 59*, 1–8. https://doi.org/10.1016/j.jenvp.2018.08.006

Joye, Y., & van den Berg, A. E. (2019). Restorative environments. In: L. Steg & J. I. M. de Groot (Eds.), *Environmental psychology: an introduction* (2nd ed.) (pp. 65–75). Wiley. https://doi.org/10.1002/9781119241072.ch7

Kang, Y., & Kim, E. J. (2019). Differences of restorative effects while viewing urban landscapes and green landscapes. *Sustainability, 11*(7), 2129. https://doi.org/10.3390/su11072129

Karaca, E. (2018). Salutogenic approach for designing restorative environments. *The Journal of Academic Social Sciences, 67*(67), 116–131. https://doi.org/10.16992/asos.13501

Karmanov, D., & Hamel, R. (2008). Assessing the restorative potential of contemporary urban environment(s): Beyond the nature versus urban dichotomy. *Landscape and Urban Planning, 86*(2), 115–125. https://doi.org/10.1016/j.landurbplan.2008.01.004

Kaw, J. K., Lee, H., & Wahba, S. (2020). *The hidden wealth of cities: Creating, financing, and managing public spaces*. World Bank.

Keyes, C. L. M. (2002). The mental health continuum: From languishing to flourishing in life. *Journal of Health and Social Behavior, 43*(2), 207–222. https://doi.org/10.2307/3090197

Keyes, C. L. M. (2007). Promoting and protecting mental health as flourishing: A complementary strategy for improving national mental health. *American Psychologist, 62*(2), 95–108. https://doi.org/10.1037/0003-066X.62.2.95

Kohn, R., Saxena, S., Levav, I., & Saraceno, B. (2004). The treatment gap in mental health care. *Bulletin of the World Health Organization, 82*(11), 858–866. Retrieved March 30, 2023, from https://www.ncbi.nlm.nih.gov/pmc/articles/PMC2623050/

Kolappa, K., Henderson, D. C., & Kishore, S. P. (2013). No physical health without mental health: Lessons unlearned? *Bulletin of the World Health Organization, 91*(1), 3-3A. https://doi.org/10.2471/BLT.12.115063

Korpela, K., De Bloom, J., & Kinnunen, U. (2015). From restorative environments to restoration in work. *Intelligent Buildings International, 7*(4), 215–223. https://doi.org/10.1080/17508975.2014.959461

Krefis, A. C., Augustin, M., Schlünzen, K. H., Oßenbrügge, J., & Augustin, J. (2018). How does the urban environment affect health and well-being? A Systematic Review. *Urban Science, 2*(1), 1–21. https://doi.org/10.3390/urbansci2010021

Kuddus, M. A., Tynan, E., & McBryde, E. (2020). Urbanization: A problem for the rich and the poor? *Public Health Reviews, 41*(1), 1. https://doi.org/10.1186/s40985-019-0116-0

Lahav, Y. (2020). Psychological distress related to COVID-19: The contribution of continuous traumatic stress. *Journal of Affective Disorders*. https://doi.org/10.1016/j.jad.2020.07.141

Lederbogen, F., Kirsch, P., Haddad, L., Streit, F., Tost, H., Schuch, P., Wüst, S., Pruessner, J. C., Rietschel, M., Deuschle, M., & Meyer-Lindenberg, A. (2011). City living and urban upbringing affect neural social stress processing in humans. *Nature, 474*, 498–501. https://doi.org/10.1038/nature10190

Lee, K. E., Williams, K. J. H., Sargent, L. D., Williams, N. S. G., & Johnson, K. A. (2015). 40 s Green roof views sustain attention: The role of micro-breaks in attention restoration. *Journal of Environmental Psychology, 42*, 182–189. https://doi.org/10.1016/j.jenvp.2015.04.003

Lin, Y.-H., Tsai, C.-C., Sullivan, W. C., Chang, P.-J., & Chang, C.-Y. (2014). Does awareness effect the restorative function and perception of street trees? *Frontiers in Psychology, 5*(906). https://doi.org/10.3389/fpsyg.2014.00906

Lindal, P. J., & Hartig, T. (2013). Architectural variation, building height, and the restorative quality of urban residential streetscapes. *Journal of Environmental Psychology, 33*, 26–36. https://doi.org/10.1016/j.jenvp.2012.09.003

Lindal, P. J., & Hartig, T. (2015). Effects of urban street vegetation on judgments of restoration likelihood. *Urban Forestry & Urban Greening, 14*(2), 200–209. https://doi.org/10.1016/j.ufug.2015.02.001

Lindström, B., & Eriksson, M. (2005). Salutogenesis. *Journal of Epidemiology & Community Health, 59*(6), 440–442. https://doi.org/10.1136/jech.2005.034777

Luís, S., Dias, R., & Lima, M. L. (2020). Greener schoolyards, greener futures? Greener schoolyards buffer decreased contact with nature and are linked to connectedness to nature. *Frontiers in Psychology, 11*, 567882. https://doi.org/10.3389/fpsyg.2020.567882

Luo, X., Estill, J., Wang, Q., Lv, M., Liu, Y., Liu, E., & Chen, Y. (2020). The psychological impact of quarantine on coronavirus disease 2019 (COVID-19). *Psychiatry Research*, 113193. https://doi.org/10.1016/j.psychres.2020.113193

Macintyre, S., Ellaway, A., & Cummins, S. (2002). Place effects on health: How can we conceptualize, operationalize, and measure them? *Social Science & Medicine, 55*(1), 125–139. https://doi.org/10.1016/s0277-9536(01)00214-3

Mandeli, K. (2019). Public space and the challenge of urban transformation in cities of emerging economies: Jeddah case study. *Cities, 95*, 102409. https://doi.org/10.1016/j.cities.2019.102409

Martin, S. (2012). Our health at risk. *Monitor Staff, 43*(3), 18. Retrieved January 28, 2023, from https://www.apa.org/monitor/2012/03/stress

Martín-Martín, A., Orduna-Malea, E., Thelwall, M., & Delgado López-Cózar, E. (2018). Google scholar, web of science, and scopus: A systematic comparison of citations in 252 subject categories. *Journal of Informetrics, 12*(4), 1160–1177. https://doi.org/10.1016/j.joi.2018.09.002

Martín-Martín, A., Thelwall, M., Orduna-Malea, E., & Delgado López-Cózar, E. (2021). Google scholar, Microsoft academic, Scopus, dimensions, web of science, and Opencitations' COCI: A multidisciplinary comparison of coverage via citations. *Scientometrics, 126*, 871–906. https://doi.org/10.1007/s11192-020-03690-4

Mayer, F. S., Frantz, C. M., Bruehlman-Senecal, E., & Dolliver, K. (2009). Why is nature beneficial? The role of connectedness to nature. *Environment and Behavior, 41*(5), 607–643. https://doi.org/10.1177/0013916508319745

McCay, L., Bremer, I., Endale, T., Jannati, M., & Yi, J. (2017). Urban design and mental health. In N. Okkels, C. B. Kristiansen, & P. Munk-Jørgensen (Eds.), *Mental health and illness in the city* (pp. 421–444). Springer Science + Business Media.

McCay, L. (2020). Built environment changes during the COVID-19 pandemic: A photo-essay (compilation). *Journal of Urban Design and. Mental Health, 6*(4). Retrieved January 30, 2023, from https://www.urbandesignmentalhealth.com/journal-6-covid-photoessay.html

McEwen, B. S. (1998). Protective and damaging effects of stress mediators. *New England Journal of Medicine, 338*(3), 171–179. https://doi.org/10.1056/NEJM199801153380307

McMichael, A. J. (2000). The urban environment and health in a world of increasing globalization: Issues for developing countries. *Bulletin of the World Health Organization, 78*(9), 1117–1126. Retrieved January 14, 2023, from https://apps.who.int/iris/handle/10665/268216

Melosi, M. V. (2000). *The sanitary city: Urban infrastructure in America from colonial times to the present*. Johns Hopkins University Press.

Milgram, S. (1970). The experience of living in cities: Adaptations to urban overload create characteristic qualities of city life that can be measured. *Science, 167*(3924), 1461–1468. https://doi.org/10.1126/science.167.3924.1461

Mitchell, R. J., Richardson, E. A., Shortt, N. K., & Pearce, J. R. (2015). Neighborhood environments and socioeconomic inequalities in mental well-being. *American Journal of Preventive Medicine, 49*(1), 80–84. https://doi.org/10.1016/j.amepre.2015.01.017

Montgomery, C. (2013). *Happy city: Transforming our lives through urban design*. Penguin Random House.

Montgomery, M. R., Stren, R., Cohen, B., & Reed, H. E. (Eds.). (2003). *Cities transformed: Demographic change and its implications in the developing world*. The National Academic Press.

Moreno, C., Wykes, T., Galderisi, S., Nordentoft, M., Crossley, N., Jones, N., et al. (2020). How mental health care should change as a consequence of the COVID-19 pandemic. *The Lancet Psychiatry, 7*, 813–824. https://doi.org/10.1016/S2215-0366(20)30307-2

Mouratidis, K. (2021). Urban planning and quality of life: A review of pathways linking the built environment to subjective well-being. *Cities, 115*, 103229. https://doi.org/10.1016/j.cities.2021.103229

Nasar, J. L., & Terzano, K. (2010). The desirability of views of city skylines after dark. *Journal of Environmental Psychology, 30*(2), 215–225. https://doi.org/10.1016/j.jenvp.2009.11.007

Neiva, S. S., Deggau, A. B., & de Andrade Guerra, J. B. S. O. (2020). Urbanization for the promotion of sustainable development. In W. L. Filho, A. M. Azul, L. Brandli, P. G. Özuyar, & T. Wall (Eds.), *Zero hunger*. Part of the book series *Encyclopedia of the UN sustainable development goals* (ENUNSDG) (pp. 1–12). Springer. https://doi.org/10.1007/978-3-319-69626-3_54-1

Newman, P., & Kenworthy, J. R. (1999). *Sustainability and cities: Overcoming automobile dependence*. Island Press.

Nordh, H., Hartig, T., Hagerhall, C. M., & Fry, G. (2009). Components of small urban parks that predict the possibility for restoration. *Urban Forestry & Urban Greening, 8*(4), 225–235. https://doi.org/10.1016/j.ufug.2009.06.003

Nordh, H., & Østby, K. (2013). Pocket parks for people: A study of park design and use. *Urban Forestry & Urban Greening, 12*(1), 12–17. https://doi.org/10.1016/j.ufug.2012.11.003

Nordh, H., Evensen, K. H., & Skår, M. (2017). A peaceful place in the city: A qualitative study of restorative components of the cemetery. *Landscape and Urban Planning, 167*, 108–117. https://doi.org/10.1016/j.landurbplan.2017.06.004

Núñez-González, S., Delgado-Ron, J. A., Gault, C., Lara-Vinueza, A., Calle-Celi, D., Porreca, R., & Simancas-Racines, D. (2020). Overview of "Systematic Reviews" of the built environment's effects on mental health. *Journal of Environmental and Public Health, 2020*, 1–10. https://doi.org/10.1155/2020/9523127

Okkels, N., Kristiansen, C. B., & Munk-Jørgensen, P. (Eds.). (2017). *Mental health and illness in the city*. Part of the book series *Mental health and illness worldwide* (MHIW). Springer Science + Business Media. https://doi.org/10.1007/978-981-10-2327-9

Ouellette, P., Kaplan, R., & Kaplan, S. (2005). The monastery as a restorative environment. *Journal of Environmental Psychology, 25*(2), 175–188. https://doi.org/10.1016/j.jenvp.2005.06.001

Panagopoulos, T., González Duque, J. A., & Bostenaru Dan, M. (2015). Urban planning with respect to environmental quality and human well-being. *Environmental Pollution, 208*, 137–144. https://doi.org/10.1016/j.envpol.2015.07.038

Patel, V., Saxena, S., Lund, C., Thornicroft, G., Baingana, F., Bolton, P., Chisholm, D., Collins, P. Y., Cooper, J. L., Eaton, J., & UnÜtzer, J. (2018). The lancet commission on global mental health and sustainable development. *The Lancet, 392*(10157), 1553–1598. https://doi.org/10.1016/S0140-6736(18)31612-X

Pearsall, H. (2010). From brown to green? Assessing social vulnerability to environmental gentrification in New York City. *Environment and Planning c: Government and Policy, 28*(5), 872–886. https://doi.org/10.1068/c08126

Peen, J., Schoevers, R. A., Beekman, A. T., & Dekker, J. (2010). The current status of urban-rural differences in psychiatric disorders. *Acta Psychiatrica Scandinavica, 121*(2), 84–93. https://doi.org/10.1111/j.1600-0447.2009.01438.x

Pelgrims, I., Devleesschauwer, B., Guyot, M., Keune, H., Nawrot, T. S., Remmen, R., Saenen, N. D., Trabelsi, S., Thomas, I., Aerts, R., & De Clercq, E. M. (2021). Association between urban environment and mental health in brussels, Belgium. *BMC Public Health, 635*(2021), 1–18. https://doi.org/10.1186/s12889-021-10557-7

Packer, J., & Bond, N. (2010). Museums as restorative environments. *Curator: The Museum Journal, 53*(4), 421–436. https://doi.org/10.1111/j.2151-6952.2010.00044.x

Penkalla, A. M., & Kohler, S. (2014). Urbanicity and mental health in Europe: A systematic review. *European Journal of Mental Health, 9*(2), 163–177. https://doi.org/10.5708/EJMH.9.2014.2.2

Polajnar Horvat, K., & Ribeiro, D. (2023). Urban public spaces as restorative environments: The case of Ljubljana. *International Journal of Environmental Research and Public Health, 20*(3), 2159. https://doi.org/10.3390/ijerph20032159

Portegijs, E., Keskinen, K. E., Tuomola, E.-M., Hinrichs, T., Saajanaho, M., & Rantanen, T. (2021). Older adults' activity destinations before and during COVID-19 restrictions: From a variety of activities to mostly physical exercise close to home. *Health & Place, 68*, 102533. . https://doi.org/10.1016/j.healthplace.2021.102533

Porter, D. (1999). *Health, civilization and the state: A history of public health from ancient to modern times*. Routledge.

Ramirez-Rubio, O., Daher, C., Fanjul, G., Gascon, M., Mueller, N., Pajín, L., Plasencia, A., Rojas-Rueda, D., Thondoo, M., & Nieuwenhuijsen, M. J. (2019). Urban health: An example of a "Health in all Policies": Approach in the context of SDGs implementation. *Globalization and Health, 15*(1), 87, 1–21. https://doi.org/10.1186/s12992-019-0529-z

Rice, L. (2020). After COVID-19: Urban design as spatial medicine. *Urban Design International*, 1–6. https://doi.org/10.1057/s41289-020-00142-6

Rigolon, A., Browning, M., Lee, K., & Shin, S. (2018). Access to urban green space in cities of the global South: A systematic literature review. *Urban Science, 2*(3), 67. https://doi.org/10.3390/urbansci2030067

Rigolon, A., Browning, M. H. E. M., McAnirlin, O., & Yoon, H. (Violet). (2021). Green space and health equity: A systematic review on the potential of green space to reduce health disparities. *International Journal of Environmental Research and Public Health, 18*(5), 2563. https://doi.org/10.3390/ijerph18052563

Roe, J., & McCay, L. (2021). *Restorative cities: Urban design for mental health and wellbeing*. Bloomsbury Visual Arts. https://doi.org/10.5040/9781350112919

Rose, N. (2020). Mental health, stress and the contemporary metropolis. In M. Keith, & A. A. de Souza Santos (Eds.), *Urban transformations and public health in the emergent city* (pp. 35–54). Manchester University Press. https://doi.org/10.7765/9781526150943.00007

Rutter, H., Savona, N., Glonti, K., Bibby, J., Cummins, S., Finegood, D. T., & White, M. (2017). The need for a complex systems model of evidence for public health. *The Lancet, 390*(10112), 2602–2604. https://doi.org/10.1016/S0140-6736(17)31267-9

Ryan, R. M., & Deci, E. L. (2001). On happiness and human potentials: A review of research on hedonic and eudaimonic well-being. *Annual Review of Psychology, 52*(1), 141–166. https://doi.org/10.1146/annurev.psych.52.1.141

Rydin, Y., Bleahu, A., Davies, M., Dávila, J. D., Friel, S., De Grandis, G., Groce, N., Hallal, P. C., Hamilton, I., Howden-Chapman, K. L., Lim, C. J., Martins, J., Osrin, D., Ridley, I., Scott, I., Taylor, M., Wilkinson, P., & Wilson, J. (2012). Shaping cities for health: Complexity and the planning of urban environments in the 21st century. *The Lancet, 379*(9831), 2079–2108. https://doi.org/10.1016/S0140-6736(12)60435-8

San Juan, C., Subiza-Pérez, M., & Vozmediano, L. (2017). Restoration and the city: The role of public urban squares. *Frontiers in Psychology, 8*. https://doi.org/10.3389/fpsyg.2017.02093

Sartorius, N. (2006). The meanings of health and its promotion. *Croatian Medical Journal, 47*(4), 662–664.

Seligman, M. E. P. (2011). *Flourish: A visionary new understanding of happiness and well-being*. Free Press.

Sibarani, G. A. H., Kusuma, H. E., & Riska, A. S. (2023). Three conceptual models of restorative environments based on user's restorative experience in the COVID-19 pandemic period. *Journal of Asian Architecture and Building Engineering*. https://doi.org/10.1080/13467581.2023.2172344

Singhal, T. (2020). A review of coronavirus disease-2019 (COVID-19). *The Indian Journal of Pediatrics, 87*(4), 281–286. https://doi.org/10.1007/s12098-020-03263-6

Sluiter, J. K. (2000). Reactivity and recovery from different types of work measured by catecholamines and cortisol: A systematic literature overview. *Occupational and Environmental Medicine, 57*(5), 298–315. https://doi.org/10.1136/oem.57.5.298

Soga, M., & Gaston, K. J. (2016). Extinction of experience: Evidence, consequences and challenges of the loss of human-nature interactions. *Frontiers in Ecology and the Environment, 14*(2), 94–101. https://doi.org/10.1002/fee.1225

Staats, H. (2012). Restorative environments. In S. D. Clayton (Ed.), *The Oxford handbook of environmental and conservation psychology*. Oxford University Press. https://doi.org/10.1093/oxfordhb/9780199733026.013.0024

Staats, H., Van Gemerden, E., & Hartig, T. (2010). Preference for restorative situations: Interactive effects of attentional state, activity-in-environment, and social context. *Leisure Sciences, 32*(5), 401–417. https://doi.org/10.1080/01490400.2010.510990

Staats, H., Jahncke, H., Herzog, T. R., & Hartig, T. (2016). Urban options for psychological restoration: Common strategies in everyday situations. *Plos One, 11*(1), Article e0146213. https://doi.org/10.1371/journal.pone.0146213

Stigsdotter, U. K., Corazon, S. S., Sidenius, U., Kristiansen, J., & Grahn, P. (2017). It is not all bad for the Grey City: A crossover study on physiological and psychological restoration in a forest and an urban environment. *Health & Place, 46*, 145–154. https://doi.org/10.1016/j.healthplace.2017.05.007

Subiza-Pérez, M., Hauru, K., Korpela, K., Haapala, A., & Lehvävirta, S. (2019). Perceived environmental aesthetic qualities scale (PEAQS): A self-report tool for the evaluation of green-blue spaces. *Urban Forestry & Urban Greening, 43*, 126383. https://doi.org/10.1016/j.ufug.2019.126383

Sullivan, W. C., & Chang, C. Y. (2011). Mental health and the built environment. In A. L. Dannenberg, H. Frumkin, & R. J. Jackson (Eds.), *Making healthy places: Designing and building for health, well-being, and sustainability* (pp. 106–116). Island Press.

Takano, T. (2002). Urban residential environments and senior citizens' longevity in megacity areas: The importance of walkable green spaces. *Journal of Epidemiology & Community Health, 56*(12), 913–918. https://doi.org/10.1136/jech.56.12.913

Todorova, A., Asakawa, S., & Aikoh, T. (2004). Preferences for and attitudes towards street flowers and trees in Sapporo, Japan. *Landscape and Urban Planning, 69*(4), 403–416. https://doi.org/10.1016/j.landurbplan.2003.11.001

Thwaites, K., Porta, S., Romice, O., & Greaves, M. (Eds.). (2007). *Urban sustainability through environmental design: Approaches to time-people-place responsive urban design*. Routledge.

Thwaites, K., Simkins, I., & Mathers, A. (2011). Towards socially restorative urbanism: Exploring social and spatial implications for urban restorative experience. *Landscape Review, 13*(2), 26–39.

Tyrväinen, L., Ojala, A., Korpela, K., Lanki, T., Tsunetsugu, Y., & Kagawa, T. (2014). The influence of urban green environments on stress relief measures: A field experiment. *Journal of Environmental Psychology, 38*, 1–9. https://doi.org/10.1016/j.jenvp.2013.12.005

UD/MH. (2017). How the city affects mental health. The Centre for Urban Design and Mental Health. Retrieved January 29, 2023, from https://www.urbandesignmentalhealth.com/how-the-city-affects-mental-health.html

UN. (2015). *Resolution adopted by the general assembly on 25 September 2015, transforming our world: The 2030 agenda for sustainable development* (A/RES/70/1). United Nations. Retrieved March 30, 2023, from https://www.un.org/en/development/desa/population/migration/generalassembly/docs/globalcompact/A_RES_70_1_E.pdf

UN. (2016). *Transforming our world: The 2030 agenda for sustainable development* (A/RES/70/1). United Nations. Retrieved January 30, 2023, from https://sdgs.un.org/sites/default/files/publications/21252030%20Agenda%20for%20Sustainable%20Development%20web.pdf

UN DESA. (2019a). *World population prospects 2019: Highlights* (ST/ESA/SER.A/423). United Nations Department of Economic and Social Affairs. Retrieved October 23, 2022, from https://population.un.org/wpp/publications/files/wpp2019_highlights.pdf

UN DESA. (2019b). *World urbanization prospects: The 2018 revision* (ST/ESA/SER.A/420). United Nations Department of Economic and Social Affairs. Retrieved October 23, 2022, from https://population.un.org/wup/publications/Files/WUP2018-Report.pdf

UN DESA. (2022). *World population prospects 2022: Summary of results* (UN DESA/POP/2022/TR/NO.3). United Nations Department of Economic and Social Affairs. Retrieved January 14, 2023, from https://www.un.org/development/desa/pd/sites/www.un.org.development.desa.pd/files/wpp2022_summary_of_results.pdf

UN-Habitat. (2017). *The new urban agenda*. United Nations Conference on Housing and Sustainable Urban Development (Habitat III), on 20 October 2016. Retrieved January 30, 2023, from https://habitat3.org/wp-content/uploads/NUA-English.pdf

UN-Habitat. (2020). *New urban agenda illustrated* (HS/035/20E). United Nations Human Settlements Program. Retrieved January 14, 2023, from https://unhabitat.org/the-new-urban-agenda-illustrated

UN-Habitat & WHO. (2020). *Integrating health in urban and territorial planning: A sourcebook*. United Nations Human Settlements Program and World Health Organization. Retrieved January 14, 2023, from https://unhabitat.org/sites/default/files/2020/05/1-final_highres_20002_integrating_health_in_urban_and_territorial_planning_a_sourcebook.pdf van den Berg, A. E., Hartig, T., & Staats, H. (2007). Preference for nature in urbanized societies: Stress, restoration, and the pursuit of sustainability. *Journal of Social Issues, 63*(1), 79–96. https://doi.org/10.1111/j.1540-4560.2007.00497.x

Vardoulakis, S., & Kinney, P. (2019). Grand challenges in sustainable cities and health. *Frontiers in Sustainable Cities, 1*, 7. https://doi.org/10.3389/frsc.2019.00007

Ventriglio, A., Torales, J., Castaldelli-Maia, J., De Berardis, D., & Bhugra, D. (2021). Urbanization and emerging mental health issues. *CNS Spectrums, 26*(1), 43–50. https://doi.org/10.1017/S1092852920001236

Vigo, D., Thornicroft, G., & Atun, R. (2016). Estimating the true global burden of mental illness. *The Lancet Psychiatry, 3*(2), 171–178. https://doi.org/10.1016/S2215-0366(15)00505-2

von Lindern, E., Lymeus, F., & Hartig, T. (2017). The restorative environment: A complementary concept for salutogenesis studies. In M. B. Mittelmark, S. Sagy, M. Eriksson, G. F. Bauer, J. M. Pelikan, B. Lindström, & G. A. Espnes (Eds.), *The handbook of Salutogenesis* (pp. 181–195). Springer International Publishing.

von Lindern, E., Lymeus, F., & Hartig, T. (2022). The restorative environment and Salutogenesis: Complementary concepts revisited. In M. B. Mittelmark, G. F. Bauer, L. Vaandrager, J. M. Pelikan, S. Sagy, M. Eriksson, B. Lindström, & C. M. Magistretti (Eds.), *The handbook of Salutogenesis* (pp. 371–385). Springer International Publishing. https://doi.org/10.1007/978-3-030-79515-3_35

von Schönfeld, K. C., & Bertolini, L. (2017). Urban streets: Epitomes of planning challenges and opportunities at the interface of public space and mobility. *Cities, 68*, 48–55. https://doi.org/10.1016/j.cities.2017.04.012

Votruba, N., Thornicroft, G., & The FundaMentalSDG Steering Group. (2016). Sustainable development goals and mental health: Learnings from the contribution of the FundaMentalSDG global initiative. *Global Mental Health, 3*, E26. https://doi.org/10.1017/gmh.2016.20

Walzer, M. (1986). Pleasures and costs of urbanity. *Dissent, 33*, 470–475.

Wang, X., Rodiek, S., Wu, C., Chen, Y., & Li, Y. (2016). Stress recovery and restorative effects of viewing different urban park scenes in Shanghai, China. *Urban Forestry & Urban Greening, 15*, 112–122. https://doi.org/10.1016/j.ufug.2015.12.003

WCED. (1987). *Our common future: Towards sustainable development*. Report of the World Commission on Environment and Development. Retrieved January 24, 2023, from https://sustainabledevelopment.un.org/content/documents/5987our-common-future.pdf

Weber, A. M., & Trojan, J. (2018). The restorative value of the urban environment: A systematic review of the existing literature. *Environmental Health Insights, 12*, 1–13. https://doi.org/10.1177/1178630218812805

Weber, R., Schnier, J., & Jacobsen, T. (2008). Aesthetics of streetscapes: Influence of fundamental properties on aesthetic judgments of urban space. *Perceptual and Motor Skills, 106*(1), 128–146. https://doi.org/10.2466/PMS.106.1.128-146

WHO. (1954). *Outline for a study group on world health and the survival of the human race*. Material drawn from articles and speeches by Brock Chisholm. World Health Organization. Retrieved January 24, 2023, from https://apps.who.int/iris/handle/10665/330666

WHO. (1996). *Creating healthy cities in the 21st century*. Background paper prepared for the Dialogue on Health in Human Settlements for Habitat II, World Health Organization. Retrieved January 28, 2023, from https://apps.who.int/iris/handle/10665/61069

WHO. (1999). *Health 21: The health for all policy framework for the WHO European Region*. WHO Regional Office for Europe. Retrieved January 28, 2023, from https://www.euro.who.int/__data/assets/pdf_file/0010/98398/wa540ga199heeng.pdf

WHO. (2001). *The world health report: Mental health: New understanding, new hope*. World Health Organization. Retrieved January 28, 2023, from https://apps.who.int/iris/handle/10665/42390

WHO. (2004). *Promoting mental health: Concepts, emerging evidence, practice* (Summary Report). World Health Organization. Retrieved January 28, 2023, from https://apps.who.int/iris/bitstream/handle/10665/42940/9241591595.pdf

WHO. (2006). *Constitution of the World Health Organization*. In World Health Organization: Basic Documents. Forty-fifth edition, Supplement. Geneva. Retrieved January 28, 2023, from https://cdn.who.int/media/docs/default-source/documents/publications/basic-documents-constitution-of-who179f0d3d-a613-4760-8801-811dfce250af.pdf?sfvrsn=e8fb384f_1&download=true

WHO. (2010a). *Why urban health matters: 1000 cities 1000 lives: World health day 2010* (WHO/WKC/WHD/2010.1). World Health Organization. Retrieved January 16, 2023, from https://apps.who.int/iris/handle/10665/70230

WHO. (2010b). *Global recommendations on physical activity for health*. World Health Organization. Retrieved January 16, 2023, from http://apps.who.int/iris/bitstream/handle/10665/44399/9789241599979_eng.pdf?sequence=1

WHO. (2016a). *Health as the pulse of the new urban agenda: United Nations conference on housing and sustainable urban development*. World Health Organization. Retrieved January 14, 2023, from https://apps.who.int/iris/handle/10665/250367

WHO. (2016b). *Physical activity strategy for the WHO European Region, 2016–2025*. WHO Regional Office for Europe. Retrieved January 14, 2023, from https://www.euro.who.int/__data/assets/pdf_file/0014/311360/Physical-activity-strategy-2016-2025.pdf

WHO. (2017). *Towards more physical activity in cities: Transforming public spaces to promote physical activity—A key contributor to achieving the sustainable development goals in Europe*. World Health Organization. Regional Office for Europe. Retrieved January 14, 2023, from https://www.euro.who.int/__data/assets/pdf_file/0018/353043/2017_WHO_Report_FINAL_WEB.pdf

WHO. (2018a). *Active: A technical package for increasing physical activity*. World Health Organization. Retrieved January 14, 2023, from https://www.who.int/publications/i/item/9789241514804

WHO. (2018b). *World Health Statistics 2018: Monitoring health for the SDGs, sustainable development goals*. World Health Organization. Retrieved January 16, 2023, from https://www.who.int/publications/i/item/9789241565585

WHO. (2019). *Sustainable cities: Health at the heart of urban development*. World Health Organization. Retrieved January 14, 2023, from https://www.who.int/sustainable-development/cities/SDG_factsheet_sustainable_cities_EN_v3.pdf?ua=1

WHO & Calouste Gulbenkian Foundation. (2014). *Social determinants of mental health*. World Health Organization. Retrieved January 29, 2023, from https://apps.who.int/iris/bitstream/handle/10665/112828/9789241506809_eng.pdf

WHO & UN-Habitat. (2016). *Global report on urban health: Equitable, healthier cities for sustainable development*. World Health Organization and United Nations Human Settlements Program. Retrieved January 14, 2023, from https://apps.who.int/iris/handle/10665/204715

WHO & UNDP. (2016). *Non-communicable diseases: What municipal authorities, local governments, and ministries responsible for urban planning need to know*. World Health Organization and The United Nations Development Programme. (WHO/NMH/NMA/16.89). Retrieved January 14, 2023, from https://apps.who.int/iris/handle/10665/250228

Wirth, L. (1938). Urbanism as a way of life. *American Journal of Sociology, 44*(1), 1–24. http://www.jstor.org/stable/2768119

World Mental Health Survey Consortium. (2004). Prevalence, severity, and unmet need for treatment of mental disorders in the

World Health Organization World Mental Health Surveys. *JAMA, 291*(21), 2581–2590. https://doi.org/10.1001/jama.291.21.2581

Wu, Y.-C., Chen, C.-S., & Chan, Y.-J. (2020). The outbreak of COVID-19: An overview. *Journal of the Chinese Medical Association, 83*(3), 217–220. https://doi.org/10.1097/JCMA.0000000000000270

Xiang, Y.-T., Yang, Y., Li, W., Zhang, L., Zhang, Q., Cheung, T., & Ng, C. H. (2020). Timely mental health care for the 2019 novel coronavirus outbreak is urgently needed. *The Lancet Psychiatry*. https://doi.org/10.1016/S2215-0366(20)30046-8

Ye, B., Wu, D., Im, H., Liu, M., Wang, X., & Yang, Q. (2020). Stressors of COVID-19 and stress consequences: The mediating role of rumination and the moderating role of psychological support. *Children and Youth Services Review, 118*, 105466. https://doi.org/10.1016/j.childyouth.2020.105466

Yin, Y., Thwaites, K., & Shao, Y. (2022a). Balancing street functionality and restorative benefit: Developing an expectation-current approach to street design. *Sustainability, 14*(9), 5736. https://doi.org/10.3390/su14095736

Yin, Y., Thwaites, K., & Simpson, J. (2022b). Exploring the use of restorative component scale to measure street restorative expectations. *Urban Design International, 27*, 145–155. https://doi.org/10.1057/s41289-022-00186-w

Zhao, J., Wu, J., & Wang, H. (2020). Characteristics of urban streets in relation to perceived restorativeness. *Journal of Exposure Science & Environmental Epidemiology, 30*(2), 309–319. https://doi.org/10.1038/s41370-019-0188-4

Urban Refugees' Impact on the Urban Fabric Form of the City Structure: The Case Study of Urban Syrian Refugees in the Sixth of October City

Alaa Mohamed Mahmoud El Sayed, Adel Samy El-Menshawy, and Amira N. Elsemellawy

Abstract

Due to war and conflicts, the world recently suffered from the highest levels of displacement. Therefore the problem of refugees and immigrants all over the world is one of the most complex issues discussed in international forums with no radical solutions. Sadly, the Arab world has had the highest percentage of refugees seeking asylum, particularly in Egypt. The Syrian war is the most serious humanitarian conflict in the world causing an increasing number of refugees over time. Because Egypt has no camping policy, all refugees who live in Egypt are independent in financial and social aspects. Egypt is the only nation that does not set up refugee camps on its territory. Many statements were made by President Abdel Fattah El-Sisi, to confirm that Egypt hosts more than 5 million refugees on its lands. The primary purpose of this article is to focus on urban immigration and highlight concerns about immigrants and city identities to better understand the connection between the city and immigrants by looking at how their daily habits influence their appreciation for the city's open spaces. This study emphasizes the relevance of knowing immigrants' urban open space preferences based on push and pulls factors by questionnaires and analyzing the results. Given this background, this study explores the condition of Syrian refugees in Cairo by applying a case study on the Sixth of October City and analyzing Syrian refugee spaces there and then providing recommendations and potential future possibilities. The study used a mixed-methods approach that included a survey, group discussions with selected survey respondents who were Syrian and Egyptian, and interviews with Syrian refugees who had lived in the city for more than ten years.

Keywords

Egypt · Syrian refugees · Cairo · Sixth of October City · Migration · Refugees · Social integration · Economic insertion · Space · Place · Urban fabric · City · Neighborhood

1 Introduction

According to the global trends report from the United Nations Refugee Agency (Global Trends Report, 2021), worldwide conflicts, acts of violence, and breaches of human rights caused 89.3 million people to be displaced by the end of 2021 (see Fig. 1). The 98.3 million has broken down into four main significant categories, including the first which is the largest number of refugees ever recorded, 27.1 million, Second on the list are the 53.2 million individuals who are internally displaced; the third is 4.6 Million people seeking asylum, and the forth is 4.4 Million Venezuelans displaced abroad.

According to the (Global Trends Report, 2021) Middle East and North Africa host 26% of refugee seekers, especially from the Syrian conflict which represents the most dramatic refugee crisis of our time causing more than 5.5 million refugees (see Fig. 2). Also, Fig. 2 shows where's the world's displaced people are being hosted.

Displacement people around the world are divided into refugees and immigrants as shown in Fig. 3. Refugees are divided into urban refugees and camp refugees, and the immigrants are divided into forced migration and labor migration, as shown in Fig. 3. Academics have recently

A. M. M. E. Sayed (✉) · A. S. El-Menshawy · A. N. Elsemellawy
Department of Architecture Engineering and Environmental Design, Arab Academy for Science Technology, Maritime Transport University, Giza, Egypt
e-mail: Engalaa.mohamed123@gmail.com

A. S. El-Menshawy
e-mail: dradelsamy@yahoo.com

A. N. Elsemellawy
e-mail: a.elsemellawy@aast.edu

I. Pigliautile et al. (eds.), *Urban Resilience, Livability, and Climate Adaptation*,
Advances in Science, Technology & Innovation, https://doi.org/10.1007/978-3-031-54911-3_15

begun to pay more attention to the issue of urban migration around the world due to the fact that the forced migration has significantly increased over the past few decades. Urban migrants account for nearly 60% of all global refugees and 6.6 million people, and around 22% of the world's refugees are thought to reside in camps (Verghis & Balasundram, 2019). Urban environments, contrary to camps, can theoretically provide refugees with freedom of movement and

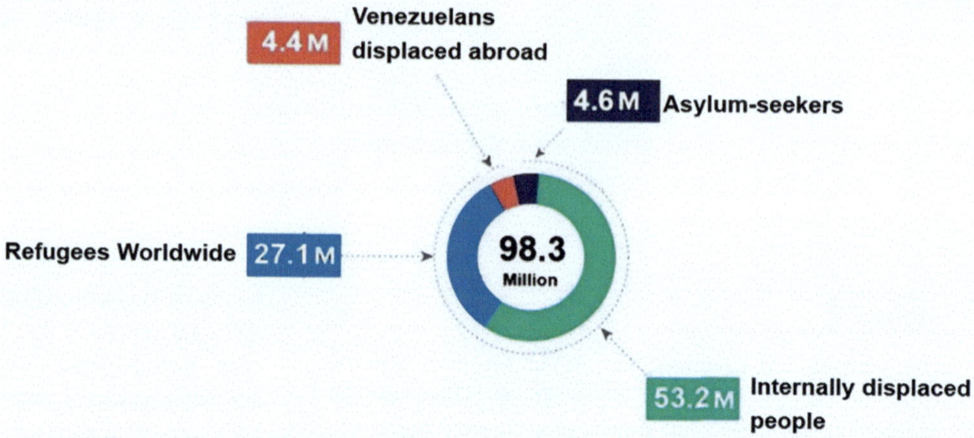

Fig. 1 Shows the total number of people displacement around the world. *Source* Global Trends Report (2021)

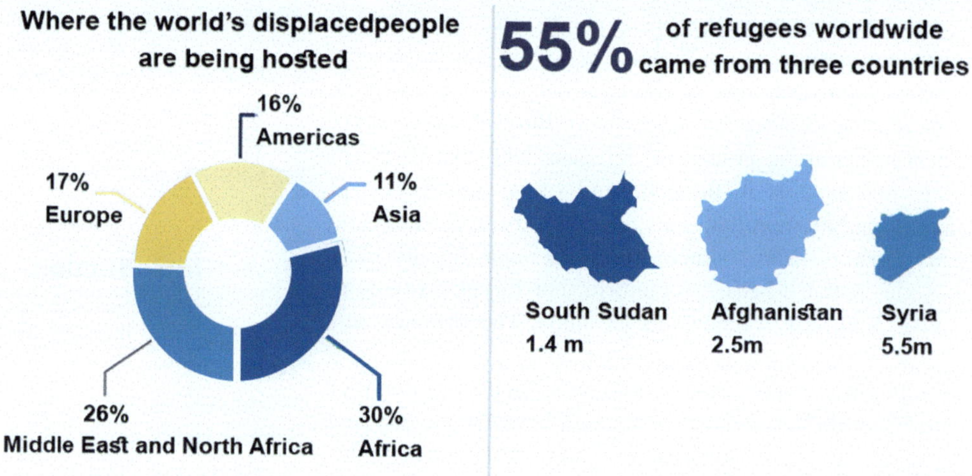

Fig. 2 Ratio of host countries for refugees. *Source* Global Trends Report (2021)

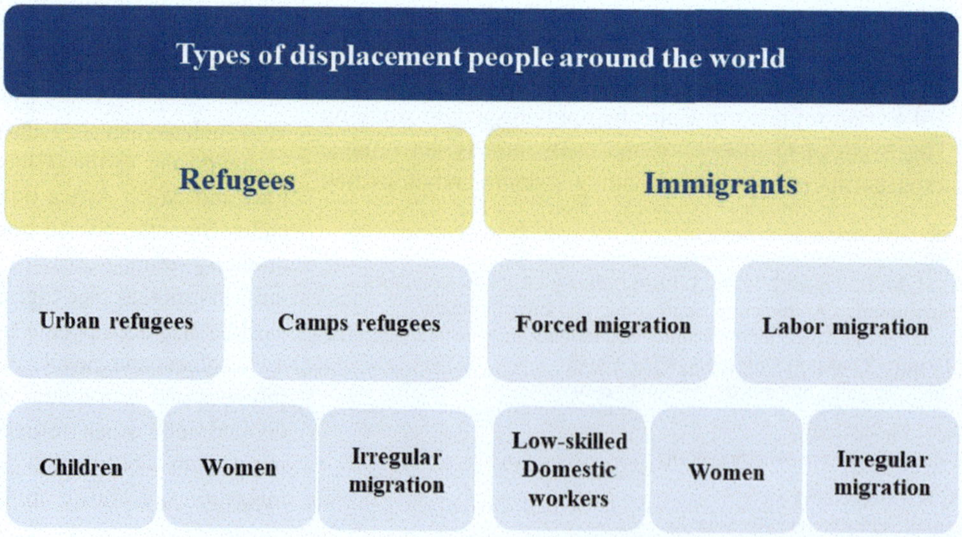

Fig. 3 Types of displacement people around the world. *Source* UNHCR global trends (2021)

career options. As a result, some refugees think that access to medical care, humanitarian relief, and even resettlement is better in cities (UNHCR, Refugee Camps Explained, 2021). As for immigrants today, the second and third generations of Turkish immigrants in Germany reflect a distinct profile of immigration that has contributed to Germany's cosmopolitan diversity as an example of labor migration. However, many forced immigrants confront challenging problems such as adaptability, as happened with Syrian refugees in Turkey (Edo et al., 2018).

People leave their country for many reasons related to push and pull forces, including economic, social, and physical factors. Attractions to settle in another country vary; some refugees are mainly seeking security. Repelling elements are also diverse such as undesirable high-priced housing and bad transit (see Fig. 4). Motivations for residing in certain areas of the city help to understand the spectrum of urban reactions to refugee influxes, which aids in forecasting likely future patterns.

Egypt is a country of migration, transit, and final destination for refugees and migrants have taken in several regional refugee groups over the past few years, particularly following the Arab Spring (Presidency, 2021), and the capital City 'Cairo' is home to one of the world's most significant urban refugee populations. Egypt was housing a total of 324,707 forcibly displaced people, including 258,391 refugees under the UNHCR's mandate, and 66,316 asylum applicants of 38 different nationalities (UNHCR, UNHCR Egypt airo: United Nations High, 2019). However, nine origin countries including; Syria, Palestine, Sudan, South Sudan, Eritrea, Ethiopia, Yemen, Iraq, and Somalia represent 99 percent of all internally displaced people in Egypt. According to the (UNHCR, UNHCR Egypt airo: United Nations High, 2019), their registered numbers are given in and illustrated in Fig. 5 and the largest percentage of refugees goes to the Syrians.

There are six million "registered" refugees in Egypt overall according to official communications from the Egyptian government (The arab republic of Egypt president, 2021) and there are 324,707 forcibly displaced people, including 66,316 asylum seekers from 38 different countries and a very small number of stateless people (UNHCR, UNHCR Egypt airo: United Nations High, 2019). On the other hand the actual unregistered refugee population is much higher 16.890.0000 according to estimates. The 'unregistered number' of Arab refugees includes Palestinians (nearly 3.000.000), Sudanese (nearly 4.300.000), Iraqi (nearly 2.270.000), Syrian (nearly 5.560.000), and Yemeni (nearly 1.760.000) refugees.

1.1 Research Problem

In 2021, conflicts, acts of violence, and natural disasters caused 53.2 million new internal displacements and this is equal to forcing one person to leave every second (IDMC, 2017). With wars growing more violent and prolonged and their underlying causes getting more complicated, global displacement is at an all-time high. At the end of 2021, 98.3 million people left their homes due to conflict, violence, and fear of persecution. This number is more than double the 42.7 million people who were forcefully displaced ten years ago and the greatest number since World War II. (UNHCR global trends, 2021). Therefore, academics have

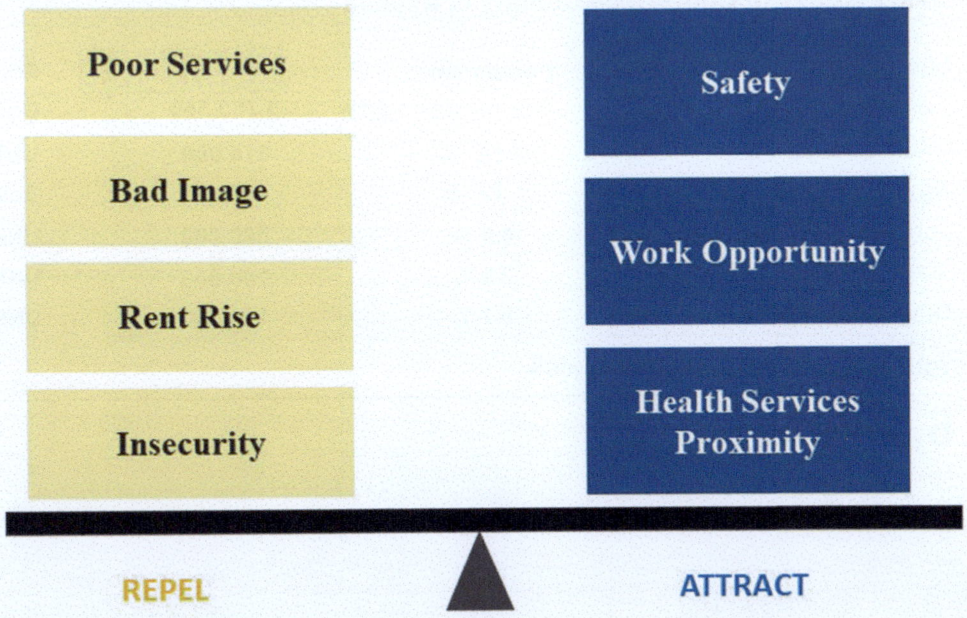

Fig. 4 The weighting of push and pull variable. *Source* By the researcher based on macro theories

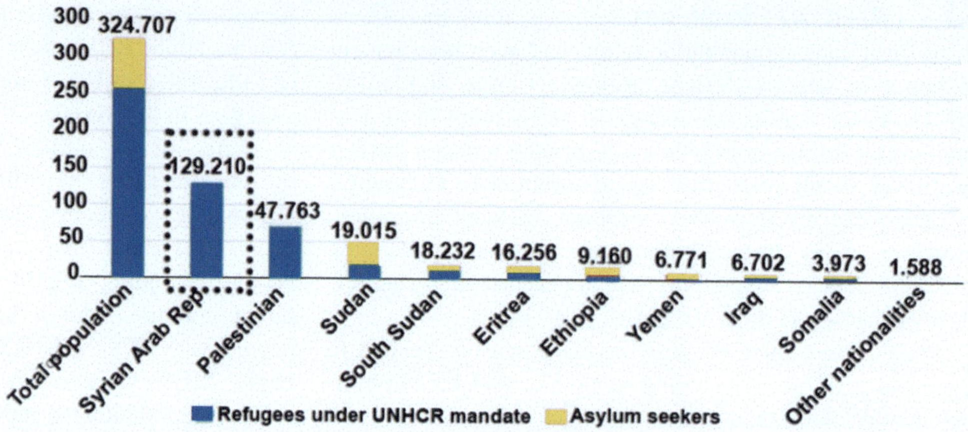

Fig. 5 Total population of refugees and asylum-seekers in Egypt by country of origin. *Source* UNHCR (2019)

recently begun to pay more attention to the issue of urban migration around the world. Syria's crisis continues to create the most refugees worldwide (6.168.565 million), as shown in Fig. 6. First, the ongoing conflict in Syria has resulted in one of the biggest humanitarian catastrophes to hit the Middle East in recent memory. The second is the internal migration of individuals, which reached 40.3 million as of the end of 2016 (ECHO, 2022). The third category is asylum seekers who have fled their country and are applying for refugee status internationally.

1.2 Aims of the Study

The main aim of this research paper is to discover how the refugees and immigrants are changing the urban fabric of the City and how they also contribute to reshaping the City's diversified urban areas. This trying to fill an academic gap in discovering the impact of refugees and immigrants in the City through many sub-objectives;

- To discover how refugees and immigrants make an impact in the urban form in different countries by exploring and analyzing international examples in Germany and Turkey as host countries examples.
- To discover how refugees and immigrants impact urban form and to explore their interaction with the host community by specifically focusing on the Syrian refugees in the Sixth of October City.
- To understand and figure out the circumstances under which refugees and immigrants reside in specific areas of the city by stressing the importance of space, or spatiality.

Total Persons of Concern by Country of Asylum

Location name	Population		Source
Turkey	65%	3.763.565	Government of Turkey, 2023
Lebanon	15 %	815.000	UNHCR, 2022
Egypt	12 %	600.000	Government of Egypt,2021
Jordan	4.6 %	560.000	UNHCR,2023
Iraq	2.5 %	280.000	UNHCR,2023
Other (North Africa)	0.7 %	150.000	UNHCR, 2022

Total Registered Syrian Refugees

6,168,565

Last updated 30 Jun 2023

Fig. 6 Registered Syrian refugees—breakdown by Country of Asylum

1.3 Research Questions

The main question of this research paper is to figure out the impact of refugees and immigrants in urban fabric form in the host nations through various elements which will introduce the following sub-questions to contribute to the findings of the main question:

- What alterations or effects have refugees and immigrants had on the city's urban form?
- What are the main factors that influence refugees' and immigrants' decisions to choose specific places of settlement in countries rather than others?
- How strongly did the flood of refugees and immigrants impact the economic labor markets?
- What is the impact of refugees in Egypt on the urban fabric through studying specific nationalities like Syrian refugees and their impact on a specific City?

2 Framework Criteria

After the researcher made a systematic review of pertinent theories related to urban refugees, the researcher conducted a conceptual developed framework shown in Fig. 7 to be applied to the global and local examples in order to examine the impact of refugees and immigrants on the urban fabric form in the City through four main measurements factors;

(1)The space responses examination that can help refugees to be present and to show their identity or not.
(2)The possibility of economic insertion.
(3)The social cohesion in the host country.
(4)The political changes during this integration process.

3 Methodology

The research that serves as the foundation for this topic is exploratory and descriptive due to it aims to describe the phenomenon of refugees in urban areas, document their impact on the urban fabric form in the City, and explain the motivations and influences that drive their movements in the City. Therefore, the methodology of this paper could be divided into four stages;

1. The first stage, gathering all the information related to the research topic, which took more than a year and a half due to this problem being uncommon in the academic field.
2. The second stage, developing a conceptual framework criteria selection after exploring Contemporary approaches and theories for urban refugees.

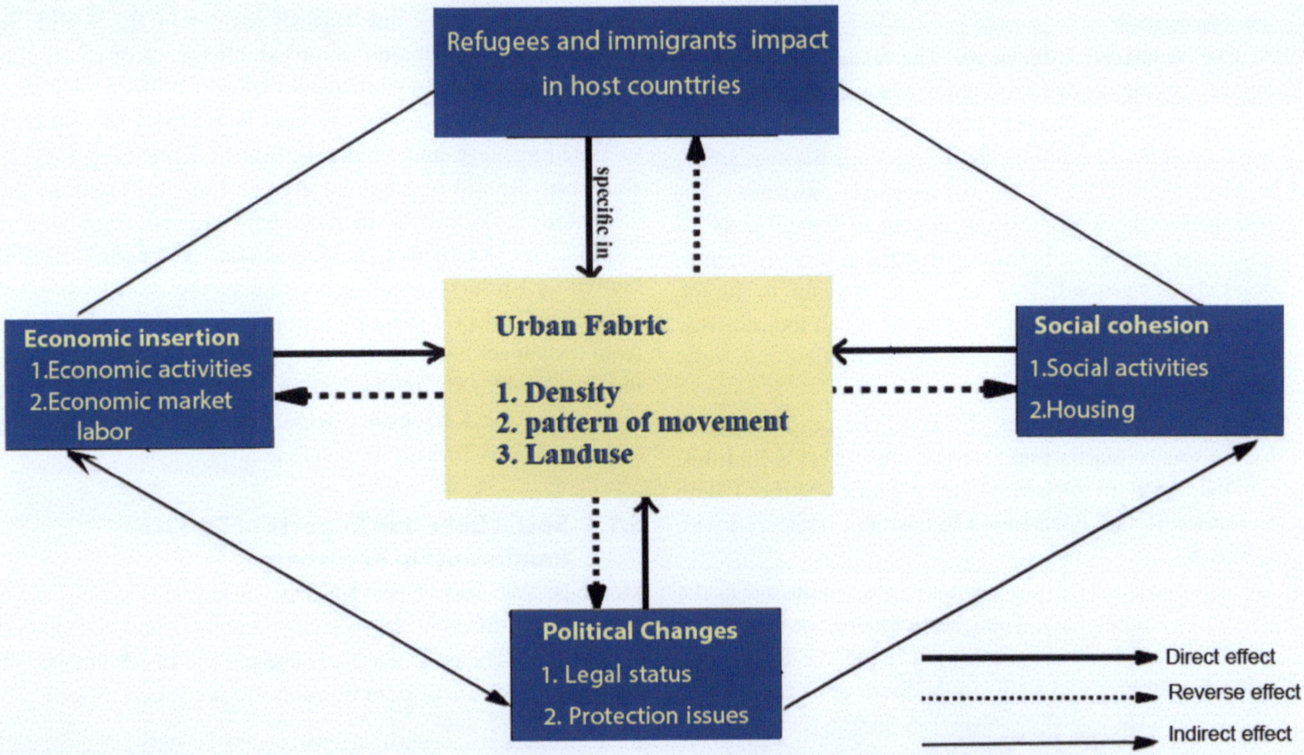

Fig. 7 The conceptual paper framework developed by the researcher

3. The third stage, conducting a comprehensive comparison between analytical examples, the researcher conducted a comprehensive review of global examples such as; Turkish immigrants in Germany and Syrian refugees in Turkey, and local examples such as Syrian refugees in Egypt on the Sixth of October City to understand how those countries dealt with refugees and how the refugees impact on the urban fabric form in the City within applying the conceptual framework criteria. Therefore, in the local example in Egypt, the researcher conducted interviews with both stakeholders and Syrian refugees in order to gain a comprehensive picture of the lives of urban refugees in Cairo and their impact on urban fabric form.
4. The fourth stage, the researcher will be able to find a discussion that will contribute to the key findings and it is going to be useful to reach good recommendations related to the impact of refugees on the urban fabric in the City based on the analytical global and local examples.

3.1 Data Collection Methods

(1) **Survey outline**: the survey shows the distribution of the sample of Syrian refugees throughout the various areas of the Sixth of October City. The researcher's ability to pinpoint the city's high concentration of Syrian refugees led to the distribution of the sample in the 1st, 2nd, 3rd neighborhoods.

(2) **Semi-Structured Interviews**: The researcher sought to balance giving study participants a list of prepared questions and allowing them to freely move between the topics they felt like talking about. Second, the researcher needed a wide range of open-ended questions. The researcher frequently did not adhere to the subjects' or questions' sequence but it covered the following topics: general background, leaving Syria, arriving in Egypt, the Sixth of October City, employment and income, distance and transportation, housing, activities, and relationships with Egyptians.

(3) **Focus group discussions (FGDs)**: The purpose of the FGDs was to clarify and interpret the survey's findings. On the Sixth of October City, a FGD was held with Syrians who had been found to have not registered with UNHCR.

(4) **Questionnaire**: The questionnaire was conducted in the Sixth of October City. The questionnaire consists of 16 questions varying between Syrian living conditions and their economic social status in the Sixth of October City.

4 Limitations of the Study

The identification of research areas of focus was made through the case study in a specific country, a specific city, and a specific nationality which took time and entailed restructuring of specific aspects. The researcher faced significant constraints like reaching out and engaging with government authorities to obtain the needed data. Furthermore, the volatile political environment in Egypt, as well as the research's dynamic character, provided the problem of regularly updating the data as refugee influxes increased day after day, consequently, the data and information changed rapidly as a result of possible changes that may occur in the Refugee Data Finder later this year. Therefore, this report should be regarded as tentative and subject to change.

5 Global Examples

5.1 Turkish Refugees in Germany

Nearly 60 years ago, the first Turks arrived in Germany as labor migrants. They were known as "Gastarbeiter" (guest laborers). Now over 3 million people with Turkish ancestry reside in Germany, and their influx has increased over the years. Turkish migrants have helped to revitalize Berlin, where they constitute the largest migrant group, and they made Berlin the second Turkish capital in the world. In addition, they contributed to the establishment of new political, economic, and social sectors (Susan Willis McFadden, 2019). The start of their movement begin from Moabit City to Wedding City and finally settled in Kreuzberg City as shown in the red highlighted area in Fig. 8. Therefore the case study area will be in Kreuzberg (Burcu Togral Koca, 2019). The Kreuzberg neighborhood in central Berlin, known as Little Istanbul, according to the world's biggest outside of Turkey. First-generation Turkish refugees resided from the outset because, in these districts, most of the housing was outdated and lacked modern utilities, and rents were low (Elie Katzenson, 2015). The streets of Kreuzberg are dotted with Turkish food sellers.

5.1.1 Social Cohesion Impacts of Turkish Immigrants in Kreuzberg

Members of this Euro-Turkish immigrant group have encountered different degrees of integration and acceptance in Germany. The concluded advantages and disadvantages of the presence of Turkish immigrants are presented in Table 1.

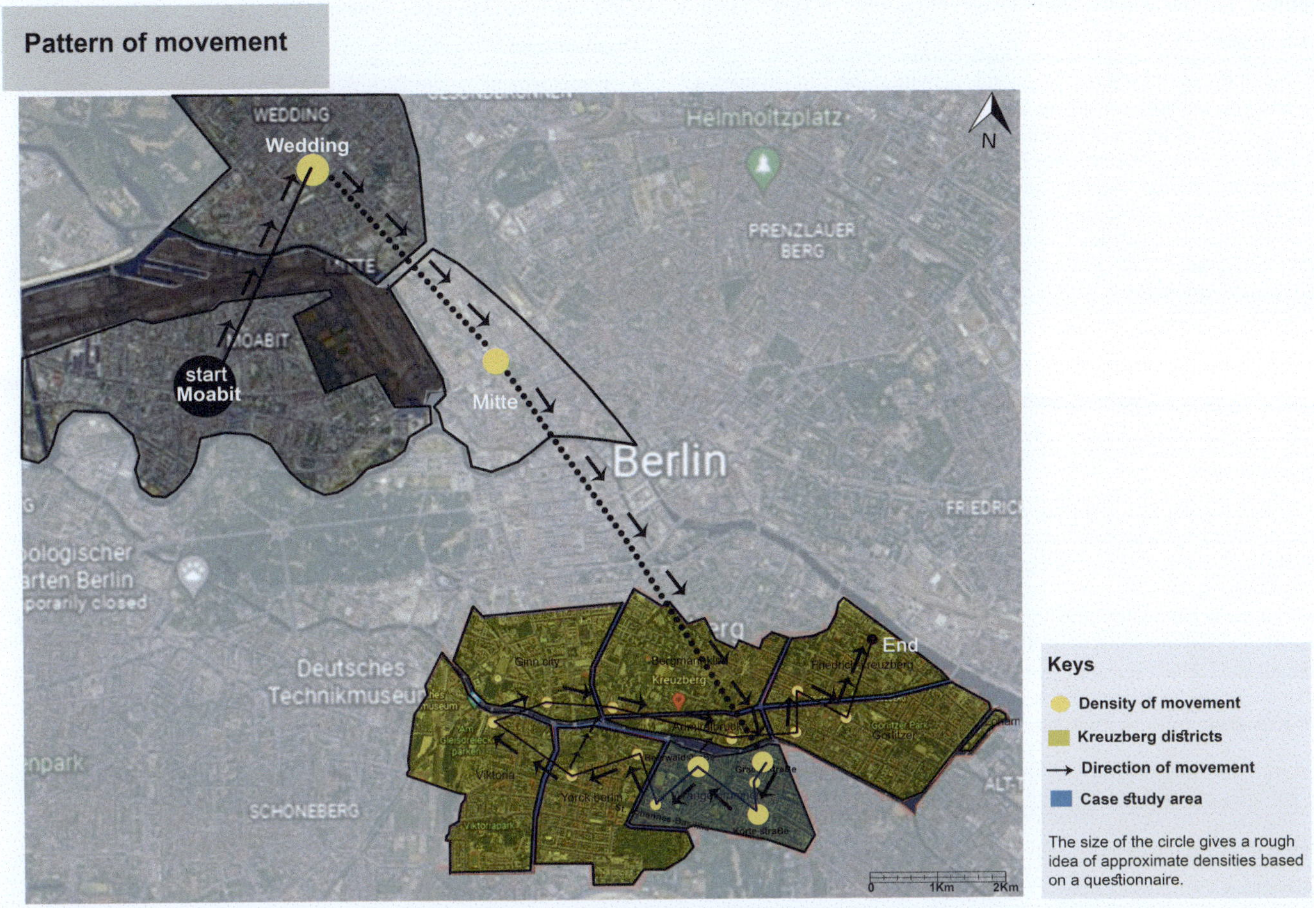

Fig. 8 Pattern of movement of Turkish immigrants in Berlin City adapted from the reviewed literature

Table 1 Advantages and disadvantages of the presence of Turkish immigrants in Kreuzberg in relation to social cohesion

Advantages	Disadvantages
• In the last years, when it has received more than 3 million Syrian refugees, Turkey has demonstrated incredible resiliency by adopting a humanist perspective and enforcing an open-door policy for migrants • Since most of the Turkish immigrants were experts and highly educated, they were welcomed at their first immigration Invalid source specified • There are many social activities the Turkish people can do in Kreuzberg neighborhoods such as cultural, sports, commercial, recreational,...etc.	• Three years after the arrival of Turkish immigrants, they started to suffer from instances of racism and marginalization, which appeared to be the driving forces for their efforts to establish unity between them and the Germans • Many of the first generations have issues of identification. They don't feel like they belong in either Germany or Turkey Invalid source specified. Therefore, many especially in Kreuzberg, mentioned how they experienced prejudice and a lack of socioeconomic and political rights

5.1.2 Economic Impacts of Turkish Immigrants in Germany

Regarding economic insertion, the rate of unemployment of Turkish immigrants is high. Consequently, many survive on state benefits and success (Topcu, 2022). In addition, conclusions concerning the advantages and disadvantages of the presence of Turkish immigrants emerge from Table 2.

5.1.3 Turkish Refugees' Impacts on Urban Fabric in Germany

Concerning urban fabric developments, Kreuzberg was full of the ugliest houses where nobody wanted to live. Later on, such an area was modernized by Turkish migrants and thus became a central and popular district after World War II. Table 3 sums up many advantages and disadvantages

Table 2 Advantages and disadvantages of presence of Turkish immigrants in Kreuzberg in terms of economic insertion

Advantages	Disadvantages
• One of the most important economic activities, the Turkish market, where you can find everything from shops to markets, sweets, perfumes …etc. is one of the most famous economic activities reflecting the Turkish identity in Kreuzberg (Akgündüz, 2021) • Germany granted residence and work permits to spontaneously enter Turkish workers to increase the volume of labor inflow from Southern Europe. Moreover, Turkish immigrants were welcomed when they first immigrated to Germany because of their being experts and highly educated individuals	• Unemployment among Turkish immigrants prevails and many survive on the benefits of the state. Hence, Turkish immigrants are primarily from the lower social classes in Germany, and their poverty rates are greater among members of the Turkish community (Akgündüz, 2021) • The emphasis on the high skill of the Kreuzberg labor market resulted in social problems of excluding Turkish immigrants and their heavy concentrations in problematic regions. This, together with the relatively high unemployment rate, is the main cause making several Kreuzberg neighborhoods susceptible regions in terms of living circumstances

Table 3 Advantages and disadvantages of the presence of Turkish immigrants in Kreuzberg in terms of urban fabric

Advantages	Disadvantages
• Kreuzberg was full of the ugliest houses where nobody wanted to live Invalid source specified. Later on, such an area was modernized by Turkish migrants and thus became a central and popular district after World War II • Enhancing the quality of living in Kreuzberg necessitated the focus on creating a new historical understanding and realizing how social, functional, and technological links depend on realizing structured examples that are not to be disregarded or let decay organically. The preservation of Kreuzberg ruins in Mietskasernen was crucial	• The Kreuzberg neighborhood's antiquated architecture had been crammed and neglected for a long time, thus making both industrial and residential activities frequently incompatible. Further, the amount to which people claimed their right to urban space has also been constrained by Berlin's neoliberal alteration of urban space, resulting in socio-spatial marginalization • In this case, Kreuzberg gentrification and the consequent increase in rent are two key factors. Because of the transition, housing costs have skyrocketed, putting a strain on services and transportation, and many Turks can no longer afford to live there

Table 4 Advantages and disadvantages of the presence of Turkish immigrants in Kreuzberg in terms of political changes

Advantages	Disadvantages
• The new citizenship legislation went into effect on January 1, 2000, by modifying the principle of descent in several ways this new citizenship rule raised the number of foreign nationals filing for naturalization • The advent of the family reunification law in 1973 permitted Turkish immigrants to bring their spouses and children who were under the age of 18 to Germany. The German government sought to limit the inflow of new foreign employees while promoting the integration of those who were already there	• As contends, as the ratio of immigration to Germany increased, other European nations expressed their desire to hire Turkish employees. As a result, Turkey's First Five Year Development Plan went into force, causing a more thorough review of migration • The first generation of immigrant Turks and many of the grandchildren and offspring, however, struggle with their identity. They are stuck between two worlds and lack a strong sense of belonging to either Germany or Turkey

that could be concluded from the presence of Turkish immigrants.

5.1.4 Political Changes of Turkish Immigrants in Kreuzberg District

Regarding political changes, Berlin has implemented new policies such as initiatives to combat marginalization and unlawful status for Turkish immigrants. Therefore, many advantages and disadvantages are detected from the presence of Turkish in Kreuzberg as Table 4 illustrates.

5.2 Syrian Refugees in Turkey

Turkey hosts over 3.7 million registered Syrian refugees (The World Bank, 2022), and the Syrian influx has increased over the years (see Fig. 8). Syrian refugees have moved to Istanbul for several reasons. One reason is the Turkish government offers housing and educational opportunities, as well as security and protection for Syrian refugee families (see Fig. 9). One case study follows the existing social networks in Syria, such as familial ties and other significant social, cultural, and religious networks (Kaya & Kıraç, 2016). The area of applying case study in Istanbul is Sultan Gazi, especially in Gaz, zubedye hanim, and 75 yil districts. 'Sultan Gazi is an Istanbul inner-city area on the European side of the Bosporus and the home to over 500,000 Turkish people and 50,000 Syrians (Syrians in Istanbul & Post-war Syrian Ghettos, 2019). Since the middle of the 1970s, the area has experienced waves of migration from eastern Turkey due to inexpensive housing and low-skilled employment prospects in textile businesses (Göksel, 2018).

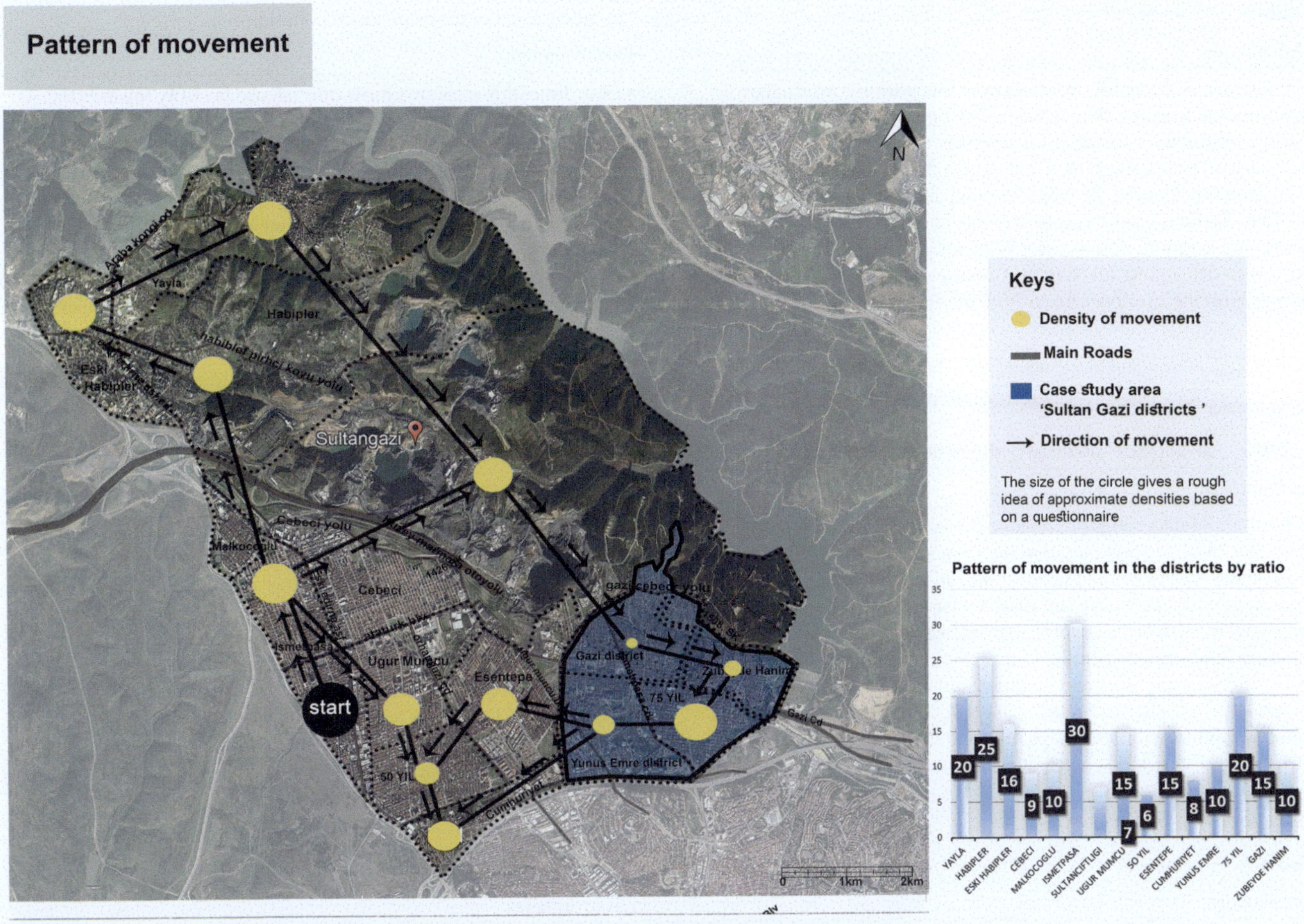

Fig. 9 Pattern of movement of Syrian refuges in Istanbul. *Source* The researcher based on the literature review

5.2.1 Syrian Refugees' Impact on Social Cohesion in Turkey

As regards social cohesion, being one of the biggest countries in the world having Syrian refugees, Turkey is a host country that received almost 4 million Syrians over the last seven years since Erdogan's implementation of the open door policy (Erdoğan, 2014). Many advantages and disadvantages are attributed to the presence of Syrian refugees in Sultan Gazi given in Table 5.

5.2.2 Economic Insertion Impacts of Syrian Refugees in Turkey

Economic insertion of Syrian refugees advantaged them through policies aiming at increasing their incentives. According to the Migration and Politics Research Center, the Syrians in Sultan Gazi have a negative influence on the socioeconomic level relative to public services like health and education due to their increasing population (Caro, 2020). Consequently, Turkish people considered them as an economic burden. The advantages and disadvantages of the presence of Syrian refugees in Sultan Gazi are summed up in Table 6.

5.2.3 Urban Fabric Impact on Syrian Refugees in Turkey

Urban spaces such as Gazi, Zubedye Hanim, and 75 Yil witnessed many urban fabric developments. For example, Syrians revitalized those areas with many facilities and services such as restaurants, bakeries, call shops, and small workshops (Adnan Doğan BULDUR, 2018). The advantages and disadvantages of the presence of Syrian refugees in Sultangazi are stated in Table 7.

5.2.4 Political Changes Caused by the Syrian Refugees in Turkey

Political conditions reflect how Turkey does not always support Syrians. Turkey sometimes denies Syrians their legal right to work and to have houses for threatening Turkish security and well-being. Turkey's asylum laws center on resettlement or repatriation (TURKSTAT's 2018 census

Table 5 Advantages and disadvantages of the presence of Syrian refugees in Sultan Gazi in terms of social cohesion

Advantages	Disadvantages
• Language is the main impediment to integration for refugees. Information learning the language also opens doors to the Sultan Gazi community's social life. Therefore free Turkish language lessons are offered to both men and women in Sultan Gazi at the Refugee Coordination Center, the Red Crescent, and the Blue Crescent • Many Syrian refugee families tend to stick to their little social networks by avoiding making new acquaintances. In Sultan Gazi, there are close ties among all Syrians. Consequently, Syrians forged intimate bonds with their fellow citizens in their neighborhoods	• Over time, this welcome turns into intense hostility towards the Syrians, as they continue to flow without returning to their homeland so that the Syrians are forced to settle in the Sultan Gazi district where they face many housing, educational, and economic issues without any rights in living and equity • Syrian refugees in Turkey caused an increase in rents and deepened residential segregation. Studies show that the Syrian refugee housing policies enacted by governments often lead to the establishment of isolated residential sites separated and disconnected from the city centers. Exercising their right to housing is one of the most challenging issues confronted by Syrian refugees

Table 6 Advantages and disadvantages of the presence of Syrian refugees in Sultan Gazi in terms of economic

Advantages	Disadvantages
• Several policies sought to remove the mobility limitations for Syrian refugees. Also, efforts targeting the expansion of the pool of qualified candidates were intended to boost incentives for Syrian refugees. Hence granting work licenses to informal Syrian workers allowed them to enjoy all employment benefits such as healthcare and retirement pensions Invalid source specified • Many Syrian vendors and facilities in the neighborhood with Arabic signboards used in restaurants, bakeries, call shops, neighborhood charities, and small workshops reflected their identity. Therefore, Syrian migrants are constantly integrating themselves into the community of Istanbul	• According to the Migration and Politics Research Center, stated that host citizens considered Syrian refugees an economic burden. Moreover, the Syrians' socio-economic impact is becoming increasingly manifest as 56% of participants believed that Syrian refugees took their jobs due to their increasing numbers in Sultan Gazi. Further, public services of healthcare and education were negatively influenced by the Syrian refugee influx • The 13.3% of the rate of adult unemployment among Syrians constitutes an intriguing part of their labor market experience. Syrian migrants are likely to encounter a somewhat unstable employment market in Sultangazi, changing jobs often

Table 7 Advantages and disadvantages of the presence of Syrian refugees in Sultan Gazi in terms of the urban fabric

Advantages	Disadvantages
• Today, Sultan Gazi is one of the top five districts in Istanbul relative to the number of Syrians hosted; they now make up more than 8% of the district's total population. Despite their struggles with poverty and isolation among other urban disadvantaged groups, Syrian refugees in Sultan Gazi are integrating into the urban fabric • The goal of governments was to assist Syrian refugees where they reside. The disadvantaged groups of refugees of unaccompanied minors, the elderly, and kids in school received psychosocial support, and health services	• Syrian immigrants typically chose to reside in Sultan Gazi's worst urban neighborhoods. Descriptive statistics reveal the negative correlation between the quality of life offered in the towns that refugees prefer to live in and their top choices for residence. There are a lot of street vendors and beggars in these spaces • The local population in Gazi, Zubedye Hanim, 75 Yil, frequently complains that the Syrian refugees prevent them from benefiting from public services like parks, health care, and schools. Also, Syrian refugees cause skyrocketing rents. Consequently, public services are already strained by the district's rapid expansion

was announced, 2018). Many advantages and disadvantages resulting from the presence of Syrian refugees in Sultan Gazi are given in Table 8.

6 Local Case Study in the Sixth of October City

6.1 Selected Background

The background of the refugees' admission into Germany, Turkey, and Egypt, as well as a general awareness of features of the urban fabric, will be discussed in the case study that follows. It will focus on the position of Syrian refugees in metropolitan areas in the Sixth of October City, as well as how they live. The presence of Syrian refugees in the Sixth of October City is distinctive enough to warrant further investigation.

6.2 Reasons for Selection of Location

The Sixth of October City has been selected as the case study since it is one of the cities of Cairo with the largest population of Syrians that has seen many economic and social changes. Most modern writing evaluates refugees' requirements and their circumstances in their new home countries, without much said about their important feelings, effects, and attitudes as mentioned above in the introduction part related to the previous research gap. Further study

Table 8 Advantages and disadvantages of the presence of Syrian refugees in Sultan Gazi in terms of political changes

Advantages	Disadvantages
• Syrian refugees have more than quadrupled in number in the Sultan Gazi area since the start of the Syrian conflict in 2011, and they have been warmly welcomed there by giving them a variety of financial and humanitarian aid • Over 3.1 million Syrians live in Turkey with the vast majority having temporary protection (TP) status. At present, approximately 82,000 Syrians are "preregistered" in Turkey seeking TP certification (TURKSTAT's 2018 census was announced, 2018)	• Turkey's asylum laws focus on resettlement or repatriation. Therefore they are still treating the presence of 2.1 million Syrian refugees as a temporary situation rather than a longer development and legal problem • It was anticipated that the battle would end quickly in Syria when started in 2011 but it is last till now. Therefore, the government referred to the Syrian refugees who were entering Turkey as "guests," which gave them no legal rights and implied a "temporary and brief" stay

on the microeconomic effects of Syrian refugees in Egypt is needed to specify how they affect the Sixth of October City in Cairo and its daily commercial and social activities. The selected area is one of the satellite communities that the New Urban Communities Administration (NUCA), which was a part of Egypt's economic reform and structural adjustment, established in Cairo in 1981 (Wahdan, 2012). The region is known as an industrial zone that offers many housing alternatives (from gated enclaves to public housing) and is thought to be a tidy and orderly environment for individuals who seek to escape Cairo's traffic (Sims, 2011). Thus the paper investigates how the city absorbed the Syrian asylum, what the social and economic repercussions of their arrival are, and what actions are taken to deal with it. In Cairo, researchers are weighing the benefits and drawbacks of the social integration of Syrian migrants. In the Sixth of October city, Syrians are distributed into districts according to their socioeconomic status, and social class. According to (Ayoub & Khallaf, 2014), greater Cairo is home to four neighborhoods with the most vulnerable Syrians found in Faisal and Omraneya, Obour, and Masaken Othman as shown in Fig. 10.

6.3 Mapping Findings of Analysis

6.3.1 Pattern of Movement of Syrian Refugees

According to the questionnaire, Syrian refugees tend to settle around well-known faith organizations i.e. around mosques like El-Hosary mosque and Abu Bakr Al-Siddiq Mosque where charities and NGOs help them. Therefore they prefer to dwell in nearby spaces similar to their former metropolitan experiences, creating a pull factor for community cohesion. When any Syrian family moves from the blocks of Bayt Al-Ayla to the neighborhoods of the Sixth of October City, their movement to these neighborhoods is not random as they consider their economic and social conditions (see Fig. 11). Another factor affecting their decision to reside in the 5th, 6th, and 4th districts is their need to be closer to services. It is worth noting that some of them moved to look for cheaper housing in the 10th, 11th, and 12th districts, where residences of young people affiliated with the government. However, a large number did not find safety and stability there. Despite the higher rents in those areas, they moved to the second and third neighborhoods for the many services offered to its dwellers such as ease of mobility.

The city of Sixth October has been a settlement for many Syrians, who arrived in families and gathered around the Beit Al-ayla neighborhood which is known as a very poor neighborhood. After this situation in beit al-Ayla Syrian communities moved to and found around the first district, the second district, the third district, (see Fig. 10) and in less concentration in the 7th district, the 4th district, the 8th district, and the sixth district (where poorer groups) where there is a big similarity in buildings between Beit Al-ayla, and 6th district. The area of applying case study covered the 1st, 2nd, and 3rd districts of the Sixth of October city and El-Housary area as shown in Fig. 12. This area, referred to as the Syrian belt area, is considered the most accommodating area for Syrians. This is due to several reasons, for example, the settlement of their relatives, and a lot of Syrian families in this place many years ago. Additionally, conforming to the migration network theory, the rest of the area outside the Syrian belt is divided into studios for rent for students of the high recorded rate of occupancy by Sudanese, Iraqis, Eritreans, and others. It resulted in raising the rate of rent that the Syrian families would not bear. Therefore Syrians settled in the aforementioned Syrian-built area, where the rate of occupancy at that time was very low due to the few offered services. However, after some time the Syrians revived this area, and it became crowded with services and occupancy.

6.3.2 Syrian Refugees' Social Cohesion Impact in the Sixth of October City

In terms of social cohesion, the advantages of the existence of Syrians were reflected very clearly in social activities. The researcher noted through a survey in main paths of districts that Syrian social activities vary between religious gatherings around mosques or commercial gatherings

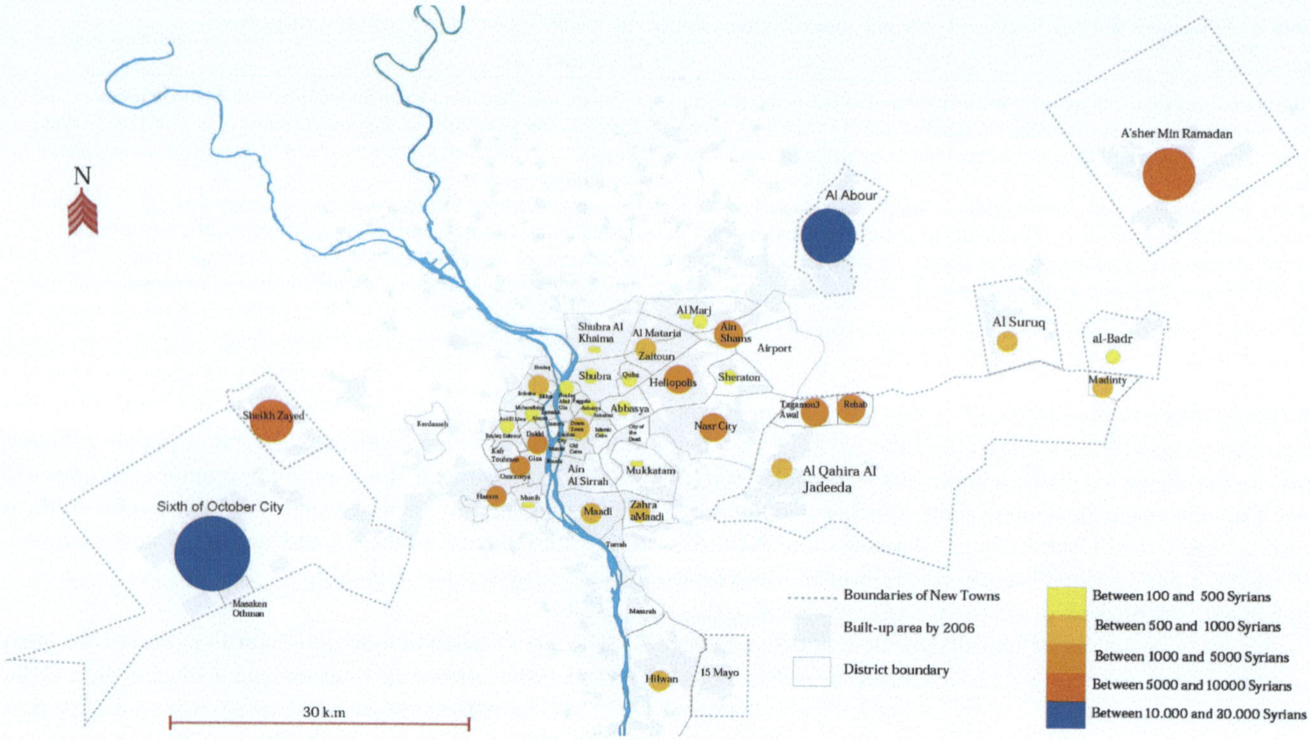

Fig. 10 Places of residence and approximate densities of the Syrian refugees. *Source* Ayoub and Khallaf (2014)

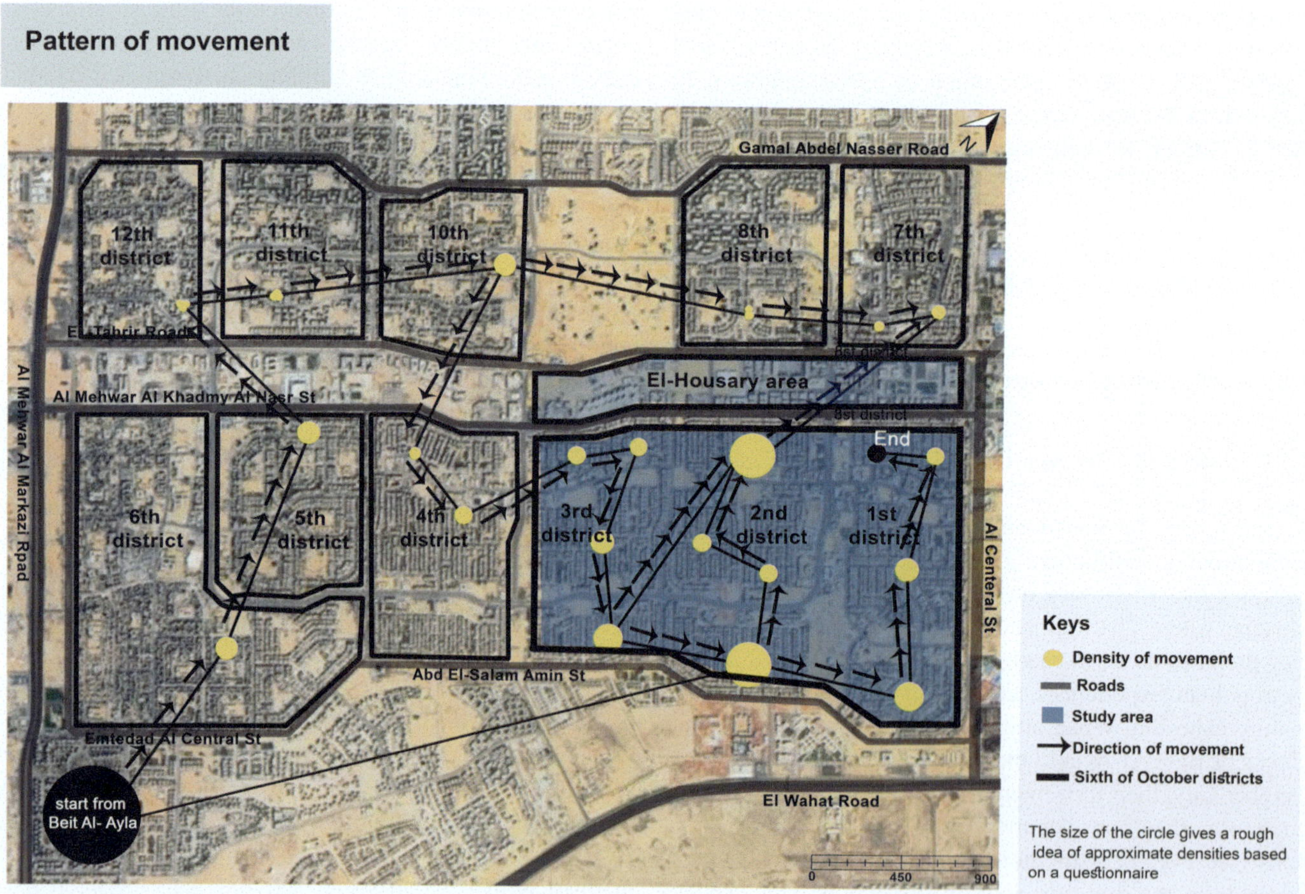

Fig. 11 Pattern of movement of Syrian refugees in the Sixth of October districts. *Source* The researcher based on survey and questionnaire

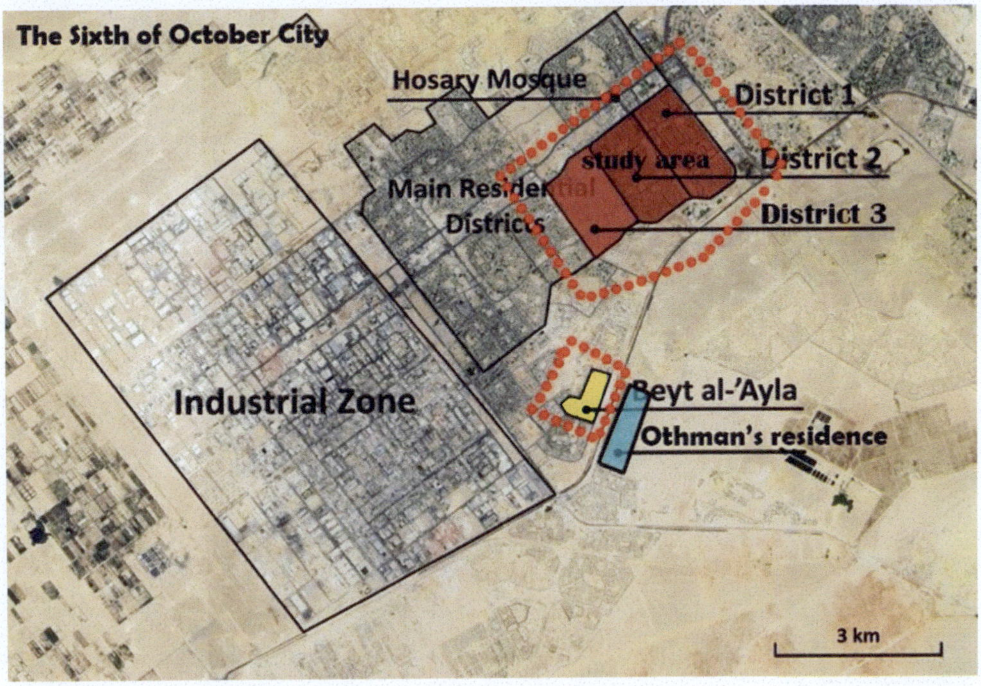

Fig. 12 Shows the main urban area of the Sixth of October City. *Source* Google Earth Pro. Additions by the researcher

around restaurants, coffee shops, and shops and others for all residential buildings in various neighborhoods as shown in Fig. 13.

Therefore some many advantages and disadvantages could be concluded from the presence of Syrian refuges in the Sixth of October city in Table 9.

6.3.3 Economic Insertion Impact of Syrian Refugees in the Sixth of October City

Figure 14 illustrates how the majority of Syrian refugees work in Egypt in many fields. Therefore, the Syrians in Egypt had an enormous effect not only on the business sector and industry but in many other fields. Therefore, there are many advantages and disadvantages that could be concluded from the presence of Syrian refuges in Sixth of October city in Table 10.

6.3.4 Urban Fabric Impact of Syrian Refugees on the Sixth of October City

In terms of urban fabric developments, the advantages of the existence of Syrians were reflected very clearly in the urban fabric and this revitalized the city of the Sixth of October city in general. Syrian-built districts in particular generated new uses for public spaces and enlarged existing ones. As shown in Fig. 15). Therefore many advantages and disadvantages could be concluded from the presence of Syrian refuges in the Sixth of October city in Table 11.

6.3.5 Political Changes Caused by Syrian Refugees on the Sixth of October City

In terms of political changes, the strength point here is, that the first arrival of Syrian refugees in Egypt was met with open arms by Egyptians, which was the prime reason why Syrians chose Egypt over other countries that are so close to Syria. Therefore some many advantages and disadvantages could be concluded from the presence of Syrian refugees in the Sixth of October city in Table 12.

7 Discussion of Findings

Germany's post-World War II reconstruction owes foreign laborers who came from Turkey much. The majority of Turkish immigrants worked in the construction, technical, and automotive industries lately sectors of tourism and services are included as well. Therefore, the highest percentage of the occupational labor market in Germany goes to Turkish immigrants with 59% as shown in the bio chart of the occupational market in Fig. 16. Also, as a result of immigration to Germany, Turkey's unemployment rates stayed within acceptable bounds. Despite this, Turkish immigrants' contributions to Germany's social and political life are overlooked as shown in the bio chart of housing income categories in Fig. 16, the Turkish immigrants settling in low-income housing with 52% which is characterized by low

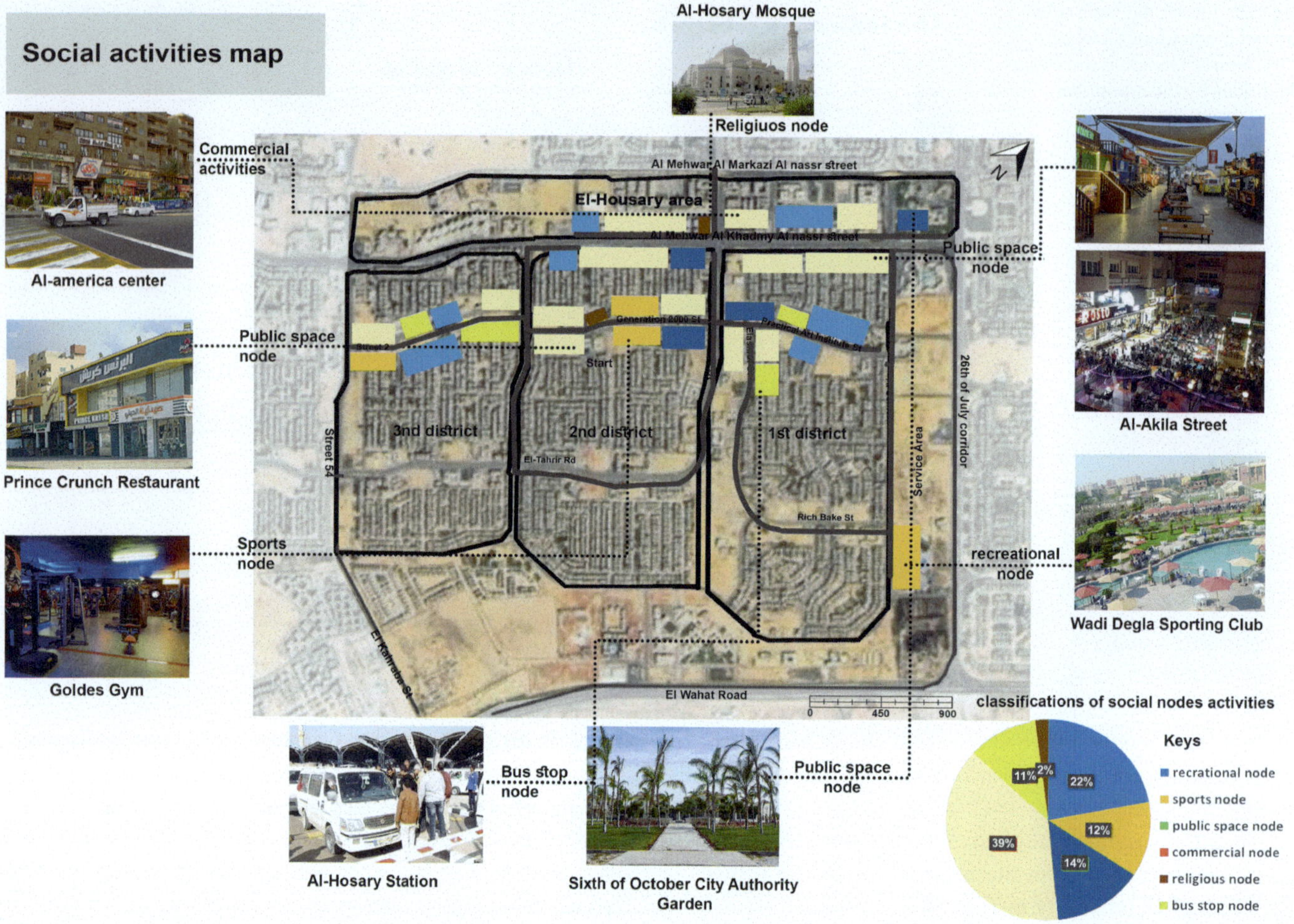

Fig. 13 Social activities of Syrian refugees in main paths in 1st, 2nd, and 3rd districts and el housari area. *Source* Developed by the researcher based on a survey and questionnaire

Table 9 Advantages and disadvantages for the presence of Syrian refugees in the Sixth of October city in terms of social cohesion

Advantages	Disadvantages
• Through the interviews Syrian refugees in Egypt claimed that the warm welcome they received from Egyptians was the main factor in their decision to travel to Egypt rather than other nearby nations that are physically closer to Syria • The advantages of the existence of Syrians relevant to social cohesion were evident in social activities. The researcher noted through a survey that Syrian social activities vary between religious gatherings around mosques or commercial gatherings around restaurants, coffee shops, and shops and others for all residential buildings in various neighborhoods as shown in Fig. 13	• Questionnaire results showed that a large majority of Syrian refugees noted a sharp change in Egyptian attitudes towards Syrians, due to the increased influx of Syrian refugees, and their extended duration of stay in Egypt • Through the interviews, numerous Syrian reported having psycho-social issues, many of which were attributed to long-term psychological stress brought on by the horrific experiences they had before departing Syria and which they claimed were made worse by the challenges they had to overcome in order to survive in Egypt

prices in rent and low recreational areas with 7% as shown in bio chart of land use classifications in Fig. 16.

Regarding Syrian refugees in Turkey, Turkey's initial refugee influx was viewed as a one-time event, and Syrian migrants were treated as "guests," but after the flood of 2014–2015, Turkish officials realized that a sizeable portion of these migrants were there to stay permanently. As a result, the emphasis has shifted from immediate protection of Syrian refugees to long-term social and economic integration. On the other hand, according to migration theory, temporary migration can easily turn into a permanent settlement in Germany. Many of the "Gastarbeiter" (guest workers) who became "Euro Turks" permanently remained in Germany. Based on the fact that Germany's economy

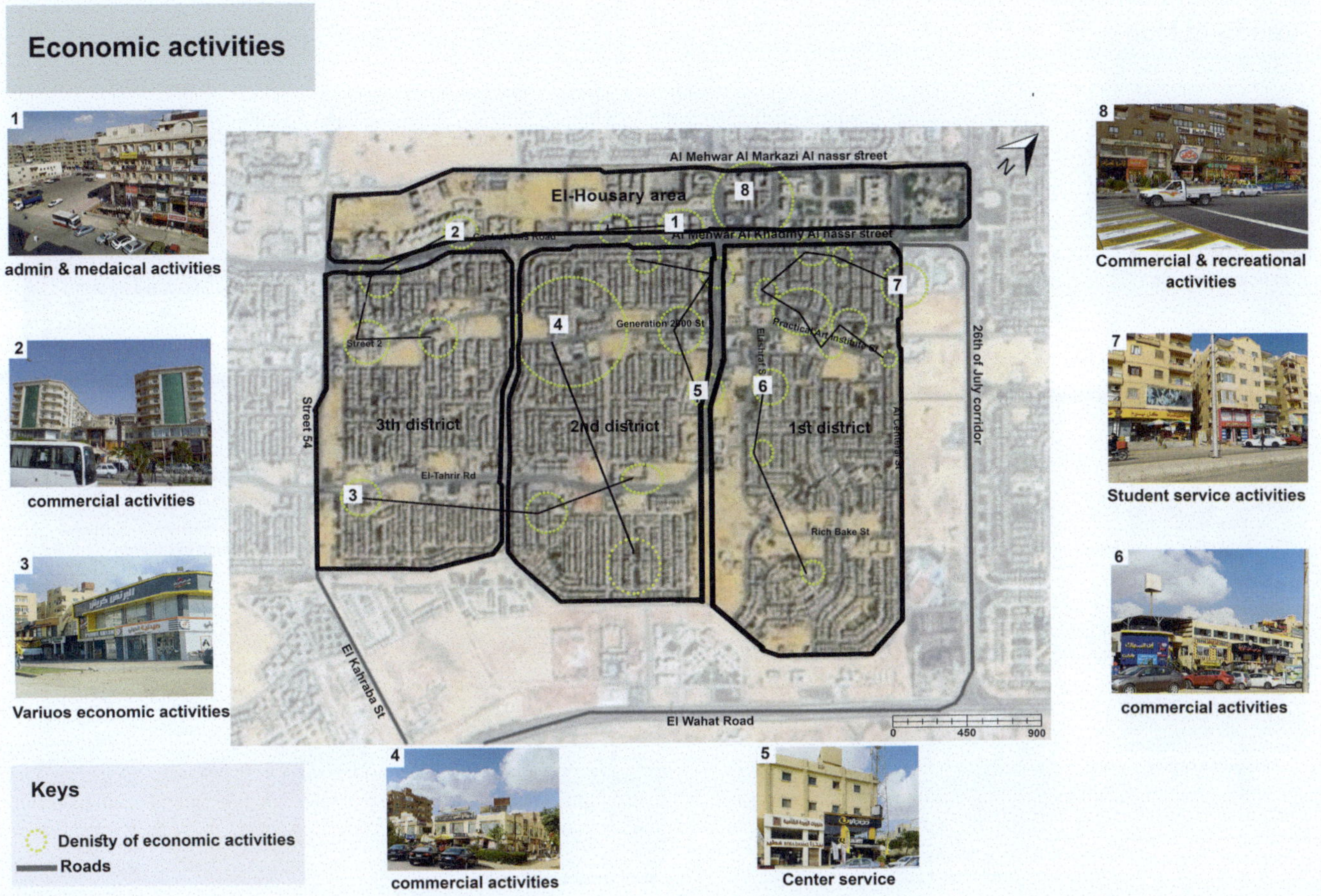

Fig. 14 Economic activities of Syrians in in 1st, 2nd, and 3rd districts and el housari area. *Source* The researcher based on a survey and questionnaire

Table 10 Advantages and disadvantages of the presence of Syrian refugees in Sixth of October city in terms of economic insertion

Advantages	Disadvantages
• The majority of Syrian refugees work in Egypt in many fields as shown in Fig. 18. Therefore the Syrians in Egypt had an enormous effect not only on the business sector and industry but in many fields • The survey of economic insertion reflected the advantages of the participation of Syrians in economic activities by highlighting the tendency to advertise their identity in economic activities. The great majority of the new shops, restaurants, and other economic activities carry names referring to Syria as a marketing strategy to differentiate them from Egyptian businesses	• Through the review of the questionnaire, increasing economic activities within the neighborhoods; led to crowding in service places, especially et al.-Hosary Square, Also, The Egyptian government is unable to offer a positive legal path to the labor market • Many other Syrians chose to leave Egypt for Europe due to the difficult economic situation there due to the government's inability to provide a favorable legal route to the job market • For employment, almost all Syrian refugees turn to the unofficial economy which exposes them to potential abuse exploitation, and instability

has grown to be among the biggest, most competitive, and most dynamic in the world. Turkey has economically benefited from emigration. Particularly in the 1970s, remittances from employees to their families in Turkey were a significant source of foreign cash that was used to balance trade imbalances and economic shortfalls. The highest percentage of the occupational labor market in Germany goes to Turkish immigrants with 59% as shown in the bio chart of the occupational market in Fig. 16. Despite this, a high refugee population frequently complains that the Turkish people prevent them from using public services like parks, health care, and education. In addition to that the skyrocketing rents, therefore as shown in the bio chart of housing income categories in Fig. 16, were the Turkish immigrants settling in low-income housing at 46% and middle-income housing at 41% with law recreational areas at 10% as shown in the bio chart of land use classifications in Fig. 16. Therefore, public services are already strained by the

Fig. 15 Land uses in in 1st, 2nd, and 3rd districts and el housari area. *Source* Developed by the researcher based on a survey

Table 11 Advantages and disadvantages of the presence of Syrian refugees in Sixth of October city in terms of urban fabric

Advantages	Disadvantages
• As a result, this led to urban and social development through socio-economic and cultural activities in the previously underutilized satellite cities of Cairo and higher demand for services. They converted many buildings from residential to residential-commercial buildings which represents their identity in the trade • As more than a year has passed since the arrival and settlement of large numbers of Syrian refugees in the Sixth of October City, some changes in the built environment have started to show, Syrian refugees revitalized the city of Sixth October in general, and in the Syrian built area they generated new uses for public spaces and enlarged existing ones they converted many buildings from residential to residential commercial buildings which represent their identity in the trade	• The changes in the built environment have a direct impact on the rental market Also the scarcity of green areas and gardens is a result of the profusion of commercial spaces due to the high demand for them • Despite the presence of many malls, such as Dolphin Mall, zumordah Mall, and others, commercial residential buildings appeared, and this indicates the intensity of consumption in these neighborhoods related to the high occupancy rate and the high demand for more services scarcity of green areas and gardens is a result of the profusion of commercial spaces due to the high demand for them • These changes in the built environment have a direct impact on the rental market and raise real estate values in these areas. These neighborhoods are characterized by the availability and diversity of services and ease of access

district's rapid expansion. These were determined to be the driving forces behind the Aleppo family's eventual departure to Europe.

De facto integration of Syrian refugees has emerged as the long-term alternative in Egypt, This is because resettlement and repatriation are dependent on factors beyond the control of UNHCR and the Egyptian government. Egyptian society is considered an essential part of this integration process through the push and pulls factors for Syrian refugees. For example, the pull factors for Syrian refugees in Egypt were the great welcome that the Syrians received upon their arrival, which encouraged others to come to

Table 12 Advantages and disadvantages of the presence of Syrian refugees in Sixth of October city in terms of political change

Advantages	Disadvantages
• Research with Syrian refugees highlighted an initial welcoming reception by Egyptians and cited that as the prime reason for choosing Egypt rather than going to other neighboring countries as well as the functional or professional distribution of the refugees in an effective way. It also depends on the willingness of the state to embrace the refugees and integrate them into their countries • Due to the reduced cost of living, good safety conditions, and the fact that certain members of the Syrian population already had familial links in Egypt, even though Egypt does not border Syria, it continues to be a favorite destination for Syrian refugees	• The questionnaire reveals that the Egyptian government's policy-making is fundamentally unclear on Syrian refugees in the country. It is not clear how long the visa requirement will be imposed on Syrian refugees, the criteria upon which Syrian refugees will be granted different types of visas, or how some Syrians were able to enter Egypt without meeting visa requirements • Many Syrians reported a substantial shift in Egyptian attitudes against them, which was linked to three major factors: an increase in the number of Syrian refugees, and their prolonged stay in Egypt

Egypt in general. As more than ten years have passed since the entrance and settlement of significant numbers of Syrian refugees in the Sixth of October city, they have brought socio-economic and cultural revivalism to the underutilized, under-served city as shown in the bio chart of land use classifications in Fig. 16, there are a variety of percentages of different land use areas. As a result, Syrian refugees contribute to urban and social development in the Sixth of October city through socio-economic and cultural activities that they have established there. Their presence has led to an increase in the use of public space and land in the city. The highest percentage of the occupational labor market in Germany goes to Turkish immigrants with 54% as shown in the bio chart of the occupational market in Fig. 16. These changes in the built environment and economic insertion made Syrian refugees move to middle-class income with the highest percentage at 52% as shown in the bio chart of housing income categories in Fig. 16. These neighborhoods are characterized by the availability and diversity of services and ease of access.

8 Conclusions and Final Remarks

This thesis goes into great length on the direct impacts of refugees on the City's urban fabric form and discusses the backdrop of hosting refugees in various nations. This Chapter's literature review is broken up into two parts: Part 1 is titled International Examples, which is divided into two sections, forced migration which is represented by the migration of Syrian refugees to Turkey, and labor migration which is represented by the migration of Turkish workers to Germany. Most refugees in Egypt are unable to return home, but at the same time, Egypt, like many other developing countries, is opposed to permanent settlements due to the country's economic issues. However, resettlement and repatriation depend on factors that are beyond the control of the Egyptian government. Therefore, local integration has become the de facto status for the refugee requirements, but they face numerous obstacles to meeting their basic needs due to the restrictive legal environment in which they live and receive services.

Inflows of both voluntary and involuntary migrants from inside Egypt and from around the world have altered the landscape of Cairo. It demonstrates how the functions of a city's spaces vary depending on its inhabitants and differ much more noticeably for refugees. As a result, Cairo has an intra-city displacement problem. This displacement affects impoverished populations and highlights the need for better planning and management. Therefore, the cityscape is disputed and complicated, making certain areas of Cairo appear distinct, dispersed, and more linked to areas outside the city than to areas inside it. Therefore, the history of the city's hosting and resettling immigrants and refugees and the purposes of its institutions are important in this respect. Analyzing refugee spaces in Cairo requires a comprehensive understanding of the city's socioeconomic and political systems, urban fabric, and urban compositions. Early multidisciplinary investigations have reflected the theoretical framework, but further in-depth investigation is needed to fully understand the factors influencing these spaces.

Immigrants, migrants, and refugees make significant contributions to cities like Cairo. The experience of living in the mega-metropolis, even if it wasn't always pleasant, has also improved the lives of these immigrants. The presence of hundreds of thousands of Syrians in Cairo at this time in history may have profound implications for the political, economic, and social ties and relationships between the two countries in the coming ten years, even though the current unrest in Egypt and the civil war in Syria make future predictions difficult. Due to recent and ongoing instability, shifts in public opinion, and shifting government views, the future of Syrian refugees in Cairo is unknown. Egypt's new strategy is reducing the number of Syrians coming to the country, and as a result, the Syrians are caught in a vicious cycle of unrest in Egypt as they flee a bloody conflict and adjust to an uncertain future. In order to deal with all of these difficulties, the Syrian and other refugee populations must proactively build their self-organized

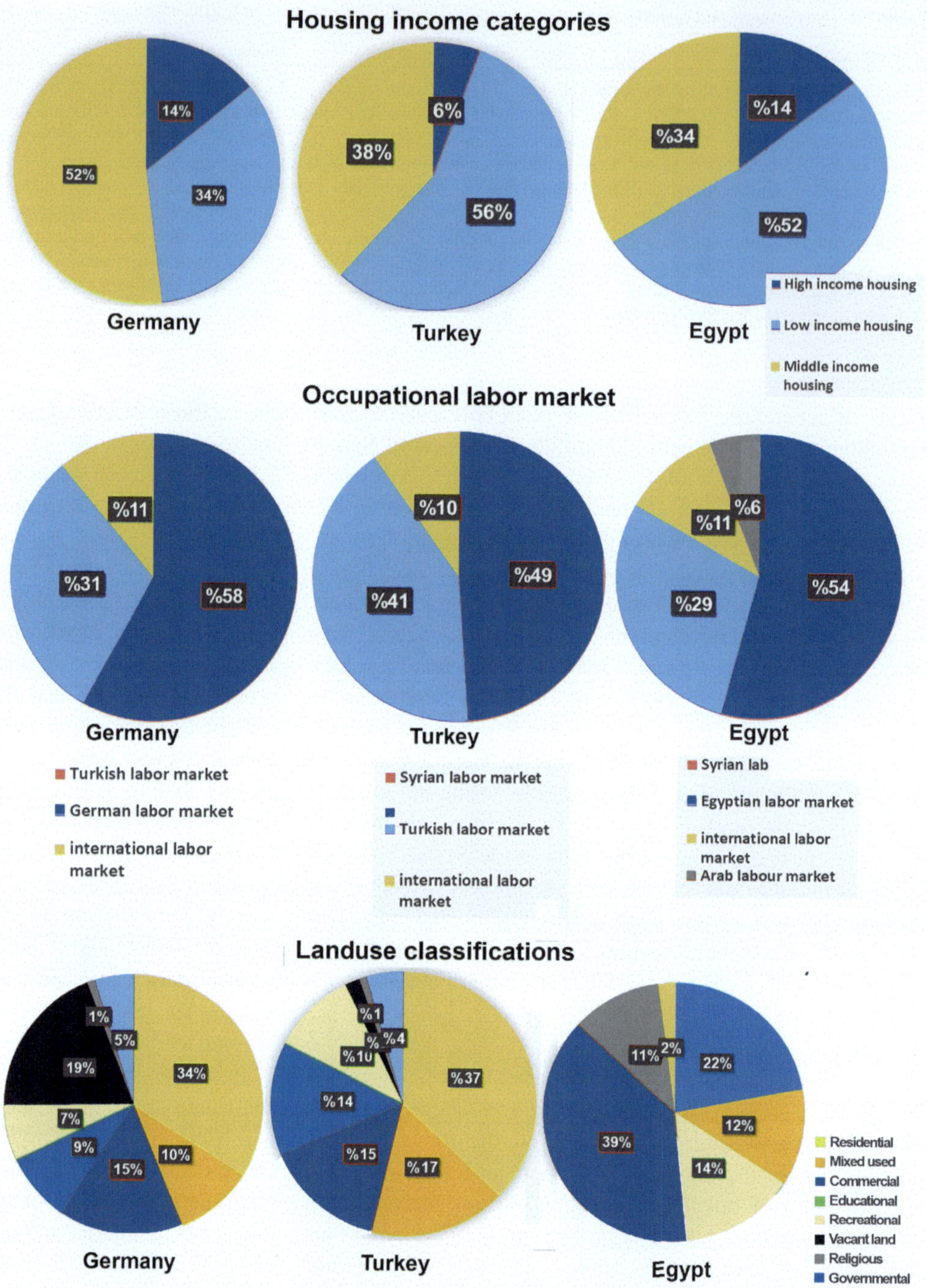

Fig. 16 Bio charts for host countries in terms of housing income categories, occupational labor market, and land use classifications. *Source* Developed by the researcher

methods. Therefore, the following Fig. 17 concludes the approaches and theories for urban refugees from the selected examples and the case study.

The integration of refugees into society improves conditions for them and benefits the host countries because it helps open up new markets and create many investment opportunities. Host countries are experiencing many economic and political problems, but let's look at the refugees from a different point of view as a great opportunity and strength to build and grow the country. Conversely, refugees can improve the economies of the nation's hosting them. Although they are frequently seen negatively from the standpoint of the host nations, labor market contacts can have favorable economic effects.

After a comprehensive review of the developed conceptual paper framework in each of the host countries, the researcher found that each country has overwhelmingly succeeded in a certain part and failed in other parts. Therefore, on the economic level, as shown in Fig. 18. It can take advantage of the experience of Turkish

Theories		Turkish immigrants in Germany	Syrian refugees in Turkey	Syrian refugees in Egypt
Urban fabric Theories	Landau Theory	Appling to Landau's theory, Turkish immigrants have a direct impact the receiving state and that's what happened exactly. Germany's post-World War II reconstruction owes foreign laborers who came from Tur-key much.	Appling to Landau's theory, Turkey has economically ben-efited from emigration. Particularly in the 1970s, remit-tances from employees to their families in Turkey were a significant source of foreign cash that was used to balance trade imbalances and economic shortfalls.	According to Landau's theory, As more than ten years have passed since the entrance and settlement of sig-nificant numbers of Syrian refugees in the Sixth of Oc-tober city, they have brought socio-economic and cul-tural revivalism to the underutilized.
	Applicability	✓	✓	✓
	Dreier et al Theory	According to Dreier theory before immigrating to the West, about half of Turkish migrants had worked in agriculture in the corresponding home country. Therefore, one of the most similar famous economic activities in Kreuzberg is the Turkish market It ex-presses Turkish identity.	According to Dreier theory, thousands of Syrians, work in the informal textile, shoemaking, and small-sized indus-trial sites. Most shoes produced in Turkey are manufac-tured by shoemaking workshops in Sultan Gazi, therefore Syrians chose Sultan Gazi to settle in contrast in their work in trade in Syria	According to Dreier theory, Syrian refugees turned the residential buildings in 1St, 2Nd, 3Th districts to the residential commercial buildings to work in trade as they used to in Syria and they can reside close to their places of employment
	Applicability	✓	✗	✓
Social cohesion Theories	international migration Theory	Appling to international migration theory, Kreuzberg has a long history as a working-class neighborhood the region saw many immigrants, in-cluding Turkish people and those from the Middle East and Europe. Therefore, many people of Turkish ancestry dwell in the Little Istanbul-dubbed Kreuzberg area in downtown Berlin.	Appling to international migration theory, there is no long history between Turkey and Syria so Syrian refugees fre-quently decide where to reside based on areas in which they already have connections or relatives. The places where Syrian refugees live typically have several charac-teristics: widespread poverty, high levels of cultural con-serve.	Appling to international migration theory, relations between Egypt and Syria have a long history. When the United Arab Republic was founded in 1958, these connections were at their best (Amelia, 2015). Before the in-flow in 2012, there were two additional sizable waves of Syrian immigration to Egypt that's what en-courage Syrians to come.
	Applicability	✓	✗	✓
	Migration network Theory	Appling to migration network theory, the Turkish immigrants settled in Kreuzberg district which is the highest density of Turkish immigrants in the whole of Berlin districts, it called as the Little Istanbul be-cause of its sizable Turkish population, outside of Turkey.	Appling to migration network theory, network Theory is perhaps the one that applies most to the situation of Syrian refugees residing in Istanbul. Family networks provide re-cently arrived refugees a sense of warmth and familiarity while also assisting them in maintaining their culture and close links with the family.	Appling to migration network theory, the Sixth of Oc-tober City has been a haven for the first arrivals from the Syrian crisis and had come in extended family groups and clustered around beit Al-Ayla but they failed to settle their due to the shortage of safety and so they moved to the rest of Sixth of October districts.
	Applicability	✓	✓	✗
Political changes Theories	Kunz's Theory	Appling to Kunz's theory, much of Northern Europe experienced a severe labor shortage after World War II. Therefore, Europe created bilateral recruitment agreements with Germany to address this labor shortage but The German government sought to limit the inflow of new foreign employees.	Appling to Kunz's theory, Turkey has economically bene-fited from emigration. Particularly in the 1970s, remit-tances from employees to their families in Turkey were a significant source of foreign cash that was used to balance trade imbalances and economic shortfalls.	Appling to Kunz's theory, Syrians were drawn to Egypt because of the country's established Syrian community, which includes companies and social net-works. Many Syrian families were thriving in Egypt's commercial world and excelled in various professions, such as; cotton trade, and restaurateurs.
	Applicability	✗	✓	✓
	Jacobsen's Theory	Appling to Jacobsen's theory, as the ratio of immi-gration to Germany increased, other European na-tions expressed their desire to hir e Turkish employ-ees. As a result, Turkey's First Five Year Develop-ment Plan (1963–67) went into force, causing a more thorough review of migration.	Appling to Jacobsen's theory, Turkey established an open-door policy for anyone needing refuge and took a strong stance against the Assa d administration. It was anticipated that the battle would end quickly when it first started. However, the government referred to the Syrian refugees entering Turkey as "guests,"	Appling to Jacobsen's theory, Egypt is considered to have the second-highest political participation for the Syrian opposition after Turkey. Syrian migrant's ori-gins and the political leanings influence the city's ge-ographical regions. Therefore, the Egyptian govern-ment's is fundamentally unclear on Syrian refugees.
	Applicability	✓	✓	✗
Economic insertion Theories	Ager & Strange's Theory	Appling to Ager's theory, the highest percentage of the occupational labor market in Germany goes to Turkish immigrants with 59% and Also, as a result of immigration to Germany, Turkey's unemploy-ment rates stayed within acceptable bounds.	Appling to Ager's theory, Syrian refugee families tend to stick to their little social networks by avoiding making new acquaintances with local Turkish people. Conse-quently, Syrians stuck to find employment and jops and forced to intimate bonds with their fellow citizens in their neighbor-hoods to find any work.	Appling to Ager's theory, the survey of economic in-sertion reflected the advantages of the participation of Syrians in economic activities by highlighting the ten-dency to advertise their identity in economic activities. Therefore the Syrians in Egypt had an enormous effect not only on the business sector.
	Applicability	✓	✗	✓
	Macro's theory	Appling to Macro's theory, economic and socio-po-litical factors in both sending and target countries can become the source of the migration, where the attractive elements in Germany are considered as "pull factors" such as higher income and "push fac-tors" such as; the poverty, higher unemployment rate high-priced housing, and poor services.	Appling to Macro's theory, families and individuals from Syria who are refugees have numerous reasons to leave Syria and relocate to big cities like Istanbul. Syrian refu-gees in Istanbul have spoken about pull factors such as; the prospects for job opportunities, social net-works in urban areas, and safety and security and that's what they need it in their country.	Appling to Macro's theory, Egyptian society is consid-ered an essential part of the push and pulls factors for Syrian refugees pull factors for Syrian refugees such as; the welcome the Syrians received upon their arri-val, which encouraged others to come to Egypt in gen-eral. As for the push factors, for example, the Sixth of October authority, removing of street vendors
	Applicability	✓	✓	✓

Fig. 17 Conclusion of contemporary approaches and theories for urban refugees in host countries. *Source* By the researcher

immigrants in Germany due to the economic integration that took place between the two countries and the success of the employment agreement between the two countries. In the process of the transformation of 'Gastarbeiter' (guest worker) to 'Euro Turk', many settled in Germany permanently. Based on that Germany became one of the largest, most competitive, and most dynamic economies in the world.

As for the urban fabric level, as shown in Fig. 18. It can take advantage of the Syrian refugees in Turkey, as they went to places that suit their identity and conditions from the beginning where Syrians tend to congregate in particular neighborhoods where poverty is more pronounced, conservatism and religion are part of daily life, social networks, and living costs are lower. Therefore, they did not have to change a large part of it, as they did in Egypt in the Sixth of October City, where it was a new residential area in need of many services, so they converted the residential buildings into residential and commercial buildings.

As for the social cohesion level, as shown in Fig. 18. It can take advantage of the Syrian refugees in Egypt because to De facto integration of Syrian refugees has emerged as the long-term alternative in Egypt, and Egyptian society is considered an essential part of this integration process through the push and pull factors for Syrian refugees. For example, the pull factors for Syrian refugees in Egypt were the great welcome that they received upon their arrival, which encouraged others to come to Egypt in general, and Syrian social activities vary between religious gatherings around El-Housary mosque or commercial gatherings such as; coffee shops, and shops and others in the 1st, 2nd, and 3rd neighborhoods. Therefore, the large influx of Syrian refugees contributed to the revival of many social and economic activities on the Sixth of October City.

As for the political changes level, as shown in Fig. 18. Unfortunately, none of the hosting countries succeeded in recognizing refugees and immigrants on their lands, except Germany. It succeeded in a small part when it legalized

Refugee's country	Refugee nationalities	Political changes		Economic insertion		Social cohesion		Urban fabric	
		Legal Status	Protection issues	Economic activities	Economic market labor	Social Activities	Housing	Urban spaces	Land use
Germany	Turkish	✓ Yes, it exists for a temporary period during bilateral recruitment agreements with Germany causing a more thorough review of migration.	✗ The first generation of immigrant Turks struggle with their identity. They are stuck between Turkey and Germany.	✓ Germany granted residence and work permits to spontaneously enter Turkish workers to increase the volume of labor inflow from Turkey.	✓ Turkish entrepreneurs are well-equipped to do international business in Kreuzberg. Therefore Germany and Turkey outstanding partners.	✗ Turks have the lowest level of social integration in Kreuzberg culture and have less luck finding employment there than local people.	✗ Housing costs have skyrocketed, putting a strain on services and transportation, and many Turks can no longer afford to live there.	✗ Kreuzberg neighborhoods were neglected for a very long time and that the amount to which people exercise their right to urban space by it has been con-strained.	✓ Kreuzberg land uses underwent modernization by the Turkish immigrants and developed into a vital and thriving region and desirable locations.
Turkey	Syrian	✗ Uncertainty about their status as refugees (specifically regarding work permits and citizenship), dire working circumstances such as; low salaries.	✗ The government referred to the Syrian refugees entering Turkey as "guests," which gave them no legal rights and implied a "temporary stay"	✗ Syrians do informally and work in low-productivity industries in low-skilled roles such as; street vendors, call shops, resturants, and small workshops.	✗ Host citizens considered Syrian refugees an economic burden. Moreover, the Syrians' socio-economic impact is becoming negative increasingly on hosting country.	✓ Many Syrian refugee families stick to their little social networks. They had many social activities such as; hung out with other Syrians after working hours.	✗ Syrian refugees in Turkey increased rents at high-quality units and deepened residential segregation. Because the refugee influx increased the demand.	✓ Syrians tend to congregate in particular neighborhoods where poverty is more pronounced, conservatism and religion are part of daily life, social networks, and living costs are lower.	✓ Municipal land uses, and services for Syrian refugees vary greatly between Gazi, zubedye hanim, 75 yil, and districts. it from residential to commercial , etc
Egypt	Syrian	✗ The Egyptian government's policy is fundamentally unclear on legal status of Syrian refugees in the country. Thus, legal status of them is un stable.	✗ Many Syrians reported a substantial shift in Egyptian attitudes against them, which was linked to increase in the number of them.	✓ The majority of Syrian refugees work in Egypt in many fields. Therefore the Syrians in Egypt had an enormous effect on the business sector and industry.	✓ The Syrian community in Egypt significantly impacted publishing, journalism, and the economic sector in Egypt and revitalized the housing market on the sixth of October City.	✓ Syrian social activities vary between religious gatherings around mosques or commercial gatherings around restaurants, coffee shops, and shops.	✓ Housing rents vary between the cheapest to the expensive even if the concentration of Syrians in particular regions, generally increased the rental market.	✗ Syrian refugees revitalized the Sixth of October City in general and they generated new uses for public spaces and enlarged existing ones. Activities in the previously underutilized.	✗ Syrian refugees converted many buildings from residential to residential commercial buildings which represent their identity in the trade.

Fig. 18 Applicability of conceptual thesis framework on the hosting countries. *Source* By the researcher

the legal status for Turkish immigrants in the country for a temporary period during bilateral recruitment agreements between Germany and Turkey causing a more thorough review of migration and other European nations (Dinbabo & Nyasulu, 2015) expressed their desire to hire Turkish employees. But after this period they felt uncertainty about their status as immigrants (specifically regarding work permits and citizenship), dire working circumstances (low pay, informal, precarious, and jobs), and discrimination were the primary barriers to their ability to integrate.

References

Adnan Doğan BULDUR. (2018). Urbanization process of Sultangazi District (Istanbul). *Turkish Studies Journal*.

Akgündüz, A. (2021). *Germany's recruitment of workers from Turkey, 1960 to 1973—Some remarks*. University of Amsterdam.

Ayoub, M., & Khallaf. (2014). Syrian refugees in Egypt: Challenges of a politically changing environment. *Cairo Studies on Migration and Refugees*.

Caro, L. P. (2020). Syrian refugees in the Turkish labour market. *ResearchGate*.

Dinbabo, M., & Nyasulu, T. (2015). Macroeconomic immigration determinants: An analysis of 'Pull' factors of international migration to South Africa. *African Human Mobility Review, 1*(1), 27–53.

ECHO. (2022). Syria factsheet.

Edo, A., Ragot, L., & Rapoport, H. (2018). The effects of immigration in developed countries: Insights from recent economic research. *CEPII—Policy Brief*.

Erdoğan, M. (2014). Syrians in Turkey: Social acceptance and integration research. *UNHCR*.

Global Trends Report. (2021).

Göksel, N. (2018). Integrating Syrian refugees in Istanbul's "District of Victimhood". *Crisis Group*.

IDMC. (2017). *Global report on internal displacment*.

Katzenson, E. (2015). *The patterns and impacts of T atterns and impacts of Turkish Immigr urkish immigration t ation to the E o the European*. University of California.

Kaya, A., & Kıraç, A. (2016). Vulnerability assessment of syrian refugees in istanbul. *Support to Life*.

Koca, B. T. (2019). Urban citizenship and the spatial encounter between Turkish Migrants and Syrian refugees in Berlin. *Sciendo yte*.

Presidency, T. A. (2021). President El-Sisi in an interview with "Die Welt".

Sims, D. (2011). *Understanding Cairo: The logic of a city out of control*. American University in Cairo Press.

McFadden, S. W. (2019). German citizenship law and the Turkish diaspora. *German Law Journal*.

Syrians in Istanbul and Post-war Syrian Ghettos. (2019).

The arab republic of Egypt president. (2021). *President El-Sisi in an interview with "Die Welt"*.

Topcu, E. (2022). Germany: Dramatic increase in number of Turkish refugees. *DW Journal*.

The World Bank. (2022). *Turkey's response to the Syrian refugee crisis and the road ahead*.

TURKSTAT's 2018 census was announced. (2018).

UNHCR global trends. (2021).

UNHCR. (2019). UNHCR Egypt airo: United Nations High.

UNHCR . (2021). Refugee camps explained.

Verghis, S., & Balasundram, S. (2019). 7 Urban refugees: The hidden population. *OXFORD Academic*.

Wahdan, D. (2012). *Transport thugs: Spatial marginalization in a Cairo Suburb*. American University Press.

Evaluating Users' Satisfaction on Urban Railway Based on Service Quality Model: The Study on KLIA Express in Malaysia

Yuhan Du and Kun Sang

Abstract

There are many countries around the world facing an increasing number of transportation challenges due to rapid urbanization. Urban railways are an important part of transportation systems. Their optimization can be an effective solution to urban problems, contributing to the improvement of urban livability and environmental sustainability. With the development of urban railways in Malaysia, the changing market environment has prompted the Kuala Lumpur International Airport (KLIA) Express to look for opportunities to optimize its services and increase users' satisfaction. Previous studies mainly investigated the influential factors on trains based on several common models, such as the satisfaction-loyalty model. The quality of urban railway services can be reflected by the service quality (SERVQUAL) model by measuring customers' perceived satisfaction. In this research, the SERVQUAL model was applied to the KLIA Express. Firstly, six dimensions, including tangibility, reliability, responsiveness, assurance, empathy, and pandemic, were developed as indicators for the assessment. Secondly, it introduced users' purchase intention as the dependent variable and the six dimensions as independent variables to verify their relationship based on the linear regression method. The findings concluded that reliability and pandemic significantly affected visitors' purchase intentions for the Express. Finally, the results can be referenced to provide suggestions for further actions to optimize the railway service.

Keywords

Users' satisfaction · KLIA express · SERVQUAL · Malaysia · Covid-19

1 Introduction

Urban railways around the world are closely linked to the daily economies of ordinary people. Kuala Lumpur (KL), Malaysia's capital city, has witnessed rapid economic growth accompanied by a boom in private transport. But a lack of patronage due to the use of private cars and other modes of transport diverting users from rail transport has become a challenge for railways to further develop, which also affects the urban development of KL to some extent. Urban railways can directly raise the attractiveness of destinations in terms of their service quality (Pineda-Jaramillo & Pineda-Jaramillo, 2022). As a country with a strong tourism industry, Malaysia benefits from the expansion of the tourism economy (WTO, 2022). At the same time, the dependence between Malaysian tourism economy and its railway development is increasing. With the influence of Covid-19 pandemic and changes in Malaysian policy, commuters' needs are expanding, and urban railways are thereby facing some promising opportunities. Among all the railways in Malaysia, the KLIA Express has a high profile and is known as the fastest and most convenient way to travel between destinations. It is also uniquely represented as one of the airport railways due to its diverse functions. According to a national survey, it served 388,949 passengers in 2020 (MTM, 2022). Though it was briefly suspended during the Malaysian Movement Control Order (MCO), the KLIA Express still has a very high patronage thanks to the crowds from all over the world that move between the destinations after its reopening in 2020. Thus,

Y. Du · K. Sang (✉)
School of Communication, Xiamen University Malaysia, 43900 Sepang, Malaysia
e-mail: kun.sang@xmu.edu.my

Y. Du
e-mail: adt2009054@xmu.edu.my

I. Pigliautile et al. (eds.), *Urban Resilience, Livability, and Climate Adaptation*, Advances in Science, Technology & Innovation, https://doi.org/10.1007/978-3-031-54911-3_16

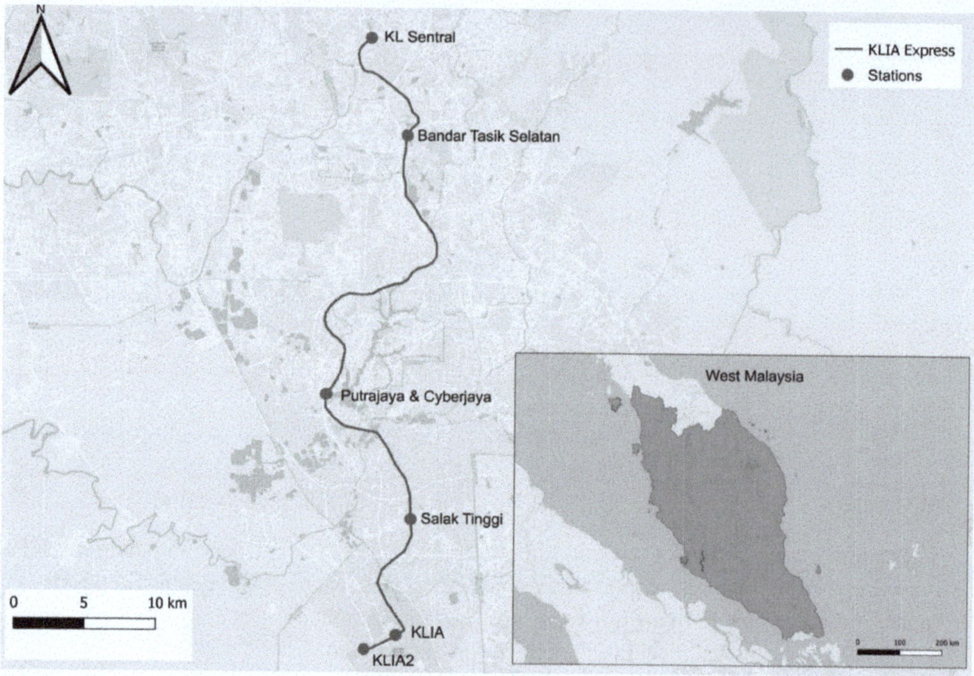

Fig. 16.1 KLIA express in Malaysia (Author)

this express was selected as the research object in this paper (Fig. 16.1).

Urban railways are a fast and high-capacity form of urban public transport. Their presence can help reduce the dependence of commuters on private transport, thereby improving commuting patterns and relieving traffic pressure. In addition to assessing the quality of service for railways, this study also contributes to the development of KL's tourism economy by identifying the needs of users for urban rail transport. Academically, in recent years, there has been little research into the users' satisfaction of KLIA Express travelers and the comprehensive assessment of KLIA Express as a case. On the other hand, hardly any studies have been conducted to assess satisfaction with the KLIA Express using the SERVQUAL model. Thus, this study not only fills a research gap in this field but also assesses satisfaction with the Express, and the indicators can be used as a reference for other similar studies for railway satisfaction assessment.

2 Literature Review

The significance of urban rail transport in the daily commutes has been widely recognized. According to Majumdar et al. (2021), commuting between origin and destination is one of the most critical components of an individual's daily routine, and the associated satisfaction with travel significantly affects commuters' well-being. From the perspective of satisfaction, previous studies have shown that commuter's satisfaction is correlated with their choice of transport, and the desire and intention to use a particular mode of travel also influence final customers' choices. The level of satisfaction with the service quality of urban rail systems plays a significant role in determining transport mode selection (De Vos et al., 2021). Chakrabarti (2017) has highlighted that promoting existing potential demand and creating new demand can increase the frequency of public transport used by travelers. Therefore, enhancing the quality of urban rail services can effectively improve their competitiveness. Scholars have been exploring ways to improve urban rail transport from an efficient, economical, sustainable, and technological perspective (Ibrahim & Borhan, 2021; Ibrahim et al., 2021; Kwan et al., 2018). As part of a city's infrastructure, urban rail is also an environmentally friendly mode of transport. Its high carrying capacity and low emissions contribute to the sustainable development of cities.

Although customer satisfaction is an important concept, it can be ambiguous and often measured using criteria like those used to assess service quality, taking into account service perceptions (Miller, 1976; Oliver, 1981). Recent studies on railway services as a tourism resource are limited, highlighting the need for further research to reflect visitor satisfaction and to develop the rail resource and tourism economy (Mohamad, 2022). In the customer-oriented public transport industry, behavioral intention refers to a visitor's intention to use a service, which can trigger future consumption behavior. Previous studies exploring the relationship between satisfaction and service quality in this field have introduced the concept of behavioral intention. In addition, Gotlieb et al. (1994) concluded that service quality is a prerequisite for satisfaction, and that behavioral intentions are influenced by satisfaction. De Ona et al. (2016) showed that users' satisfaction predicts their intention to use the service again. Recently, there has been renewed interest in the relationship between customer satisfaction and purchase intention (Boubker & Douayri, 2020; Tran, 2020).

In terms of the model for evaluating service quality, Parasuraman et al. (1988) introduced the basic SERVQUAL model, including five dimensions: tangibility, reliability, responsiveness, assurance, and empathy. The SERVQUAL model conceptualized perceived quality as a way of capturing actual service quality by measuring perceived quality. Zeithaml (1988) defined perceived quality as the consumer's judgment of the overall superiority of a good. Later, the SERVQUAL model has been a popular method for assessing customer satisfaction with transportation, and this has served as a basis for extending a few specific models. For example, Ok et al. (2018) further demonstrated that service quality is expressed through the customer's evaluation and perception of the service, which implies that whether a consumer is satisfied with the service or not determines the quality of the service. During the Covid-19, scholars also included "pandemic" as one of the dimensions for the evaluation model, known as Pandemic-SERVQUAL 4.0 (P-SERVQUAL), in addition to the five dimensions developed (Tumsekcali et al., 2021). Therefore, the prerequisite for providing quality services is to fully consider market needs and improve the level of service that customers can perceive. With the development of time and the renewal of the industry, the concept of service quality has been widely applied to various emerging industries. However, current studies have not fully developed indicators for railway studies, which presents an opportunity for future studies to develop new assessment indicators based on the specific railway's actual situation.

Based on the literature review, this study aims to (1) quantify users' satisfaction value using the SERVQUAL model with selected indicators; (2) explore the relationship between KLIA users' satisfaction and their purchase intention. Six hypotheses (Fig. 16.2) contribute to achieving these two objectives, establishing the KLIA express users' satisfaction assessment system, and providing recommendations for the KLIA express services.

H1: Tangibility has a significant influence on users' purchase intention for the service of KLIA express.

H2: Reliability has a significant influence on users' purchase intention for the service of KLIA express.

H3: Responsiveness has a significant influence on users' purchase intention for the service of KLIA express.

Fig. 16.2 Theoretical framework (Author)

H4: Assurance has a significant influence on users' purchase intention for the service of KLIA express.

H5: Empathy has a significant influence on users' purchase intention for the service of KLIA express.

H6: Pandemic has a significant influence on users' purchase intention for the service of KLIA express.

3 Methodology

This study constructed a theoretical framework based on the SERVQUAL model for evaluating the users' satisfaction with the KLIA Express. Considering the influences of Covid-19, an evaluation system was developed with reference to the P-SERVQUAL 4.0 model to evaluate user satisfaction with urban rail transport in Malaysia. To test the hypotheses above, this paper used Statistical Package for the Social Sciences (SPSS) software for data analysis. The data needs to pass the validity and reliability tests. The validity of the survey was examined using the Kaiser–Meyer–Olkin Measure of Sampling Adequacy (KMO) and Bartlett's test. Then, this study used multiple linear regression (MLR) to examine the effects of six dimensions on users' satisfaction and intention. In this analysis method, R-square represents the fitness of model to the data; the coefficient (F-value) represents the overall significance, and the p-value is the criterion to test the hypotheses. In addition, the coefficient B-value is a predictor that shows how the dependent variable is affected by changes in the independent variable. Based on the evaluation system, the details are shown in the following steps (Fig. 16.3).

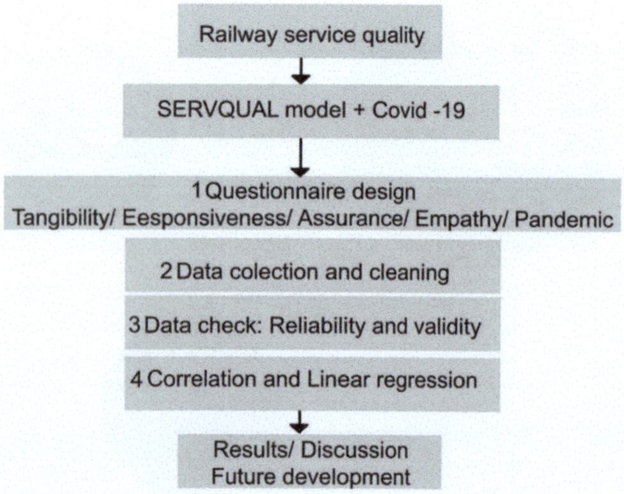

Fig. 16.3 Workflow (Author)

3.1 Data Collection

The sampling method used in this study was purposive sampling (a non-probability sample). The KLIA Express has a significant diversity of customers from all over the world. To ensure the accuracy of the responses, the study designed a sampling criterion: visitors who have taken the KLIA Express in the past and at least two stops should pass. According to Long & McMellon (2004), the sample size should be five times the scaled questions. Thus, the sample size for this research should be at least 105 eligible questionnaires. The questionnaires were distributed online and offline in November 2022. The online questionnaires were distributed via social media platforms such as WeChat and Instagram, while offline questionnaires were given to eligible users at three stations (KLIA, KLIA 2, and KL Sentral) along the Express.

Fieldwork was carried out during this period to collect data. The questionnaire was pilot-tested, with 30 responses to check its feasibility. The overall reliability analysis of the scale used in this study was 0.919 (Cronbach's Alpha), indicating high reliability. After importing the data into the software, it was found that all 185 respondents had answered the questions without any missing values. Among these, 169 responses were identified as inliers and retained, while 16 extreme values were excluded using the Local Outlier Factor method in Orange. The response rate for the questionnaire collection was 91.5%, which was considered high.

3.2 Evaluation System

Using dimensions as a primary level and indicators as a secondary level, a questionnaire was developed (Table 16.1). In this questionnaire, a 7-point Likert scale allowed for a more granular description of each level (from 1 to 7, the values mean the degree of agreement from "strongly disagree" to "strongly agree"). In this research, the dependent variable (Y) measured was users' purchase intention for the service of KLIA express, and the independent variables (X) were tangibility (X1), reliability (X2), responsiveness (X3), assurance (X4), empathy (X5), and pandemic (X6) (Loo et al., 2016; Sam et al., 2018; Shi et al., 2021).

4 Results

4.1 Descriptive Analysis

The proportion of female respondents was more than three times that of male respondents, with females exceeding males by 60%. Most respondents, 158 (93%), were aged between 18 and 25. Regarding occupation, 152 (90%)

Table 16.1 Indicators and questions of the system

Dimensions	Indicators	Questions
X1. Tangibility	I1. Modernity	Q1. The facilities at KLIA express are modern
	I2. Aesthetic	Q2. The facilities at KLIA express are aesthetically pleasing
	I3. Matching	Q3. The facilities at KLIA express match the services offered
X2. Reliability	I4. Waiting Time	Q4. KLIA express provides services during your waiting time at the station
	I5. Accessibility	Q5. You can easily access to the KLIA express by other transports
	I6. Punctuality	Q6. The KLIA express service is provided in a timely manner and the trains are basically not late
X3. Responsiveness	I7. Reminder	Q7. The KLIA express can inform users timely when the service is late
	I8. Helpfulness	Q8. The help center of KLIA express can answer users' request in a timely manner
	I9. Ticketing	Q9. Ordering KLIA express tickets (online or offline) is easy and guaranteed
X4. Assurance	I10. Reassuring	Q10. The KLIA express travel is reassured that personal injury is largely unlikely to occur during travel
	I11. Emergency	Q11. The KLIA express has emergency equipment (like window hammer)
	I12. Belongings Security	Q12. Your personal belongings are safe during the journey with KLIA express
X5. Empathy	I13. Personalization	Q13. The KLIA express are providing you with personalized attention
	I14. Users' Interests	Q14. The KLIA express care about the interests of the users
	I15. Convenience	Q15. The KLIA express service is available at the same time when you need
X6. Pandemic	I16. Hygiene	Q16. KLIA express vehicle and stations are clean and hygienic
	I17. Social distance	Q17. KLIA express maintains a social distance between customers, which makes you feel safe
	I18. Travel time	Q18. KLIA express has a shorter travel time, leaving you feel safe
Y. Purchase intention	B1. Continuity	BQ1. You intend to continue using the KLIA express in the future
	B2. Priority	BQ2. You consider using the KLIA express as your first choice when you need
	B3. Recommendation	BQ3. You will highly recommend others to use the KLIA express

respondents were students, followed by 14 (8%) white-collar workers, one (1%) blue-collar worker, and two (1%) freelance workers. Students made up the largest proportion of respondents. Concerning the frequency of passing stations, Salak Tinggi and KL Sentral were two of the most frequently visited stations, with 109 respondents. Putrajaya and Cyberjaya, which was chosen by only 47 respondents, was the least frequently visited station overall. Since the questionnaire was distributed both online and offline, with irrelevant responses excluded, little variability was observed between online and offline answers.

The most frequently cited reason for choosing the KLIA Express was its convenience and accessibility, chosen by 113 respondents. This indicated that the KLIA Express provided easy and convenient access to their destinations. Compared to other public transportation, it had a faster frequency and more frequent departures during peak traffic periods, making it more suitable for some respondents' schedules than buses, which was why 49 respondents chose it. Regarding the frequency of different respondents traveling with the KLIA Express, 77 (46%) of the respondents used the express with variable frequency when they needed it. Eight respondents (5%) used it three or more times a week. The largest number of respondents, apart from the variable frequency of use, took the KLIA Express between one and three times a month, with 39 (23%) respondents.

The results of the study showed that most respondents had a neutral attitude towards their relative satisfaction with the KLIA Express service. The mean values of the six dimensions measured their satisfaction levels, with assurance (5.32) and responsiveness (5.30) scoring the highest, followed by pandemic (5.15), tangibility (5.05), reliability (5.01), and empathy (4.95). Among all the factors, the highest satisfaction scores were achieved by punctuality (5.71), travel time (5.48), and hygiene (5.42), while the three lowest satisfaction scores were for aesthetics (4.84), personalization (4.81), and waiting time (4.40). The final section of the questionnaire was an open-ended question asking users for suggestions to improve the service. The responses were analyzed using Orange, and a word cloud was generated to identify the most common suggestions. According to Fig. 16.4, users were most concerned about price, stops, announcements, safety, punctuality, maintenance of facilities, and frequency. These areas could be prioritized for future development of the express service.

4.2 Reliability and Validity

This study examined the consistency of all items on the scale by assessing the Cronbach's Alpha value. Table 16.2 demonstrated that the Cronbach's Alpha value for each variable was above 0.6, indicating acceptable reliability for each individual variable. The overall Cronbach's Alpha value of the questionnaire was 0.935, indicating high reliability and consistency (Heo et al., 2015). KMO and Bartlett's test were conducted to check the validity of each dimension in the questionnaire, and the result was 0.861 (significant value 0.0), indicating high validity (Table 16.3).

Table 16.2 Reliability analysis

Variables	Number of items	Cronbach's alpha
Tangibility	3	0.781
Reliability	3	0.675
Responsiveness	3	0.772
Assurance	3	0.765
Empathy	3	0.735
Pandemic	3	0.775
Purchase intention	3	0.876
Total	21	0.935

Table 16.3 KMO and Bartlett's test

KMO		0.861
Bartlett's test of sphericity	Approx. chi-square	702.908
	df	21
	Sig	0.000

4.3 Correlation and Regression

After data pre-processing, valid data from 169 respondents was obtained, which exceeded the minimum requirement of 105. The measurements used in the questionnaire

Fig. 16.4 Word cloud (Author)

Table 16.4 Pearson's correlation

Variables	X1	X2	X3	X4	X5	X6	Y
X1	1.000						
X2	0.605	1.000					
X3	0.522	0.710	1.000				
X4	0.566	0.448	0.646	1.000			
X5	0.646	0.675	0.648	0.573	1.000		
X6	0.631	0.474	0.530	0.660	0.632	1.000	
Y	0.470	0.462	0.529	0.513	0.523	0.645	1.000

for multiple linear regression (MLR) were continuous data derived of the computed mean from the sample, meeting the requirement of data format. Furthermore, Pearson's correlation in SPSS was used to examine the correlation between each independent variable and the dependent variable (Table 16.4). The correlation values among independent variables were all lower than 0.8, indicating no multicollinearity (Shrestha, 2020). Additionally, the correlation value between each independent variable and the dependent variable was higher than 0.45, suggesting a moderate to high correlation.

4.4 Regression Formula

Using SPSS, Table 16.5 showed the overall coefficient of MLR. In terms of significance value, it showed that only pandemic and reliability were the variables that significantly affect purchase intention (Y), while the other independent variables do not significantly affect Y. Thus, the findings validated H6 and H3, but H1, H2, H4, and H5 were not supported.

Thus, the regression equation was reconstructed to show the relationship between X6 and X3 and Y. The coefficients (Table 16.6) represented that both X3 and X6 positively influenced Y. Overall, the adjusted equation was as followed:

$$\begin{aligned} \textit{Purchase Intention} &= 0.283 \times \textit{Reliability} \\ &+ 0.548 \times \textit{Pandemic } 0.989 \end{aligned} \quad (1)$$

Within the dimension of reliability and pandemic, a new regression was done to compare the weight of each factor (Table 16.7). Among them, helpfulness (Reliability), and social distance and travel time (Pandemic) had a significantly positive influence on Purchase Intention. Among the six factors, Travel time was the most influential one, followed by Helpfulness, Social distance, Ticketing, and Hygiene.

5 Discussion and Conclusion

In conclusion, it can be summarized that KLIA Express visitors were basically satisfied with the services provided in six aspects, with reliability, assurance, and responsiveness being the most satisfying, while empathy, pandemic, and tangibility were the least satisfying. Among all the factors, matching, punctuality, and belongings security had the highest satisfaction scores, while hygiene, personalization, and waiting time had the lowest scores. The study also revealed that reliability and pandemic were the two key factors that influenced visitors' purchase intentions.

Table 16.5 Coefficients for six dimensions.

Model	Unstandardized		Standardized	t	Sig.	Collinearity statistics	
	B	Std. error	Beta			Tolerance	Variance inflation
Constant	2.640	1.247		2.117	0.036		
X1	−0.011	0.097	−0.010	−0.118	0.906	0.442	2,264
X2	0.044	0.101	0.041	0.436	0.664	0.370	2.703
X3	0.217	0.106	0.199	2.057	0.041	0.349	2.861
X4	0.031	0.103	0.026	0.298	0.766	0.417	2.400
X5	0.066	0.103	0.059	0.639	0.524	0.379	2.641
X6	0.509	0.094	0.472	5.419	0.000	0.433	2.309

Table 16.6 Coefficients of adjusted formula

Model	Unstandardized		Standardized	t	Sig.	Collinearity statistics	
	B	Std. error	Beta			Tolerance	Variance inflation
Constant	0.989	0.379		2.610	0.010		
X3	0.283	0.073	0.260	3.877	0.000	0.719	1.392
X6	0.548	0.072	0.508	7.582	0.000	0.719	1.392

Table 16.7 Coefficients of the selected factors

Model	Unstandardized		Standardized	Sig
	B	Std. error	Beta	
Constant	1.099	0.421		0.010
Q7. Reminder	−0.036	0.066	−0.039	0.580
Q8. Helpfulness	0.189	0.072	0.184	0.009
Q9. Ticketing	0.091	0.065	0.098	0.163
Q16. Hygiene	0.075	0.074	0.078	0.316
Q17. Social distance	0.138	0.063	0.164	0.030
Q18. Travel time	0.339	0.068	0.358	0.000

Based on the observations and an online survey, the KLIA Express was found to be a fast and convenient rail service that significantly reduced travel time between destinations, and visitors were particularly satisfied with the well-maintained facilities and fast-moving trains, which provided a comfortable and efficient travel experience. Punctuality was also a key factor in user satisfaction, as the express experienced fewer delays compared to other modes of transportation such as cars, buses, and flights. Additionally, visitors appreciated the luggage storage facilities above the seats, which ensured their belongings were safe and secure during the journey. However, visitors were dissatisfied with the hygiene of the train, as solid waste was not cleaned regularly, and the air circulation was a concern during the pandemic. Personalization options were also reduced due to the pandemic, which resulted in visitors being less satisfied. Finally, waiting time was a significant issue, as visitors sometimes had nowhere to wait and no one to communicate with. Future developments should maintain high levels of satisfaction in areas such as matching, punctuality, and belongings security while improving factors such as hygiene, personalization, and waiting time to increase overall satisfaction levels.

The MLR analysis revealed that reliability and pandemic had a significant positive impact on visitors' purchase intentions, confirming hypotheses H3 and H6. Conversely, tangibility, responsiveness, assurance, and empathy did not have a significant effect on visitors' purchase intentions. The study's results on tangibility were consistent with Loo et al.'s findings, likely because the express maintained consistency in service over time. Upgrading amenities was expensive and time-consuming, making it challenging to improve this service. Responsiveness did not significantly affect visitors' purchase intentions, which differed from Loo et al.'s previous findings. Visitors experienced less waiting time at the express, and most preferred to enter the terminal at the appointed time, so the service offered during the waiting period was not as significant. Additionally, it was difficult to alter the express's accessibility to alternative transportation options. Assurance did not significantly influence visitors' decision-making, as metropolitan public railroads usually had fewer traffic safety issues, and KLIA Express had sufficient measures, such as a safety hammer, to ensure passengers' safety. The empathy service was primarily reflected in the online website and App of KLIA, while other public transportation in KL had attendants, giving KLIA users less access to interactive empathy services. The study's results also demonstrated that pandemic factors significantly influenced users' purchase intentions. People cared more about factors such as hygiene, social distance, and travel time. KLIA express emphasized these factors more due to the growing number of users' concerns about their health and carriage cleanliness during travel. The pandemic-induced economic rebound in Malaysia and the resurgence of tourism activities also contributed to it.

Based on the findings, future research will investigate the satisfaction of KLIA Express from a variety of angles. Firstly, a larger sample size during the data collection period will make the results more accurate and reliable. Additionally, more stations should be investigated to study as many visitors as possible (Niu et al., 2019). Therefore, more time should be devoted to studying visitors' satisfaction with KLIA Express. Secondly, while this study used the SERVQUAL model to select six satisfaction-related dimensions of purchase intention, further studies should focus on perspectives from multiple disciplines to select more useful indicators and factors for railroad satisfaction studies (Tripp & Drea, 2002). For example, from a geographic perspective, factors outside of the trains can be an important aspect that influences the railway experience. Future studies will also consider how to use more landscape and geographic factors to combine with the SERVQUAL model for a more comprehensive evaluation.

Acknowledgements This research was funded by Xiamen University Malaysia Research Fund: Digital heritage: spatial technologies for railway heritage documentation in Asia (XMUMRF/2022-C9/IART/0013).

References

Boubker, O., & Douayri, K. (2020). Dataset on the relationship between consumer satisfaction, Brand Attitude, brand preference and purchase intentions of dairy product: The case of the laayoune-sakia el hamra region in Morocco. *Data in Brief, 32*, 106172.

Chakrabarti, S. (2017). How can public transit get people out of their cars? an analysis of transit mode choice for commute trips in Los Angeles. *Transport Policy, 54*, 80–89.

De Ona, J., De Ona, R., Eboli, L., Forciniti, C., & Mazzulla, G. (2016). Transit passengers' behavioural intentions: The influence of service quality and customer satisfaction. *Transportmetrica A Transport Science, 12*(5), 385–412.

De Vos, J., Singleton, P. A., & Gärling, T. (2021). From attitude to satisfaction: Introducing the travel mode choice cycle. *Transport Reviews, 42*(2), 204–221.

Gotlieb, J. B., Grewal, D., & Brown, S. W. (1994). Consumer satisfaction and perceived quality: Complementary or divergent constructs? *Journal of Applied Psychology, 79*(6), 875–885.

Heo, M., Kim, N., & Faith, M. S. (2015). Statistical power as a function of Cronbach alpha of instrument questionnaire items. *BMC Medical Research Methodology, 15*(1), 1–9.

Ibrahim, A. N., & Borhan, M. N. (2021). Sex disparity in satisfaction and loyalty towards urban rail transit: A survey of Light Rail Transit (LRT) passengers in Kuala Lumpur, Malaysia. *International Journal of Integrated Engineering, 13*(3), 223–228.

Ibrahim, A. N., Borhan, M. N., Mat Yazid, M. R., Rahmat, R. A. & Yukawa, S. (2021). Factors influencing passengers' satisfaction with the light rail transit service in Alpha Cities: Evidence from Kuala Lumpur, Malaysia using structural equation modelling. *Mathematics, 9*(16).

Kwan, S. C., Sutan, R., & Hashim, J. H. (2018). Trip characteristics as the determinants of intention to shift to rail transport among private motor vehicle users in Kuala Lumpur, Malaysia. *Sustainable Cities and Society, 36*, 319–326.

Long, M., & McMellon, C. (2004). Exploring the determinants of retail service quality on the internet. *Journal of Services Marketing, 18*(1), 78–90.

Loo, C. L., Chua, S., Foong, H. Z., Loke, K. & Pang, E. C. (2016). Service quality on behavioural purchase intentions in the Public Land Transportations service in Kuala Lumpur, Malaysia. Bachelor thesis of UNIVERSITI TUNKU ABDUL RAHMAN.

Majumdar, B. B., Jayakumar, M., Sahu, P. K., & Potoglou, D. (2021). Identification of key determinants of travel satisfaction for developing policy instrument to improve quality of life: An analysis of commuting in Delhi. *Transport Policy, 110*, 281–292.

Miller, J. A. (1976). Exploring some alternative measures of consumer satisfaction. *Marketing, 1776*, 661–664.

Ministry of Transport Malaysia (MTM). (2022). Quarterly statistics of rail transport. Retrieved September 30, 2022, from https://www.mot.gov.my/en/land/reports/quarterly-statistics-of-rail-transport

Mohamad, N. (2022). The influence of service quality on tourist loyalty: A case of electric train services. *International Journal of Knowledge Management in Tourism and Hospitality, 2*(4), 315.

Niu, H., Yao, J., Zhao, J., & Wang, J. (2019). SERVQUAL model based evaluation analysis of railway passenger transport service quality in China. *Journal on Big Data, 1*(1), 17–24.

Ok, S., Suy, R., Chhay, L., & Choun, C. (2018). Customer satisfaction and service quality in the marketing practice: Study on literature review. *Asian Themes in Social Sciences Research, 1*(1), 21–27.

Oliver, R. L. (1981). Measurement and evaluation of satisfaction processes in retail settings. *Journal of Retailing, 57*(3), 25–48.

Parasuraman, A., Zeithaml, V. A., & Berry, L. L. (1988). SERVQUAL: A multiple-item scale for measuring consumer perceptions of service quality. *Journal of Retailing, 64*(1), 12–40.

Pineda-Jaramillo, J., & Pineda-Jaramillo, D. (2022). Analysing travel satisfaction of tourists towards a metro system from unstructured data. *Research in Transportation Business & Management, 43*, 100746.

Sam, E. F., Hamidu, O., & Daniels, S. (2018). SERVQUAL analysis of public bus transport services in Kumasi Metropolis, Ghana: Core user perspectives. *Case Studies on Transport Policy, 6*(1), 25–31.

Shi, R., Feng, X., Li, K., & Tao, Z. (2021). Evaluation of passenger service within the area of Beijing West railway station. *Smart and Resilient Transportation, 4*(1), 2–11.

Shrestha, N. (2020). Detecting multicollinearity in regression analysis. *American Journal of Applied Mathematics and Statistics, 8*(2), 39–42.

Tran, V. D. (2020). The relationship among product risk, perceived satisfaction and purchase intentions for online shopping. *The Journal of Asian Finance, Economics and Business, 7*(6), 221–231.

Tripp, C., & Drea, J. T. (2002). Selecting and promoting service encounter elements in Passenger Rail Transportation. *Journal of Services Marketing, 16*(5), 432–442.

Tumsekcali, E., Ayyildiz, E. & Taskin, A. (2021). Interval valued intuitionistic fuzzy AHP-WASPAS based public transportation service quality evaluation by a new extension of SERVQUAL model: P-SERVQUAL 4.0. *Expert Systems with Applications, 186*, 115757.

WTO. (2022). Global and regional tourism performance. Retrieved September 30, 2022, from https://www.unwto.org/global-and-regional-tourism-performance

Zeithaml, V. A. (1988). Consumer perceptions of price, quality, and value: A means-end model and synthesis of evidence. *Journal of Marketing, 52*(3), 2–22.

Sport as a Tool for the Development of Healthy and Sustainable Cities: A Strategic Documentation Review

Aleksander Orlowski and Narek Parsamyan

Abstract

The rapidly changing cities and their environment are causing new challenges for which actions and solutions must be sought. Most of all, a major challenges facing cities are adverse environmental changes and issues related to public health and citizens well-being. Sport is present in the lives of most of us and is an essential component of urban infrastructure, while in relation to this, to what extent is it being used to support health and sustainable goals, shaping cities of tomorrow. The area of relevance of sport for healthy and sustainable cities has not been explored in depth so far. The study aims to examine the sport dimensions and sport factors that contribute to healthy and sustainable city frameworks, through a literature review and analysis of international institutions' policy documents and metrics. The study will contribute to broader understanding of the role of sport within the framework of healthy and sustainable cities and creates the paradigm of future considerations of the relevance of sport from the perspective of city management and policy maker. The novelty of this study is the use of primary sources (strategic documents) for analysis, being a relevant step for the further process of identifying the sport role towards healthy and sustainable cities.

Keywords

Healthy city · Sustainable city · Sport · City planning · Sustainable Development Goal (SDG)

A. Orlowski · N. Parsamyan (✉)
Faculty of Management and Economics, Gdansk University of Technology, Gdansk, Poland
e-mail: narek.parsamyan@pg.edu.pl

A. Orlowski
e-mail: aorlowsk@zie.pg.gda.pl

1 Introduction

With the greatest migration in history happening around us, we are at an unprecedented moment in human evolution. Less than a decade ago, most people still lived in the countryside, but today most people live in cities. By the 2030, the deadline of the new Sustainable Development Goals (SDGs), 60% of the population will live in cities, double the number in 1950 (Salim, 2013). Due to the significant increase in the number of people living in urban areas, aspects of urban health and sustainable development are becoming increasingly important and receiving increasing attention among scholars (Gortmaker et al., 2011; Jararweh et al., 2020; Maller et al., 2006). City planners and policymakers are seeking for solutions to increase urban density while maintaining city liveability. However, when cities expand, they also encounter a range of problems, among which are brought on by population growth. Numerous difficulties can result from excessive population, such as environmental degradation driven by increased pollution and more waste on the streets, difficulty in gaining equitable access to public services, and crowded living conditions that can exacerbate health issues and poor quality of life. There is no single answer to the issue of urban overpopulation, but coordinated action by city planners and policy makers is necessary if these areas are to continue to progress in the coming years (Ramaswami et al., 2016).

In their widely acclaimed book 'Soft City' Sim (2019) noted that people must occupy ever-smaller places as urban density and housing costs grow across the world, it is becoming increasingly difficult to achieve a balance between personal and work activities. Spending most of our time in mechanically ventilated, artificially illuminated rooms and commuting everywhere by automobile, our health will go to suffer, hence spending more time outside, interacting, and moving around is more important than ever. Sim (2019) discusses also the need of implementing soft solutions to cities using simple, affordable, low-tech,

I. Pigliautile et al. (eds.), *Urban Resilience, Livability, and Climate Adaptation*,
Advances in Science, Technology & Innovation, https://doi.org/10.1007/978-3-031-54911-3_17

and human oriented, to the problems caused by increasing urbanization instead of looking for a solution in complex, modern technologies. Perhaps these solutions will make our lives simpler, more attractive in the city, and more convenient, and thus softening the approach may be a wise way out. This begs the question whether sport could promote greater physical activity and a more balanced and friendly way of living in cities.

In recent years, sport and its contribution to health and sustainability have become increasingly intertwined (Cornax-Martín et al., 2020; Dickson & Zhang, 2020; Millington et al., 2022). Particularly, scholars are interested in urban areas where sporting activities can play an important role in promoting physical activity and reducing the risk of chronic diseases (WHO, 2020). Sport has been shown to have a significant impact on overall health and well-being, both physically and mentally; provide opportunities to socialize and build cohesion within communities, while promoting a sense of identity and belonging (Bruner et al., 2021). Among other things, sport can help improve fitness levels, discipline, teamwork skills, self-confidence, and public health awareness. In the same vein, the study by Parsamyan and Orłowski (Parsamyan & Orłowski, 2022) has emphasized the great sport contribution of the living conditions of city inhabitants affecting their health and well-being, and underlines the need to measure the impact of sport on the smartness of the city, in the context of the urban planning process and complex city management. Our previous study (Parsamyan & Orłowski, 2022) explores the role of sport on smart cities through systematic literature review, which identified 38 (out of 764 articles) contextually relevant journal articles indicate the relevance of sports to the development of smart cities, including the majority of the reviewed literature indicate the significant impact of sports on the areas of health (physical and mental activity—living domain) and sustainability (water and energy management—natural resources and energy domain). Therefore, the results of the previous study leads to a need for further consideration of the role of sports in tomorrow's cities, particularly on its direct impact on urban health and sustainability. However, the aforementioned literature review did not consider primary sources, but only secondary ones, due to literature inclusion criteria, thus the strategic documents as urban agendas, reports, guidelines, action plans, and policies may be omitted, so that, consequently, an important source of information was not considered in the previous review. To fill this gap, the intention is to review primary sources (strategic documentation) regards strategic concepts of healthy and sustainable cities which will going to provide the direct access to the subject of the research. Thus, provide an opportunity to review current and relevant strategic documents in this area.

This study delves further into the fundamental aspects of how sports influence the development of cities, with a particular emphasis on how sports contribute to healthy living and sustainability, completing the shortcomings of a prior our research. With these considerations in mind, public engagement in sports is critical for the overall development and well-being of a community. Cities authorities can be engage by providing resources and infrastructure (e.g., parks, fields), encouraging physical activity (e.g., through advertising campaigns), participating in joint ventures with sporting entities (e.g., developing youth leagues). However, it is also important to remember that residents must be given opportunities to choose sport as an activity—excluding any sport from a community will limit its potential benefits for development. Given these points, sport can be used as a tool for urban governance as one of the factors that contributes to improving the health and well-being of citizens and taking measures to mitigate unfavorable climate change.

The purpose of this article is to provide an overview of sport influence on the development of healthy and sustainable cities and try to understand how sport can contribute to fulfil the cities goals and agendas. Specific research questions are as follows: Q1: Is sport included in strategic documents related to health and sustainable development, if so, in what context? Q2: How can city governments use the power of sport to build a healthier and more sustainable city? The research objective will be achieved through the analysis of public strategic documents related to the initiative of a healthy and sustainable cities. The examples and reflections on the sports implications of the concept of a healthy and sustainable city presented in this article attempt to fill the gap in the context of the current state of the art in the field of the contribution of sport to shaping cities of tomorrow.

2 Literature Review

2.1 Healthy Cities

Various social and physical structures, conditions, and factors that shape the urban environment are interconnected and influence our quality of life. Urban governance should consider all relationships and interconnections between different urban domains and features in order to maintain balance and harmony, in order to establish a foundation for the development of a cohesive agenda for healthy cities (Healthier & Communities, 2022). Considerations regarding the need for creating healthy cities were already emphasized by the following scholars, Northridge (Northridge et al., 2003) point out the need of introducing policies and actions for improving the health of citizens. Another example is the research by Handy (Handy et al., 2002), which states that to build a healthier environment is the need to understand human behaviours and adjust the city planning to their needs, thereby doing so in a way that encourages physical activity. The relevant reason for raising the discourse around health cities is the urge to fight against obesity pandemic, as Gortmaker et al. (2011) noted, there is an unquestionable need to take concerted action to create a complex framework for policy makers. Similarly, Barton and Grant (2013) indicates that one of the major challenges for city management is to create framework for policy interventions that are going to improve health and well-being.

The idea of a healthy city is influenced by sport, which offer an enjoyable and stimulating way to maintain regular exercise, making it essential for physical activity. Cities that invest in sports facilities and programs are supportive of residents' physical fitness, thus promoting a healthy lifestyle and sustainable living (Rocha et al., 2019). Furthermore, engaging urban citizens in physical activity is a remedy for reduce chronic illnesses such as obesity, diabetes, and cardiovascular diseases, which are major public health problems for many cities (Giles-Corti et al., 2016). The promotion of social and community involvement through sport can have a positive impact on mental health and well-being (Lindsey et al., 2020). The inclusion of sport in urban areas and their availability can foster a culture of health and wellness that is beneficial to all.

2.2 Sustainable Cities

Sustainable cities can be linked to the notion of healthy city because of its framework for action for improving health and well-being. According to UN-Habitat (Cities Report, 2020), cities currently account for 80% of the world's CO_2 emissions and are also the places where climate change will have the most impact. Rapid urbanization will result in an increase in urban people and carbon emissions, increasing the likelihood that they may be exposed to climate threats. Therefore, city authorities should react and develop solutions to minimize the negative impact on the environment, and the concept of the sustainable city was created in this respect. The following definition has been given by Bibri and Krogstie (2017): "a sustainable city can be described as an urban environment designed with the primary aim of contributing to improved environmental quality and protection and social equity and well–being over the long run … this entails working strategically towards mitigating the environmental impacts derived from the intensive consumption of energy, while promoting social justice, safety, and stability". Nevertheless, a sustainable city can be defined in a variety of ways (Bibri & Krogstie, 2017). Jabareen (2006) work has made a significant contribution to the field to distinguish four dimension of urban sustainability forms: the neotraditional development, the urban containment, the compact city, and the eco-city. It is worth noting that the concept of sustainable city is often combined with the popular smart city concept, a combination known as the sustainable smart city. Instead of just being concerned with technology (smart city), the effectiveness of smart solutions, intelligence in this context should be primarily directed toward the objectives of sustainable development, thus cities should carefully prioritize goals regardless smartness and sustainability ones (Marsal-Llacuna et al., 2015).

Sport and sustainability are two seemingly unrelated topics, but in reality, they impact each other. The potential of sport has power to influence sustainability in many ways, and it is essential that we recognize this and take action to ensure a sustainable future for our planet (Glibo et al., 2022; Tangen, 2021). From reducing the carbon footprint of sports events to promoting eco-friendly practices among athletes and fans, there are many ways in which sports can help promote sustainability, for instance, many sports organizations are now incorporating sustainable practices into their operations, such as using renewable energy sources and reducing waste (Sustainability in sport, 2023). Furthermore, athletes can use their platform to raise awareness about environmental issues and promote sustainable living (Hayes et al., 2019).

3 Methodology

The methodology of the study includes a systematic process of document analysis and review (Wolfswinkel et al., 2013). The process of document analysis involves evaluating electronic and physical documents to interpret them, understand their meaning, and further develop the information provided. In order to look for the presence of sport in the strategic documents, a broad overview of public records will be conduct, involving documents such as urban agendas, reports, guidelines, action plans, and policies related to a healthy and sustainable city. The analysis process consists of reading and studying the document thoroughly, looking at the goals and objectives, and, above all, looking for the sports contribution in these documents. The following steps are included in the document analysis and review process:

- Step 1: Identification
  Search for publicly available documents thematically related to the development of healthy and sustainable cities by searching the browser engine for search terms containing the following phrases: "healthy cities agenda", "sustainable cities agenda", "healthy

cities programme", "sustainable cities programme", "healthy cities guidelines", "sustainable cities guidelines", "healthy cities report", "sustainable cities report", "healthy cities policy", "sustainable cities policy", "healthy cities action plan", and "sustainable cities action plan". The search takes into account the foremost international documents, but also consider relevant national and regional ones.

- Step 2: Screening
  Exclusion of documents which that have been superseded by a recent edition or version. Only official documents approved by the respective organization were selected for review.
- Step 3: Inclusion
  The selection of documents takes into account all documents that make strategic reference to the planning, construction, the development of healthy and sustainable cities, as well as the best practices in this area.
- Step 4: Analysis and insights
  Search for keywords in the documents: "sport", "play", "game", "movement", "activity", and "leisure". The selected keywords were chosen on the basis of their proximity to the word "sport" and their overall thematic validity. In addition, the search will include search for sport-related words, thanks to which the scope of searches will be broadened. By incorporating terms related to sports, it is possible to expand the range of search results. After searching for phrases, specific sentences will be extracted and their context will be analysis and then interpreted.

The analysis performed is not a standard literature review approach, as the search for strategic documents required going beyond the standard content analysis of articles. The searched collection of documents did not come from a literature review (academic research databases), but from online sources available to the public, published by international organizations, institutes, government departments, municipal organizations, associations, and partnership initiatives. Only primary documents were considered in the analysis process. The significance of primary sources lies in their originality and authenticity, which enables to gain a deeper understanding of the topic. In searching for content, a web browser engine was used to identify strategic documents related the concept of healthy and sustainable cities. As a result of the search, based on the level of relevance, seven documents related to the healthy city concept and eight documents related to the sustainable city concept were selected for further study. The methodology presented documents exclusion process is depicted in Fig. 17.1. In the following fourth chapter, a detailed analysis of the results are presented.

4 Results

The value of strategic documents is not only relevant for city governance and decision-makers but also for the stakeholders who want to be engaged in the process of improving the quality of urban life, such as private and non-profit sectors, individuals, and local community inclusively (WHO, 2021). It is important to take a well-thought-out and multifaceted approach to the production of such strategic documents, as they later become the basis for, among other things, the construction of urban strategies and city's agendas. To better illustrate the role of sports in strategic documents related to the idea of a healthy and sustainable city, a collective table (Tables 17.1 and 17.2) has been prepared, showing selected strategic documents and their inclusion of the sports factor.

Seven strategic documents related to the development of healthy cities were analyzed for the presence of the sport factor and sport connection to the development of the healthy city concept (Table 17.1). In most documents, connections were found and are primarily focused on the aspect of promoting and encouraging physical activity (Healthier & Communities, 2022; Lafond, 2015; WHO, 2016a, 2020, 2022). Sport is an activity source that is an indisputable factor for a physical and mental health and overall well-being. Moreover, sport and physical activities not only improves fitness but also reduces the risk of chronic diseases and minimizes the risk of obesity (Healthier & Communities, 2022; WHO, 2016b), which is essential in the vulnerable group of children and adolescents (Obesity, 2021). Creating compact cities is one way that city planning can encourage physical activity, people can easily access the locations they need to go, reduces intra-urban distances (WHO, 2016b). Additionally, it has been found that interventions such as the development of walkable streets, connecting walkways, and proper lighting and visibility can promote regular physical activity by up to even 160% (WHO, 2016b). What is noticeable here is that sport may be used for raising the awareness of the necessity of a healthy lifestyle (Lafond, 2015; WHO, 2020), sport can promote health-oriented behavior and encourage locals to get active through sports education programs and initiatives. Another important aspect from the perspective of health is to ensure the possibility of access to sports facilities in the city (WHO, 2016a), where one can practice sports, but also simply relax, walk, and enjoy the fresh air. The location and surroundings of the sports facilities should be greenish (WHO, 2016a), which would further encourage residents to spend time in the midst of nature. A study found that being close to parks was one of the most reliable predictors of children's physical activity (WHO, 2016a). Green spaces and recreational places provide local residents with

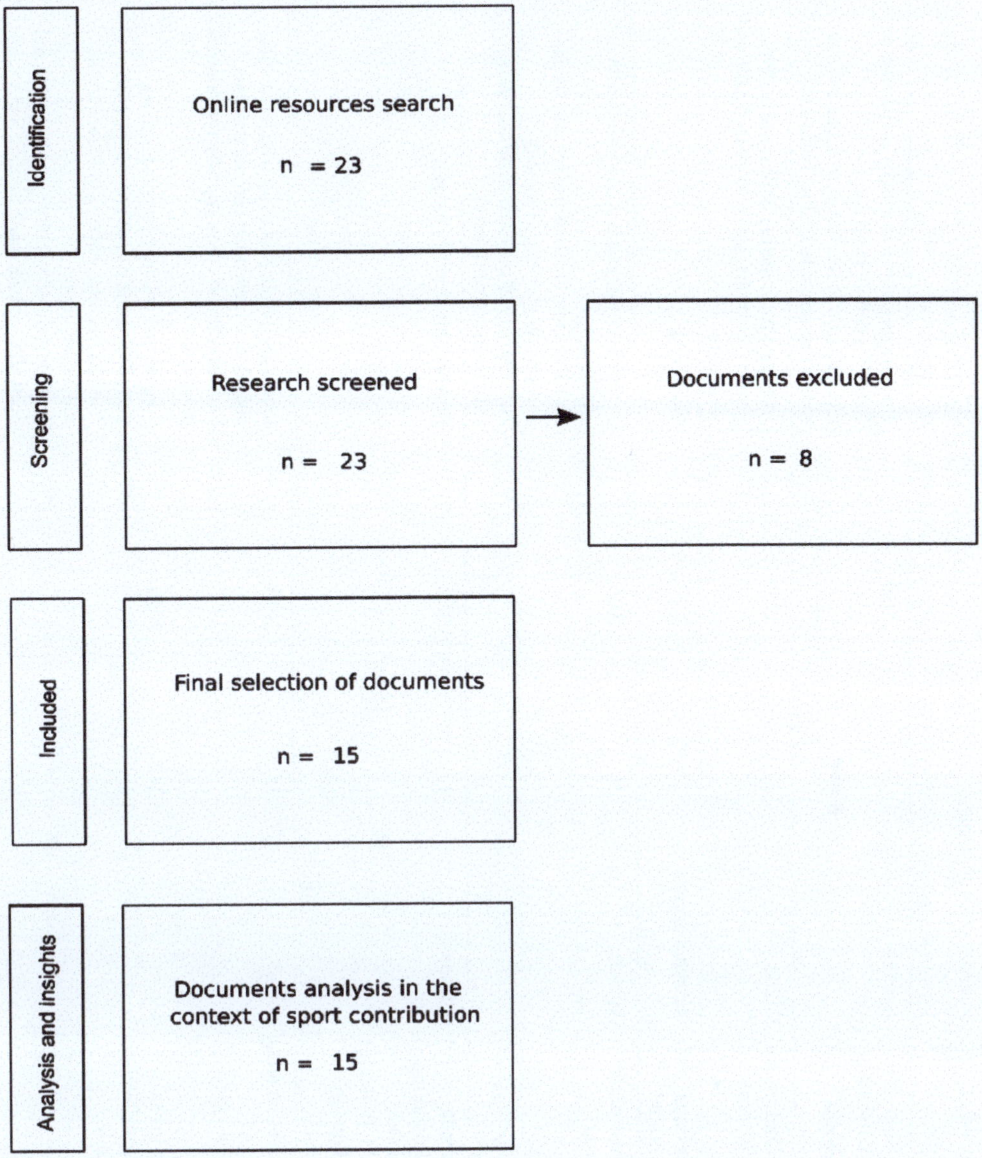

Fig. 17.1 Documents exclusion process. *Source* Own elaboration based on Wolfswinkel et al. (2013). The search was conducted at the end of December 2022

the opportunity to be physically active and to support mental health (WHO, 2016b). Yet, one more important role of sports is to create a child-friendly city, providing them with opportunities for fun, play, and social interaction (Healthier & Communities, 2022; WHO, 2016a) which is so vital in the era of pandemic COVID-19.

Table 17.2 reveals a relationship between sport and strategic documents regarding the concept of sustainable cities. In most of documents, the sport presence is noticeable, only in one document out of eight investigated, no connections were found. One of the important reasons for considering sports in the context of sustainable cities is its flexibility and ability to influence on many levels impacting a social, economic, and environmental sphere. Sport provides possibilities for people with disabilities to be empowered, creating awareness, and fostering respect, thereby sport can so effectively be used to include everyone, regardless of age, gender, race, ethnicity, origin, sexual orientation, gender identity, religion, or economic or other status (Sport & the SDGs, 2015). Sport popularity and a favorable attitude toward it make it an appropriate instrument for addressing inequality in hard-to-reach locations and groups, it can help combat prejudice and bigotry by encouraging tolerance and pro-social behavior (Sport & the SDGs, 2015). Also, cities must offer the facilities necessary for higher living standards, such as social services, education, health care, leisure activities, safety, and security, to promote urban sustainability (*Reference Framework for SC*, 2016). These will make it possible for people to live healthy, fulfilling lives, and to the fullest extent of their ability (Urban Sustainability Framework, 2018). Ensure access to high-quality schools, health services, and other public facilities in the public

Table 17.1 The strategic documentation of a healthy city initiative and the presence of sport within them

No	Organizations	Name of document	Type of document	No. of pages	Year of publishing	Sport presence
1	World Health Organizations (2020)	"Healthy Cities—Effective approach to a rapidly changing world"	Agenda	20	2020	Encouraging physical activity and active living for people in all age groups Contribution to the design of urban environments that promote healthy behavior
2	World Health Organizations—European Healthy Cities Network (WHO, 2022)	"How to develop and sustain healthy cities in 20 steps"	Guide	62	2022	Increased participation in physical activity Promotion of physical activity for health and wellbeing
3	World Health Organizations (WHO UH, 2022)	"Setting global research priorities for urban health"	Research agenda	24	2022	Not found
4	World Health Organizations (New Urban Agenda, 2016)	"Health as the Pulse of the New Urban Agenda"	Agenda	56	2016	Promotion of physical activity for health and wellbeing Physical sport facilities use for play, leisure and general greening of the city Accessibility of play and leisure areas that enhance healthier lifestyle
5	Big cities health coalition (Healthier & Communities, 2022)	Urban Health Agenda "Protecting and Promoting the Well-being of Our Nation's Big Cities Building Healthier, More Equitable Communities"	Agenda	12	2022	Provide opportunities for play
6	Impact on Urban Health (2013)	"Global perspectives on urban health"	Report	35	2013	Increasing physical activity which reducing obesity and preventing chronic diseases Creating playful cities, friendly for children perspective
7	World Health Organizations, United Nations-Habitat (WHO, 2016b)	"Global report on urban health: equitable, healthier cities for sustainable development."	Report	242	2016	Support and enhance physical activity for reducing the risk of chronic diseases. Promote cycling, walking to reduce traffic and air pollution Development of walkable streets, connected sidewalks, and adequate lighting and visibility increase regular physical activity Active and compact city design can increase physical activity that is a key to preventing overweight and obesity The relevance of green and recreational spaces are crucial additions to mixed land use planning for cities. Green spaces and recreational areas give city residents the opportunity to be physically active and promote mental health

Source Own elaboration

Table 17.2 The strategic documentation of a "sustainable city" and the presence of sport within them

No	organizations	Name of document	Type of document	No. of pages	Year of publishing	Sport presence
1	United Nations—office on sport for development and peace (Sport & the SDGs, 2015)	"An overview outlining the contribution of sport to the SDGs" Goal 11. Make cities and human settlements inclusive, safe, resilient and sustainable	Guidelines	17	2015	The value of social and cultural diversity can be promoted through sport, which can also help to advance equality Leveraging the potential of sports events for promote respect, tolerance and human rights Sports are believed to have a positive impact on people and particular on vulnerable communities. The prevention of current inequalities can be achieved through the establishment of policies, agendas and programs related to sport, which aim to address existing injustice
2	United Nations Habitat (New Urban Agenda, 2017)	New Urban Agenda	Agenda	66	2017	Improving physical and mental health and promoting attractive and liveable cities
3	United Nations (UNIDO, 2016)	"Sustainable Cities Hubs of Innovation, Low Carbon Industrialization and Climate Action"	Action plan	24	2016	Not found
4	Global platform for sustainable cities world bank (Urban Sustainability Framework, 2018)	"Urban Sustainability Framework: 1st ed"	Guidelines	160	2018	Promote an active and healthy lifestyle that of all ages can live, work, learn, and play in neighborhoods Providing recreation is an important cities responsibilities, ensuring work-life balance Develop inclusive policies, provide equal access to public services, including recreation facilities
5	UN-Habitat (Cities Report, 2020)	"World cities report, 2020: the value of sustainable urbanization"	Report	418	2020	Investments in sports and recreational facilities will make cities more attractive and healthier for youth while encouraging prosocial behavior Urban planning principles can lead the way to child-friendly cities: creating vibrant play spaces
6	World Resources Report (Mahendra et al., 2021)	"Seven Transformations for More Equitable and Sustainable Cities."	Report	224	2021	Use of drones for acquiring geospatial data regards utilization of land and open spaces. The governance may monitor, for example, areas where people engage in sports and physical activity

(continued)

Table 17.2 (continued)

No	organizations	Name of document	Type of document	No. of pages	Year of publishing	Sport presence
7	United Nations (People-Smart Sustainable Cities, 2020)	"People-Smart Sustainable Cities"	Guidelines	96	2020	Increased support for sports clubs and the rehabilitation of outdoor public spaces to make them more attractive Invest in places or facilities that are designed to facilitate a recreation or leisure activity Access to public spaces, green spaces (parks, forests) as well as blue spaces (water bodies) has proved important for physical and mental wellbeing
8	The reference framework for sustainable cities (Reference Framework for SC, 2016)	"5 dimensions and 30 objectives for a European vision of tomorrow's cities"	Guidelines	160	2016	Cities should provide places for rest and leisure: in playgrounds and sport facilities Sport competition fosters the local economy and the tourism sector. Cities may benefit from creating synergies between health and sport activities Ensure access to high quality public facilities including play and sport grounds Use urban public space as open venues for leisure activities and sport. Public spaces can be places for biodiversity in cities and for sustainable urban drainage

Source Own elaboration

domain, such as green spaces (People-Smart Sustainable Cities, 2020), play areas, and sports fields (*Reference Framework for SC*, 2016), for example, through regeneration programs are particularly meaningful for the child's development (Cities Report, 2020). Children's experiences and observations about the functionality and accessibility of urban space can be highly relevant to decision-makers in the process of planning and designing sustainable cities of the future as they take into account children's needs by including them in holistic urban considerations, consequently, which will then subsequently reflected in urban agendas and policies (Cities Report, 2020). In addition to providing critical services for the city to function properly as a city, the city's social infrastructure, including public spaces and green areas, bring citizens together and meet the needs of different urban social groups (People-Smart Sustainable Cities, 2020). Therefore, if used strategically, it can help address key pressing issues facing cities such as inequality, social disintegration, and promote physical and mental health; Mahendra et al., 2021; New Urban Agenda, 2017; People-Smart Sustainable Cities, 2020). Moreover, sport competition can foster the local economy and the tourism sector, cities may benefit from creating synergies between health and sport activities (*Reference Framework for SC*, 2016).

Surprisingly, the direct sport connotations in the analyzed documents were rather weak. Generally, only a few documents include the phrase "sport/s," which may be surprising given that playing sports is one of the most popular activities toward activeness. What could be the reason for this, perhaps sport is such an obvious issue that international organizations and committees working on those strategic documents do not see the need for additional anchoring of sports in the general framework as a determinant of healthy and sustainable cities.

5 Discussion

The knowledge gained from organizations that are responsible for preparing and implementing strategies for healthy and sustainable cities is crucial in terms of a broader understanding of the area under study. The juxtaposition of sports and strategies in terms of the development of healthy and sustainable cities has not been considered to date. Thus, obtained results from documents analysis allow positioning the sport in the line with health and sustainable cities and to identify areas where sports are relevant and truly related to the topic. To have so far, this type of relationship has not been investigated.

The analysis shows that the presence of sport in strategic documents for a healthy and sustainable city agenda. Sport appears as a factor in similar areas in these two different concepts, which by definition, in principle, deal with different areas and are aimed at solving different problems. Thus, it shows that sport is multidimensional and has a wide range of possibilities to influence our lives.

The compilation of strategic documents and their analysis made it possible to identify the role of sports, in the context of the implementation of strategies derived from these documents and possible further development of the concept of healthy and sustainable cities. Based on the analysis of the strategic documents (Tables 17.1 and 17.2), the comparative summary was prepared indicating the sports factors and, on this foundation, the areas of the city's operation affected by the sports factor were identified (Table 17.3). The classification of areas in which cities should engage to develop and use the potential of sports are:

- **Improve communication and enhance for active participation**—a commitment to promote healthy lifestyles and raise awareness of the role of physical and mental health. Consequently, create a strategy to make city residents more active and participate more in the planning and implementation process.
- **Provide access to facilities and services**—a commitment to creating a sports city, with access to infrastructure for physical activity, recreation, play, and the availability of green and blue spaces.
- **Reorganize the urban planning and development approach**—involvement in the design and planning process of the city in a way that incorporates physical activity and urban compactness, creating a street where people are drawn to walk and where fast and heavy car traffic is not a priority. Moreover, providing a play area for children, created and tailored to their needs—that is important for ensuring the mental health of children.
- **Ensure social inclusion and equity strategy**—involvement in the creation of sports projects targeting specific areas of social life, eliminating discrimination, and promoting equality and diversity.
- **Provide respective resource allocation**—a commitment to reasonable resource planning, taking into account all areas of urban life, including the development of sports clubs and rehabilitation centers, which are an important nexus between physical activity and the health of residents.

It is important to note the similarity of the areas in which sports have been identified for both healthy and sustainable cities in the three areas of city engagement. In contrast, interestingly, in the sustainable cities strategic documents, the broader use of sports are outlined as platforms for promoting social equality and the exclusion of discrimination, and moreover, is underlining the need to include funds for the implementation of projects related to the activities of

Table 17.3 The role of sport in healthy and sustainable city strategic documents—a comparison

City's engagement area	Sport factors	Healthy city	Sustainable city
Improve communication and enhance for active participation	Promoting physical and mental health, and well-being	X	X
	Increased participation in physical activities	X	X
	Raising awareness about healthy lifestyle	X	X
	Promoting active transportation (walking, cycling)	X	X
Provide access to facilities and services	Access to sport facilities	X	X
	Access to leisure and recreational spaces	X	X
	Access to green spaces	X	X
	Access to high quality public services		X
Reorganize the urban planning and development approach	Compact and active city design	X	
	Developing of walkable streets	X	
	Developing a child-friendly city (improve mental health and reduce obesity among children)	X	X
Ensure social inclusion and equity strategy	Developing sport programs and events		X
Provide respective resource allocation	Investments in the activities of sports clubs and rehabilitation's centers		X

Source Own elaboration

sports clubs and rehabilitation centers. An explanation for the greater number of parallels found with sport within the concept of sustainable cities than in the context of healthy cities may be due to the broader definition of sustainable cities, which include not only the broad aspects of health but also take into account: social development, economic development, environmental sustainability, peaceful, just, and inclusive societies, and partnership. In contrast, the strategies identified in strategic documents related to healthy cities emphasize the importance of planning and designing active and compact cities, promoting health and physical activity and the need for life work balance in the urban environment. It is important from the point of view of the city to ensure that residents can benefit from various types of urban functions (work, habitation, relaxation, recreation, physical activity).

First, a review of strategic documents indicated that the proposed activities do not usually indicate the specific groups around whom we want to address the solution (e.g., the elderly, children with obesity, people with disabilities, etc.). Explicit identification of target groups can significantly facilitate the implementation of goals in a measurable and genuinely life-changing way. What's more, as part of the implementation of the strategy, cities should identify the target group to which activities will be directed prior to undertaking such activities (Shirazi et al., 2015; Varming et al., 2018). In this context, the planning and information acquisition phase may be important, cities should use information from their partners organizations, and if there is no such information then they should undertake activities to gain such information, e.g., surveys, interviews, observations, field work.

Second, highlight the role of participation and collaboration in promoting and developing societal and urban synergy. Participation refers to the process through which individuals, groups, organizations, or institutions are given an opportunity to express their views on matters that concern them or affect their interests. Collaboration refers to the process by which two or more parties work together toward a common goal. Both participation and collaboration can be seen as complementary because they both contribute to promote synergy between individuals, groups, organizations, or institutions. This collaborative approach should be a good practice for cities and a way to manage sports projects as early as the planning stage and include the parties involved (Leino & Puumala, 2021). A review of strategic documents has revealed a lack of links in the area of cooperation between the city and other partners. The importance of health is so important that it is in everyone's interest to take action in this area. On the other hand, the importance and role of health is so important that it is in everyone's interest to take action in this area and consider individuals, institutions and governments' points of view.

Third, strategic documents about sustainability do not mention how sports affect the environment, e.g., the environmental damage caused by the sports industry such as massive energy consumption, waste and transportation issues. Sports facilities such as sports arenas, football, and

Olympics stadiums, are often very energy demanding and their maintenance costs sometimes exceeds their operations revenue. Realizing what a negative impact mega-sports events can have, it would be appropriate to point this out as a threat to sustainability and how city authorities can deal with it (Vanwynsberghe, 2015). What type of technology, or sustainable building materials, should be used to minimize the negative impact of sports business on the environment. These types of solutions should be developed as part of the planning and design of the buildings themselves but also of plans for land use, and analyzing the placement of such facilities in terms of their impact on the surrounding area.

6 Conclusions

This paper identified the sport factors that were present in the strategic documentation regarding the concepts of healthy and sustainable cities. These results emphasize the importance of sport as a source of a source for initiating change in our environment, including affecting our health and a sustainable stance. Moreover, examination of the content of strategic documents that are guidelines for policy-makers in the decision-making process serves as the ground for further urban policies evaluation (measurable indicators), therefore becomes a valuable measure. The results of the review of strategic documentation related to the concept of a healthy and sustainable city, allowed to answer the research questions formulated beforehand:

- Q1: Is sport included in strategic documents related to health and sustainable development, if so, in what context?
  Sport was identified in most of the strategic documentations, only in two documents (each concerning the concept of a healthy and sustainable city) there were no mention about sport at all. The founding's connections were mainly related to the: (a) promotion of physical and mental health, and well-being (b) increasing participation in physical activities (c) sport facilities and green areas (d) urban and spatial planning (e) sports programs and agendas (f) investments in sports clubs and rehabilitation centers
- Q2: How can city government can be engaged to build a healthier and more sustainable city through sport?
  The city authorities can demonstrate their commitment to the following tasks: (a) improve communication and enhance for active participation, (b) provide access to the facilities and services, (c) reorganize urban planning and development approach (d) ensure social inclusion and equity strategy, (e) provide the respective resource allocation.

The primary source strategic documents, serve as the foundation of knowledge for the further analysis of a given subject, and are means for exploring the relationship between the role of sports within healthy and sustainable cities. The above-mentioned results provide a significant first step toward filling the gap regarding the current insufficient consideration of the role of sport in shaping the cities of the future. Moreover, the results of this article indicate the importance of paying attention to the important aspect of sport, the potential of which seems unused by cities, often by not taking into account sports initiatives as a factor influencing the implementation of strategic goals of cities, such as the development of health and sustainable development of cities. Stakeholders involved in building and designing cities should pay attention to the possibility of using sport as a multidimensional platform, which is a kind of bridge between the city and their inhabitants.

References

Barton, H., & Grant, M. (2013). Urban planning for healthy cities a review of the progress of the European healthy cities programme. *Journal of Urban Health, 90*(1), 129–141. https://doi.org/10.1007/s11524-011-9649-3

Bibri, S. E., & Krogstie, J. (2017). Smart sustainable cities of the future: An extensive interdisciplinary literature review. *Sustainable Cities and Society, 31*, 183–212. https://doi.org/10.1016/j.scs.2017.02.016

Bruner, M. W., McLaren, C. D., Sutcliffe, J. T., Gardner, L. A., Lubans, D. R., Smith, J. J., & Vella, S. A. (2021). The effect of sport-based interventions on positive youth development: A systematic review and meta-analysis. *International Review of Sport and Exercise Psychology*. https://doi.org/10.1080/1750984X.2021.1875496

Cities Report. (2020). United nations human settlements programme. *Unpacking the value of sustainable urbanization*. https://doi.org/10.18356/c41ab67e-en

Cornax-Martín, M., de Salazar, N. N. G., Rosa-Jiménez, C., & Luque-Gil, A. (2020). Healthy cities, new technologies and sustainability: A collaborative mapping of informal sport activity in the public space of cities as an innovative tool for understanding city sport phenomena. *Sustainability (Switzerland), 12*(19), 1–20. https://doi.org/10.3390/su12198176

Dickson, G., & Zhang, J. J. (2020). Sports and urban development: An introduction. *International Journal of Sports Marketing and Sponsorship*. https://doi.org/10.1108/IJSMS-11-2020-0194

Glibo, I., Misener, L., & Koenigstorfer, J. (2022). Strategic sustainable development in international sport organisations: A Delphi study. *Sustainability (Switzerland), 14*(16), 1–18. https://doi.org/10.3390/su14169874

Giles-Corti, B., Vernez-Moudon, A., Reis, R., Turrell, G., Dannenberg, A. L., Badland, H., Foster, S., Lowe, M., Sallis, J. F., Stevenson, M., & Owen, N. (2016). City planning and population health: A global challenge. *The Lancet, 388*(10062), 2912–2924. https://doi.org/10.1016/S0140-6736(16)30066-6

Gortmaker, S. L., Swinburn, B. A., Levy, D., Carter, R., Mabry, P. L., Finegood, D. T., Huang, T., Marsh, T., & Moodie, M. L. (2011). Changing the future of obesity: Science, policy, and action. *The Lancet, 378*(9793), 838–847. https://doi.org/10.1016/S0140-6736(11)60815-5

Handy, S. L., Boarnet, M. G., Ewing, R., & Killingsworth, R. E. (2002). How the built environment affects physical activity: Views from urban planning. *American Journal of Preventive Medicine, 23*(2 SUPPL. 1), 64–73. https://doi.org/10.1016/S0749-3797(02)00475-0

Healthier & Communities. (2022). *Urban health agenda, protecting and promoting the well-being of our nation's big cities building healthier, more equitable communities*. Retrieved December 13, 2022, from https://www.bigcitieshealth.org/wp-content/uploads/2022/06/BCHC_UrbanHealthAgenda.pdf

Hayes, M., Filo, K., Riot, C., & Geurin, A. (2019). Athlete perceptions of social media benefits and challenges during major sport events. *International Journal of Sport Communication, 12*(4), 449–481.

Jabareen, Y. R. (2006). Sustainable urban forms: Their typologies, models, and concepts. *Journal of Planning Education and Research, 26*(1), 38–52. https://doi.org/10.1177/0739456X05285119

Jararweh, Y., Otoum, S., & Ridhawi, I. A. (2020). Trustworthy and sustainable smart city services at the edge. *Sustainable Cities and Society, 62*, Article 102394. https://doi.org/10.1016/j.scs.2020.102394

Lafond, L. J. (2015). National health cities networks in the WHO European region. *WHO Regional Office for Europe, 1*, 136.

Leino, H., & Puumala, E. (2021). What can co-creation do for the citizens? Applying co-creation for the promotion of participation in cities. *Environment and Planning c: Politics and Space, 39*(4), 781–799. https://doi.org/10.1177/2399654420957337

Lindsey, I., Chapman, T., & Dudfield, O. (2020). Configuring relationships between state and non-state actors: A new conceptual approach for sport and development. *International Journal of Sport Policy and Politics, 12*(1), 127–146. https://doi.org/10.1080/19406940.2019.1676812

Mahendra, A., King, R., Du, J., Dasgupta, A., Beard, V. A., Kallergis, A., & Schalch, K. (2021). Towards a more equal city: Seven transformations for more equitable and sustainable cities. *World Resources Institute*. https://doi.org/10.46830/wrirpt.19.00124

Maller, C., Townsend, M., Pryor, A., Brown, P., & St Leger, L. (2006). Healthy nature healthy people: "contact with nature" as an upstream health promotion intervention for populations. *Health Promotion International, 21*(1), 45–54. https://doi.org/10.1093/heapro/dai032

Marsal-Llacuna, M. L., Colomer-Llinàs, J., & Meléndez-Frigola, J. (2015). Lessons in urban monitoring taken from sustainable and livable cities to better address the Smart Cities initiative. *Technological Forecasting and Social Change, 90*(PB), 611–622. https://doi.org/10.1016/j.techfore.2014.01.012

Millington, R., Giles, A. R., van Luijk, N., & Hayhurst, L. M. C. (2022). Sport for sustainability? The extractives industry, sport, and sustainable development. *Journal of Sport and Social Issues, 46*(3), 293–317. https://doi.org/10.1177/0193723521991413

New Urban Agenda. (2017). United nations habitat III. Retrieved December 12, 2022, from https://habitat3.org/the-new-urban-agenda/

Northridge, M. E., Sclar, E. D., & Biswas, P. (2003). Sorting out the connections between the built environment and health: A conceptual framework for navigating pathways and planning healthy cities. *Journal of Urban Health, 80*(4), 556–568. https://doi.org/10.1093/jurban/jtg064

Obesity. (2021). Overweight and obesity. Retrieved November 26, 2022, from https://www.who.int/health-topics/obesity#tab=tab_1

Parsamyan, N., & Orłowski, A. (2022). *The role of sport in Smart City concept*. Unpublished manuscript. Gdańsk University of Technology, Faculty of Management and Economics

People-Smart Sustainable Cities. (2020). People-smart sustainable cities. United Nations. Retrieved December 5, 2022, from https://unece.org/housing/publications/people-smart-sustainable-cities-E

Ramaswami, A., Russell, A. G., Culligan, P. J., Rahul Sharma, K., & Kumar, E. (2016). Meta-principles for developing smart, sustainable, and healthy cities. *Science, 352*(6288), 940–943. https://doi.org/10.1126/science.aaf7160

Reference Framework for SC. (2016). *Reference framework for sustainable cities "5 dimensions and 30 objectives for a European vision of tomorrow's cities"*. Retrieved December 5, 2022, from http://www.rfsc-community.eu/

Rocha, N. P., Dias, A., Santinha, G., Rodrigues, M., Queirós, A., & Rodrigues, C. (2019). Smart cities and public health: A systematic review. *Procedia Computer Science, 164*, 516–523. https://doi.org/10.1016/j.procs.2019.12.214

Salim, Z. (2013). Global perspectives on urban gating. *In City, 17*(1), 119–121. https://doi.org/10.1080/13604813.2012.754187

Shirazi, M., Engelman, K. K., Mbah, O., Shirazi, A., Robbins, I., Bowie, J., Popal, R., Wahwasuck, A., Whalen-White, D., Greiner, A., Dobs, A., & Bloom, J. (2015). Targeting and tailoring health communications in breast screening interventions. *Progress in Community Health Partnerships: Research, Education, and Action, 9*, 83–89. https://doi.org/10.1353/cpr.2015.0030

Sim, D. (2019). *Soft city: Building density for everyday life*. Island Press.

Sport and the SDGs. (2015). An overview outlining the contribution of sport to the SDGs. Retrieved November 28, 2022, from https://www.sport-for-development.com/imglib/downloads/unosdp2014-sport-and-the-sustainable-development-goals.pdf

Sustainability in Sport. (2023). *Sustainability in sport events from environmental, economic, and social perspectives*. Retrieved January 7, 2023, from https://aists.org/sustainability-in-sport-events/

Tangen, J. O. (2021). Is sport sustainable?—It depends! *Frontiers in Sports and Active Living, 3*, 679762. https://doi.org/10.3389/fspor.2021.679762

UNIDO. (2016). *Sustainable cities: Hubs of innovation, low carbon Industrialisation and climate action*. Retrieved December 19, 2022, from https://www.unido.org/sites/default/files/2017-03/CITIES_22.06.2016_WEB_0.PDF

Urban Health. (2013). Global perspectives on urban health. Retrieved December 20, 2022, from https://urbanhealth.org.uk/wp-content/uploads/2021/04/Global-perspectives-on-urban-health.pdf

Urban Sustainability Framework. (2018). Retrieved December 15, 2022, from https://www.ptonline.com/articles/how-to-get-better-mfi-results

Vanwynsberghe, R. (2015). The Olympic Games Impact (OGI) study for the 2010 winter Olympic games: Strategies for evaluating sport mega-events' contribution to sustainability. *International Journal of Sport Policy, 7*(1), 1–18. https://doi.org/10.1080/19406940.2013.852124

Varming, A. R., Torenholt, R., Andersen, T. H., Møller, B. L., & Willaing, I. (2018). Targeting "hardly reached" people with chronic illness: A feasibility study of a person-centered self-management education approach. *Patient Preference and Adherence, 12*, 275–289. https://doi.org/10.2147/PPA.S148757

WHO. (2016a). Health as the Pulse of the New Urban Agenda: United Nations conference on housing and sustainable urban development. In *United Nations Conference on Housing and Sustainable Urban Development Quito*. Retrieved December 12, 2022, from https://www.who.int/publications/i/item/9789241511445

WHO. (2016b). *Global report on urban health: equitable, healthier cities for sustainable development*. Retrieved December 15, 2022, from http://www.who.int/about/licensing/copy-right_form/index.html

WHO. (2020). *Healthy cities effective approach to a rapidly changing world*. Retrieved December 7, 2022, from https://www.who.int/publications/i/item/9789240004825

WHO. (2021). *A healthy city is an active city: A physical activity planning guide.* Retrieved December 12, 2022, from https://apps.who.int/iris/handle/10665/341088

WHO. (2022). National healthy cities networks in the WHO European Region: How to develop and sustain healthy cities in 20 steps. World Health Organization. Regional Office for Europe. Retrieved December 10, 2022, from https://apps.who.int/iris/handle/10665/364675

WHO UH. (2022). Global research priorities for urban health. *Bulletin of the World Health Organization, 100*(12), 750–750A. https://doi.org/10.2471/BLT.22.289135

Wolfswinkel, J. F., Furtmueller, E., & Wilderom, C. P. M. (2013). Using grounded theory as a method for rigorously reviewing literature. *European Journal of Information Systems, 22*(1), 45–55. https://doi.org/10.1057/ejis.2011.51

SPRINGER NATURE

GPSR Compliance

The European Union's (EU) General Product Safety Regulation (GPSR) is a set of rules that requires consumer products to be safe and our obligations to ensure this.

If you have any concerns about our products, you can contact us on ProductSafety@springernature.com

In case Publisher is established outside the EU, the EU authorized representative is:

Springer Nature Customer Service Center GmbH
Europaplatz 3
69115 Heidelberg, Germany

The manufacturer's authorised representative in the EU is Springer Nature Customer Service Centre GmbH, Europaplatz 3, 69115 Heidelberg, Germany. If you have any concerns regarding our products, please contact ProductSafety@springernature.com

Printed and bound by CPI Group (UK) Ltd, Croydon, CR0 4YY
01/10/2026
02225993-0001